Photometry and Polarization
in Remote Sensing

Photometry and Polarization in Remote Sensing

Walter G. Egan

Research Department
Grumman Aerospace Corporation

Lamont-Doherty Geological Observatory
Columbia University

Elsevier
New York • Amsterdam • Oxford

Elsevier Science Publishing Co., Inc.
52 Vanderbilt Avenue, New York, New York 10017

This book has been registered with the Copyright Clearance Center, Inc.
For further information, please contact the Copyright Clearance Center,
Salem, Massachusetts.

Distributors outside the United States and Canada:

Elsevier Science Publishers B.V.
P.O. Box 211, 1000 AE Amsterdam, The Netherlands

Library of Congress Cataloging in Publication Data

Egan, Walter G.
    Photometry and polarization in remote sensing.

    Bibliography: p.
    Includes index.
    1. Remote sensing. 2. Photometry. 3. Polarization (Light) I. Title.
G70.4.E38   1985   621.36'78   84-10304
ISBN 0-444-00892-6

Manufactured in the United States of America

Contents

Preface

This book is a practical study of optical remote sensing in the wavelength region from 0.185–12 μm, with particular focus on the spectral region between 0.4 and 1.0 μm wavelength. The areas covered range from ground truthing in hydrology, agriculture, forestry, and atmospheres, to remote sensing from a few feet distance up to aircraft, satellite, and astronomical distances. Great emphasis is placed on absolute photometric referencing as the basis of the establishment of a library of photometric signatures of ground features free from extraneous effects such as atmospheric absorption and scattering, whereby ground truthing would be unnecessary to establish recognition in remote sensing. The term *absolute photometric reference* is a new one, and denotes the bidirectional reflectance (scattering) of a surface (ground feature) independent of the effects of atmospheric absorption and scattering. The bidirectional reflectance is expressed in the most general fashion by the four optical Stokes parameters: intensity of radiation, amount of plane polarization, the angle of the plane polarization, and the amount of circular polarization.

A number of practical treatments of the problem of atmospheric absorption and scattering are offered in order that these extraneous effects be elucidated for aircraft and satellite remotely sensed data. Simultaneously, the overall aerosol absorption and scattering properties of the atmosphere are obtained.

An appendix is included that lists the optical complex indices of refraction of various photometric standards, minerals, and aerosol components, in order that surface and atmospheric modeling may be accomplished readily. The photometric standards include Nextel paints (of various colors), $BaSO_4$, $MgCO_3$, sulfur, and various types of beach and desert sands. The atmospheric aerosols include those collected by the author at an invited "First International Workshop on Light Absorption by Aerosol Particles," as well as those collected as "acid rain" over a six-month period in suburban New York.

Polarization is then considered as an outgrowth and a sequel to highly accurate photometric reference photometry because the plane polarization components are in reality the difference between precision photometric measurements in two mutually perpendicular directions. The angle of the plane of polarization arises from the quantitative determination of the plane polarization. Circular polarization is then approached as a final refinement in the complete specification of the optical properties of surfaces and of the atmosphere in terms of Stokes parameters.

Tried and proven techniques are described for aircraft measurements to determine the photometry and polarization of ground features and the atmosphere. Techniques for satellite photometry are summarized and projected techniques for satellite measurements of polarization are described.

The level of treatment of the subject is that of undergraduate college, but the text may be used to advantage by advanced high school students. The book also serves as a useful reference for graduate level college students.

The potential audience is students and workers in aircraft and satellite remote sensing, in areas of agriculture, forestry, marine biology, oceanography, hydrology, pollution (marine and atmospheric), natural resources, the environment, estuarine research, limnology, and data analysis.

The book is divided into three sections plus an appendix; each section contains a generous number of figures to illustrate points made in the text. The Introduction sets the tone of the book. Part I, Mathematical Bases (consisting of three chapters), furnishes the mathematical background relevant to photometry and polarization and their representation in terms of Stokes parameters.

Part II, Optical Fundamentals (consisting of seven chapters), deals with the photometric and polarimetric properties of targets, sensor systems (imaging and nonimaging), contrast (signal to noise: S/N), calibration, atmospheric effects (scattering and absorption), data handling and analysis, and culminates in a chapter on interpretation and information.

Part III, Applications (consisting of ten chapters), presents specific photometric, polarimetric, and Stokes parameter determinations from laboratory measurements and remote sensing. The effect of the atmosphere on polarization and photometry is described, as is the determination of the absorption and scattering properties of the atmosphere given the aerosol and molecular loading. The Applications Section could be used independent of Parts I and II to design an experiment, but the interpretation of the data acquired requires a full knowledge of the photometric and polarization mechanisms, and an appreciation of the significance of calibration.

The specific applications treated are, in sequence: hydrology, marine biology and water quality, agriculture, forestry, planetary astronomy, stellar astronomy, atmospheric constituents, oceanography, depolarization, and radiative transfer.

In all fairness, a separate book could be written covering the material contained in each of the chapters; nevertheless, sufficient information and references are provided to give the remote sensing investigator sufficient background to undertake work in any one of the fields presented.

Further, the first portion of the book (Parts I and II) builds up the important correlation between remote sensing system concepts that ultimately relate to the applications.

Another book has not yet been published that covers the material included here in such detail, written by, or in consultation with, acknowledged experts in the field. In particular, the use of absolute photometric references for ground feature recognition in areas where ground truthing is unavailable has been discussed in the literature only by the author. There is a distinct need for a book of this type for the emerging era of photometrically precise remote sensing.

Acknowledgments

The specialists whose talents were available and were utilized for consultation in the preparation of text material include Ted Hilgeman, coauthor of *Optical Properties of Inhomogeneous Materials*, and John Selby, collaborator on LOWTRAN Atmospheric modeling. The meteorite photometry and polarimetry was proposed by Carl Sagan. The author was assisted by J. Ververka and M. Noland of Cornell University. E. Fireman furnished the Bruderheim Meteorite sample. The Adelphi University Department of Marine Sciences, and in particular H. Brenowitz, M. Hair, J. Cassin, and A. Cok, collaborated in the estuarine and wetlands work. The Martian atmospheric modeling followed suggestions of A. Dollfus, of the Observatory at Meudon, Paris, and Pic du Midi; the staffs of Pic du Midi and Mauna Loa Observatory (MLO), Hawaii, were cordial and hospitable during the author's visits. Special thanks are given to K. Coulson, Director of MLO, for his extensive help. The overview material was prepared with the help of E. Nowatzki, of the University of Arizona. R. Hargraves, of Princeton University, suggested and furnished samples of hematite and nontronite for Martian surface simulation; J. Pollack, of Ames Research Center, furnished experimental data. V. V. Salmonson, U.S.G.S., R. Byrnes, of EROS Data Center, and J. L. Engel, of Santa Barbara Research Center, helped in the acquisition of material describing the LANDSAT-D. Samples of Mt. St. Helens' ash were obtained by L. Radke and P. Hobbs, of the University of Washington. The National Weather Service was helpful in supplying essential meteorological data about Newark Airport, Hawaii, Alabama, and Atlantic City.

The St. Thomas, Virgin Islands program benefited greatly from the field coordination provided by E. Medlicott.

Special thanks are due to the NASA photographic laboratories at Houston and the Grumman Photo Laboratory (D. Rice) for cooperation in the film calibration programs.

My co-workers at Grumman, in particular W. Coulbourn, M. D'Agostino, A. Favale, W. Fischbein, S. Fishburne, K. Foreman, J. Grusauskas, J. Halajian, H. Hallock, T. Hilgeman, J. Krassner, N. Milford, W. Muench, E. Nowatzki,

J. Reichman, D. Reid, J. Selby, L. Smith, were endlessly enthusiastic and helpful.

A. Speidel furnished superb design engineering support. My technicians, H. Wheeler, C. Beneke, J. Augustine, and K. Simco, with additional shop coordination by E. Knoflicek and C. Clamser, were essential to the successful implementation of the field programs.

Computer programming assistance was furnished by T. McGivney, A. Kaercher, and L. Supakoff.

Typing by M. Sudwischer, B. Baldwin, A. McPhillips, and J. K. Egan, the author's patient and understanding wife, was sincerely appreciated, as was report production coordination by J. Compton.

Finally, the unstinting editorial help of K. M. Silverio was instrumental in bringing the book to fruition.

1

Introduction

Photometry and Polarization in Remote Sensing may appear to be an intriguing, but perhaps initially elusive, title. Polarization may be unfamiliar in association with remote sensing. It has been acknowledged in astronomy as a technique for remote sensing, but it is not used as consistently for terrestrial observations. More frequently, polarization effects are associated with atmospheric optical phenomena.

As will be seen in Section I, polarimetry is highly accurate photometry, requiring absolute photometric standards as a reference for observations. Exactness is presently demanded in certain aspects of remote sensing; cartography, stereo interpretation, urban analysis, forest and agricultural inventories, cloud and weather analysis, sea surface temperature measurement, and range inventories, for example, produce quantitative information. However, these analyses concern the contrast between adjacent areas in the imagery, as well as concurrent ground truthing[1], and are relatively independent of the absorption and scattering effects of the atmosphere and visibility, particularly if there is contrast enhancement of the imagery. This text is intended not to downgrade the present approaches in remote sensing, but to augment and enhance image analysis by the application of state-of-the-art photometric calibration technology.

The deactivated LANDSAT 2 has a multispectral scanner that digitizes (quantizes) the scene brightness of a picture element (pixel) representing a ground area of 79×79 m into 64 levels. By data processing that averages adjacent pixels, the brightness levels may effectively be increased to 128. The next generation of LANDSAT sensors (i.e., the multispectral scanners) has a digitizing capability of 256 scene brightness levels, which may be processed to yield 512 levels by averaging with adjacent pixels. This higher scene brightness resolution is more accurately described as higher photometric resolution. Photometry is the science of light

---

[1] The term "ground truthing" denotes on site verification on the surface of the earth of the exact properties of the material or structures located there.

measurement, and the measurement of the brightness of a scene (in appropriate photometric units) is essentially a photometric measurement.

One may naively ask why, if the photometric measurement shows a contrast difference, and we can emphasize the differences by computer enhancement, one need establish contrast levels quantitatively. After all, we then can easily see contrast in an enhanced photograph or image, and relate it to ground truth.

Perhaps the best answer lies in a remark by a New York State official responsible for environmental monitoring and the enforcement of pollution control regulations. At a meeting in New York City of the Regional Planning Commission, he said in essence that remote sensing imagery from uncalibrated "black boxes" is useless for assessing atmospheric or water pollution, much less for enforcing regulations on these problems. Of course, if there is evidence of pollution in remote sensing imagery, it might still be possible to mobilize quickly a ground truthing team, even working without the aid of quantitative photometric remote sensing.

The fact is that the quantitative determination of physical phenomena is required. We readily accept the need for weights and measures even in day-to-day activities. Photography itself recognizes the need to quantify the picture-taking process with light meter readings (which may be accomplished automatically within the camera) in order to produce pictures that are neither under- nor overexposed. Picture development and film production are also subject to tight quality controls.

In remote sensing, as in other fields, quick success is politically expedient and publicly appreciated; we are all subject to this rule in life. If a group relying on remote sensing is shown quick and startling results in imagery, whose value is clear even to those lacking scientific background, the results are recognized as good. The subtlety of this reasoning will be elaborated on subsequently; the basic question is whether an image is more convincing (see, for instance, Fig. 1.1) than tabular or computer data arrays (see, for instance, Fig. 1.2). The general answer is that the choice of presentation technique depends on the nature of the audience. An image (or picture) may impress and easily be understood by a nontechnical audience, whereas data arrays, although harder to understand, are generally of more use to a technical group.

Figure 1.1 Image representation of water depths in St. Thomas Harbor, U. S. Virgin Islands, and adjacent land areas produced from LANDSAT 1 multispectral scanner using computer-compatible tape. Elevation of 4600 ft, a viewing height of 5500 ft, viewing angle north to south.

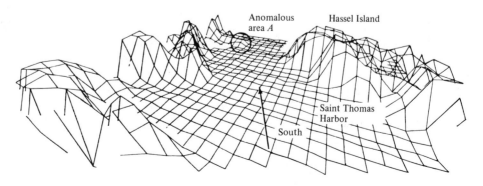

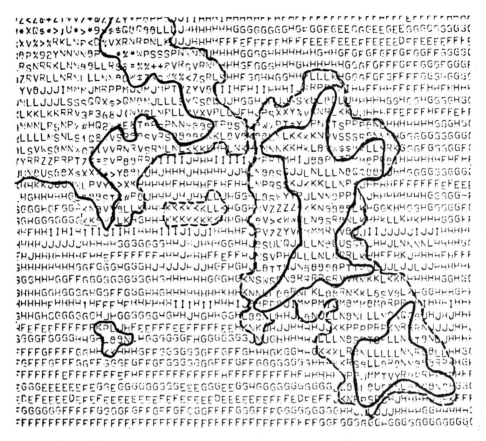

Figure 1.2 Land/water boundaries, water depths and land heights in a digital display whereby letters of the alphabet represent various brightness levels (developed from multispectral scanner bands 4 and 7). Contour lines separate scanner values $\leq J$ and values $\geq K$.

Scientific analyses, from which we develop engineering systems, depend on numerical results. Combustion and chemical processes, for instance, are specified in detail by mathematical equations representing certain physical phenomena. This should also be a goal of remote sensing. Remote sensing ultimately must stand on its own, ideally with little or no ground truthing, making use only of a photometric library to evaluate imagery, just as astronomers have made extensive use of photometric and polarization calibration techniques in the observations of stars and planets.

Absolute photometric calibration (i.e., photometry referred to a standard) is important for another reason: Atmospheric absorption and scattering properties are disregarded in those techniques that depend only on relative scene contrast. With relative photometric measurements, it is difficult to remove the effect of the atmosphere above a terrestrial scene; however, if the atmospheric scattering and absorption can be determined absolutely on a global basis (and the effects are large near urban areas; see Fig. 1.3) and related to the aerosols in the atmosphere, then a significant input may be added to an analysis of the global energy balance (including weather-related phenomena). It has been conjectured that atmospheric aerosols reduce the solar energy reaching the earth's surface, and the net long-term effect would be an average cooling of the earth's surface. Even though one observation of the scattering and absorption of aerosols would not mean much, on a terrestrial synoptic basis much can be gleaned from multiyear observations. Further, such terrestrial events as Sahara or Gobi desert sandstorms can be monitored quantitatively without detailed air truth. Industrial atmospheric pollution too can be remotely sensed and monitored.

LANDSAT remote sensing involves viewing the earth's surface perpendicularly. As will be seen in Chapter 5, the brightness of a surface strongly depends on both the solar illuminating and sensor viewing angles. A future LANDSAT is projected to use a high resolution pointable imager (HRPI), which permits viewing terrestrial surfaces at angles up to 45° from the vertical (perpendicular) direction. Viewing angles away from the vertical reduce the scene brightness as a result of two factors:

Figure 1.3 Optical depths (τ_{aer}) of aerosol at various sites in the Northern Hemisphere (Rahn, 1980). (Attenuation: 63% for τ_{aer} of 1.0, 39% for 0.5, and 10% for 0.1.)

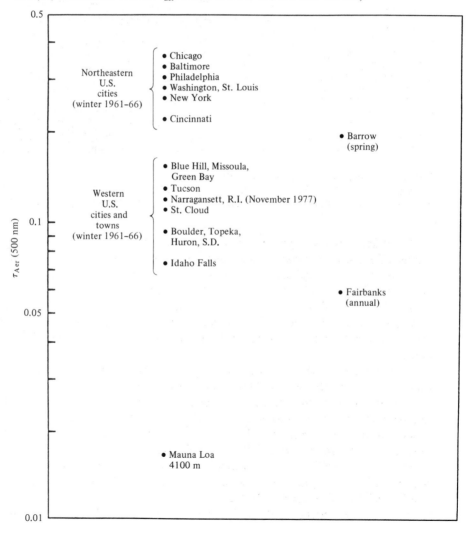

(1) The atmospheric path length is increased, and (2) the surface brightness is apparently reduced in nonnormal viewing (in a nonperpendicular direction). This latter effect is sometimes called a *phase angle effect*, in analogy to the phases of the moon; the phase angle is the angle between the incident solar direction and the direction to the viewer on the earth.

The foregoing paragraphs serve as an introduction to and motivation for the remainder of this book. What follows is an extensive in-depth presentation of the phenomena involved and their analyses, evaluation, and application to remote sensing problems.

Remote sensing has been conceived as the study of distant phenomena using various sensing techniques. The distant phenomena may involve stars, the galaxy, planetary surfaces, the moon, the earth, the interstellar medium, or the earth's atmosphere. In its modern usage, remote sensing requires sensors; data-processing and display equipment; communications and recording systems; information theory; satellites, aircraft, and even ground-located platforms; and much computer software. This text will consider remote sensing in the wavelength range 0.185–12 μm (from ultraviolet through the visible, near infrared, and far infrared).

Most treatments of remote sensing consider terrestrial atmospheric effects as troublesome artifacts in conjunction with the essential objective of remote sensing of the earth's surface from aircraft or spacecraft. In contrast, we shall consider the atmospheric transmission and scattering effects as yielding useful and essential information permitting the establishment of the absolute photometric levels of ground targets. In fact, the effect of the atmosphere can be characterized to some degree by ground-based observations. The atmosphere that blankets the earth has great importance in the heat balance of the earth surface; the circulation patterns cause weather, lead to the transport of anthropogenic pollutants as well as dust from sandstorms, and influence atmospheric photochemistry. Clouds obscure surface remote sensing, but there is interest in the processes that cause the clouds to form (temperature, humidity, pressure, and condensation nuclei).

More strictly interpreted, remote sensing may involve distances of only a few meters (or centimeters) if no physical contact is made with the object being sensed. This near-type remote sensing will not be emphasized in this text.

Remote sensing imagery in itself may be delightful to view. Initially the imagery may be in the form of a tape (to be subsequently converted to an image, either photographic or video), or already be a photograph (in true color or false-color infrared, or narrow-wavelength band, or in black and white). The image, of whatever sort, is an array of data, conventionally either one or two dimensional, although it may be higher dimension. The third dimension may be a physical one, time, or some other variable. The data contained in remote sensing imagery is of astronomical proportions. One set of four images from the LANDSAT 2 multispectral scanner contains 31,000,000 data points (pixels) each having a possible 64 digitized levels. Adding to this the 44,500 daytime scan frames that occur during the LANDSAT 18 day cycle, we have 1.38×10^{12} data points with 64 times this (8.8×10^{13}) possible digital levels. Thus each satellite produces a formidable amount of data.

The data, however, are *not* information; information must be obtained *from* the data by some suitable processing technique. What constitutes information depends on the use to which the remote sensing imagery is to be put. Information appropriate to one application may be useless in another. The user may be furnished LANDSAT imagery (data) from the EROS (Earth Resources Observation Systems) Data Center (in Sioux Falls, South Dakota), for instance, which he subsequently must adapt to his own application. Sometimes he may use the imagery as received for the observation of large photometric differences (contrasts); these differences can easily be seen without enhancement of contrast. Enhancement can bring out subtle contrasts but ones that are uncalibrated with respect to an absolute photometric reference level. If they are calibrated at all, the calibration depends on the degree of enhancement, and this usually varies from scene to scene.

Information may depend on correlations with factors not in the imagery, such as the sediment load in water, insect damage to crops or forest areas, or soil moisture.

Figure 1.4 Tape dump of TIROS-N satellite data.

TAPE TYRO RECORD #1*****
----

*M RECORD 1 (%1)
000000: 0130 9E60 01C5 1B52 11D9 9E60 01D0 RA96 3032 3436 3636 3600 0001 11DA 0026 0063
000020: 0RF3 5C00 0000 0000 0000 0000 0000 0000 0000 0000 0000 0000 0000 0000 0000 0000
000040: 0000 0000 0000 0000 0000 0000 0000 0000 0000 0000 0000 0000 0000 0000 0000 0000
000060: 0000 0000 0000 0000 0000 0000 0000 0000 0000 0000 0000 0000 0000 0000 0000 0000
000100: 0000 0000 0000 0000 0000 0000 0000 0000 0000 0000 0000 0000 0000 0000 0000 0000
000120: 0000 0000 0000 0000 0000 0001 0000 0000 016A 02F0 016A 009A 0700 0000 RR03 4A7C
000140: 0403 4E7C ADBB 9100 AFEB 91E1 CB85 0001 F213 4E7D AFFC 8523 CAR0 00FF CBR5 0002
000160: F203 4EA7 0ED0 0002 AFER 9101 CA00 00FF CRR5 0004 FF03 4FRR 0F40 0650 F203 4FD5
000200: 6DC1 6FC1 2D70 0002 AFF3 9205 93C3 4F6F FER3 506D CRR0 0006 D7RR 4E7R ADRR 9100
000220: C903 0038 AFEB 912B D783 4E6C FF03 4F19 6FC2 2DF0 FR6F F9RC D0FF E98R R4R3 5610
000240: D4R3 55F0 FRR3 4089 0000 0008 0013 55F0 0203 560R D793 55F0 ADRR 9100 C903 003R
000260: AFER 912A D783 4E6C 6FC2 2DF0 CBR0 FFFF FRFF F9RC ADRR 9100 AECR 9101 CAR5 0002
000300: F203 4F49 FREF E98C ADR3 4E6C 6DC2 0002 D0FF F98D R4R3 5610 D4R3 55F0 D5R3 55F4
000320: FRR3 4089 0000 0008 0013 55F0 0203 560C D793 55F0 ADR3 55F4 AFFF E98C F203 4FA1
000340: D0EF E98D R4R3 5610 D4R3 55FR FRR3 4089 0000 0008 0013 55FR 0203 560R D793 55FR
000360: ADRR 9100 AFEB 91E1 EF03 4FD1 FR6R 91F1 C903 0009 6DC2 0002 D0F3 92BC R4R3 5610
000400: FRR3 4061 D783 4E6C AF03 4E6C FR93 9151 ADRR 9100 C903 003R ADFR 912C 6DC2 0002
000420: D0EF E98D R4R3 5610 D4R3 55F0 FRR3 4089 0000 0008 0013 55F0 0203 5604 D7F0 0002
000440: EF03 5039 ADRB 9100 AFFC 84FF D783 4F6C 2D70 0002 C903 003R ADR3 4E6C AFFR 912A
000460: D7CC R501 6D41 0002 ADRR 9100 6DC1 0002 AFF3 9209 D7C4 R503 ADRR 9100 D0FR 9101
000500: R4R3 5610 D4R3 55F0 D0FR 91D1 R4R3 5610 D4R3 55F4 FRR3 4RR9 0000 000C 000R 4F7R
000520: 0013 55F0 0013 55F4 FC13 4E7D ADRR 9100 6DC2 0002 AFF3 3R24 FF03 50R6 0R0R 0009
000540: D78R 4E78 FR63 3824 FC03 4EDD ADRR 9100 AFFR 9101 CRR5 0002 F203 50C1 C903 002C
000560: 6DC1 0002 AFE3 91DR D7F3 9203 ADRR 9100 C903 0016 6DC2 0002 AFF3 91DC D7F3 9204
000600: ADRB 9100 AFEB 9101 CRR5 0004 FE03 5131 D0R3 5615 FRR3 564D ADRR 9100 C903 002C
000620: 6DC1 0002 D0E3 91D3 R4R3 5610 D5R3 55F0 ADR3 55F0 D7F3 91D3 ADRR 9100
000640: C903 002C 6DC1 0002 D0E3 91DR R4R3 5610 D5R3 55F4 FRR3 5RC9 ADR3 55F4 D7F3 91DR
000660: ADRR 9100 AFEB 9101 CRR5 0002 F203 51R1 D0R3 5617 FRR3 564D ADRR 9100 C903 002C
000700: 6DC1 0002 D0E3 91FR R4R3 5610 D5R3 55F0 ADR3 55F0 D7F3 91FR ADRR 9100
000720: C903 002C 6DC1 0002 D0F3 9203 R4R3 5610 D5R3 55F4 FRR3 5RC9 ADR3 55F4 D7F3 9203
000740: CRR0 0001 D4R3 4E68 ADRR 9100 C903 0016 3990 6DC2 AFF3 91C4 D7F3 4D5C CRR1 0001
000760: CRR5 0016 EF03 5185 ADRR 9100 C903 0009 6DC2 0002 D0F3 92C0 R4R3 5610 D4R3 55F0
001000: FRR3 4089 0000 0008 0213 55F0 0203 5EFC D793 55F0 ADRR 9100 C903 0009 6DC2 0002
001020: AC8R 9100 CR03 0016 6CC2 0002 D0R3 4D68 R4R3 5610 D5R3 4F6C FRR3 4D61 D7R3 55F4
001040: FRR3 4089 0000 0008 0203 55F4 0203 5610 ADR3 4E6C FRF3 92C0 ADRR 91RR C903 0009
001060: 6DC2 0002 D0E3 92BC R4R3 5610 FRR3 4RR3 4F03 4F6C FRR3 9145 ADRR 9100
001100: 6DC1 0002 CBR0 0001 EBE3 9201 ADRR 9100 C903 0009 6DC2 0002 D0F3 92C0 R4R3 5610
001120: D4R3 55F0 FRR3 4089 0000 0008 0213 55F0 0203 55FC D793 55F0 ADRR 9100 C903 0009
001140: 6DC2 0002 AC8R 9100 CR03 0003 6CC2 0002 D0R3 4D9C R4R3 5610 D5R3 55F4 FRR3 4D61
001160: D783 55FR FRR3 4089 0000 0008 0203 55FR 0203 5610 ADR3 55F4 FRF3 92C0 AF03 4F6C
001200: FR93 9149 AF03 4E6C FR93 9151 ADRR 9100 FR6R 91E1 C903 003R AFFR 912R D7R3 4F6C
001220: EF03 5349 6FC2 2DF0 D0EF E98D R4R3 5610 D4R3 55F0 D5R3 55F4 FRR3 4D89 0000 000R
001240: 0013 55F0 0203 560C D793 55F0 ADRR 9100 AFEC 8523 CAR0 00FF CRR5 0002 F203 5323
001260: 0FD0 0002 AFEB 9101 CAR0 00FF CRR5 0002 FF03 5337 0F40 0650 FF03 5340 ADR3 55F4
001300: CRR0 FFFF EREF F98C ADRR 9100 C903 003R AFFR 912A D7R3 4F6C ADRR 9100 AFFC R523
001320: C900 00FF CRR5 0002 F203 536F 0R20 0002 AFFR 9101 C9R0 00FF CRR5 0002 EE03 5383
001340: 0DR0 05A0 EE03 5395 6FC2 2DF0 CRR0 FFFF FRFF F9RC ADR3 4F6C 6DC2 0002 D0FF F98D
001360: R4R3 5610 D4R3 55F0 FRR3 4D89 0000 0008 0013 55F0 0203 560C D793 55F0 ADRR 9100
001400: C903 003R AFEB 912B EE03 53D1 D7R3 4E6C ADRR 9100 C903 0009 6DC2 0002 AFE3 92C4
001420: CRR5 018R ED03 5449 FRR3 4D89 0000 0008 000F E9F6 0203 5600 D7RF F9F6 FC03 5449
001440: FRR3 4071 0000 0008 000F E9F6 0203 5600 D7RF F9F6 FC03 5449 ADRR 910R AD6C R4FF
001460: AF83 4E6C D7CC R501 6D41 0002 D503 5EF0 C903 0009 6DC2 0002 AFF3 92C4 CRR1 0001
001500: AD83 55F0 D7E4 R503 CRR0 0006 D783 4F6C ADRR 9100 C903 003R AFFR 912R F203 5469
001520: CRR0 0004 D783 4E6C ADRR 9100 C903 001C R9R3 4F6C 6DC1 0002 ADRR 9100 6D41 0002
001540: AFC3 9201 D7E3 9127 FR43 9201 F843 9205 CBR0 0007 D78R 4E78 D0R3 5615 FRR3 59ED

```
CHANNEL 1 DATA BEING STORED IN FILE

HRPT DATA FOR SCAN LINE   851. DAY  96. YEAR 79. TIME 29836560  -  QUALITY INDICATORS = 00000010 00000000 00000000 00000000
CALIBRATION DATA -      CHANNEL 1             CHANNEL 2             CHANNEL 3             CHANNEL 4             CHANNEL 5
            SLOPE      .10713041D+00         .10502547D+00        -.17338991D-02        -.19916749D+00        -.19954038D+00
        INTERCEPT     -.39230824D+01        -.34826603D+01         .16410799D+01         .19638461D+03         .19655576D+03
SOLAR ZENITH ANGLES:
  111.0  110.0  109.0  108.5  108.0  107.5  107.0  106.5  106.0  106.0  105.5  105.5  105.0  104.5  104.5  104.0
  104.0  104.0  103.5  103.5  103.0  103.0  102.5  102.5  102.5  102.0  102.0  101.5  101.5  101.5  101.0  101.0
  100.5  100.5  100.0  100.0   99.5   99.5   99.0   99.0   98.5   98.0   98.0   97.5   97.0   96.5   96.0   95.5
   95.0   94.0   93.5
HRPT DATA FOR SCAN LINE   852. DAY  96. YEAR 79. TIME 29836724  -  QUALITY INDICATORS = 00000010 00000000 00000000 00000000
CALIBRATION DATA -      CHANNEL 1             CHANNEL 2             CHANNEL 3             CHANNEL 4             CHANNEL 5
            SLOPE      .10713041D+00         .10502547D+00        -.17338991D-02        -.19916749D+00        -.19954038D+00
        INTERCEPT     -.39230824D+01        -.34826603D+01         .16410799D+01         .19638461D+03         .19655576D+03
SOLAR ZENITH ANGLES:
  111.0  110.0  109.0  108.5  108.0  107.5  107.0  106.5  106.0  106.0  105.5  105.5  105.0  104.5  104.5  104.0
  104.0  104.0  103.5  103.5  103.0  103.0  102.5  102.5  102.5  102.0  102.0  101.5  101.5  101.5  101.0  101.0
  100.5  100.5  100.0  100.0   99.5   99.5   99.0   99.0   98.5   98.0   98.0   97.5   97.0   96.5   96.0   95.5
   95.0   94.0   93.5

    RECORD IS1450
```

683	682	684	684	685	691	693	689	686	685	685	685	682	680	683	686
686	686	688	685	681	678	659	648	656	656	654	657	651	649	648	660
671	667	670	680	691	697	701	709	714	718	709	701	699	702	705	708
712	708	701	701	710	715	717	720	724	728	727	728	731	731	732	731
730	726	727	730	730	727	726	728	728	728	727	730	729	726	724	720
721	725	727	724	720	714	707	708	712	715	716	716	712	704	699	699
699	701	698	690	689	692	695	696	695	700	705	705	712	712	708	707
706	705	699	691	690	687	687	687	689	686	684	683	686	690	693	697
700	701	703	705	707	705	704	708	712	712	708	702	698	699	702	707
708	704	697	695	691	692	689	685	683	683	682	680	679	678	680	688
690	686	686	681	679	681	677	673	675	675	679	686	686	678	674	675
671	668	672	678	672	673	677	677	681	684	678	680	684	692	693	695
697	697	697	696	690	688	694	697	697	694	692	688	689	693	696	696
690	689	691	691	693	693	696	700	705	704	705	707	708	709	709	711
714	712	710	710	712	713	710	707	709	710	716	711	712	712	712	711
709	712	715	721	724	726	731	732	733	736	738	742	741	739	736	731
728	727	732	732	731	727	728	731	735	733	732	734	734	736	736	734
729	722	722	731	736	734	731	733	734	736	737	737	735	736	736	736
735	734	733	732	731	728	729	729	726	725	726	728	732	731	733	734
732	733	734	733	732	731	732	731	731	729	729	726	719	719	728	731
732	730	726	718	719	707	687	689	689	692	700	709	716	721	718	722
724	729	730	725	725	734	729	712	699	705	717	717	718	724	717	711
708	708	705	698	695	690	689	679	665	665	674	674	669	670	669	667
666	665	667	674	677	675	677	676	682	678	663	657	665	661	661	676
684	682	684	688	690	694	688	683	681	682	680	680	682	693	704	700
672	667	682	685	690	691	691	693	705	725	732	731	725	713	708	709
721	739	749	753	753	749	746	748	752	756	765	774	774	772	767	763
758	757	757	762	771	770	761	757	759	761	756	753	755	749	739	728
724	723	716	711	705	703	694	680	674	669	658	648	637	628	624	622
618	615	610	607	604	602	600	601	603	605	606	605	604	600	602	602
604	604	604	603	602	605	609	611	612	612	612	611	608	605	605	605

Figure 1.5 Tape of TIROS-N satellite data (header plus data).

In such cases ground truthing is necessary to establish the physical, biological, or chemical relationships. Subsequently, when an adequate library or file is built up, absolute photometry can be used further to interpret this imagery (be it tape or photographic) in terms of the ground cover or structure.

Correlations may best be established with the aid of a computer; thus the data must be in some form of computer-compatible tape format. Simply having data on magnetic tape is inadequate; the data must be in a format (with a header) so that it may be read into an appropriate computer for analysis. Satellite imagery is in a *packed* format for best use of the data link from the satellite to the ground station. For analysis and display purposes, the data must be unpacked, formatted, and printed out for convenient display on a line printer or electronic display. The use of already unpacked satellite data in photographic form is grossly inadequate for absolute photometric analysis. Such factors as unequal sensitivity of the detectors in the LANDSAT multispectral scanner produce striping, and the nonlinear nature of the photographic printing and duplication process results in uncontrolled variables; some form of microdensitometry is necessary to read from the photograph or transparency to a computer-compatible tape from which all the imagery was produced in the first place.

Examples of various types of data displays are shown in Figs. 1.4–1.6. Figure 1.4 is a copy of a section of advanced high resolution radiometer (AVHRR) channel 4 packed TIROS-N data. The format is such as to make maximum use of the satellite–ground radio link; all the data representing the image are crammed

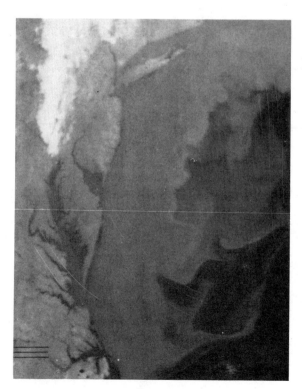

Figure 1.6. Thermal TIROS-N data from channel 4 showing the Gulf Stream east of the Delaware–New Jersey Coast. The darker areas represent higher temperatures.

together, with no spaces or blanks. When unpacked, the data have the form shown in Fig. 1.5, with a header describing the imaging conditions and the numbers representing the uncorrected photometric brightnesses. A black and white photographic representation of this digital data set is shown in Fig. 1.6. The effect of atmospheric emission on the photometry has not been eliminated, and the data are thus termed *uncorrected*.

The TIROS-N channel 4 data are thermal, or far-infrared, photometric data (10.4–12.5 μm in wavelength); the output of the sensor represents the temperature of the ocean as modified by the radiative effect of the atmosphere. Thus, photometry is concerned not only with visible, ultraviolet and near-infrared data, but with far-infrared data as well; the image in Fig. 1.6 is thus an uncorrected thermal image, with the darker areas representing higher temperatures.

Polarimetry, as we have stated and as will be shown in Chapter 2, is merely highly accurate photometry. Polarization is essentially the difference between two photometric measurements, made in mutually perpendicular (orthogonal) directions, of the intensity of the plane-polarized radiation reaching a sensor. Although polarimetry is not currently used in operational remote sensing, it is already used extensively in stellar and planetary astronomy. In addition, the polarimetric properties of surfaces are available as an aid in remote sensing. Polarization also furnishes a supplemental characterization of the molecular and aerosol constituents of the atmosphere. Some surface polarimetric properties are already given in the literature, including polarimetric measurements of various surfaces and foliage (see Figs. 1.7 and 1.8). The use of polarizing sunglasses to minimize sun reflection from lakes is common knowledge.

Another point to be made is that lasers almost invariably produce plane-polarized radiation and that remote sensing with lasers involves not only polarization, but also depolarization on reflection from surfaces. The literature already contains references to these depolarization effects of surfaces.

Polarization sensing is the next step in the development of remote sensors. This will require some rethinking of the design of remote sensing scanners because the scanning is generally accomplished with a scanning mirror that in itself introduces extraneous polarization by reflection from its scanning surfaces.

Figure 1.7 Percent polarization of (a) the mineral Autunite, $Ca(UO_2)_2P_2O_8 \cdot 8H_2O$, as the angle between the source and sensor (phase angle) is increased, and of (b) chrysocolla, $CuSiO_3 \cdot 2H_2O$.

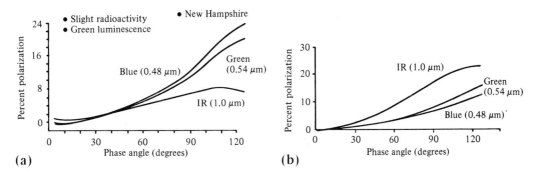

(a) (b)

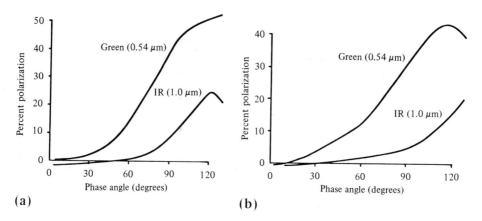

Figure 1.8 Percent polarization of (a) Rhododendron leaves as the angle between the source and sensor (phase angle) is increased, and of (b) white pine foliage.

To put quantitative photometry and polarimetry into perspective, let us consider the percent equivalents of various quantization levels now existing or possible:

Quantization level	Equivalent (%)
1 in 64	1.6
1 in 128	0.0078
1 in 256	0.0039
1 in 512	0.0020
1 in 1024	0.0010

Thus, to increase the photometric accuracy of sensors, as is currently being done, one must emphasize the need to assess quantatively the effect of the intervening atmosphere, as well as the effects of the illuminating and viewing geometry. Looking with a magnifying glass at a poor (low-resolution) photograph or image does not put information into the picture that was not there originally. However, there are techniques for enhancing a signal relative to noise by repetitive scans of the noisy scene or by determination of the appearance of specific spatial frequencies associated with a particular target of interest; calibration, by determining and "calibrating out" the scattering and absorption effects of the atmosphere and the illuminating and viewing geometries, also serves these objectives.

I

MATHEMATICAL BASES

In order to understand and apply the photometric (and polarization) corrections required to achieve absolute calibration, it is necessary to review the fundamental optics involved. These concepts are mathematical in nature but are not difficult to comprehend, being based on relatively simple geometrical relationships.

Photometry will be considered in Chapter 2, followed in Chapter 3 by a discussion of polarization as an outgrowth of highly accurate photometry. Subsequently, in Chapter 4, the mathematical representation of the optical properties of surfaces, the atmosphere, and illuminating sources will be described with Stokes parameters.

Recognition of a particular surface or material (or atmospheric aerosol) by its optical properties depends upon the interaction of radiation with the particular surface or material. The interaction is determined by the radiation wavelength (color), the incident and viewing angles to the surface, and, most important of all, the fundamental optical properties of the surface or material. The fundamental optical properties are embodied in the complex index of refraction (as a function of wavelength). There is nothing "complex" about the index of refraction; the refractive and absorptive portions of the index are simply represented mathematically as a complex number (composed of a real part and an imaginary part). The real part designates the refractive portion, and the absorptive portion is preceded by the letter i or j representing $\sqrt{-1}$. The origin of this convention lies in vector notation used in mathematics, physics, and engineering, whereby graphically the x axis is taken as the real axis and the y axis is designated the imaginary axis. The convention follows through in the characterization of scattering and absorption of electromagnetic radiation through Maxwell's equations. Essentially, scattering is produced by the refractive portion and absorption by the imaginary portion of the optical complex index of refraction.

The optical appearance of a surface depends on its scattering and absorptive properties, which in turn depend on the geometrical structure of the surface as well as the optical complex index of refraction. The most general way of characterizing a surface is through a model. However, in the absence of a model, a family of parametric curves would serve as a substitute.

Recognition of an object implies contrast with its background, and as a result the scattering and absorption properties of the background also affect optical recognition. It frequently happens that the angular dependence of the radiation scattered from one surface differs from that from the adjacent (background) surface, and these properties may be used to advantage in the process of recognition.

2

Photometry

Radiative energy is carried through space (including the earth's atmosphere) by invisible electromagnetic waves. The evidence of the presence of these waves is given when they interact with a sensor. In the optical region, the sensor may be the eye, a phototube, a solid state photosensor, or some combination of one or more of these devices in a system. The system may be a telescope, a scanner, or a mosaic system.

Electromagnetic radiation travels through free space with the velocity of light $(3 \times 10^{10}$ cm/sec). Since electromagnetic radiation has a wave property, it has a wavelength, and concurrently a frequency. The relationship between the wavelength and frequency is given by the relationship

$$\lambda = c/\nu, \tag{2-1}$$

where c is the velocity of light (radiation) in a vacuum $(3 \times 10^{10}$ cm/sec), λ is the wavelength (in centimeters), and ν is the frequency, in hertz (sinusoidal cycles per second).

When electromagnetic radiation passes through a nonabsorbing atmosphere, or any lossless dielectric, it is slowed down by a factor equal to the refractive portion of the index of refraction of the medium. This slowing down can cause a bending or refraction of the wave. The frequency ν remains the same in the dielectric, but the wavelength decreases, in accordance with the expression given above for the wavelength. The electromagnetic wave consists of two field components, an electric field $\mathbf{E}$ and a magnetic field $\mathbf{H}$; these components are mutually perpendicular and follow the same amplitude variations. In the simplest form, the amplitudes are represented as sine waves of one frequency (Fig. 2.1).

This sine wave disturbance $\boldsymbol{\xi}$ may be expressed mathematically

$$\boldsymbol{\xi} = \mathbf{A} \sin(\omega t - 2\pi z/\lambda), \tag{2-2}$$

where the quantities are defined as follows: $\mathbf{A}$ is the vector amplitude of the wave (half the peak-to-peak deviation), ω is the angular frequency, and z is the distance

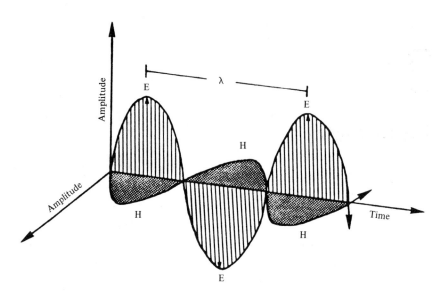

Figure 2.1 Electric (**E**) and magnetic (**H**) vectors of a plane electromagnetic wave.

along the direction of propagation. An alternative representation in complex notation is

$$\xi = \mathbf{A}e^{i(\omega t - 2\pi z/\lambda)} \tag{2-3}$$

or, alternatively,

$$\xi = \mathbf{A}e^{i\phi}, \tag{2-4}$$

where ϕ is termed the *phase angle* (not to be confused with the phase angle in astronomy, mentioned later in this chapter); here

$$\phi = (\omega t - 2\pi z/\lambda) \tag{2-5}$$

A is a vector quantity (i.e., it has a direction and an amplitude $|\mathbf{A}|$), and may represent the **E** or **H** field. When **A** is directed along the x axis (the real axis), a unit vector **i** lying on the x axis and directed in the positive sense is used as a prefix as $\mathbf{i}|\mathbf{A}|$; however, if the unit vector prefix **j** is used (i.e., $\mathbf{j}|\mathbf{A}|$), the direction of the absolute magnitude is along the positive y-axis direction. Similarly, the unit vector **k** is directed along the positive z axis.

The intensity (the physically measurable quantity) is the amplitude squared (i.e., $|A|^2$). The plot of Fig. 2.1 is versus time (or position). The wavelength is shown, and the direction of the **E** field is taken conventionally as the plane of polarization (historically, though, the **H** field was once taken as the polarization direction). Figure 2.1 represents a plane polarized wave, and the polarization direction may assume any direction in a plane perpendicular to the direction of propagation. Unpolarized radiation would be represented by an infinite number of plane polarized waves of equal infinitesimal amplitude A oriented in all directions perpendicular to the direction of propagation. Another representation would be two mutually orthogonal plane polarized waves not having phase coherence.

The techniques for producing and analyzing polarized radiation will be described in the chapter following; the mathematical representation of the phenomena will be outlined in Chapter 4.

Table 2.1 Radiometric and Analogous Photometric Units

Quantity	Symbol	Description	Defining equation	Representative units
Radiometric units				
Radiant energy[a,b]	Q_e		$Q_e = \int_0^\tau \Phi_e\, dt$	joules $\left(1\text{ J} = 10^6\text{ erg} = \dfrac{1}{4.186}\text{ cal}\right)$
Radiant density[a,b]	w_e	Radiant energy per unit volume	$w_e = \dfrac{\partial Q_e}{\partial V}$	joules per cubic meter
Radiant flux[a]	Φ_e	Radiant energy per unit time	$\Phi_e = \dfrac{\partial Q_e}{\partial t}$	watts ($1\text{ W} = 1\text{ J sec}^{-1}$)
Radiant intensity[a]	I	Radiant power per unit solid angle	$I = \dfrac{\partial \Phi_e}{\partial \omega}$	watts per steradian
Radiant emittance[a]	M_e	Radiant power emitted per unit area of source	$M_e = \dfrac{\partial \Phi_e}{\partial A}$	watts per square meter
Irradiance[a]	E_e	Radiant power impacting per unit area of receiver	$E_e = \dfrac{\partial \Phi_e}{\partial A}$	watts per square meter
Differential irradiance (directional radiance)	dE_e	Elemental contribution of flux at a point from a single direction (θ, ϕ)	$dE_e = L_e \cos\theta\, d\omega$	watts per square meter
Radiance	L_e	Radiant power per unit solid angle per unit projected area of source	$L_e = \dfrac{\partial^2 \Phi_e}{\partial\omega\,(\cos\theta\,\partial A)}$	watts per steradian square meter
Photometric units				
Luminous energy[a,b]	Q_v	Portion of radiant energy in visible region	$Q_v = \int_0^\tau \Phi_v\, dt$	lumen second, talbot
Luminous density[a,b]	w_v	Luminous energy per unit volume	$w_v = \dfrac{\partial Q_v}{\partial V}$	lumen seconds per cubic meter
Luminous flux[a,b]	Φ_v	Luminous energy per unit time	$\Phi_v = \dfrac{\partial Q_v}{\partial t}$	lumen
Luminous intensity[a]	I_v	Luminous power per unit solid angle; candlepower	$I_v = \dfrac{\partial \Phi_v}{\partial \omega}$	lumen per steradian, candela
Luminous exitance[a]	M_v	Luminous power emitted per unit area of source	$M_v = \dfrac{\partial \Phi_v}{\partial A}$	lumen per square meter
Illumination[a] (illuminance)	E_v	Luminous power impacting per unit area of receiver	$E_v = \dfrac{\partial \Phi_v}{\partial A}$	lumen per square meter
Differential illuminance (directional illuminance)	dE_v	Elemental contribution of flux at a point from a single direction (θ, ϕ)	$dE_v = L_v \cos\theta\, d\omega$	lumen per square meter
Luminance	L_v	Luminous power per unit solid angle per unit projected area from a source; brightness	$L_v = \dfrac{\partial^2 \Phi_v}{\partial\omega\,(\cos\theta\,\partial A)}$	lumen per steradian square meter

[a] Radiometric and photometric quantities may be limited to a narrow-wavelength band by adding the word *spectral* and indicating the wavelength; the corresponding symbols may be modified by adding the subscript λ.

[b] Radiant and luminous energy and flux can be expressed as areal densities at a surface (quantity per unit area at a given point on a surface, e.g., $\partial Q / \partial A$) or as intensity (quantity per unit solid angle in a given direction, e.g., $\partial Q / \partial \omega$), or as specific intensity (quantity per unit solid angle in a given direction per unit projected area at a given point, e.g., $\partial^2 Q / \partial\omega(\cos\theta\,\partial A)$). The spectral quantities can be described in a similar manner.

Quantitative Descriptors

The radiant energy Q (in ergs or joules; see Table 2.1) is transported by electromagnetic waves; the energy is proportional to the product of the amplitudes of **E** and **H**. The magnitudes of **E** and **H** are related by a constant of proportionality, the intrinsic impedance η of the medium:

$$\mathbf{E} = \eta\mathbf{H}. \tag{2-6}$$

When radiation enters a medium other than a vacuum, it is slowed down. The index of refraction m is the ratio of the velocity v of energy flow in the medium to that in free space (a vacuum), c:

$$m = v/c. \tag{2-7}$$

The rate of flow of this energy (in ergs per second, joules per second, or watts) is termed the radiant flux Φ (Table 2.1), and this flux striking a surface is the irradiance E_e (watts per square meter; or BTU per square foot; Table 2.1). A flux being emitted from a surface is described by the radiant emittance M_e. The radiation occupies a volume, with a radiant density w_e.

The ideal of all radiation moving in parallel rays is not met in practice; both radiation sources and sensors are at finite distances, and as a result the radiation will diverge radially from sources and converge radially on sensors. Therefore, this concept must be brought into the mathematical representation by way of solid angles. A sphere encloses a solid angle ω of 4π steradians in space; the solid angle ω enclosed by *part* of a sphere is the sphere area divided by the sphere radius squared (Fig. 2.2) or, mathematically,

$$\omega = 4\pi \sin^2(\theta/4), \tag{2-8}$$

where θ is the vertex angle.

Figure 2.2 The concept of the solid angle in angular measure. (Copyright 1975 by the American Society of Photogrammetry. Reproduced with permission.)

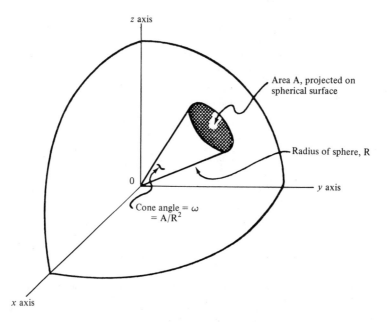

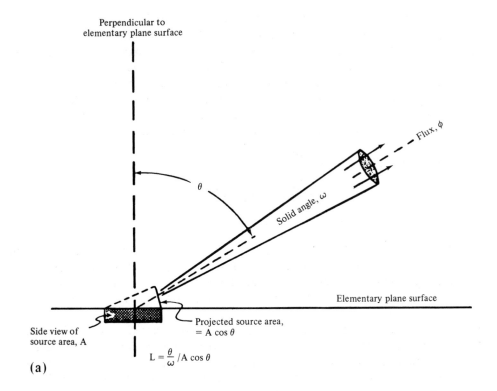

Perpendicular to
elementary plane surface

Flux, ϕ

Solid angle, ω

θ

Elementary plane surface

Side view of
source area, A

Projected source area,
= A cos θ

$L = \dfrac{\theta}{\omega} / A \cos \theta$

(a)

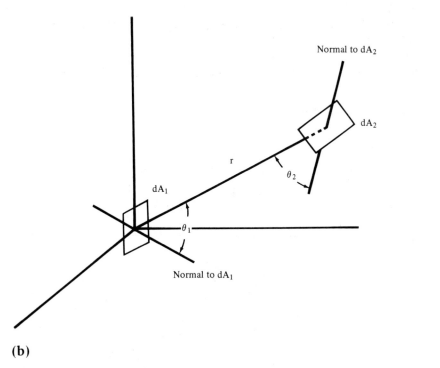

Normal to dA_2

dA_2

r

θ_2

dA_1

θ_1

Normal to dA_1

(b)

Figure 2.3 (a) The concept of radiance. (Adapted from *Manual of Remote Sensing*, Vol. 1, Chapter 3. Reproduced with permission.) (b) Geometrical relationships.

A point source of radiant flux will radiate into a solid angle of 4π steradians (a sphere) and produce a radiant intensity I of $\Phi/4\pi$ (Table 2.1) in units of watts per steradian. But since the actual radiant surfaces have a finite area, as well as sensors, the radiant intensity I per unit source surface area (the radiance L in Fig. 2.3a; also Table 2.1) is the quantity to be reckoned in source–sensor calculations. For a diffusely radiating source of area A_1 and radiance L_1, the radiation intercepted by a sensor of area A_2, located at an angle θ_1 relative to the surface normal of A_1 (having the surface normal at angle θ_2 relative to A_1), at a distance r is

$$P_2 = L_{el}A_1\cos\theta_1\frac{A_2\cos\theta_2}{r^2}. \tag{2-9}$$

(See Fig. 2.3b.) The differential irradiance at the sensor A_2 will be

$$dE_2 = L_{el}\cos\theta_1\left(dA_2/r^2\right). \tag{2-10}$$

In terms of photometry (Table 2.1), the analogous symbolism applies, with the exception that the luminous flux Φ_v is limited to that in the spectral region to which the eye is sensitive. However, in the more general sense (as used in this text), the term photometry characterizes radiation in the spectral range from the ultraviolet to the infrared.

In remote sensing photometry, the brightness of surfaces is the basic parameter of interest; the brightness of a surface is the product of the bidirectional reflectance and the irradiance. Bidirectional reflectance of a surface is the ratio of radiant flux reflected or scattered in a particular direction (measured relative to the surface normal, θ') to that incident at an angle θ (also measured relative to the surface normal) (Fig. 2.4). Both θ and θ' may assume any value from 0 to $\pm90°$; further, the reflected or scattered ray may be in a direction other than in the projection of the incident direction on the surface. It may thus take any value between 0 and 360°, the azimuthal variation ϕ, measured counterclockwise from above. Another angle of interest is the phase angle α, measured in a plane including the incident and

Figure 2.4 Geometrical relationships.

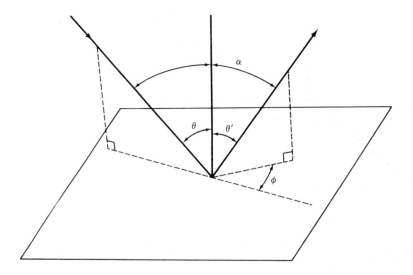

reflected or scattered rays. The bidirectional reflectance yields the specular reflectance when $\theta = \theta'$ and $\phi = 180°$; the sum (integral) of the bidirectionally reflected and scattered flux over a hemisphere enclosing the surface yields the total diffuse reflectance. Specular reflectance is of importance in the remote sensing of water surfaces, or smooth man-made surfaces. Diffuse reflectance is of no interest because remote sensors do not measure it.

Absolute Photometric Measurements

The irradiance of a surface is usually composed of two components; (1) a directional incident (collimated) flux and (2) a diffuse incident flux. For instance, the incident solar radiation is directionally incident (or at least spatially collimated to 0.5°, this being the angular diameter of the sun as seen from earth) and the incident flux from the sky (and clouds, when present) is diffuse. Accurate measurement of these specular and diffuse components may be accomplished by various types of photometers. Several kinds are manufactured by The Eppley Laboratory, Inc. Newport, Rhode Island. Figure 2.5 shows normal incidence pyrheliometer (with a solar tracker), used for measurement of direct solar irradiance (in watts per square meter); the instrument is mounted on a solar tracker to permit aiming the sensor directly at the sun throughout the day without manual attention. Filters may be mounted on the optical system when a particular spectral range is to be measured. The acceptance angle of the sensor system is 5°43′30″, which is about 10 times the sun angular diameter. Thus, the system also will sense the "halo" around the sun caused by

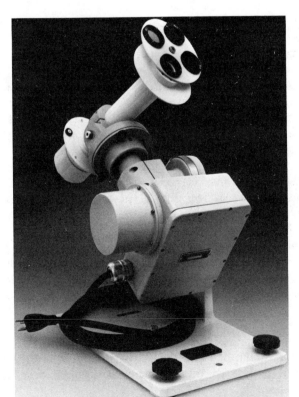

Figure 2.5 Eppley normal incidence photometer mounted on solar tracker.

Figure 2.6 Eppley black and white pyranometer.

scattering of the particulate aerosols in the earth atmosphere (Egan and Shaw, 1981). The irradiance on a horizontal surface will be the projection, i.e., the irradiance multiplied by the cosine of the angle that the sun's radiation makes with the perpendicular to the surface (i.e., the solar zenith angle, which is the complement of the solar elevation angle).

For the measurement of total incident radiation (direct plus diffuse), a pyranometer, such as shown in Fig. 2.6, may be used. The diffuse radiation is the difference between the radiation on the pyranometer and normal incidence pyrheliometer.

For both the normal incidence pyrheliometer and the pyranometer, as well as other related instruments, a certification of the calibration is furnished (Fig. 2.7): traceability to primary standards is noted, as well as the calibration factors.

The diffuse radiation alone may be measured if a paddle (a circular disk) is held at an appropriate distance in front of the pyranometer so as to form a shadow of the sun on the pyranometer; the object is to have this disk block the direct solar rays from striking the pyranometer. In addition, a pyranometer must be located free of nearby walls, buildings, or obstacles in order to prevent radiation scattered off these objects from causing erroneous contributions to the sensed irradiance.

The spectral variation of the incident solar radiation may be measured by using narrow band filters of the thin film interference type. The filters transmit a narrow wavelength band nominally given at 566 nm as shown in Fig. 2.8. The peak transmission of the filter is 44.0% at a wavelength of 569 nm, and the bandwidth (FWHM or full width half-maximum) of 19 nm is shown.

The irradiance transmitted by this filter is the wavelength by wavelength product of the incident radiance and the filter transmission (solid line). Equivalently, when the incident radiance does not vary significantly over the filter pass band, a 100% (or 50%) transmitting bandwidth, hypothetical filter may be chosen. This hypothetical filter will have the same transmission–wavelength product as the actual filter. The determination is made using a planimeter or other area measuring device on the actual transmission curve shown, and the result is the dashed line (Fig. 2.8). If the incident radiance varies significantly over the transmission band of the filter, the "equivalent filter" is then broken up into as many as required to adequately represent the overall irradiance.

EPLAB

THE EPPLEY LABORATORY, INC.
SCIENTIFIC INSTRUMENTS
NEWPORT, R.I. 02840 U.S.A.

STANDARDIZATION
OF
EPPLEY BLACK AND WHITE PYRANOMETER

(horizontal surface receiver - 180°)

Model 8-48   Serial Number 16209   Resistance 345 ohm at   26°C

Temperature Compensation

Range -20 to + 40   °C

This radiometer has been compared with the Eppley group of reference standards, under radiation intensities of about 700 watts meter^{-2} (roughly one-half a solar constant), the adopted calibration temperature is   25 °C.

As a result of a series of comparisons, it has been found to develop an emf of:

11.91 x10^{-6} volts/watt meter^{-2}

8.30 millivolts/cal cm^{-2} min^{-1}

The calculation of this constant is based on the fact that the relationship between radiation intensity and emf is rectilinear to intensities of 1400 watts meter^{-2}. This pyranometer is linear to within ± 1.0 percent up to this intensity.

The calibration of this instrument is traceable to standard self-calibrating cavity pyrheliometers in terms of the Systems Internationale des Unites (SI units), which participated in the Fourth International Pyrheliometric Comparisons (IPCIV) at Davos, Switzerland in October 1975.*

Useful conversion facts: 1 cal-cm^{-2} min^{-1} = 697.3 watts/meter^2

1 BTU/ft^2·hr^{-1} = 3.153 watts/meter^2

Date of Test: August 19, 1977   IN CHARGE OF TEST

The Eppley Laboratory, Inc.   *Richard H. Hatch*

By: *W. J. Scholes*   S. O.   35206

Newport, R. I.   Date   August 26, 1977

Figure 2.7 Typical photometer certification.

A variant of the pyranometer and pyrheliometer, with spectral filters included all in one instrument is the Volz photometer. This versatile device has been widely used as a tertiary standard for sun and sky measurements with purported ±1% to 2% accuracy. The device is portable, self-contained, and intended for hand-held measurements of direct and scattered solar radiation for determination of atmospheric aerosol extinction and precipitable water. Four interference filters are normally used (0.865, 0.500, 0.380 μm as recommended by the World Meteorology Organization and EPA/NOAA, and a supplementary one at 0.945 μm for water vapor). Additional bands are furnished, on special request, at 0.342 and 1.67 μm. The field of view is less than 2°, and an air mass diopter is furnished to correct for solar elevations other than directly at the zenith.

By using calibration factors (furnished with each instrument), the effect of ozone and Raleigh absorption is theoretically eliminated. The aerosol extinction may then be calculated for 1 air mass (i.e., sun at the zenith) at the wavelengths sensed. At 0.945 μm, both aerosol and water vapor extinction are sensed; by ratioing the extinction at 0.945 μm to that at 0.865 μm, the precipitable water vapor may be

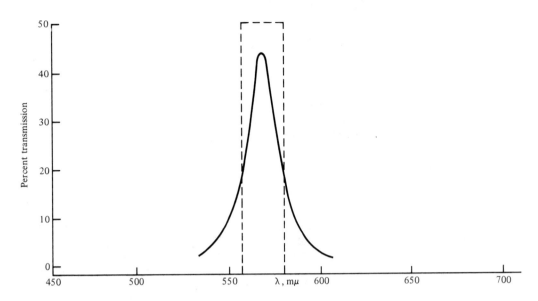

Figure 2.8 Typical interference filter transmission curve and equivalent (shown dotted).

deduced. By shifting the viewing area off the incident solar direction and maintaining the same elevation angle, the brightness of the solar aureole may be determined. This measurement is actually along the solar almucantar, and the sky (aerosol) brightness as a function of displacement angle is a function of aerosol size. This sky-brightening near the sun is caused by the diffraction of solar radiation by aerosol particles. To a first approximation, it is similar to Fraunhofer diffraction by a disk of radius r. The diffracted radiation is mainly within a scattering angle θ_0, where

$$\theta_0 = \sin^{-1}(0.6\lambda / r). \tag{2-11}$$

For instance, for radiation of wavelength 0.5 μm ($= \lambda$) and particles of 10 μm radius, the brightening region (solar aureole) has a radius of 2°, whereas for 1 μm radius particles, the radius is 20°.

Photometric calculations with polarized radiation sometimes can cause difficulties. For instance, when 100% plane polarized radiation is incident on a perfectly diffusing surface, the scattered radiation will be completely depolarized and the geometrical factor that describes the spatial distribution of radiation ($1/\pi$) is identical to that for unpolarized incident radiation, providing a nonpolarization sensitive radiation detector is used. If the detector is polarization sensitive (i.e., having a polaroid in front of it to detect plane polarized radiation), the corresponding factor is $1/2\pi$. Since most surfaces are imperfect diffusers, the dominant plane polarized scattered (reflected) radiation will lie in the same plane of polarization as the incident radiation. In this case, a calculation of the scattered radiation must take into account the phase function of the surface (Chapter 10).

The sun itself is not exactly a blackbody radiator. The internal nuclear furnace, which supplies the solar energy by transforming hydrogen into helium, is enclosed

Table 2.2 Solar Spectral Distribution

λ	B_λ	B'_λ	$I_\lambda(0)$	$I'_\lambda(0)$	f_λ	$\dfrac{I_\lambda(0)}{I'_\lambda(0)}$	$\dfrac{B'_\lambda}{I'_\lambda(0)}$	$\dfrac{B_\lambda}{I'_\lambda(0)}$
μ		10^{10} erg sr^{-1} cm^{-2} μ^{-1} sec^{-1}			erg cm^{-2} Å^{-1} sec^{-1}			
0.20	0.018	0.04	0.03	0.06	1.2	0.5	0.60	0.30
0.22	0.066	0.13	0.10	0.21	4.5	0.5	0.62	0.31
0.24	0.094	0.19	0.14	0.29	6.4	0.5	0.63	0.32
0.26	0.19	0.39	0.30	0.60	13	0.5	0.64	0.32
0.28	0.37	0.85	0.66	1.30	25	0.51	0.66	0.29
0.30	0.87	1.65	1.30	2.45	59	0.53	0.67	0.36
0.32	1.25	2.24	1.88	3.25	85	0.58	0.687	0.39
0.34	1.68	2.68	2.37	3.77	114	0.63	0.710	0.44
0.36	1.69	3.03	2.56	4.13	115	0.62	0.730	0.42
0.37	1.87	3.15	2.45	4.23[a]	127	0.58	0.737	0.43
0.38	1.78	3.35	2.59	4.63	121	0.56	0.710	0.39
0.39	1.69	3.54	2.62	4.95	115	0.53	0.711	0.34
0.40	2.35	3.71	3.24	5.15	160	0.63	0.720	0.46
0.41	2.75	3.82	3.68	5.26	187	0.70	0.728	0.52
0.42	2.78	3.90	3.85	5.28	189	0.73	0.736	0.53
0.43	2.69	3.91	3.73	5.24	183	0.71	0.743	0.51
0.44	2.95	3.90	4.14	5.19	201	0.80	0.750	0.57
0.45	3.13	3.87	4.38	5.10	213	0.86	0.757	0.61
0.46	3.16	3.82	4.35	5.00	215	0.87	0.763	0.63
0.48	3.13	3.71	4.16	4.79	213	0.87	0.774	0.65
0.50	3.00	3.57	3.96	4.55	204	0.87	0.783	0.66
0.55	2.90	3.25	3.72	4.02	198	0.92	0.803	0.72
0.60	2.75	2.86	3.42	3.52	187	0.97	0.817	0.78
0.65	2.46	2.54	3.00	3.06	167	0.98	0.832	0.80
0.70	2.19	2.25	2.65	2.69	149	0.985	0.843	0.82
0.75	1.90	1.93	2.25	2.28	129	0.986	0.853	0.83
0.8	1.68	1.73	2.01	2.03	114	0.987	0.863	0.83
0.9	1.32	1.35	1.54	1.57	90	0.98	0.878	0.85
1.0	1.08	1.11	1.25	1.26	74	0.99	0.887	0.87
1.1	0.90	0.91	1.00	1.01	61	0.99	0.895	0.89
1.2	0.73	0.74	0.80	0.81	50	0.99	0.900	0.90
1.4	0.48	0.48	0.52	0.53	33	0.99	0.910	0.91
1.6	0.328	0.33	0.36	0.36	22.3	0.993	0.918	0.92
1.8	0.218	0.220	0.235	0.238	14.8	0.994	0.925	0.92
2.0	0.150	0.151	0.159	0.160	10.2	0.996	0.930	0.93
2.5	0.073	0.073	0.78	0.078	4.97	0.998	0.940	0.94
3.0		0.039		0.041	2.63	1.0	0.947	0.95
4		0.013 7		0.014 2	0.93	1.0	0.957	0.96
5		0.006 0		0.006 2	0.41	1.0	0.963	0.96
6		0.003 1		0.003 2	0.21	1.0	0.968	0.97
8		0.000 93		0.000 95	0.063	1.0	0.976	0.98
10		0.000 34		0.000 35	0.023	1.0	0.983	0.98
12		0.000 18		0.000 18	0.012	1.0	0.985	0.98

Note: Adapted from Allen (1963).

[a]Hypothetical continuum between Balmer lines: 5.6.

by a shell of hot gases. These gases have absorption and produce the Fraunhofer absorption spectrum. The solar output as a function of wavelength between 0.2 and 12 μm (above the earth's atmosphere) is shown in Table 2.2. This table is from an astronomy text, and the radiance per unit wavelength (i.e., μ^{-1}) is termed *intensity*. Also, 10^{10} erg sr^{-1} cm^{-2} μ^{-1} sr^{-1} = 10^7 watts m^{-2} sterad μ. The quantity B_λ is the average solar intensity in the presence of the Fraunhofer absorption lines, and B_λ' is that observed between the Fraunhofer lines. Because the solar intensity decreases toward the outer rim of the sun's disk (limb darkening), the center of the sun is brighter, and the quantities $I_\lambda(0)$ and $I_\lambda'(0)$ are those representing B_λ and B_λ' at the center of the solar disk.

The appropriate absolute reference values depend on the sensor system. If the sensor measures the average output of the entire sun, or the reflected and scattered energy produced therefrom, then B_λ is correct. If a very high resolution spectrometer is used to look at the entire sun or similarly the specularly reflected or scattered energy, then B_λ' is appropriate. For atmospheric observations whereby the center of the sun is viewed (within the 0.5° sun diameter), then the $I_\lambda(0)$ and $I_\lambda'(0)$ are appropriate depending upon whether the sensor is broadband or very high resolution, respectively.

Surface Reflectance

The bidirectional reflectance of a surface is the result of the interaction of radiant flux with the surface. The radiant flux, for instance, may consist of direct sunlight plus diffuse sky light. (We are considering wavelengths shorter than about 3 μm, where surface emission is negligible.) The interaction of the incident radiant flux with the surface is through scattering and absorption of the radiation, which occurs by virtue of the microstructure of the surface and the optical complex index of refraction. Various successful surface models relate the optical complex index of refraction and the surface geometry with the reflectance. These are generally based on radiative transfer analyses such as the Kubelka–Munk or modified Kubelka–Munk. However, the fundamental property is the optical complex index of refraction of the surface material.

An example of a typical material is silica sand from Cape Hatteras, shown in the Appendix as Table A.64. Here the refractive portion (n) and the absorptive portions (k) are listed and plotted versus wavelength in micrometers. To relate these basic optical properties to observed optical phenomena requires the use of models; the model ties the refractive and absorptive index to the observations in the most general case. The structural properties of a surface affect the relative effect of the refractive portion on scattering and the absorptive portion on absorption. Because of the almost infinite range of measurable combinations, a surface model is appropriate to represent the phenomena most generally. The Appendix contains an extensive original reference listing of optical complex index of refractions in the wavelength range from 0.185 to 1.105 μm, with some extending to 2.7 μm. These indices of refraction thus are available for use in specific problems in remote sensing. It is of interest to note that chlorophyll (maple leaf) is among the indices listed.

Surfaces as they occur in nature are almost never ideal, whether diffuse or specular; the surfaces are generally some combination of specularity, diffuseness, shadowing, and retroreflecting. These properties of specularity and diffuseness have already been alluded to; interparticulate shadowing occurs even from surfaces that have a detailed small scale microstructure. The shadowing causes a diminished surface reflectance as the incident radiation angle increases; also, as the viewing angle increases, the surface reflectance decreases nonlinearly (see Fig. 12.2 and Chapter 10 for instance). Furthermore, as the incident radiation and sensor come into alignment (phase angle of 0°), there is a strong increase in reflected (scattered) radiation; this effect is termed *retroreflectance* (or in astronomy, *the opposition effect*).

The atmosphere can significantly affect the remotely sensed photometry of surfaces, sometimes by an order of magnitude, particularly visibility. The almost infinite variety of possible atmospheric effects again requires the use of modeling to represent them. Basically, atmospheric effects broadly include scattering and absorption: Rayleigh scattering, molecular absorption, aerosol scattering, and absorption. The aerosol scattering and absorption again require a knowledge of their optical complex index of refraction (see Appendix) for modeling. The Rayleigh scattering may be readily represented theoretically, but molecular absorption is particularly tricky, because the absorption depends on the molecular line shapes; these in turn depend on the pressure and temperatures near the molecules. These latter vary throughout the earth's atmosphere, particularly vertically, but also to some extent horizontally. The modeling of molecular absorption bands (which are the sums of the effects of many lines) is dependent on the line parameters and the combination techniques, and these techniques are still being developed and evaluated. More specifically, water vapor, carbon dioxide, ozone, nitrous oxides, methane, carbon monoxide, nitrogen, and oxygen are the strong molecular absorbers of interest. A thorough discussion of the present state of knowledge is embodied in a section on atmospheric effects.

3

Polarization

Polarization was initially introduced and characterized in Chapter 2, on photometry (see Fig. 2.1). Polarization is quite common in nature. However, its use in remote sensing has been limited to astronomy, where every bit of information available in the optical spectral region must be utilized in analyses because most astronomical objects will not be reached for *in situ* measurements within our lifetime.

Terrestrial polarization occurs in sky and sunlight reflection and scattering from water and land surfaces, roads and highways, windows, and vehicle bodies, to name a few. Contrast enhancement of clouds may be achieved by using a polaroid filter on a camera (with color film). Glare from polarized reflection from water surfaces may be minimized by using polarizing eyeglasses. The radiation scattered from the moon's surface is partially polarized, and this property had been used to advantage to model it prior to the lunar landing; the object was to determine the surface bearing strength in order to design the lunar lander appropriately (the LM, the lunar module). All gas lasers with Brewster angle mirrors produce 100% polarized radiation. In fact, every surface produces some polarization of the scattered radiation, even if only very slight.

This chapter will continue with the description of polarization, its production, and detection by virtue of the interaction of radiation with matter. The emphasis will be on the spectral region between 0.185 and 3.0 μm where reflected and scattered radiation dominate; polarization of emitted radiation will be mentioned briefly. The present status of surface and atmospheric modeling will be described.

Radiation Sources

A single atom emitting a photon of radiation would be expected to produce plane polarized waves, though of random orientation. Many atoms of different elements clustered together, with pressure broadened spectral lines—as in the sun or stars—would produce blackbody radiation (Fig. 3.1). The radiation behaves in accordance with the Planck function, with the radiation intensity plotted as a

28

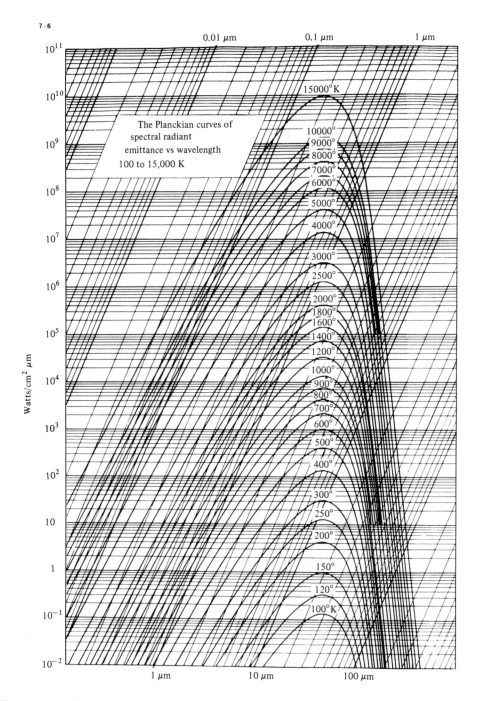

Figure 3.1 An interesting and very useful diagram of Planckian curves is obtained by inclining the vertical lines of the log–log representation so that the peaks of the maximum spectral emittance are aligned vertically. The entire range of temperatures from 100 to 15,000 K can thus be shown most conveniently with good separation between the curves, including the long-wavelength area of the spectral band.

function of wavelength. The radiation from the sun would be expected to be unpolarized; however, the existence of magnetic fields causes preferential orientation of atoms located therein (Zeeman Effect), and plane polarized or circularly polarized radiation occurs depending upon the field direction.

Terrestrially, blackbody radiation sources are incandescent or heated solids. In solids, the atoms are linked to each other by lattice forces and the resulting effect is the smearing out of the sharp spectral lines that would normally be emitted by the individual atoms. The result is a continuous radiation emission, although not usually quite equal to that of a blackbody. In this case, the radiation is less, and the emission is termed graybody (i.e., the emissivity of the surface is less than 1). For solids (and liquids) where a well-defined surface exists, a directional radiation property exists whereby polarized radiation is emitted. The polarization is described by Fresnel's equations. These expressions arise from the application of Maxwell's equations to surfaces, with the appropriate boundary conditions; the fundamental optical properties of the surfaces (i.e., the complex index of refraction) determine the fields that exist within and outside the surface. In reality, surfaces are not perfectly smooth, and modifications are necessary to Fresnel's equations to account for the surface roughness.

However, for a smooth surface, the emissivity ϵ is made up of two components relative to the plane of emission, one parallel ($\parallel$) and one perpendicular ($\perp$);

$$\epsilon = \tfrac{1}{2}(\epsilon_\parallel + \epsilon_\perp), \tag{3-1}$$

$$\epsilon_\parallel = \left[\frac{2\sin\theta\cos\phi}{\sin(\theta + \phi)\cos(\phi - \theta)} \right]^2, \tag{3-2}$$

$$\epsilon_\perp = \left[\frac{2\sin\theta\cos\phi}{\sin(\theta + \phi)} \right]^2, \tag{3-3}$$

where θ is the angle of emission from surface normal and ϕ is the angle of incidence for internally generated blackbody radiation. Snell's law is obeyed:

$$\sin\theta = m\sin\phi, \tag{3-4}$$

where m is the optical complex index of refraction:

$$m = n - ik_0. \tag{3-5}$$

Equations (3-1) through (3-4) may, in principle, be combined to give the total emissivity ϵ as a function of the angle of emission from the surface normal (θ), and the optical complex index of refraction m:

$$\epsilon = \epsilon(\theta, n, k_0). \tag{3-6}$$

Typically, the total normalized emissivity, as well as the parallel and perpendicular components, are shown versus emission angle in Fig. 3.2. It can be seen that the parallel component is stronger than the perpendicular component. The emission is Lambertian (cosine function) near normal up to about $\theta = 55°$, for instance, for a glass radiator having a refractive index of 1.5 (Fig. 3.3). For larger angles of emission θ, the directional spectral emissivity decreases; the ∞ curve is for a thick plate of glass, and the other curves are for thinner plates having the dimensionless product of thickness and spectral absorption coefficient equal to 2, 1, 0.3, and 0.1. The reason for the decrease in emissivity with increased angle of emission is that the directional transmittance of the glass–air surface decreases at the larger angles.

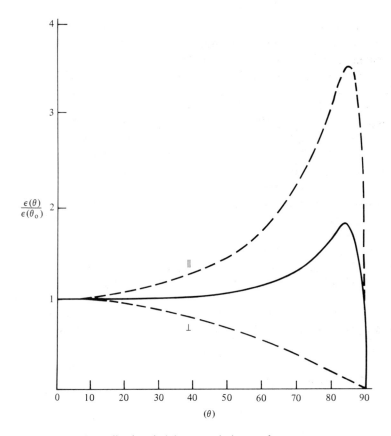

Figure 3.2 Normalized emissivity vs emission angle.

Figure 3.3 Angular variation of the transmissivity τ' of a glass–air surface calculated for $n_{\text{glass}} = 1.5$ (right-hand side), and the emissivities of various metals (left), after Schmidt and Eckert (1935): I, glass; II, clay; III, alumina (coarse); IV, copper oxide.

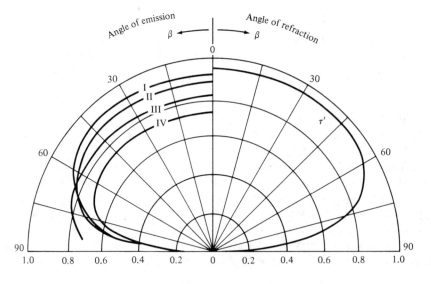

Thus, tungsten incandescent filaments can be expected to produce some polarized radiation at all wavelengths depending on their physical structure. A convenient optical structure that serves to remedy this situation is the Lyot depolarizer. It is placed in the optical path between the tungsten lamp source and the object to be illuminated. The depolarizer (Lyot, 1929) is made of two quartz or gypsum plates cut parallel to their optic axes. Their thicknesses are in the ratio of $2:1$, and they are glued together with Canada balsam so that their optic axes make an angle of $45°$. Because noncoherent radiation of varying wavelength undergoes varying retardations passing through the assembly, any broadband polarization in the incident radiation will be eliminated. It should be noted that the Lyot depolarizer will not be effective for monochromatic, coherent radiation.

Mathematical Representation

Plane polarized radiation was depicted in Fig. 2.1 of this section and represented as Eq. (2-2), (2-3) or (2-4). The plane of polarization of this wave could also rotate continuously in its plane, resulting in a circularly polarized wave. It could rotate clockwise (viewed end on from the left), producing right circularly polarized radiation, or counterclockwise, producing left circularly polarized radiation. It is to be noted that both the $\mathbf{E}$ and $\mathbf{H}$ fields do not necessarily occur concurrently in phase (maximum $\mathbf{E}$ occurs with maximum $\mathbf{H}$) in free space. In some crystals and at metal surfaces, the $\mathbf{E}$ and $\mathbf{H}$ amplitudes can be out of phase. In another variant, the $\mathbf{E}$ vector could simultaneously rotate and periodically change its amplitude, resulting in elliptically polarized radiation.

These preceding characteristics may be expressed mathematically. A circularly polarized wave $\boldsymbol{\xi}$ may be represented as the sum of two equal amplitude sine waves in mutually perpendicular directions, with the second one phase-displaced by $90°$ $(\pi/2)$:

$$\boldsymbol{\xi} = \mathbf{i}|\mathbf{A}|e^{i\phi} + \mathbf{j}|\mathbf{A}|e^{i(\phi \pm \pi/2)}, \tag{3-7}$$

where the $+$ and $-$ signs refer to positive and negative circular polarization, respectively.

Elliptically polarized radiation is represented by a wave consisting of nonequal mutually perpendicular components A_1 and A_2:

$$\boldsymbol{\xi} = \mathbf{i}A_1 e^{i\phi} + \mathbf{j}A_2 e^{i(\phi \pm \pi/2)} = \mathbf{E}_1 + \mathbf{E}_2. \tag{3-8}$$

If the A_1 and A_2 plane polarized waves are out of phase by some angle ϕ_2, then the representation is

$$\mathbf{E}_1 = \mathbf{i}A_1 e^{i\phi},$$
$$\mathbf{E}_2 = \mathbf{j}A_2 e^{i(\phi + \phi_2)}. \tag{3-9}$$

The phase angle ϕ_2 can be the result of scattering (or reflection) of the second plane wave off a material having a significant absorption (see, for instance, Egan and Hilgeman, 1979; Eq. I-19).

When temporally incoherent radiation is partially plane polarized, and the intensities (absolute amplitudes squared) measured in mutually perpendicular directions are I_x and I_y, a convenient representation of plane polarization is percent plane polarization:

$$\% \text{ polarization} = (I_x - I_y)/(I_x + I_y) \times 100. \tag{3-10}$$

The plane of polarization θ_0 is given by

$$\theta_0 = \tan^{-1}(I_x/I_y).\tag{3-11}$$

If circular polarization exists, these expressions represent only the kind resulting from plane polarization. It is thus seen [Eq. (3-10)] that polarization is generally highly accurate photometry, since it may be the small difference between two terms $|I_x|$ and $|I_y|$, which in themselves may each be large.

Polarization is produced by Fresnel reflections of radiation at a surface; the vector expressions for the parallel and perpendicular reflected components are

$$\hat{R}_{\parallel} = \left\{ \frac{(\hat{m}_{i'}/\hat{m}_i)^2\cos\hat{\theta}_i - \left[(\hat{m}_{i'}/\hat{m}_i) - \sin^2\hat{\theta}_i\right]^{1/2}}{(\hat{m}_{i'}/\hat{m}_i)\cos\hat{\theta}_i + \left[(\hat{m}_{i'}/\hat{m}_i)^2 - \sin^2\hat{\theta}_i\right]^{1/2}} \right\}^2,\tag{3-12}$$

$$\hat{R}_{\perp} = \left\{ \frac{\cos\hat{\theta}_i - \left[(\hat{m}_{i'}/\hat{m}_i)^2 - \sin^2\hat{\theta}_i\right]^{1/2}}{\cos\hat{\theta}_i + \left[(\hat{m}_{i'}/\hat{m}_i)^2 - \sin^2\hat{\theta}_i\right]^{1/2}} \right\},$$

Figure 3.4 Reflectances for a dielectric having $n = 1.50$ (Jenkins and White, 1957).

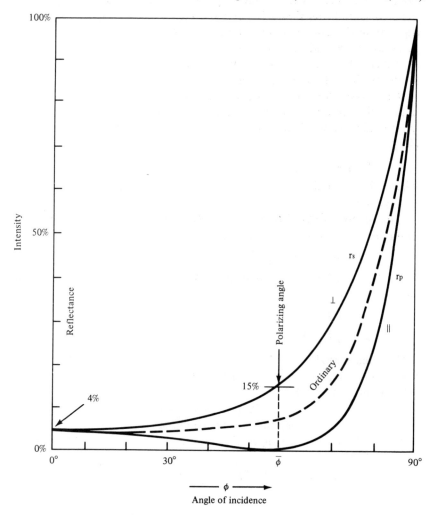

where $\hat{\theta}_i$ is the angle of incidence of wave to surface normal, $\hat{m}_i$ is the complex index of refraction of incident media, and $\hat{m}_{i'}$ is the complex index of refraction of the second refractive media.

Figure 3.4 shows the typical properties of the reflected intensities as a function of incident angle for dielectric reflection, where both $\hat{m}_i$ and $\hat{m}_{i'}$ are real. The Brewster angle occurs at ϕ, where the smaller parallel component goes to zero, and there is 100% polarization with the perpendicular component. The tangent of ϕ yields the index of refraction of the dielectric.

For metallic reflection, where $\hat{m}_{i'}$ is complex, a typical set of reflectance curves is shown in Fig. 3.5; here it can be seen that the trends in the reflecting power are analogous to the dielectric case. However, the perpendicular component does not go to zero, but achieves a minimum at ϕ, termed in this instance the *principal angle of incidence*.

From the application of Fresnel's equations to dielectric and metallic reflection, it is found that a phase change occurs for the electric vector. This is shown in Fig. 3.6 typically for the dielectric case and in Fig. 3.7 for metallic reflection. For pure dielectric reflection (Fig. 3.6) a phase change (δ_s) of 180° (π) occurs for the

Figure 3.5 Reflectances for plane polarized white light off gold and silver mirrors (Jenkins and White, 1957).

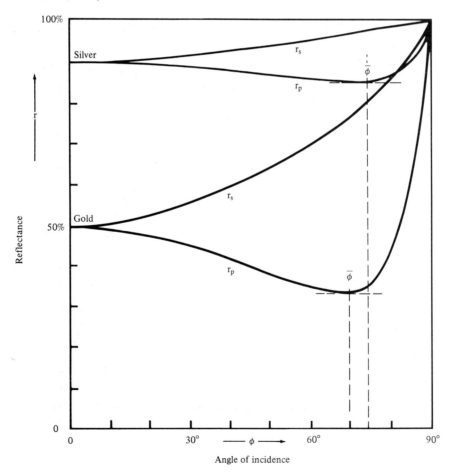

Angle of incidence

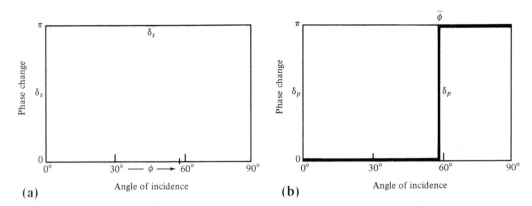

(a)

(b)

Figure 3.6 Phase change of the electric vector of plane polarized light externally reflected from a dielectric (Jenkins and White, 1957).

perpendicular component, but the parallel component has no phase change (δ_p) up to the Brewster angle (ϕ), when a phase change of π occurs with increased angle of incidence. A phase change of π merely means a polarity reversal of the wave. However, for metallic reflection (Fig. 3.7), the phase change is more involved; of particular interest is the difference δ between δ_p and δ_s. This difference is always 90°

Figure 3.7 Phase changes of the electric vector for reflection from a metallic steel mirror (Birge, 1934).

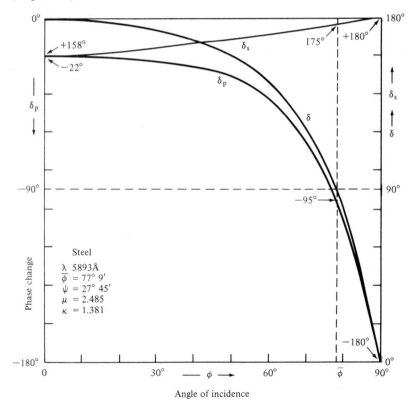

Angle of incidence

at the principal angle of incidence ϕ. The principal azimuth ψ is given by

$$\tan \psi = R_{\parallel}/R_{\perp}. \tag{3-13}$$

For metallic reflection at angles other than ϕ, the phase change δ will be some other value. This will result in $\delta = \phi_2$ in Eq. (3-9) and the amplitudes A_1 and A_2 will be given by the Fresnel reflectivity.

For rough surfaces, an approach frequently used is to assume that an assemblage of facets may be employed to represent the irregularities. Then the Fresnel laws of reflectance may be applied statistically to characterize the reflectance (and scattering). However, in these models, as carefully as they may be constructed, an empirical factor is always needed somewhere to represent certain uncalculable effects.

Detection and Measurement of Polarized Radiation

In order to sense polarization in an electromagnetic wave, an optical element is necessary that responds nonuniformly to the vector directional property of the incident radiation. Such devices are polaroids, wire grid polarizers, dichroic polarizers, piles of plates, double refraction prisms (Glan–Foucault, Nicol, Wollaston, or Rochon, for instance), or asymmetrical scatterers. Various degrees of mechanization and computational facilities have been described in the extensive literature on polarization analysis. This section will describe some of the convenient polarization analyzers, the basic reasoning for the analysis of polarized radiation, and a few mechanized versions.

The most antique instrument for analyzing polarization is the wire grid; Heinrich Hertz was the first to use the wire grid as a polarizer in 1888 to evaluate the properties of the then recently discovered radio waves. The grid, consisting of 1 mm copper wires spaced 3 cm apart, was used to check the properties of 66 cm radio waves. By orienting the grid parallel and perpendicular to the electric vector of the radiation, the electric field was either "short circuited" or permitted to pass through, respectively. This property of asymmetric absorption is termed *dichroism*.

This same principle was extended into the infrared by duBois and Rubens in 1911 with 25 μm diameter noble metal wires for polarization analyses of radiation from 24 to 314 μm. The most recent application was by Bird and Parrish in 1960; wire grids were made by evaporating gold onto a plastic replica of a diffraction grating

Figure 3.8 Evaporation of metal on a plastic grating replica.

Evaporating metal

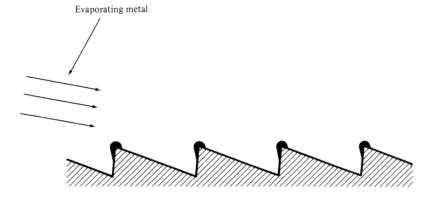

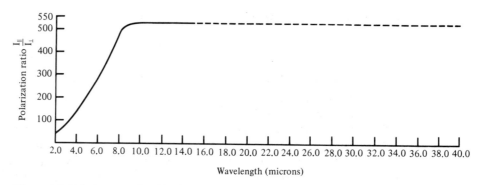

Figure 3.9 Polarization of gold wire grid polarizer on silver bromide substrate.

(Fig. 3.8). Such grids are available commercially from Perkin–Elmer with transmission in the wavelength range from 2.5 μm to well beyond 333 μm. The wavelength range from 2.5 to 35 μm uses a silver bromide substrate, and the region between 20 and 333 μm employs a polyethylene substrate. The polarization properties of these wire grid polarizers are shown in Fig. 3.9; the plots are the ratios of the parallel component of intensity to the perpendicular versus wavelength. The larger the ratio, the better the performance of the wire grid as an analyzer. There is some inherent attenuation of the wire grid analyzer (64% at 2.6 μm) because of the obscuration of the optical path through it by the wires.

As a result of research by E. H. Land, in 1938 the chemical analog of the wire grid was invented. Long, thin polymer molecules aligned nearly parallel to each other, with high conductivity produced by free electrons associated with the iodine atoms in the molecules, served to absorb the electric field parallel to the molecules; the perpendicular component then passes through with a small absorption.

Various sheet polarizers are available from the Polaroid Corporation, Norwood, Massachusetts 02062. The usable wavelength ranges cover the region from 275 nm to 2.0 μm. Also, Minneapolis Honeywell manufactures ultraviolet polarization analysers. Table 3.1 lists some significant Polaroid linear polarizers, their region of operation, and the reference curve for the performance.

The preceding list is not all inclusive, but represents practical, inexpensive polarization analyzers that are available in a 2 in. (51 mm) diameter (some mounted between glass disks) and appropriate for use in laboratory and field polarimeters.

More exotic, laboratory-type polarization analyzers are piles of plates, double refraction prisms, and asymmetrical scatterers. These devices are described in the

Table 3.1 Polaroid Linear Polarizers, Operating Range, and Reference Curves

Polarization analyzer	Wavelength range (μm)	Performance curve
105 UV	0.18–0.28	Fig. 3.10
HNP'B	0.28–0.450	Fig. 3.11
HN 22	0.450–0.760	Fig. 3.12
HR	0.8–2.0	Fig. 3.13

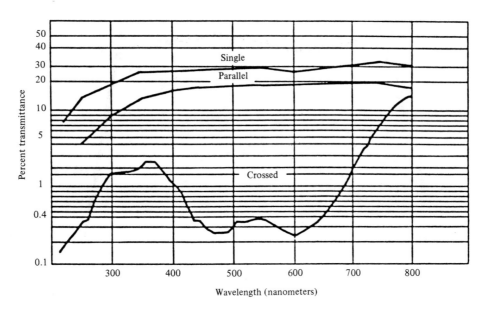

Figure 3.10 105 UV transmittance.

Figure 3.11 Spectral curve: HNP'B.

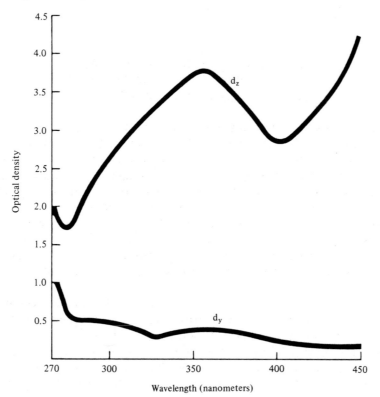

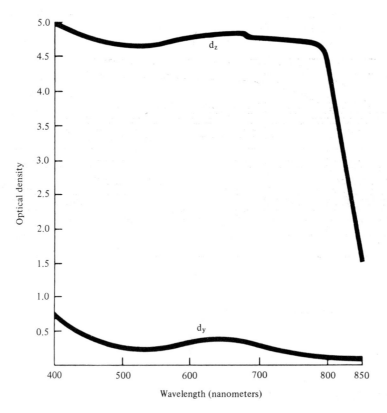

Figure 3.12 Spectral curve: HN22.

vast literature on polarization and generally cause mounting and alignment problems, particularly when the analyzer is to be rapidly rotated in an automated polarization analyzer, and maximum accuracies of $\pm 0.1\%$ polarization [Eq. (3-10)] are required.

In order to sense polarized radiation, much less radiation itself, an appropriate detector is necessary. The human eye itself is one of the first polarization sensors, but rather inadequate for quantitative work. The eye has been used quite extensively in photometry to judge matching brightness, but the range of brightnesses occurring in polarimetry and the need for quantification require some form of accurate photometric detector. Some types are phototubes, photomultipliers, photodiodes, thermocouples, thermopiles, pneumatic bolometers, and photoconductive or photovoltaic semiconductors. The chapter on sensor systems (Chapter 6) will discuss detectors in detail, because the choice of appropriate detector is tied into system considerations such as wavelength sensitivity of the detector, the frequency response, and the sensor area.

In order to characterize polarization, a series of logical steps are necessary; these steps are presented in Fig. 3.14. The steps require a polarization analyzer and a device which serves to retard one radiation component relative to the component perpendicular to it; this device is termed a *retardation plate* or *wave plate*. We will digress to discuss the device. The retardation plate, when appropriately chosen, can efficiently convert any form of polarization into any other form. More specifically, a

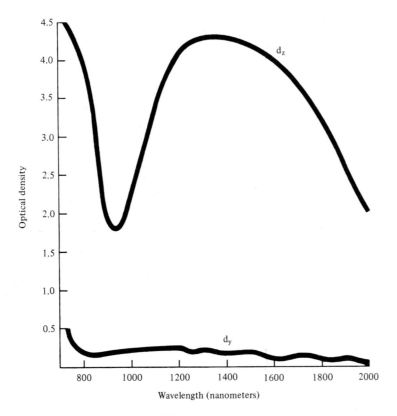

Figure 3.13 Spectral curve: HR.

primary function of retarders is to produce and analyze nonusual circular and elliptical polarization.

Physically, a retarder is a thin crystalline plate, such as calcite, with its optic axis parallel to the plane of the plate. Depending upon whether the incident ray from the left is polarized parallel to the x or y axis, the index of refraction that it sees differs. The index of refraction along the y axis (o ray) is larger than that along the x axis (e ray), resulting in a wave that travels faster when polarized along the x axis. Thus, when a wave is plane polarized at an angle of 45° (half way between the x and y axis), it is split into two components along the respective axes. If the relative delay of the fast and slow waves is an integral multiple of a wavelength, there is no effect on the ray emerging from the retarder except for some slight loss on external reflection from the left side of the plate, internal reflection on the right side of the plate, and absorption within the plate (and possibly scattering from molecules within the plate). However, if the delay is some fraction of a wavelength, various things may happen. Thus, if the delay is 90° (i.e., $\pi/2$ or $\lambda/4$), it may be seen [Eq. (3-7)] that circularly polarized radiation is produced. Conversely, circular polarization may be converted to linear polarization using a $\lambda/4$ (quarter wave) plate. If the incident plane polarized radiation were at an angle other than 45°, elliptical polarization would be produced, the ellipticity depending upon the angle.

Most quarter wave plates have chromatism; that is, the phase shift depends on the wavelength of the radiation: a quarter wave plate cut for one wavelength would

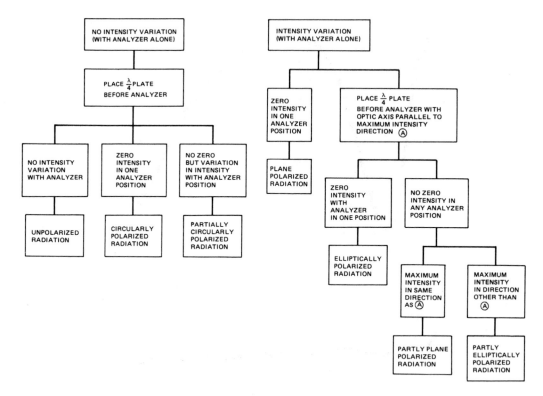

Figure 3.14 Characterization of polarization.

produce a different phase shift at another wavelength. Because of the need for a fixed phase shift ($\lambda/4$, for instance) in order to analyze polarized radiation, adjustable phase shifters are available (made by Continental Optics, Hauppauge, NY), transmitting the range from the ultraviolet to the long wavelength infrared. However, the Polaroid Corporation makes a variety of circular polarizers, restricted to producing a quarter wave phase shift at one wavelength.

Returning to Fig. 3.14, and using the quarter wave plate, we can characterize the polarization of an incident beam of radiation. The quantitative specification depends upon a mathematical description, a convenient one being Stokes parameters.

4

Stokes Parameters

The general representation of radiation—unpolarized, partially polarized, completely polarized, as well as circularly or elliptically polarized—is achieved through the use of the Stokes vector. The Stokes vector consists of a set of four numbers arranged in a matrix form defining a particular form of radiation:

$$\begin{bmatrix} 1 \\ 0 \\ 0 \\ 0 \end{bmatrix}: \qquad \text{unpolarized radiation of unit intensity;} \qquad (4\text{-}1)$$

$$\begin{bmatrix} 1 \\ \pm 1 \\ 0 \\ 0 \end{bmatrix}: \qquad \begin{array}{l}\text{horizontally }(+)\text{ or vertically }(-)\text{ polarized radiation of} \\ \text{unit intensity;}\end{array} \qquad (4\text{-}2)$$

$$\begin{bmatrix} 1 \\ \pm 0 \\ 1 \\ 0 \end{bmatrix}: \qquad \begin{array}{l}\text{radiation of unit intensity polarized linearly at }\pm 45°\text{ to} \\ \text{the reference axes;}\end{array} \qquad (4\text{-}3)$$

$$\begin{bmatrix} 1 \\ 0 \\ 0 \\ \pm 1 \end{bmatrix}: \qquad \begin{array}{l}\text{right }(+)\text{ or left }(-)\text{ circularly polarized radiation of} \\ \text{unit intensity;}\end{array} \qquad (4\text{-}4)$$

$$\begin{bmatrix} I \\ Q \\ U \\ V \end{bmatrix}: \qquad \begin{array}{l}\text{general form of Stokes vector, where }I,\ Q,\ U,\text{ and }V\text{ are} \\ \text{the Stokes parameters.}\end{array} \qquad (4\text{-}5)$$

The four Stokes parameters are defined by

$$I = \langle A_x^2 + A_y^2 \rangle = A^2,$$

$$Q = \langle A_x^2 - A_y^2 \rangle,$$

$$U = \langle 2 A_x A_y \cos \gamma \rangle,$$

$$V = \langle 2 A_x A_y \sin \gamma \rangle,$$

(4-6)

where A_x, A_y are the amplitudes of the electromagnetic waves in mutually perpendicular directions, A^2 is the intensity, γ is the phase angle between A_x and A_y, $\langle \ \rangle$ indicates time averaging, and $I^2 = Q^2 + U^2 + V^2$.

Briefly, the quantity I, the intensity of radiation, is the result of photometric measurement. The quantity Q is the difference between the intensities of radiation in the mutually perpendicular directions used to specify A_x and A_y. If the mutually perpendicular directions used in the remote sensing program are those perpendicular and parallel to the plane of vision (i.e., the plane defined by the sun, the viewed point on the remotely sensed surface, and the sensor), then it has been found that, because of symmetry, $U = 0$. The quantity U indicates the excess of radiation in the $+45°$ direction over that in the $-45°$ direction relative to the plane of vision; V is a quantity that indicates the amount of circularly polarized radiation. The quantity Q is also measured by a plane polarization sensor; Q/I is then proportional to the percent polarization P (%) [Eq. (3-10)]:

$$P\ (\%) = 100 \frac{\langle A_x^2 - A_y^2 \rangle}{\langle A_x^2 + A_y^2 \rangle} = 100 \frac{Q}{I}.$$

(4-7)

There have been a number of apparatuses described that perform automated analyses of polarized radiation, some generating values for the Stokes parameters. Plane polarized radiation is the most commonly occurring form in remote sensing of natural objects. Strong circular or elliptical polarization occurs with metallic reflection, but to a slight degree may occur in atmospheric scattering by particulate aerosols, especially if they are anisotropic.

A spectropolarimeter/photometer that has been used by the author for laboratory and aircraft measurements is described in Chapter 8. The system utilizes quartz optics, and observations may be made in the spectral range from 0.185 to 2.7 μm. In operation, the spectral region is chosen by the appropriate filter–sensor–polarization-analyzer combination. The polarization analyzer is rotated by an electric motor at 6.5 Hz producing a fluctuating voltage, for polarized incoming radiation, of 13 Hz. Simultaneously, a sine/cosine resolver electrically indicates the angular position of the polaroid on an oscilloscope. The voltage from the output anode of the photomultiplier may be simultaneously fed to both an ac and dc recorder. Provision must be made for elimination of dark current from the photomultiplier or PbS system sensor. The dc and ac outputs are proportional to the Stokes parameters I and Q. The plane of polarization is read from the oscilloscope from a Lissajous figure. The analysis of circularly polarized radiation of a particular wavelength requires the use of a quarter wave plate before the analyzer in accordance with the procedures outlined in Fig. 3.14.

Conventional ellipsometers, on the other hand, are designed to determine the polarization of radiation reflected from thin films or metallic surfaces, where elliptically polarized radiation is normally produced. Plane polarization and photometry are generally available within the system, but are of little interest. However, astronomers are interested in both photometry and polarization, and their polarimeters are designed to function in the manner of the spectropolarimeter/photometer just described, to produce Stokes parameters. There are also polarimeters, based on Fresnel's equations, that yield only polarization, the photometry being submerged into a calibration procedure.

A short description of these three representative types of polarimeters will be given.

Ellipsometer

The conventional Gaertner ellipsometer (1968) consists of a light source, a polarizer, a compensator or a quarter wave plate, experimental surface, analyzer, and detector. The intent of the ellipsometer is the measurement of the polarizing properties of surfaces (i.e., ultimately the Stokes parameters of the surface); the polarimetric properties of remotely scattered or reflected radiation is not determined with this type of instrument. There is a need to characterize surfaces in the laboratory in terms of Stokes parameters for subsequent use in mathematical models. A measurement with the Gaertner ellipsometer involves the manual adjustment of the polarizer and analyzer positions (or the compensator could be adjusted) to provide a null position. The azimuthal positions of the polarizer and analyzer yield the phase retardation Δ and amplitude attenuation ψ that represent the phase relationship change of the $\parallel$ and $\perp$ components upon reflection from the experimental surface (i.e., elliptical polarization). The plane polarized components are obtained for the polarizer and analyzer both oriented $\parallel$ or $\perp$ (with no compensator), and the cross-polarized components for the polarizer $\parallel$, analyzer $\perp$, and polarizer $\perp$, analyzer $\parallel$. These plane polarization measurements imply a reference plane, usually taken as the plane of vision, as indicated previously. The ellipsometric procedure requires a double balancing, which is time consuming.

At this point, it is appropriate to digress again and bring in the matrix representation of the Stokes parameters for a general surface that has both Fresnel reflectivity plus scattering. In essence, the surface performs a transformation of the properties of the radiation incident upon it, just as a polarizer, analyzer, or quarter wave plate does. For instance, a linear polarizer with the transmission axis horizontal ($\theta = 0°$) would be represented in either of the following two ways by the Mueller matrix,

$$\frac{1}{2}\begin{bmatrix} 1 & 1 & 0 & 0 \\ 1 & 1 & 0 & 0 \\ 0 & 0 & 0 & 0 \\ 0 & 0 & 0 & 0 \end{bmatrix}$$

or the Jones matrix

$$\begin{bmatrix} 1 & 0 \\ 0 & 0 \end{bmatrix}. \tag{4-8}$$

In order to discuss these matrices in general, the following representations are used:

$$\begin{bmatrix} A_{11} & A_{12} & A_{13} & A_{14} \\ A_{21} & A_{22} & A_{23} & A_{24} \\ A_{31} & A_{32} & A_{33} & A_{34} \\ A_{41} & A_{42} & A_{43} & A_{44} \end{bmatrix},$$

$$\begin{bmatrix} A_{11} & A_{12} \\ A_{21} & A_{22} \end{bmatrix}.$$

(4-9)

The various properties of surface reflection and scattering are then given in terms of the matrix coefficients. Thus A_{11} represents $\parallel$ polarizer, $\parallel$ analyzer conditions; A_{22} for $\perp$ polarizer, $\perp$ analyzer; and A_{12} and A_{21} are the cross-polarization components. The coefficients in the other positions in the Mueller matrix (as well as admitting the possibility of the elements in the Jones matrix assuming values of $A_{12} = A_{21} = \pm i$ and $A_{12} = -i$, $A_{21} = i$) permit assuming a reference plane that is not the plane of vision (i.e., at $\pm 45°$), and right circular polarization respectively. This brief exposure will serve to introduce the concept of matrix representation that will be developed more fully in later chapters.

We now return the discussion to the ellipsometer. Ord has constructed a null-balancing ellipsometer with the polarizer and analyzer positions controlled by a computer. However, the system requires a high signal-to-noise ratio, achieved only with a laser. The first rotating polarizer unit was suggested by Kent and Lawson, and elaborated upon by Van der Meulen and Hien. The Van der Meulen and Hien device samples the light passing through a rotating analyzer at 512 equally spaced intervals, and subjects the data to Fourier analysis to obtain the ellipse parameters. A more recent embodiment employs a double modulation configuration of Stobie, Rao, and Dignam; double modulation is achieved by the use of a polarizer and analyzer at fixed azimuths and a rotating polarizer either before or after the reflecting surface. The system has the advantage that it is not a null system in the conventional sense, and does not involve signal amplitude measurements. However, in order to obtain the Stokes parameters for a surface, the plane polarized components must be determined.

Astronomical Polarimeter/Photometer

A typical astronomical polarimeter/photometer has been used by Gehrels (1960) at the University of Arizona for measurements of planetary photometry and polarization. It consists of a cryostat (ice box) containing photomultiplier sensors preceded by a Wollaston prism polarization analyzer. Dual photomultipliers are used with integrators, one set with an S-1 response (RCA-7102) and one set with S-20 response (RCA-7265)—see Chapter 6 for description of response curves. Dry ice is used to cool the photomultipliers, although liquid nitrogen could also be used. Precautions must be taken to assure no moisture accumulates on the photomultiplier tube terminals. The purpose of the dual photomultipliers is measurement reliability because the sensitivity of end window photomultipliers can vary greatly over the sensitive area. The cryostat is positioned at the telescope focal plane and can be rotated to determine the polarization properties of the incoming radiation. A telescope guiding eyepiece is included; the photometer system contains a Lyot depolarizer (see Chapter 3), a field viewing eyepiece, a focal-plane diaphragm slide, a centering eyepiece, and a filter holder. Although originally set up as a U, B, V photometer (see Chapter 17), narrow band interference filters may also be used.

During operation, the integration times usually were between 10 and 40 sec, although photon counting could be used for weak sources. The Wollaston prism gave full transmission down to a wavelength of 0.290 μm and to the infrared at 1.0 μm. Frequent calibration assured the accuracy and precision of the measurements. The removable depolarizer allows determination of the first Stokes parameter with the same apparatus.

Lyot Polarimeter

The Lyot polarimeter is a visual device that measures only polarization; the photometry is canceled out by the calibration technique inherent in its operation. The polarimeter has the advantage that no electronic instrumentation is necessary for the operation, but the considerable disadvantages are that it is limited to the visual range, is nonrecording, and does not supply photometry. The device directly measures the percent polarization and is an outgrowth of the Savart polariscope. The amazing sensitivity of the unit is that a polarization of 0.001 can be measured. It has been used to measure planetary and terrestrial polarization.

The principle of operation consists of slightly polarizing the light rays that enter the polariscope alternately parallel and perpendicular to the principal axis of the analyzer. With this artifice, interference fringes become quite visible. By rotating the polarimeter around its axis to assume orthogonal positions, the equality of the fringe intensity is achieved through the adjustment of a compensating plate. The percent polarization is then read directly from a calibrated sector.

A few words are now in order about the Stokes vector as a tool for interpreting the photometric and polarimetric properties of surfaces.

Assume that the exact nature of the surface scatterer is known. Then the scattered field could, in principle, be determined from the incident field

$$\begin{pmatrix} E_{\perp} \\ E_{\parallel} \end{pmatrix} = \begin{pmatrix} A_2 & A_3 \\ A_4 & A_1 \end{pmatrix} \begin{pmatrix} E_{0\perp} \\ E_{0\parallel} \end{pmatrix};$$

where $E_{\perp \, (\parallel)}$ are the components of the scattered fields parallel (perpendicular) to the plane of vision and $E_{0\perp \, (\parallel)}$ represents the same quantities for the incident field. The elements of the matrix (A) depend on the detailed nature of the scatterer, as well as the propagation vectors of both the incident and the scattered fields. The analysis of a scattering problem consists essentially of determining the matrix (A). In the optical region, the features of the incident and scattered beams that are usually observed are the energy intensities in various polarization modes. The most complete description of such a beam is given by the four-element Stokes vector that can be determined by experiment.

The Stokes vectors of the incident and scattered beams are also related by a matrix transformation

$$
\begin{pmatrix} I \\ Q \\ U \\ V \end{pmatrix} = \begin{pmatrix} F_{11} & F_{12} & F_{13} & F_{14} \\ F_{21} & F_{22} & F_{23} & F_{24} \\ F_{31} & F_{32} & F_{33} & F_{34} \\ F_{41} & F_{42} & F_{43} & F_{44} \end{pmatrix} \begin{pmatrix} I_0 \\ Q_0 \\ U_0 \\ V_0 \end{pmatrix},
$$

where I, Q, U, and V provide information about the intensity of light, the intensity of polarized light, the position of the plane of polarization, and the extent of circular polarization present. The elements of the matrix (F) can be expressed in terms of the elements of the matrix (A). Thus, given a knowledge of the scatterer, one can, in principle, determine the properties of the scattered beam, i.e., the Stokes vector.

However, in the problem of interpreting a remotely sensed signature, we are given the Stokes vector and seek to determine the properties of the scatterer. This requires (1) determining the elements of the matrix (F) from the observational data, (2) inverting the relationship between (A) and (F) so that the elements of (A) can be determined from the elements of (F), and (3) determining the properties of the scatterer from a knowledge of the matrix (A). The last of these is the mathematically difficult inverse scattering problem. A determination of the matrix (A) provides more insight than is available from the Stokes vector alone. Before this can be done, the first of the requirements above must be completed, namely, the determination of the matrix (F).

The elements of (F) are functions of wavelength and the propagation vectors of the incident and scattered fields. It is clear that to determine (F) for one wavelength and at each phase angle, one must perform four experiments, each of which measures the Stokes vector of the scattered beam. The incident beam for the four experiments must be, respectively, (1) linearly polarized parallel to the plane of vision, (2) linearly polarized perpendicular to the plane of vision, (3) linearly polarized at some angle to the plane of vision other than 90° or 0°, and (4) circularly polarized.

When an atmosphere intervenes, which in itself is a scatterer and absorber, a second scattering matrix operates both on the source components and the surface scattered components. The ensuing requirement for the determination of the Stokes parameters of the intervening atmosphere as well as the underlying scattering surface is even more difficult mathematically.

A procedure for determining certain optical properties of the atmosphere (such as the aerosol particle size distribution) is by constrained linear inversion (Egan and Shaw, 1981). In essence, the atmospheric absorption and scattering problem is set up in terms of certain significant variables which are permitted to vary within specified bands in order that singularities not occur in the possible solutions.

II

OPTICAL FUNDAMENTALS

This section deals with the optical and data processing technology that ties the spectral characteristics of a ground, atmospheric, or space target to the final information desired or required about that target. The following are included within the scope of this section: (1) the optical photometric and polarimetric characteristics of selected targets; (2) both imaging and nonimaging sensor systems that translate the optical properties of the target into electrical impulses, whether analog or digital; (3) electrical data transmission systems that carry these analog or digital signals to the final product, whether magnetic tape, photo imagery, or a computer listing; the concept of signal-to-noise ratio (S/N), and the preservation through the data handling system will be discussed; (4) the concept of calibration, introduced in Section I, will be extended to the sensor systems, including photometry, polarimetry, and geometrical registration; (5) introduction to the atmospheric effects of photometry and the polarization on the target signature sensed, in particular the effects of atmospheric scattering and absorption by aerosols as well as molecules; (6) techniques of data handling and analysis, both optical and computer approaches; and lastly (7) the obtaining of the final product of information about the target and the interpretation.

Not included in this section are the relationships between the optical characteristics of the target and the coexisting chemical, physical, biological, geological, or other nonoptical properties. These relationships are more subtle and will be considered in specific applications in the following section (IV).

In anticipation of the detailed sensor system descriptions to follow, a brief outline of the essential characteristics of a remote sensing system is necessary; the characteristics of a target that can be sensed depend upon the sensor system that views the target. The conceptual relationship between the elements of a remote sensing system is shown in Fig. II.1; the ultimate product of the system is information that either may be visually perceived by a viewer or may be presented in some graphic or digital form by a computer. The intervening elements will be indicated briefly.

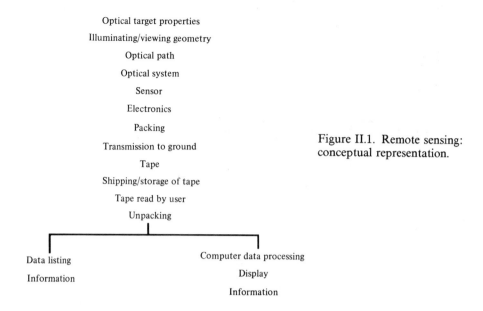

Optical target properties
Illuminating/viewing geometry
Optical path
Optical system
Sensor
Electronics
Packing
Transmission to ground
Tape
Shipping/storage of tape
Tape read by user
Unpacking

Data listing
Information

Computer data processing
Display
Information

Figure II.1. Remote sensing: conceptual representation.

The target itself has certain fundamental physical (as well as perhaps biological and chemical) characteristics. Incident radiation from the sun or sky optically interacts with the target to produce scattered or reflected radiation. The mechanism of the interaction is via the optical complex index of refraction and the inherent geometry of the target. The incident radiation has a spectral distribution of energy, which may be the solar spectrum (Table 2.2), or the solar spectrum modified by transmission properties of the earth's atmosphere (Fig. II.2). The conditions of Fig. II.2 are listed in Table II.1. The scattered radiation from a target has spectral properties, and these spectral properties depend not only on the inherent optical properties of the target but also on the illuminating and viewing geometries.

The optical path (essentially the atmosphere through which the target is viewed) can affect the spectral, photometric, and polarimetric properties of a target as well as the apparent size (i.e., because of haze, the boundaries may no longer be distinct).

Because of its physical size, the optical system has a limit to the number of resolution elements that it views; if it is a scanning system (as most are), it also has a

Table II.1 Conditions for Fig. II.2

Curve	Path length	Date (1956)	Time (P.M.)	Temperature (°F)	Relative humidity (%)	Precipitable water (mm)	Visual range (miles)
A	1000 ft	20 March	3	37	62	1.1	22
B	3.4 miles	20 March	10	34.5	47	13.7	16
C	10.1 miles	21 March	12	40.5	48	52.0	24

Window definitions

I: 0.72–0.94 μm   V: 1.90–2.70 μm
II: 0.94–1.13 μm   VI: 2.70–4.30 μm
III: 1.13–1.38 μm   VII: 4.30–6.0 μm
IV: 1.38–1.90 μm   VIII: 6.0–15.0 μm

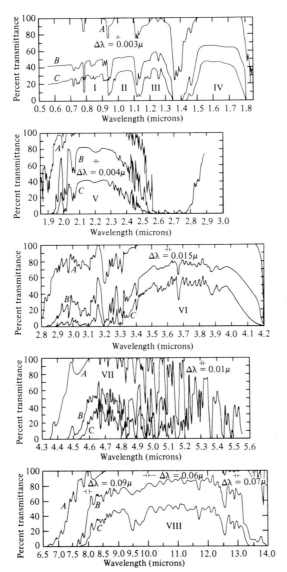

Figure II.2. Atmospheric transmission. (After Taylor and Yates, 1956).

finite time dwell on each target element in the field of view. Large targets are then constructed of a multitude of elements over many scans. The optical system also may introduce extraneous polarization from reflecting off scanning mirrors.

The sensor, which has a finite size and a given spectral sensitivity, may have polarization sensitivity.

The electronics connected with the operation of the sensor are not perfectly linear and may introduce distortion, if not just simply limiting the frequency response of the system.

The electronic data are then packed to optimize the use of the microwave data transmission link to the ground station. The packed data are then recorded on tape and shipped, generally through a center such as the one at Sioux Falls, to the user. The tape is then unpacked, put into a format appropriate for a listing or computer processing. The display or the listing can then furnish the required information.

5

Target Characteristics

A convenient way to lead into a discussion of the characteristics of ground targets is by way of aerial photography interpretation. Since aerial photography involves qualities of man/land features that can be seen visually, and mentally correlated with previously perceived factors derived from experience, the extension to machine (sensor) perceived features and their correlation (via computer with appropriate software) to factors also perceived by machine (chemical, physical, biological properties) are quite easily accomplished. Further, machine processing of remote-sensed data has led to inductive type reasoning, whereby unexpected correlations among data can occur, leading to subsequent inferences about the causative mechanisms.

There will be an emphasis in this text on aspects of target characteristics that are amenable to computer recognition. Thus, for instance, a photointerpreter uses size and shape as recognizable features of a ground target. A computer recognizes boundaries between adjacent areas; the boundary may be sharp or gradual. To recognize a shape requires a correlation be made over an area.

The optical properties of targets are most generally described in terms of spectral photometry, with polarization as a subdivision of photometry. Linked to spectral photometry is the spatial and temporal resolution of the sensor system. Therefore, in the following discussion of target characteristics, the aerial photography features will be described in relation to spectral photometry, polarization, and spatial and temporal resolution of the sensor system that views the target. Sensor systems will be covered in the section following.

One material that has been studied in some detail is volcanic ash from Haleakala on the island of Maui, Hawaii. The reason for the analysis was rooted in prelunar landing preparations, whereby the physical properties of the lunar surface were being inferred from the remotely sensed photometric, polarimetric, and thermal properties. Because the surface of the moon appears to consist of pristine, unweathered soil, it was logical to investigate the optical properties of volcanic material of relatively recent terrestrial origin as a first approximation. Most terrestrial

Figure 5.1 Haleakala volcanic ash, ground and sorted.

materials have been subject to intense weathering and do not really fulfill the requirements of simulated lunar rock. However, volcanic ash samples are readily available materials that permit investigation of the optical properties in question.

The Haleakala volcanic ash as obtained was sorted into different size ranges by mechanical sieving, the reason being that it had been hypothesized that the lunar surface was composed of dust. Thus it was reasonable to determine the photometric and polarimetric properties of the simulated lunar material of various particle sizes. Figure 5.1 shows the general physical appearance of the subdivided samples; it can be seen that the finer powders appear brighter. The actual microscopic appearance of the subdivided samples are shown in Figs. 5.2a–5.2g. The larger particles (Fig. 5.2a) have a vesicular, porelike structure, which gradually degrades to a simple rough surface as the particles become smaller (Figs. 5.2b–5.2g). For the finest particles (Fig. 5.2g), the particles are simply angular, with edges and asperities.

The optical scattering properties of materials ranging from centimeter-size chunks to fine-grained powders require an instrumentation that accurately depicts the actual sensor physical situation (i.e., a telescope). Thus, if one used a very small "scale" system[1], such as a microscope to measure the photometric scattering properties of the Haleakala volcanic ash, the result would be what a microscopic sensor would see. But since the remote sensing used for lunar surface analyses was by means of telescopes, this constitutes a very large scale measurement. To represent the telescopic measurements of the lunar surface properly, a large scale photometer is necessary; such an instrument is shown in Fig. 5.3. This spectrophotometer/polarimeter permits measurements of the scattering properties of samples up to 10 cm in diameter and is suitable for granular samples with large particles up to a few

[1]*Large scale* and *small scale* are used here colloquially, in a sense opposite to that of the corresponding terms in geology and cartography.

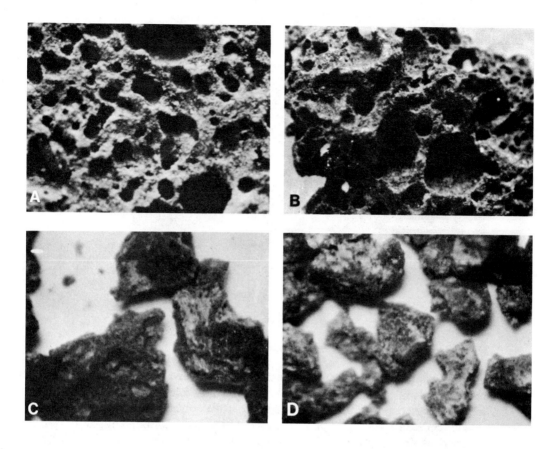

centimeters in size. Where rocks or boulders exist, on the order of meters in size, this apparatus would no longer be suitable. The reason for the large size instrument is the measurement of shadowing produced by the particles on each other from different incident illuminating geometries; a small size viewing area instrument will not measure this effect. Further, the spectrophotometer/polarimeter of Fig. 5.3 is a nonimaging type, i.e., it integrates the photometric/polarimetric response over the entire sample area, and does not form an image of it. Also the device permits spectral measurements at wavelengths from the blue (0.400 μm wavelength) to the infrared (1.0 μm wavelength).

The geometrical relationships possible with the photometer/polarimeter are shown in Fig. 5.4; here the incident ray direction is variable, the source of illumination being on a moving arm. Three sensors are permanently positioned at angles to the vertical of 0°, 30°, and 60°. The angle between the incident direction and the sensor direction is termed the *phase angle*, analogous to the phase of the moon. The plane defined by the incident direction and viewing direction is called the *plane of vision*, and it does not always contain the perpendicular line to the surface observed.

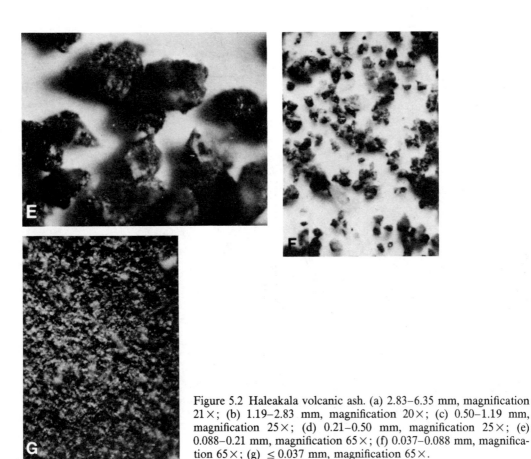

Figure 5.2 Haleakala volcanic ash. (a) 2.83–6.35 mm, magnification 21×; (b) 1.19–2.83 mm, magnification 20×; (c) 0.50–1.19 mm, magnification 25×; (d) 0.21–0.50 mm, magnification 25×; (e) 0.088–0.21 mm, magnification 65×; (f) 0.037–0.088 mm, magnification 65×; (g) ≤ 0.037 mm, magnification 65×.

The results of measurements of the radiation scattered from the various particle sizes of Haleakala volcanic ash with the photometer/polarimeter of Fig. 5.3 are presented in Fig. 5.5. Here the relative brightness as seen by the 0°, 30°, and 60° sensors is plotted versus both angle of incidence of the illumination and the corresponding phase angle. The average normal to the sample surface lies in the plane of vision. The physical properties of the volcanic ash are listed in Table 5.1 (i.e., grain sizes, weight and volume of solids, apparent and real densities, porosity, and albedo versus viewing angle). (The albedo is the average reflectance of the sample relative to a freshly prepared $MgCO_3$ block used as a reference standard). The shaded areas in Figs. 5.5a, 5.5b and 5.5c represent the range of photometric properties of the lunar surface as indicated by Minnaert-Van Diggelen, and Fedoretz-Orlova (for continents and maria).

It is seen that in all cases, the brightness tails off at the higher phase angles, drastically differing from the idealized Lambertian (cosine) dependence. The reason for this extreme difference is surface shadowing of one particle by adjacent particles.

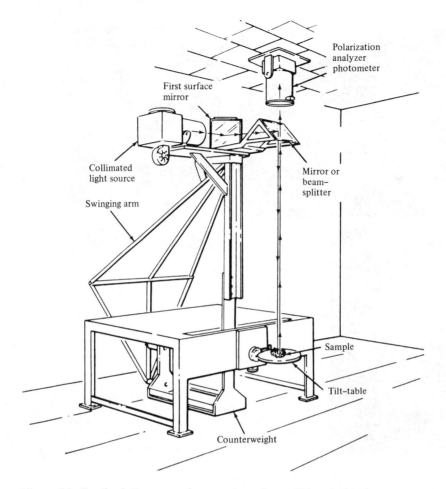

Figure 5.3 Sketch of Grumman photometric analyzer (30° and 60° photometers not shown).

The discontinuity near 0° phase angle is caused by the system mirror or beam splitter getting in the optical path to the respective sensors. The peaking of the curves near 0° phase angle is called the *opposition effect*, and results in a higher surface brightness than that of a Lambertian surface. The observed representation of this opposition effect is reduced by the attenuation of the beam splitter, the peaking actually producing a photometric result greater than 1.0 at 0° phase angle. All of the curves are normalized to 1.0 at a 5° phase angle for convenience of comparison, even though the smaller particles are brighter (higher albedo, Table 5.1).

In comparison, the corresponding polarimetric curves are shown in Figs. 5.6 and 5.7. Here the percent polarization is plotted versus phase angle for two of the three sensor positions, 0° and 60°. The standards of comparison are two lunar areas, one a darker maria (Crisium) and the other a lighter highland area (Clavius).

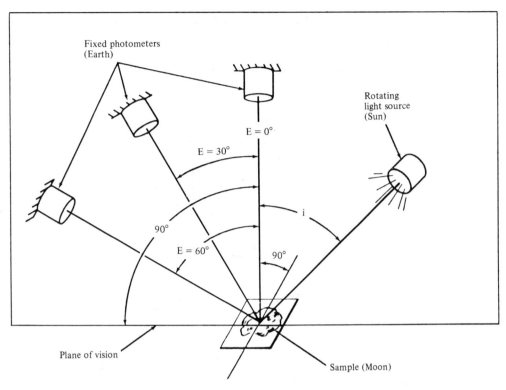

Figure 5.4 Laboratory simulation of sun–moon–earth optical relationship (intensity equator and phase angle are in plane of vision).

The percent polarization is seen to rise at the higher phase angles (Figs. 5.6 and 5.7), with a change in polarization from positive to negative at a phase angle near 22.5° (termed the inversion angle). In Fig. 5.7, the smaller, brighter surfaces composed of smaller particles have the lower maximum polarization at the highest phase angles, and the darker, rougher surfaces have the higher polarization. Neither the location of the inversion angle nor the negative polarization maximum trends are clearly related to particle size. The effect of change in viewing angle from 0° to 60° on polarization (compare Fig. 5.6 with Fig. 5.7) is negligible. Further, the effect of simulated change in lunar longitude produced negligible change on the percent polarization as a function of phase angle.

Thus, even though the Haleakala volcanic ash was chosen as an example to characterize the optical effects of all photometric surfaces, the same basic properties are apparent for all surfaces, whether they be soil, cornfields, trees, houses, buildings, etc. The effect of shadowing on apparent remotely sensed surface brightness is very significant, particularly with machine processing of data. One convenient way out of this dilemma is the use of "training areas," areas of known composition, and programming the computer algorithm to recognize a particular brightness range as

Table 5.1 Grain Size, Porosity and Albedo of Volcanic Cinder No. 4 Powders

Curve number ref. Figs. 5.1, 5.2	Grain size		Weight of solids, W_s g	Apparent volume, V_a cm^3	Volume of the solids, V_s cm^3
	Range mm	Average mm			
1		80			
2	6.35–2.83	4.6	4.9	5.8	1.8
3	2.83–1.19	2.0	4.7	4.7	1.6
4	1.19–0.50	0.85	6.4	5.5	2.1
5	0.50–0.21	0.35	5.5	4.0	1.8
6	0.21–0.088	0.15	6.3	4.8	2.1
7	0.088–0.037	0.062	6.5	5.4	2.3
8	0.037	0.02	4.4	5.5	1.5
	Average ρ_r				

[a] Based on average ρ_r

Figure 5.5 Photometry of Haleakala volcanic cinder as a function of grain size, porosity, and albedo. Hatched area is lunar standard. Specimens were volcanic cinder No. 4, curves 1–8. Refer to Table 5.1 for grain size, porosity, and albedo.

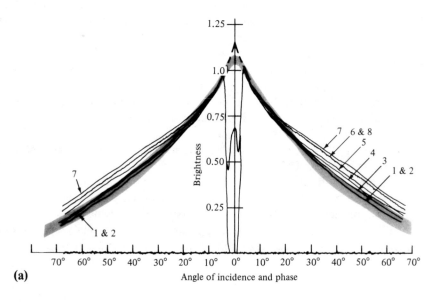

(a)

Table 5.1 (*continued*)

Apparent density, g/cm³ $\rho_a = \dfrac{W_S}{V_a}$	Real density g/cm³ $\rho_r = \dfrac{W_S}{V_s}$	% porosity[a] $n = 100\left(1 - \dfrac{\rho_a}{\rho_r}\right)$	Albedo @ viewing angle		
			0°	30°	60°
			0.14	0.14	0.14
0.85	2.72	71	0.14	0.14	0.14
1.00	2.94	64	0.14	0.14	0.14
1.16	3.05	60	0.14	0.14	0.14
1.38	3.05	53	0.14	0.14	0.14
1.31	3.00	55	0.14	0.13	0.13
1.20	2.82	59	0.14	0.13	0.12
0.80	2.93	73	0.18	0.17	0.16
	2.93				

Figure 5.5. (*continued*)

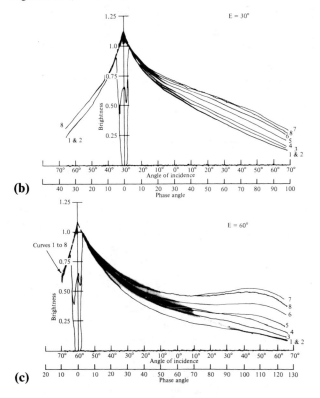

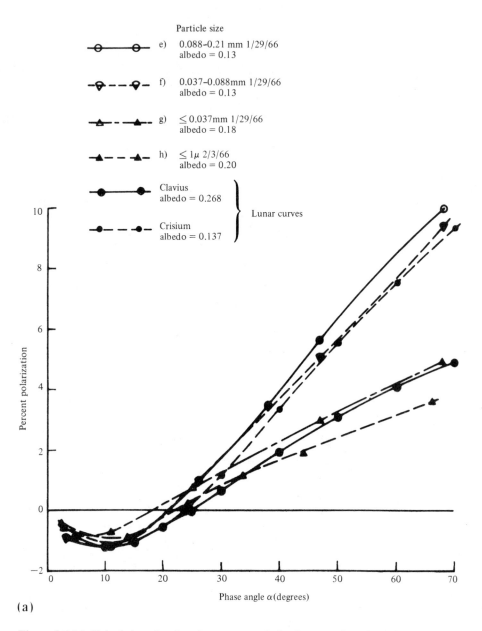

Figure 5.6 (a) Haleakala volcanic ash: percent polarization as a function of particle size for smallest particles. 0° polarimeter. (b) Haleakala volcanic ash: percent polarization as a function of particle size for largest particles. 0° polarimeter.

the known composition. However, when the ground truth is not conveniently available, or a comparison is to be made under different illuminating and viewing directions, the surface photometric function must be determined (and hopefully in the not too distant future, the polarimetric function).

The causes of the observed photometry and polarization are linked to the

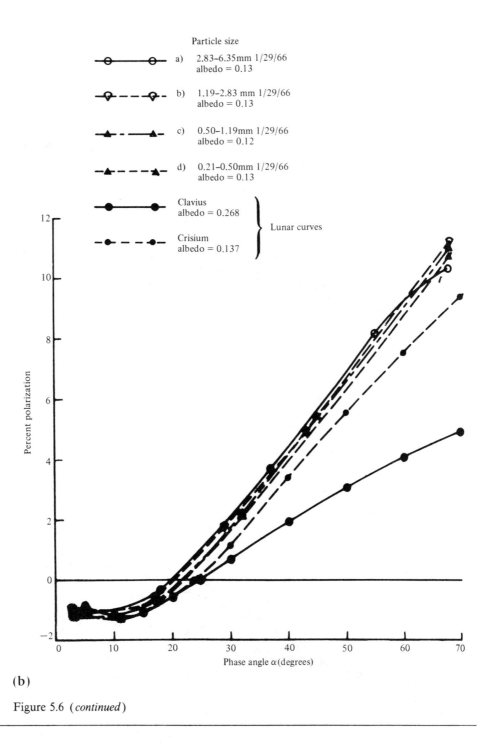

(b)

Figure 5.6 (*continued*)

geometry and optical properties of the scattering surface. For instance, the complex index of refraction of the Haleakala ash (Fig. 5.8) can be used in the modeling of a nonshadowed surface (Egan and Hilgeman, 1979). However, the photometric shadowing requires another approach. One photometric modeling approach uses large scale analytical geometrical features to represent the irregularities of the

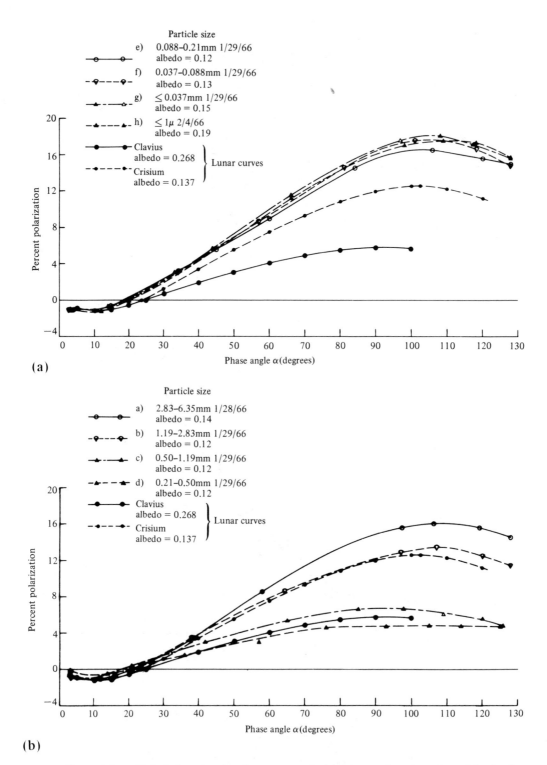

Figure 5.7 (a) Haleakala volcanic ash: percent polarization as a function of particle size for smallest particles. 60° polarimeter. (b) Haleakala volcanic ash: percent polarization as a function of particle size for largest particles. 60° polarimeter.

surface particles. Figure 5.9 represents the geometry of a "T" shaped structure situated atop the viewing plane. The "T" may be separated into a horizontal top (Fig. 5.10, top), or the vertical portion (Fig. 5.10, middle), or the composite (Fig. 5.10, bottom). The corresponding computed photometric functions are shown on the right-hand side of Fig. 5.10; the "T" model is not too different from the lunar photometric function. By adding additional features to the "T," small changes in the photometric behavior are produced. It is assumed that the elements of the model are Lambertian.

However, the polarization properties of a simple dielectric rough surface model are not so easily modeled. Various unsuccessful approaches have been tried, and it is of historical interest to examine these to gain insight into polarization mechanisms.

The Mie theory has been used to model the lunar surface, assumed composed of dust particles, admittedly not one of the more sophisticated terrestrial surfaces. Qualitatively, the curves of Fig. 5.7 may be explained for very small particles by the idea that most of the scattered light is reflected from the surfaces and tends to be positively polarized at the higher phase angles. As the particle size increases, more light penetrates the particle and upon emergence tends to be negatively polarized. Hence the curves of Fig. 5.7 may be explained qualitatively by saying that as the particle size increases, negative polarization from the interior is added to positive polarization from the surface in varying amounts.

The same reasoning applies when an imaginary component is added to the real portion of the complex index of refraction. The increase of absorption within the particle tends to suppress the contribution of the negative polarization and generally

Figure 5.8 Optical properties of Haleakala volcanic ash (see also Appendix).

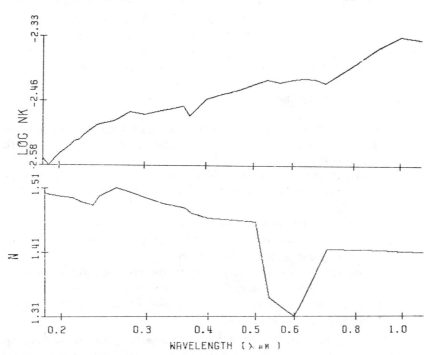

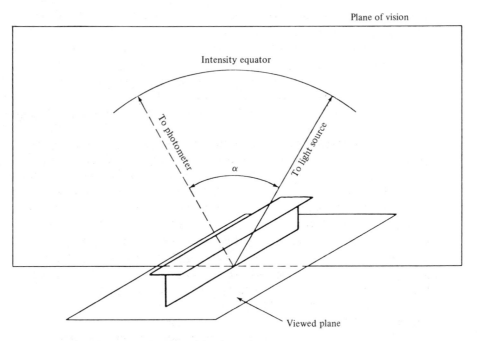

Figure 5.9 Intensity equator and model orientation.

shifts the polarization toward more positive values. This is shown quite clearly in Fig. 10 of Deirmendjian, Clasen, and Viezee (1961) and reproduced here as Fig. 5.11.

The lunar polarization curves are quite similar to the ones shown typically in Fig. 5.7. However, it is difficult to explain how all the laboratory samples, from chunks up to 1 cm in size with pores down to powders less than 1 μm in size, plus the lunar data, all seem to fit a Mie theory analysis with the Mie size parameter approximately equal to 2 for an index of refraction of 1.33 at a wavelength of 0.54 μm. This seems to indicate quite clearly the inapplicability of the Mie theory to polarization.

Criticism of the Mie theory approach is based on the following facts:

1. Crystals in lunar rock particles are composed of rough and planar surfaces and would be expected to give different scattering diagrams than spherical Mie particles of the same size.
2. Deirmendjian et al. show that spheres scatter strongly in the forward direction. This is contrary to the backscatter of light and opposition effect observed for lunar features.
3. It is doubtful whether the range of size parameter and index of refraction of the lunar surface material could be as small as would be necessary to explain negative polarization by the Mie theory.
4. The scattering efficiency (ratio of scattering cross section to geometric cross section) of a 0.8-μm-diameter Mie particle should decrease with increasing wavelength for $\lambda \geq 1$ μm. However, in contrast, the reflectivity of the moon continues to rise in the infrared.

At this point, the reader may feel that there is undue emphasis on the polarimetry of the moon, but it must be emphasized that astronomers have made detailed

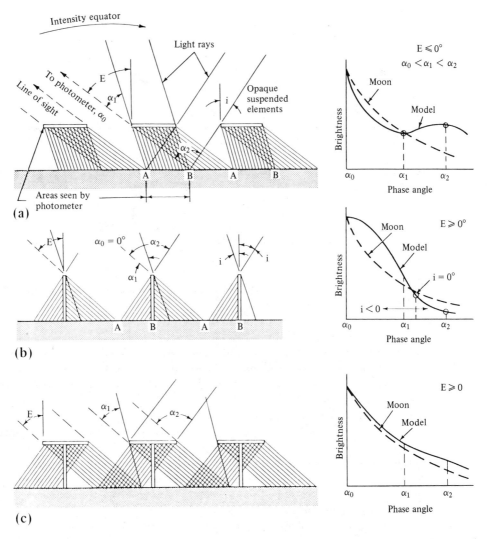

Figure 5.10 "Simple" models and their computed photometric function. All surfaces and cross sections in plane of vision. (a) Model with horizontal elements only. (b) Model with vertical elements only. (c) Model "T": horizontal and vertical elements.

polarimetric measurements of the moon for over 50 years; it is thus about the best data bank available in polarization.

Another approach for polarization modeling was the use of the Born approximation that expresses the far field scattered by a closed surface dielectric, of index of refraction close to unity, in the form of an infinite series. The result is the representation of Rayleigh scattering in the first Born approximation.

A single layer, plane facet model (Fig. 5.12), where two dielectric regions in one plane are separated by a faceted surface, has been investigated. The upper dielectric region has an index of $n_1(=1)$ and that below it, n_2. The normal to each facet (the micronormal) is different from that of the surface (the macronormal). Unfortunately, the polarization properties of this model are unsatisfactory because only the Fresnel reflection coefficients are involved, and since the parallel component of reflected

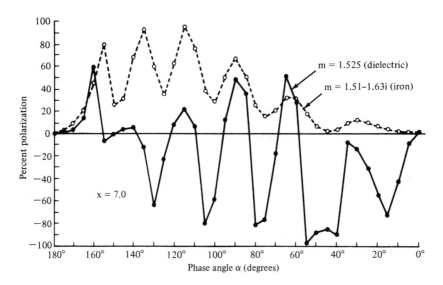

Figure 5.11 Effect of absorption on polarization curves computed from Mie Theory for $x = 7$ and $m = 1.525$ (dielectric) and $m = 1.51-1.63i$ (iron).

radiation is zero at the microsurface Brewster angle, this model predicts 100% polarization at a phase angle of twice the Brewster angle. Also, since the percent polarization is a function only of the phase angle and not the incident angle of the radiation or the sensor direction, the shadowing caused by the facets within an element or by elements of the macroscopic model will have no effect on the polarization properties.

Thus double layer models appear to be necessary, at least in a two-dimensional model. There is a requirement in the plane parallel model to have some of the parallel polarized light transmitted into the medium be scattered back out again at the Brewster angle, so that the polarization does not reach 100%. This produces a closer fit to the real physical situation because it is reasonable to assume that surface layers of the moon are inhomogeneous and light will be scattered back by inhomogenieties. However, this model requires rather restrictive values on the indices of refraction of the layers and does not produce negative polarization at low phase angles. Also the polarization maximum and phase angle at which it occurs are higher than those on the lunar surface. Further, the polarization properties of the model do not depend on the geometry of the surface.

To examine the dependence of the polarization on geometry, it is necessary to construct a model in which the rays that are refracted within the upper layer pass

Figure 5.12 Single layer-plane facet model.

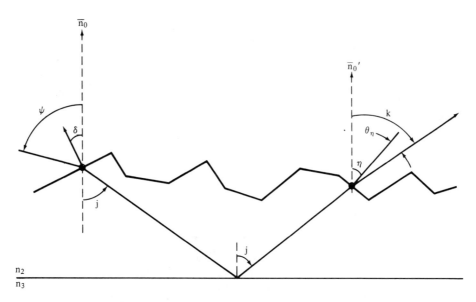

Figure 5.13 Path of a ray through the medium η_1 from the entrance facet ($\xi = \delta$) to the exit facet ($\xi = \eta$).

out through a facet other than the one through which they entered. This model is shown in Fig. 5.13. The ray enters a facet into medium of index n_2, is reflected from medium n_3 back into medium n_2, and out through another facet. The results of the modeling are shown in Fig. 5.14; negative polarization is reproduced, as required, and the lunar polarimetric and albedo properties are appropriately characterized. The model is still somewhat contrived, in that the negative polarization is dependent on the selection of n_2 and n_3.

Photometry of Composite Areas

Since most targets are composites depending on the scale of the imagery, the remotely sensed photometric and polarimetric properties (colloquially termed the *signature*) will correspondingly be a composite. Thus, the photometric properties (first Stokes parameter) will be a photometric weighting of the various identifiable elements i in a scene.

Let us define the following: $I_{i\lambda}(\theta, \varphi)_\perp$ is the normalized relative brightness (photometric function) of crop or background in perpendicular orientation in field of view of sensor; $A_{i\perp}$ is the projected perpendicularly oriented area of crop or background in field of view of sensor; $\mathscr{A}_{i\lambda\perp}$ is the geometric albedo (relative to $MgCO_3$) of crop or background in perpendicular orientation; and the following are similar terms for parallel orientation: $I_{i\lambda}(\theta, \varphi)_\|$, $A_{i\|}$, $\mathscr{A}_{i\lambda\|}$, where θ, φ are the incident and emergent light directions and $I_{T\lambda}(\theta, \varphi)$ scene brightness relative to $MgCO_3$ as a function of geometry at a given wavelength λ. Then, for a geometrical condition where the average normal to the earth's surface lies in the plane of vision (POV):

$$I_{T\lambda}(\theta, \varphi) = \frac{\sum\limits_{i=1}^{n} \left[\mathscr{A}_{i\lambda\perp} A_{i\lambda\perp} I_{i\lambda}(\theta, \varphi)_\perp + \mathscr{A}_{i\lambda\|} A_{i\lambda\|} I_{i\lambda}(\theta, \varphi)_\| \right]}{\sum\limits_{i=1}^{n} \left(A_{i\perp} + A_{i\|} \right)}. \tag{5-1}$$

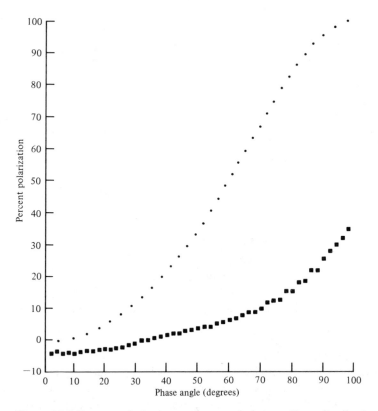

Figure 5.14 Percent polarization vs phase angle for a uniform distribution with facets at every 0.5°; viewing angle 10°, detector acceptance half-angle 1°; dots represent $m_2 = 1.2$, $m_3 = 1.4$, albedo = 0.010; crosses represent $m_3 = 1.4$, albedo = 0.19.

This calculation is made, in the most general case, with a computer and requires the determination of the photometric functions $I_{i\lambda}(\theta, \varphi)_{\perp}$ and $I_{i\lambda}(\theta, \phi)_{\|}$ as well as the geometric albedos $\mathscr{A}_{i\lambda \perp}$ and $\mathscr{A}_{i\lambda\|}$.

Thus, the effect of orientation may be explicitly determined in the laboratory.

Polarimetry of Composite Areas

An expression analogous to Eq. (5-1) applies for the combination second Stokes parameter (polarization) for a scene $[Q_{T\lambda}(\theta, \phi)]$ within the resolution element of a sensor, i.e., I is replaced by Q in Eq. (5-1), i.e., $Q_{i\lambda}(\theta, \varphi)_{\perp}$ is the relative polarization (second Stokes parameter) of crop or background in perpendicular orientation in field of view of sensor (referenced to the normalized relative brightness curves) and $Q_{i\lambda}(\theta, \varphi)_{\|}$ is the similar term for parallel orientation. Thus:

$$Q_{T\lambda}(\theta, \varphi) = \frac{\sum_{i=1}^{n} \left[\mathscr{A}_{i\lambda \perp} A_{i\lambda \perp} Q_{i\lambda}(\theta, \varphi)_{\perp} + \mathscr{A}_{i\lambda\|} A_{i\lambda\|} Q_{i\lambda}(\theta, \varphi)_{\|} \right]}{\sum_{i=1}^{n} \left(A_{i\perp} + A_{i\|} \right)}. \qquad (5\text{-}2)$$

This expression, in combination with Eq. (5-1), utilizing the second Stokes parameters $Q_{i\lambda}(\theta, \varphi)_{\perp}$ and $Q_{i\lambda}(\theta, \varphi)_{\|}$, yields additional information to relate the variables to a particular remotely sensed area.

6

Sensor Systems

A sensor, in the context of remote sensing, is a device that converts radiant electromagnetic energy from a target into some other form such as electrical, chemical (photographic), or thermal energy (thermopile or pneumatic). Sensors are characterized as imaging and nonimaging generally based on spatial resolution, although the distinction is not a clean one. Thus, depending on the scale of the viewed area as seen by a sensor, one may, in principle, distinguish between imaging and nonimaging sensors. For instance, an aircraft photograph of an agricultural area to any scale would be an imaging system; however, unless a very long focal length camera lens system (with a technique to compensate for aircraft motion) were used, one would not expect to image each leaf of the plants in a cornfield; nor would one expect to see the cell structure of the corn leaves. Scanning sensors, such as those used in satellites, are imaging only to the resolution of the sensors. As individuals, we relate to our surroundings through our senses, vision being the most important. Thus, when remote sensing systems are conceived, people look for high resolution (possibly comparable to the eye as the ultimate standard, or better); there is a desire to "see" the effects of water pollution, forest damage by disease, status of crop maturation, distinction between planted areas, etc.

On the basis of human perception, the push to higher resolution sensors and systems is easily justifiable, although just how high a resolution is open to question. However, how does one "see" polarization? The eye has certain optical properties that permit observation of special polarization effects (Haidingers brush). However, observation of natural polarization phenomena requires the aid of polarization sensing instrumentation (a polarization analyzer). The fact that we do not visually see polarization does not negate its existence. But since polarization does exist in nature and it carries information, it can at least equal the data obtainable photometrically, because for each photometric observation, a corresponding plane polarization property exists. The value can be greater in image classification, to reduce uncertainties based on photometry alone. Natural circular polarization is quite small; a large amount of circular polarization is produced only by reflection or scattering from high conductivity materials (metallic man-made structures). Circular polarization is not as readily sensed as plane polarization, because it requires a quarter wave plate for the wavelength sensed as well as a polarization analyzer.

Furthermore, most sensors, as well as the lenses or mirrors that make up the optical sensor system, may have a polarization sensitivity or a polarization bias. The fidelity of sensor response desirable for astronomical observations is a plane polarization bias of $< 0.01\%$, with a resolution of 0.01% plane polarization. Terrestrial observations should aim for these levels of operation, as borne out by field and laboratory measurements of terrestrial objects.

In addition to the aforementioned spatial resolution of an image, temporal and spectral properties of the target must be taken into consideration. The temporal aspect may be on the order of a year or more or may be as short as nanoseconds. Annual long period phenomena could be crop maturation, water pollution, urban activity, and the like, whereas shorter time span phenomena could be atmospheric electrical discharges. However, the temporal aspects of the acquired data arising from natural phenomena are generally much longer than the response time of sensors; the high speed response of sensors is ordinarily necessitated by the scanning speed of the mirrors in the remote sensing system. Photographic imagery is not subject to this limitation.

Spectral resolution of a remote sensing system may vary from a fraction of a wavenumber (wavenumber $=1/$ wavelength) to a few tenths of a micrometer wavelength. The higher resolutions are required for remote sensing of the atomic and molecular properties of atmospheric spectra, whereas the lower spectral resolutions are required for the broadband phenomena associated with water pollution, plant leaf quality, and the distinction of man-made features in terrestrial observations. The spectral resolution of a system is keyed into what is to be observed. High spectral resolution is costly to implement in terms of dollars and costly in terms of the length of time necessary for data acquisition and transmission in order to achieve an adequate signal-to-noise ratio (see Chapter 7).

The treatment of sensors in this text initially will be general, outlining the spectral, spatial, and temporal responses of various sensors with emphasis on the region between 1850-Å and 2.5-μm wavelengths. Then, a sensor of a specific satellite will be described (the LANDSAT 4), which is the most recently launched high resolution earth resources satellite. Detailed discussion of the thermal infrared is beyond the scope of this volume; one thermal infrared application of polarization will be presented for sea state classification.

Since no polarimetric remote sensing satellite has been launched to date, the next section provides a state-of-the-art projection of the LANDSAT 4 design as a representation of polarimetry with photometry in remote sensing.

Sensors may have some small polarization sensitivity; however, in general it may be neglected, except in the case of highly refined photometric analyses.

The following discussion of sensors will include photomultipliers, vidicons, and semiconductor detectors. The semiconductor detectors will be further divided into photodiodes, line and area arrays, and charge coupled devices. Then the multispectral sensor (MSS) and the thematic mapper (TM) of LANDSAT 4 will be described as a representative optical system, and modifications suggested for polarization measurements.

The various sensors will be described in terms of their principles of operation, spectral sensitivities, physical characteristics, and ruggedization. For satellite qualification, the ruggedization and reliability approximates that required for military systems. The detectors may be used in broadband or narrow band application by the use of glass or interference filters. For high spectral resolution (atmospheric molecular or atomic species), an interferometer spectrometer would be incorporated in the fore-optics system.

Spectral, spatial, and temporal response characteristics differ drastically among remote sensing detector requirements, from broadband applications of crop variety sensing to high resolution sensing of atmospheric constituents, including temperatures inferred from atomic and molecular line shapes.

Photomultipliers

Photomultipliers are photoemissive detectors ideally suited for operation in the spectral region < 0.550 μm, where their quantum efficiency is high. The signal response time may be made as short as 0.1 nsec. The radiation is sensed by a photosensitive surface partly composed of alkali metals such as cesium, potassium, or sodium. The incident quanta of electromagnetic radiation release photoelectrons from the photocathode of a diode phototube. A photomultiplier has a series of dynodes that are appropriately biased to collect the initially released photoelectrons, add kinetic energy to them, and then, by means of secondary emission, amplify the original photoelectric signal. The photoemissive surface area may vary from a fraction of an inch in diameter up to many inches, depending on application.

The spectral sensitivity depends on the composition of the photosensitive surface. Figure 6.1 shows the spectral sensitivity of various photoemitters; the range of operation extends from 100 nm (0.1 μm) to nearly 1100 nm (1.1 μm). The sensitivity in the red region is limited by thermal noise and can be improved by cooling photomultipliers.

A photomultiplier itself is basically a nonimaging detector, but may be converted into an imaging device by the use of a scanning mirror optical system preceding the sensor.

It is appropriate to specify the performance of a detector in terms of quantities that represent the response of the detector to electromagnetic radiation above the inherent noise that exists within the detector, regardless of how well it is made. Three important quantities will now be described.

The *spectral noise equivalent power* (NEP$_\lambda$) is the rms value of sinusoidally modulated monochromatic (as far as possible) electromagnetic radiant power incident upon a detector, that subsequently produces an rms signal voltage from the detector equal to the rms noise voltage (also from the detector) over a 1-Hz bandwidth (sinusoidal) frequency interval. This requires specifying the chopping (sinusoidal) frequency, electrical bandwidth, detector area, and sometimes the field of view of the detector. The units are W/Hz$^{1/2}$.

The *spectral detectivity* $D(\lambda)$ is a measure of detector sensitivity, being the reciprocal of (NEP$_\lambda$). Also, the chopping frequency, electrical bandwidth, detector sensitive area, and sometimes field of view are specified. The units are W^{-1}.

Lastly, the *spectral D star* $D^*(\lambda, f)$ is the quantity that attempts normalization of the spectral detectivity to take into account detector area and electrical bandwidth dependence. When detectors are background noise limited, the field of view must be specified. The normalization is not valid for many types of detectors.

70

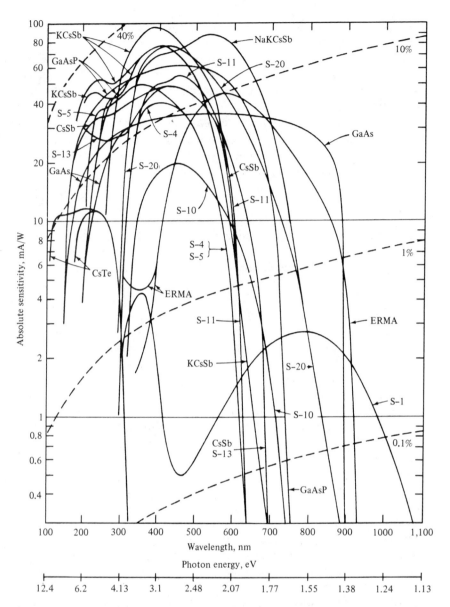

Figure 6.1. Spectral sensitivity of various photoemitters. Dotted lines indicate photocathode quantum efficiency. Chemical formulas are abbreviated to conserve space. S-1 = AgOCs with lime or borosilicate crown-glass window; S-4 = Cs_3Sb with lime or borosilicate crown-glass window (opaque photocathode); S-5 = Cs_3Sb with ultraviolet-transmitting glass window; S-8 = Cs_3Bi with lime or borosilicate crown-glass window; S-10 = AgBiOCs with lime or borosilicate crown-glass window; S-11 = Cs_3SB with lime or borosilicate crown-glass window (semitransparent photocathode); S-13 = Cs_3Sb with fused-silica window (semitransparent photocathode); S-19 = Cs_3Sb with fused-silica window (opaque photocathode); S-20 = Na_2KCsSb with lime or borosilicate glass window. ERMA = extended red multialkali (RCA; ITT uses MA for multialkali). Thus, curve is representative of several manufacturers' products. Many variations of this response are available, e.g., trade-offs between short- and long-wavelength response. (From RCA Electronic Components, PIT-701B.)

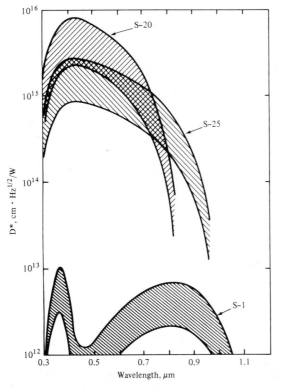

Figure 6.2. Range of D^* for uncooled photomultiplier tubes ($T = 300$ K). For abbreviations see Fig. 6.1. S-25 = same as S-20 but different physical processing. (Based on material from RCA.)

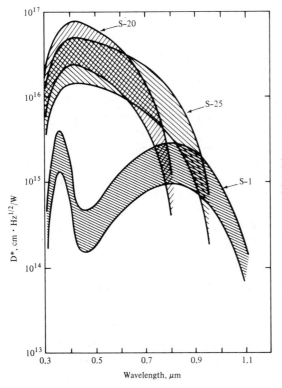

Figure 6.3. Range of D^* for cooled photomultiplier tubes ($T = 200$ K). S-25 same as S-20 but different physical processing. (Based on E. H. Eberhardt, "D^* of photomultiplier Tubes and Image Detectors," ITT Industrial Labs, 1969.)

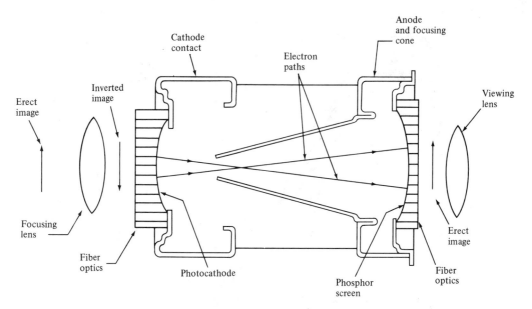

Figure 6.4 Schematic diagram of a typical single-stage first generation electrostatically focused image intensifier tube.

As an example, cooling photomultipliers can significantly increase their sensitivity. Figures 6.2 and 6.3 show a comparison of the range of D^* for uncooled and cooled photomultipliers with S-1, S-20, and S-25 surfaces.

Image Intensifiers

An image-intensifier tube (also known as an image or image converter tube) has a photosensitive surface similar to a photomultiplier. However, it is an imaging device that reproduces the image on the photocathode onto a flourescent screen (Fig. 6.4). The tube has the advantage that the output image is brighter than the input image, and that the output image may be in a different wavelength range than the input image. It thus may convert ultraviolet or infrared radiation into a visible image. The electrons emitted by the photocathode gain kinetic energy from the acceleration through the electric field within the tube to achieve a luminance gain of 50 to 100. If the image tubes are coupled, luminance gains of 10^5 to 10^6 are possible. Typical photosensitive surfaces are ERMA 6-1 and S-20 (Figs. 6.5 and 6.1); tube diameters are typically 18, 25, and 40 mm. Tubes may be electrostatically or magnetically focussed.

Silicon-Intensifier Target Tubes and Intensifier-Silicon-Intensifier Target Tubes

Both silicon-intensifier target (SIT) and intensifier–silicon-intensifier target (ISIT) tubes are imaging devices that employ photosensitive cathodes similar to a photomultiplier. However, the photoelectrons are focused onto a target that provides the gain through a semiconductor device (Fig. 6.6); this device is a tightly spaced matrix

ERMA 6–1, Multialkali
(as modified by fiber optic window)

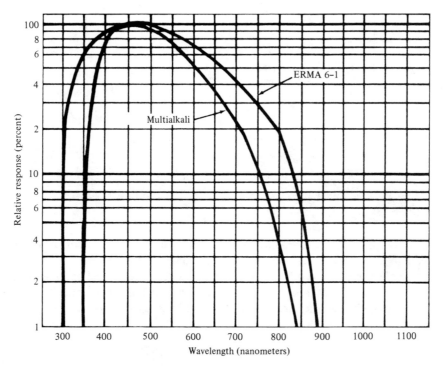

Figure 6.5 ERMA 6-1: The ERMA spectral response (extended red multialkali) is enhanced in the near infrared region of the spectrum and is tested specifically in that region.

of *p–n* junction diodes. Target is used in two senses in this book: as a ground object or area, and as an element of a sensor onto which an image is focused (as in the present section). The center-to-center spacing of the diodes is about 14 μm. The gain is achieved by imparting kinetic energy to the photoelectrons emitted by the photosensitive surface, and may be 1000 or more. Multiple dissociations of

Figure 6.6 Schematic representation of a SIT camera tube.

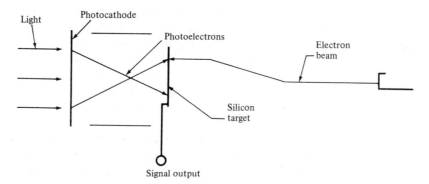

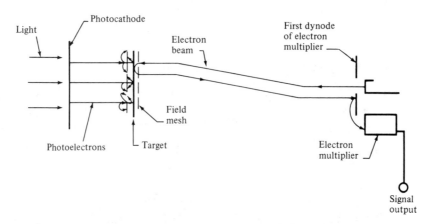

Figure 6.7 Schematic representation of an image orthicon camera tube.

electron–hole pairs occur, and the holes are collected at the p side of the diode; the charge so induced is neutralized by the scanning electron beam.

These tubes are used extensively for low light level pickup. For even greater sensitivity, the ISIT tubes employ a SIT tube that is fiber optically coupled to a single stage image intensifier stage. This results in an increase in sensitivity of about 30 times that of an SIT tube.

Image Isocons and Image Orthicons

Both tubes are imaging types, and the principle of operation is depicted in Fig. 6.7 for the image orthicon tube. The tube utilizes an S-10 photocathode as the sensitive surface. The image pattern consisting of photoelectrons is focused by an axial magnetic field onto a moderately insulating target surface. Then secondary emission produces a net positive charge on the tube target; this charge is neutralized by the scanning electron beam, and the modulated beam returns to an electron multiplier surrounding the gun. The image isocon is essentially an electrostatically focused version of the image orthicon.

Semiconductor Detectors

Strictly speaking, a semiconductor is a solid state material decreasing in conductivity (higher resistivity) as temperature decreases. Some typical materials are silicon, germanium, and carbon. The reason for the decrease in resistance with increasing temperature is that electrons (or holes) acquire enough energy thermally to pass across an energy bandgap with rise in temperature. By proper selection of impurities in the semiconductor, the bandgap may be changed to optimize the ability of the semiconductor to detect a certain range of wavelengths; as the detection wavelength is increased, the detector must be cooled accordingly to decrease the thermal excitation (noise).

Electrons (holes) may be given sufficient energy to cross the bandgap thermally or by photons of electromagnetic radiation. The movement of electrons or holes within the semiconductor from photoexcitation gives rise to a decrease in resistance or an

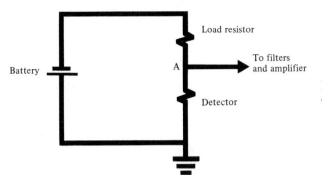

Figure 6.8 Photoconductive detector bias circuit.

electrical current; devices of the former type are termed *photoconductive* (PC), and exposure to electromagnetic radiation causes a small decrease in resistance. Typical detectors of this type are lead sulfide, lead selenide, lead oxide, antimony trisulfide, mercury–cadmium–telluride, and doped and undoped germanium silicon. These detectors must be operated with a bias circuit (Fig. 6.8). The decrease in resistance of the detector as a result of incident radiation will cause a decreased voltage across it, but results in an increased voltage across the load resistor. If the signal is taken as shown, the decrease in detector voltage will be offset by an increase in bias current if a small load resistor is used; conversely, if the voltage is taken across the load resistor, an augmentation will occur, but the increased current through the detector could destroy it.

The other type of detector is the *photovoltaic* (PV). This is a diode that generates a current proportional to the number of photons of electromagnetic radiation striking it. An operational amplifier feedback circuit is generally used with these detectors to furnish a signal at point A (Fig. 6.9). This circuit holds the voltage at A near zero, where the detector has the maximum sensitivity. Photovoltaic detectors include indium antimonide as well as mercury–cadmium–telluride.

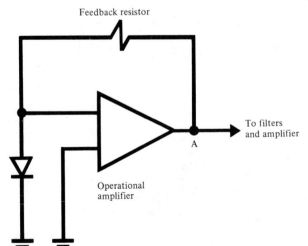

Figure 6.9 Photovoltaic detector with operational amplifier in feedback mode.

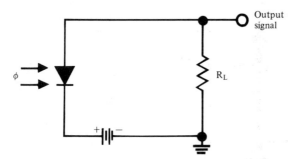

Figure 6.10 Basic solid state photodiode circuit.

Figure 6.11 Typical spectral response characteristics for $p-i-n$ photodiodes. (For Avalanche Photodiodes, multiply responsivity by the gain.)

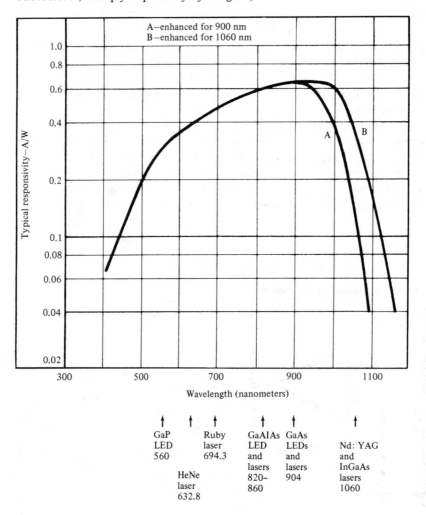

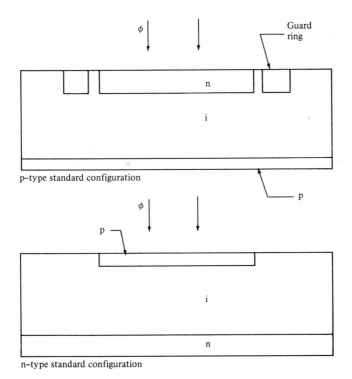

Figure 6.12 General structure of different $p-i-n$ silicon photodiodes.

For the specific optical electrical and physical characteristics of these detectors, generally used in the infrared region of the spectrum, the reader should consult the references.

For many satellite remote sensing applications in the visual spectral region, silicon diodes are used. Silicon has a bandgap of 1.1 eV. Radiant energy ϕ, with wavelengths shorter than 1.1 μm, is absorbed in the silicon to create electron–hole pairs. The electric field supplied by a reverse bias (as in Fig. 6.10) sweep the

Figure 6.13 General structure of an RCA silicon avalanche photodiode.

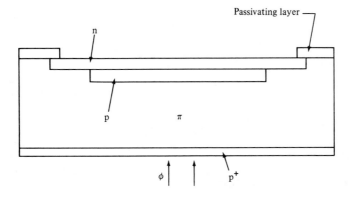

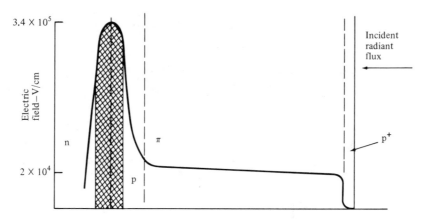

Figure 6.14 Electric field profile for a "reach-through" silicon avalanche photodiode.

electron–hole pair out of the depletion region to develop a voltage across a load resistor. The wavelength response of typical $p-i-n$ photodiodes is shown in Fig. 6.11. The general structure of $p-i-n$ silicon photodiodes is shown in Fig. 6.12, and the diodes may be configured into a monolithic array structure for remote sensing. The photodiode may be augmented with an internal gain mechanism, the counterpart of a photomultiplier, by adding a p-type substrate to the intrinsic silicon (Fig. 6.13). This double-diffused reach-through structure (Fig. 6.14) effectively provides two regions: a high field multiplying region a few micrometers wide, and a wide drift region between 30 and 110 μm, in which the incoming radiation is absorbed. The radiation ϕ enters the p^+ surface and the generated electrons are swept to the high field region, where impact ionization causes multiplication. The resulting holes produced in the high field region traverse the π region back to the p^+ surface to yield the multiplied signal. The response time is typically 0.5–3 nsec.

Vidicons are widely used in remote sensing and military applications. They are an outgrowth of using a photoconductive target in a camera tube instead of a photo-

Figure 6.15 Schematic representation of a vidicon camera tube.

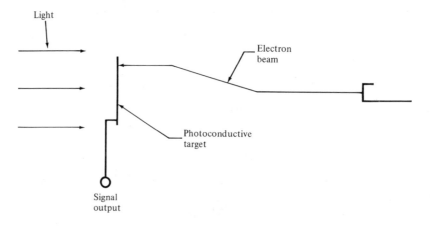

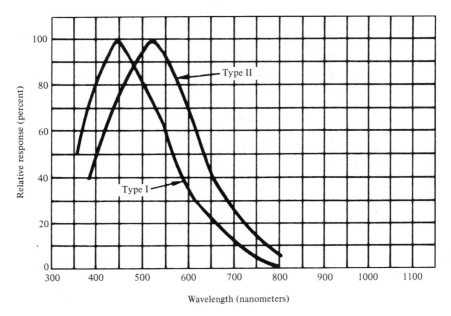

Figure 6.16 Photoconductor I: Vidicons employing this medium-sensitivity low-lag photoconductor are intended for use at the relatively high-light levels of film pickup service. Photoconductor II: Vidicons employing this high-sensitivity low-lag photoconductor are intended for use primarily in live TV pick-up service.

emissive surface (Fig. 6.15). The electron beam scans the photoconductive surface; the other side of the surface is coated with a transparent conductive layer. The signal electrode is maintained at a positive voltage with respect to the backside of the photoconductor, which is at near zero (cathode) voltage. The scanning beam charges the backside of the target to cathode potential. A light pattern focused on the photoconductor increases the conductivity proportional to brightness, increasing the positive charge on the backside. The electron beam essentially reads the signal by providing a capacitively coupled signal to the signal electrode.

The typical spectral response of a silicon target vidicon is shown in Fig. 6.16, as is also that of the improved vidicon (the ultricon) in Fig. 6.17a–6.17d. Originally, the term *vidicon* applied to antimony trisulfide tubes, but it is often applied to any of a variety of tubes with a photoconductive target. General Electric produces the FPS Epicon, which in their terminology applies only to a vidicon with a silicon target; FPS stands for focus, projection, and scanning within the same tube element, to produce a very rugged sensor tube. Vidicons are compact, rugged, partly solid state devices, but have the disadvantage of requiring an electron beam for scanning and readout, limiting the ultimate photometric linearity performance.

Typical photoconductor wavelength responses of vidicons are shown in Fig. 6.17a–6.17d. The application, as indicated in the legends, will dictate the appropriate surface.

Line and area arrays are a further development in the use of solid state silicon detectors. One such system (the Reticon) consists of 512 or 1024 elements, each of dimension 25 μm $\times$ 2.5 mm or 512 elements with the dimensions 50 μm $\times$ 0.432 mm (Fig. 6.18). There are three versions commercially available, with three response

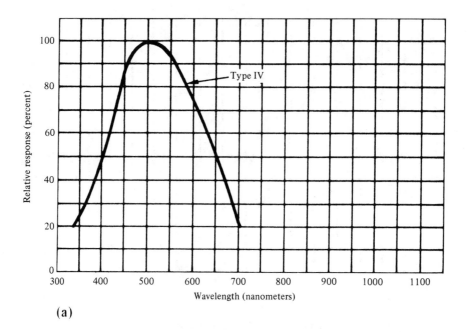

(a)

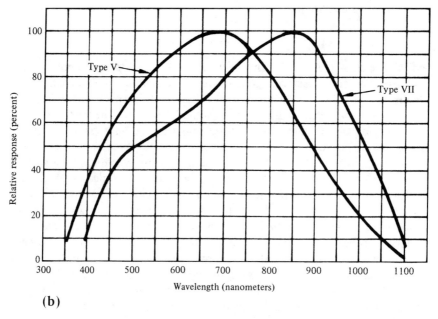

(b)

Figure 6.17 (a) Photoconductor IV: This photoconductor has very long-lag characteristics. Vidicons employing this material are intended for signal-storage applications such as the telecasting of PPI-type radar displays. (b) Photoconductor V (silicon-target diode-mosaic ultracon): Vidicons employing this extremely high-sensitivity and very low-lag photoconductor are especially suited for operation in the visible portion of the spectrum. They are controlled to provide uniformly good resolution even at the very short wavelength (blue–violet) end of the light spectrum. They are also useful in near infrared television systems. Where primary interest is in the near infrared, photoconductor VII should be specified. Photoconductor VII (silicon-target diode-mosaic vistacon): This photoconductor has very low lag with

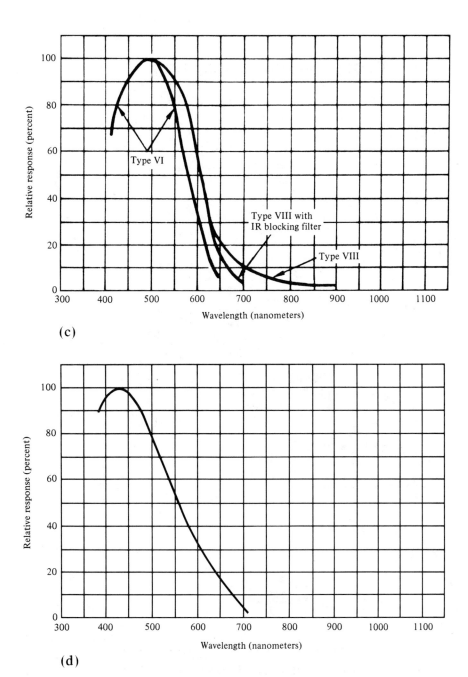

(c)

(d)

an enhanced infrared response. Where primary interest is in good detail response to visible light, photoconductor V should be specified. (c) Photoconductor VI (vistacon): This lead oxide photoconductor used in vistacons has high sensitivity and extremely low lag. Photoconductor VIII (vistacon): This lead oxide photoconductor used in vistacons has enhanced red response and is especially useful in the red channel of color broadcasting cameras. This photoconductor is also available with integral infrared blocking filters. (d) Photoconductor IX: This selenium–arsenic–tellurium photoconductor used in SATICON vidicons has a response that encompasses the entire visible spectrum. The spectral response is especially suited to color TV pickup both live and film. Spectral sensitivity is high in the blue and negligible in the infrared regions of the spectrum.

TN 6100 Detector Specifications

	TN-6111	TN-6112	TN-6121	TN-6122	TN-6131	TN-6132	TN-6133
Detector Array Type							
512 elem, 50µ x 0.432mm	X		X		X		
1024 elem 25µ x 2.5mm		X		X		X	
512 elem 25µ x 2.5mm							X
Intensification 25 mm Image Inverting, ERMA Photocathode	N/A	N/A	X	X			
25 mm Proximity Focused Broad Band S20 Photocathode					X	X	
18 mm Proximity Focused Broad Band S20 Photocathode							X
Persistance	N/A		90%-10% in 25 ms Typ		90%-10% in < 1 ms Typ		
Photon Gain (Photons Out/Photons In)	N/A		6000		4000		
Effective Active Area # Elements	510	1022	500	1000	500	1000	510
Array Length (in.)	1.0						0.5
Resolution (FWHM)	1-2 diodes		2-3 diodes	4-6	4 (typ)		
Noise*** (cnts/ch/scan)	2-3	0.7	2.5-6	0.7-2	2.5-6	0.7-2.0	
Cooling	Dual-Peltier cooling with Liquid Heat dissipation capabilities included. Normal array operating temperature is −5°C to 0°C.						
Min Gatetime (ns)	Non-Gatable				10ns		5ns
On/Off Ratio	N/A				> 10^4 (10^6 typ)		

* Minimum specified gain @525 nm (photons out/photons in)
** Design goal specifications

*** RMS Noise at minimum exposure time and at 1 second

NOTE: Other detector/intensifier combinations are available upon request. Contact factory for details.

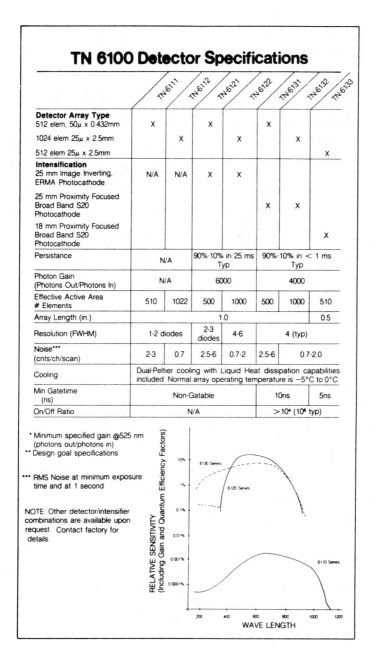

Figure 6.18 Reticon system. (Courtesy of Tracor Northern, Inc.)

characteristics. In the 6100 series, the detectors are unintensified, the 6120 series has an inverting microchannel plate image intensifier, and the 6130, a proximity focused image intensifier with a gating provision. A gating time of 5 or 10 nsec is all that is necessary to read either the 18-mm or 25-mm arrays. Cooling is required, and the arrays are rather expensive at present. The self-scanning monolithic arrays may be assembled into a two dimensional area mosaic, or may be used in linear fashion to acquire the output of a high resolution spectrograph quickly; it can be used as well for electrons or x rays.

A two dimensional charge injection device (CID), the G.E. CID-11, has been constructed and is composed of a mosaic of silicon sensors read out using the charge injection technique. There are 244 horizontal lines, each containing 248 individual pixels. Each of these pixels is 1.4×1.8 mils in size; the total active area of the array is 0.342 in. (8.7 mm) $\times 0.447$ in. (11.3 mm) with a diagonal of 0.563 in. (14.3 mm). Units having diagonals as large as 75 mm also have been constructed. Since the output signal is obtained via two shift registers, the CID is a completely solid state, addressable device that has been used in an ultraminiature camera, the GE TN 2500.

Charge coupled devices (CCD) are another version of a completely solid state sensor. The concept of the CCD is shown in Fig. 6.19. Each sensing element (pixel) in the imaging area is a metal oxide semiconductor (MOS) capacitor. There are sequential groupings of three phase transfer electrodes in the vertical direction and adjacent channel stops in the horizontal direction, both essentially serving to define the pixel size. The three phase electrodes for each row of pixels are connected in parallel with the corresponding three phase transfer electrodes of the other rows; clocking voltages applied to these electrodes accomplish charge transfer to the storage area. An optical image focused onto the image area produces a charge array of electrons during the active field interval which is subsequently transferred to the temporary storage area during the vertical blanking interval following. The transfer electrodes are made of transparent polysilicon, making the entire area light sensitive.

The storage area has the same construction as the image area because it is the temporary storage site for the entire image charge pattern for sequential readout. The three phase transfer electrodes of the storage area are clocked in unison with the image area during the vertical blanking interval to transfer the complete field of

Figure 6.19 Schematic representation of a CCD solid-state imager.

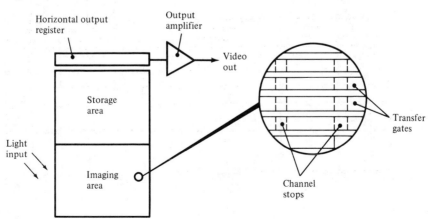

charge, representing the image, to the storage area. During a horizontal blanking interval, the charge pattern is moved toward the horizontal output register, one line at a time, to be loaded into the horizontal register and to be read out during the next active horizontal interval.

Also, the horizontal register has a three phase construction, thus receiving one line of picture information from the storage area during each horizontal blanking interval. In standard TV applications, each line is clocked out at a 6.1 MHz rate, producing a readout of all pixels in a standard active horizontal line time of 52.7 μsec.

An output gate electrode (the last CCD gate in the horizontal register) extracts the picture signal from the horizontal register, and sends it to an output transistor.

A typical response curve for a silicon CCD is shown in Fig. 6.20. There have been CCD systems that embody image intensifiers for low light level television. An outline drawing of the ITT PFCIT is shown in Fig. 6.21 (PFCIT = proximity focused channel intensifier tube). An optical image focused on the faceplate of the tube produces photoelectrons at the cathode. This electron signal is proximity focused on to the multichannel plate (MCP) input by application of 180 V (E_k) across the input section of the tube. Electron amplification occurs by the application of a variable voltage of 500 to 900 V (E_{mcp}) across the microchannel plate. Typical gains are 800

Figure 6.20 This device is a silicon imaging device that utilizes an array of charge-coupled device shift registers for photosensing and readout. The useful spectral response range extended from 420 to 1100 nm.

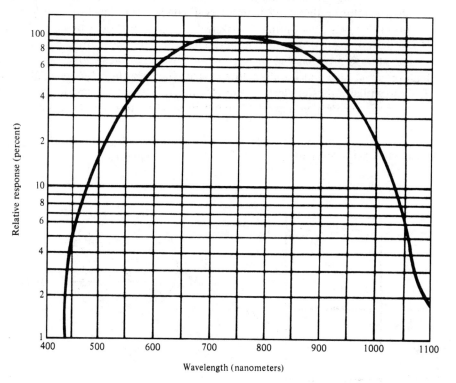

Wavelength (nanometers)

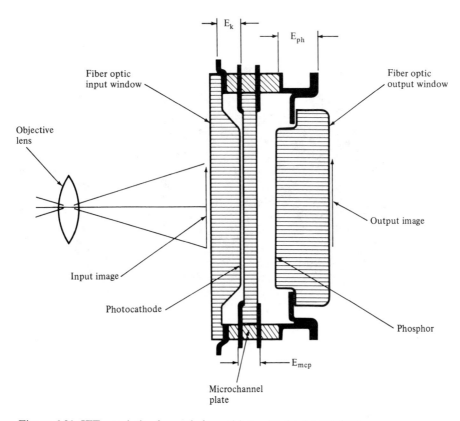

Figure 6.21 ITT proximity focused channel intensified tube (PFCIT).

to 1000. The MCP output current is proximity focused on to the phospor screen with a 5 kV accelerating voltage (E_{ph}), converted thereby into photons, and thence to the CCD or CID. With the use of tandem microchannel plate arrays, the sensitivity can be raised sufficiently to do photon counting.

Modification of LANDSAT 4 for Polarization

The LANDSAT 4 is the most recent of the Earth Resources Technology Satellite (ERTS), supported by NASA. It represents a new generation of land-observing satellites and was launched July 16, 1982. If one were to project the possibility of using a similar sensor system, by way of illustration, for a spectro-polarimetric/photometric mission, what would be a suitable configuration?

To answer this question, consider the two sensor systems aboard the spacecraft, the multispectral scanner (MSS) and the thematic mapper (TM) (See Figs. 6.22 and 6.23). The visible and near infrared system sensors are photomultipliers and photodiodes. In principle, a *simultaneous* signal is required from all sensors, one signal with the polarization sensing plane parallel to the plane of vision and the other one perpendicular. The simultaneity is of utmost importance, based on the author's experience, because, what would otherwise be small photometric temporal or spatial

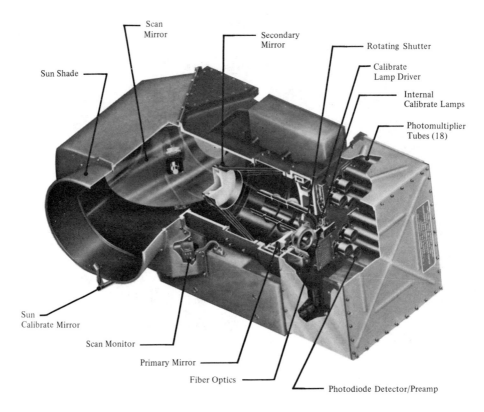

Figure 6.22 Cutaway view of multispectral scanner. (Courtesy of Santa Barbara Research Center.)

Figure 6.23 Thematic mapper.

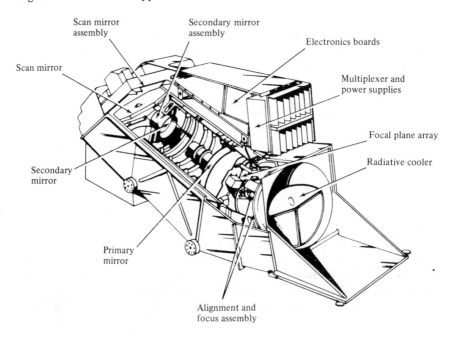

changes, become magnified in the differencing required in photometric measurements. Thus dual detection systems are required that operate through the same optical system, with beam splitters, and polarization analyzers for each detector. Where a detector array exists, one suitable plane polarization analyzer for the array would suffice.

However, beam splitters introduce significant polarization; consequently the beam splitter must be oriented relative to its subsequent polarization analyzer so that one polarization analyzer axis is parallel to the beam splitter plane of incidence and the other, perpendicular. This leaves open the not insurmountable question of how to maintain the system so that the plane of incidence of the beam splitter is kept

Figure 6.24 Relative positions of detector elements.

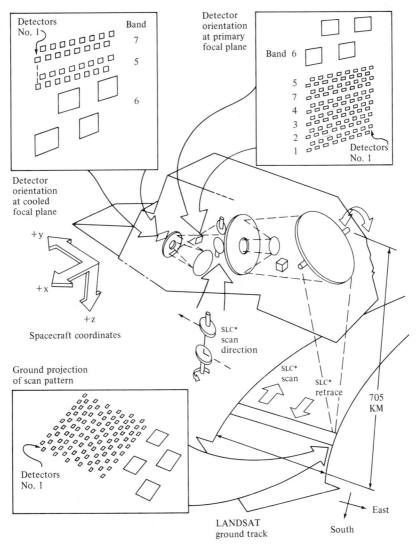

*Scan line corrector

either parallel or perpendicular to the plane of vision defined by the sun, earth target, and satellite sensor. Further, consider the relative position and number of detector elements of the TM (Fig. 6.24); each mirror reflection introduces a polarization bias; however, the optical Stokes parameters of each reflection can be measured, and matrix manipulation performed by computer, to calculate the plane polarimetric and photometric properties of a ground target as viewed through the atmosphere.

A satellite spectro-polarimetric/photometric sensor is possible with the present state of the art, but the advantages to be gained using polarization in contrast to higher resolution must be considered. This is examined in the Applications section.

7

Contrast: Signal/Noise

The application of remote sensing to ground target recognition or land use classification involves a decision process, either human or machine. This decision process is of a binary form—yes or no—in terms of classification. Noise, in one conventional electronic sense, is random, wide band "white" noise, that may interfere with the decision process. However, in more general terms, noise in remote sensing is any signal or effect that interferes with the decision process. This noise could be caused, for instance, by atmospheric scintillation or absorption, sensor mirror jitter, electronic circuit noise, sensor noise, scene noise (arising from nonhomogeneity, or from very high resolution sensors), downlink noise, tape recording noise, and possible deterioration of tapes during storage and shipment.

In a mathematical sense, noise of all types is amenable to a symbolic representation, and remote sensing systems may be designed based on optimization techniques. While system optimization to achieve maximum data transfer with minimum power consumption and system size is commendable, normal procedure is to veer toward overdesign in the interest of reliable data transfer.

This section will outline the various causes of noise, its type, effect, and minimization techniques. The subject of optimization will be treated briefly since there are many treatises on it. The system requirement imposed by polarization will be emphasized.

Scene Noise

Scene noise, a term introduced by Wiersma and Landgrebe (1978), is the spectral inhomogeneity involved in photometrically describing a land cover or remotely sensed area. The scene noise is averaged out at lower resolutions, but this reduces the mathematical size of the spectral space for a given cover class. As a result, there is less overlap with other cover classes, with ensuing higher classification accuracies. But at the boundaries between two or more cover classes or where the ground target is relatively small, there will be a high percentage of mixed pixels, containing combined optical contributions from these two or more cover classes; the result is a decrease in classification accuracy. Thus, a scene with small, relatively uniform targets would have a higher classification accuracy with higher sensor resolution, and in comparison, large heterogeneous targets would have the opposite trend.

A number of studies have been made of the relative effects of boundary pixels and scene noise. One such study by Markham and Townshend (1981) used a modular multispectral scanner (MMS) in flight lines over eastern Maryland. Of the 11 MMS channels available, a subset of 4 channels was used, 0.532–0.572, 0.653–0.692, 0.762–0.856, and 8–14 μm. These channels represented a close approximation of the bands recommended for agriculture, range, and forestry classification, with the first three bands corresponding to the thematic mapper bands 2, 3, and 4. Data was acquired at 5-m resolution (at an aircraft altitude of 2000 m) and the spatial resolution degraded to 10 , 20-, 40-, and 80-m pixel sizes. The degradation was accomplished using a filter developed by Sadowski and Sarno (1976). However, Gaussian noise had to be added to all of the lower resolution data to counteract the improvement in signal-to-noise level from applying the spatial filter.

Typical results for scene noise are presented in Fig. 7.1; the 80-m data are omitted because of insufficient pixels for analysis. Classification was achieved using a "training area" (an area of pixels of known characteristics from ground truthing), rather than from the derivation of an independent statistically derived testing set, which is usually different. In itself, this procedure introduces classification errors as distinguished from scene noise, but nevertheless "noise." The quantity of variance (S^2) or standard deviation (S) is one measure of scene noise, but variance is usually correlated with the mean ($\mathbf{X}$). Thus the coefficient of variation ($S/\mathbf{X}$) constitutes another measure of variability. The quantity transformed divergence D_{ij}^{τ} is also a useful parameter to characterize variability between classes where

$$D_{ij} = 2\left[1 - \exp\left(- D_{ij}/8\right)\right]. \tag{7-1}$$

Here D_{ij} is the ordinary divergence, and is identified in terms of some involved mathematical symbolism as follows. The probability that a vector $\mathbf{X}$, the result of a series of simultaneous spectral or polarimetric measurements of a ground target belongs to either class ω_i or ω_j, is represented by the likelihood ratio $L_{ij}(\mathbf{X})$, where

$$L_{ij}(\mathbf{X}) = \frac{p(\mathbf{X}|\omega_i)}{p(\mathbf{X}|\omega_j)}, \tag{7-2}$$

$$D_{ij} = E\left[\log_e L_{ij}(\mathbf{X}|\omega_i)\right] + E\left[\log_e L_{ji}(\mathbf{X}|\omega_j)\right]. \tag{7-3}$$

Here $E[\ ]$ denotes the expectation or average value over the variation possible in $\mathbf{X}$.

The statistical notation is useful in computer classification, and the statistical approach is covered in the work of Swain and Davis (1978).

In Fig. 7.1, it is seen that scene noise varies considerably between land cover categories (i.e., CBD is the central business district). Also, the scene noise varies significantly between different spectral bands for the same land cover categories. Further, the rate at which scene noise changes with coarser resolution also significantly varies between land cover types and different spectral bands for the same type of land cover. There is a parallel between standard deviation and coefficient of variation. The transformed divergence shows an increased separability with increased spatial resolution, but the relationship is imperfect with decline in scene noise and improvement in divergence. The classification accuracy versus spatial

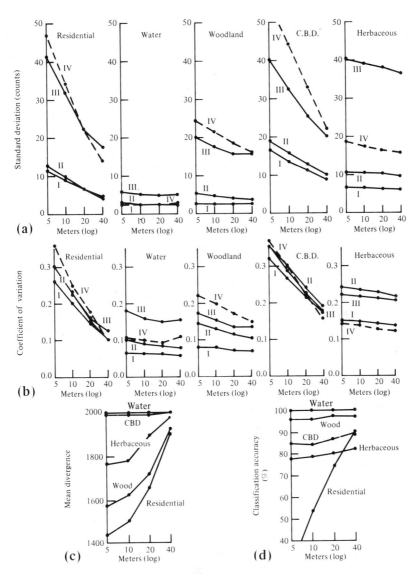

Figure 7.1 Results for the Annapolis area. (a), (b) Scene noise versus spatial resolution; (c) transformed divergence versus spatial resolution; (d) classification accuracy versus spatial resolution. Note: Channel 4 was not used in the divergence. Key to bands: I, 0.532–0.572 μm; II, 0.653–0.692 μm; III, 0.762–0.856 μm; IV, 8.0–14.0 μm.

resolution reveals that the classification accuracy is not only a function of scene noise, but of the category position in feature space relative to other categories.

As a result of the study, it was found that boundary and scene noise may offset each other almost exactly when scene noise varies, resulting in the same classification accuracy through a wide resolution range. For target areas of moderate scene noise, boundary effects tended to dominate over scene noise at resolutions poorer than 40 m. Classes with higher variances absorbed more mixed pixels, and tended to fare better in classification, except when the boundary pixels did not fit into a constituent class.

Polarization, when added to the photometric characteristics of a scene, has the effect of multiplying the number of vector descriptors by 2; this makes analysis at least twice as involved, but can reduce the uncertainties of classification by more than a factor of 2, because of the unique polarimetric properties associated with ground targets.

Atmospheric Optical Noise

Atmospheric "noise" results in the distortion of the wavefront from an image caused by nonhomogeneity of the earth's atmosphere. For sensors of high resolution, the noise effects can be significant. The noise arises from atmospheric turbulence. The fundamental property of the atmosphere that determines the propagation of radiation is the optical complex index of refraction. The index of refraction is proportional to atmospheric density, which in turn is inversely proportional to temperature. Thus, atmospheric temperature changes become changes in the atmospheric index of refraction. There is usually an air flow over the surface of the earth that varies with height, resulting in atmospheric shear, causing turbulence.

Thus, the focused image may be displaced, distorted, or broadened relative to a diffraction limited case. For severe turbulence, or over long path lengths, an image of a target can be broken up or bear little resemblance to the original beam or object.

Turbulence may be expressed in terms of the structure function C_N^2. The quantity C_N is a property at a point in the atmosphere that affects the optical properties in terms of a path integrated value along the optical path. Values of C_N can be determined from measurements of the temperature and humidity fluctuations or measurement of the fluctuation of light over a path. The real portion of the index of refraction of the atmosphere, n, is given by (Crittenden et al., 1978)

$$n - 1 = 79 \times 10^{-6} p / T, \tag{7-4}$$

where p is the pressure in millibars and T is the absolute temperature.

The average of the square of the difference of the index of refraction at two points $\mathbf{r}_1$ and $\mathbf{r}_2$ separated by distance $\mathbf{r}$ is related to C_N by

$$\left\langle \left[n(\mathbf{r}_1) - n(\mathbf{r}_2) \right]^2 \right\rangle = r^{2/3} C_N^2, \qquad l_0 \ll r \ll L_0, \tag{7-5}$$

where l_0 and L_0 are the inner and outer scales of turbulence, respectively; l_0 taken as 1 mm and L_0 as the path length of interest.

Also, C_N is related to the temperature turbulence structure constant C_T:

$$C_N = 79 \times 10^{-6} \left(p / T^2 \right) C_T. \tag{7-6}$$

An example of the variation of C_N^2 with altitude is shown in Fig. 7.2. The curve represents the functional relationship of Hufnagel and Stanley (1964) for a mean square wind speed of 27 m/sec.

As a result of temperature and pressure variations along an optical path, a point on an advancing wave front through the atmosphere may fluctuate in magnitude and phase. The magnitude of this fluctuation is termed *scintillation*. The atmospheric turbulence can be expressed in terms of a two dimensional image of a point source called the "point spread" function, and the corresponding two dimensional Fourier transform, the optical transfer function (OTF). Because the calculation of a two dimensional Fourier transform is quite large, the point spread function is simplified into a one dimensional "line spread" function by scanning the image of a point

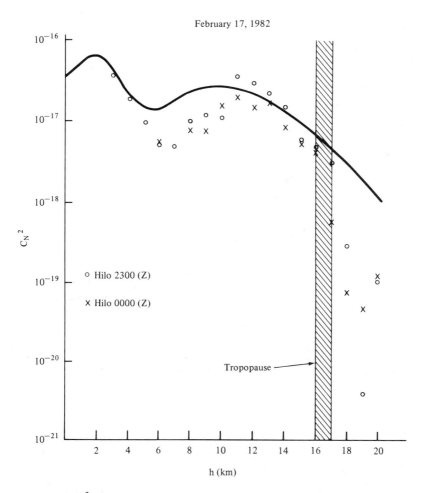

Figure 7.2 C_N^2 as a function of altitude.

source with a slit. The OTF requires that the images have circular or elliptical symmetry. The final results can be inverted with the Abel transformation to give the two dimensional representation.

An example of the long-term atmospheric OTF as a function of spatial frequency (the number of resolution elements per milliradian) is shown in Fig. 7.3. The example uses a 0.6328 μm laser, and it is seen that the higher spatial frequency elements are not transmitted by the atmosphere.

Polarization analyses generally would require a two dimensional Fourier transform because of the nonsymmetry of the transmission of radiation through the atmosphere along a slant path. The one dimensional OTF would be a good approximation.

The effects of atmospheric turbulence can also be expressed in terms of the mutual coherence function (MCF), which is the OTF expressed in terms of different variables. The modulation transfer function (MTF) is the modulus of the OTF; the OTF is the fundamental quantity that contains the effects of atmospheric turbulence although the MTF may also be used. The MTF lacks information on phase distortion.

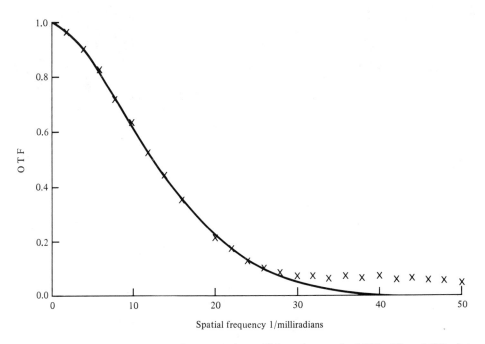

Figure 7.3. Long-term OTF of the atmosphere (Crittenden et al., 1978). Time 1417; date, December 15, 1976; wavelength = 0.6328 μm; range = 14150 m; diameter of optics = 0.4572 m; C_N = 5.36E-08; C_{NSQ} = 2.87E-15.

Detector Noise

The ultimate limit in terms of signal-to-noise ratio for any detector is set by photon arrival times. There are three limiting cases:

1. photon detector where the photon arrival level far exceeds that of background;
2. photon detector where background photon level exceeds the signal photon rate;
3. thermal detector that is background limited.

For the strong signal case (1), the noise equivalent power (NEP) is given by

$$\text{NEP} = \left(2N_s A/\eta\right)^{1/2} h\nu, \tag{7-7}$$

where N_s is the rate of arrival of photons of energy $h\nu$ per unit area, A is the detector area, and η is the quantum efficiency.

For case (2):

$$\text{NEP} = \left(\frac{2\langle N_B\rangle A}{\eta}\right)^{1/2} h\nu, \tag{7-8}$$

where $\langle N_B\rangle$ is the average rate of background photon arrival and $D^* = (n/2N_B)^{1/2}(1/h\nu)$; N_B is the background photon rate.

Figure 7.4 shows the photon noise limited D^* at a peak wavelength assumed to be the cutoff wavelength; the parameter is the background temperature in degrees Kelvin. It applies to photovoltaic and photoemissive detectors. The D^* for photoconductors is reduced by $\sqrt{2}$ because of recombination noise.

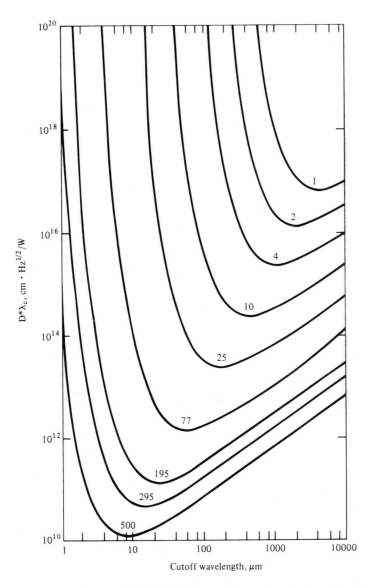

Figure 7.4. Photon-noise limited D^* at peak wavelength assumed to be cutoff wavelength, for background temperatures $1,2,4,10,25,77,195,295,$ and 500 K (assumes 2π fov and $\eta=1$). [From Jacobs and Sargent (1970).]

For Case (3), assuming no cutoff wavelengths,

$$D^* = \left(4.0\times10^{16}\epsilon^{1/2}\right)\big/\left(T_1^5 + T_2^5\right)^{1/2}, \qquad (7\text{-}9)$$

where ϵ is the detector emissivity, T_1 is the detector temperature, T_2 is the background temperature, σ is the Stefan–Boltzmann constant, and k is the Boltzmann constant.

Photoemissive detectors (photomultipliers) can detect single photons with an efficiency of ~ 30% with a time resolution as short as 0.1 nsec. The limiting noise for a photomultiplier (PMT) depends upon the background illumination level. For low level detection, dark current (i_{dark}) shot noise i_n is the limitation

$$i_n = (2ei_{\text{dark}}\Delta f)^{1/2},$$

where Δf is the bandwidth and e is the electron charge. At high illumination levels, the shot noise is

$$i_n = (2ei_{\text{signal}}\Delta f)^{1/2}. \tag{7-10}$$

The dark current can be reduced by operating the PMT at a lower temperature; the improvement upon cooling depends on the particular photocathode in the PMT.

Diffused silicon diodes are available with a variety of characteristics, as indicated in Section III.2. There are five general types of silicon detectors: type I, high sensitivity; type II, faster but less sensitive; type III, large area, guard ring type; type IV, fast with red response; type V, avalanche.

The detectors are subject to thermal noise and shot noise. Thermal rms noise voltage $V_{n,\text{rms}}$ is given by the expression

$$V_{n,\text{rms}} = (4kTR\Delta f)^{1/2} \tag{7-11}$$

Figure 7.5 D^* vs λ for small-area type I silicon photodiodes: curves A, B, and C correspond to areas of 0.02, 0.2, and 1 cm². The lower D^* for smaller-area detector performance is due to amplifier limitations rather than intrinsically poorer D^* for small-area detectors. (Texas Instruments, Infrared Devices, SC-8385-366; courtesy of Texas Instruments, Inc.)

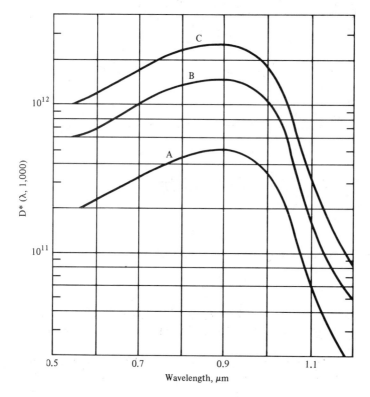

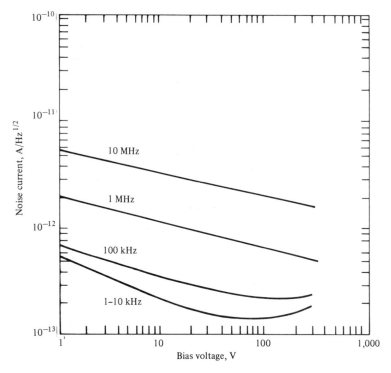

Figure 7.6 Typical noise current vs bias voltage and frequency for type III silicon photodiode ($A = 0.05$ cm^2). (Electronuclear Laboratories, *source*: IR Industries, Inc.)

Figure 7.7 Example of noise vs frequency for PbS detectors (1×1 mm area). (Santa Barbara Research Center, Brochure.) (Courtesy of Hughes Aircraft Company.)

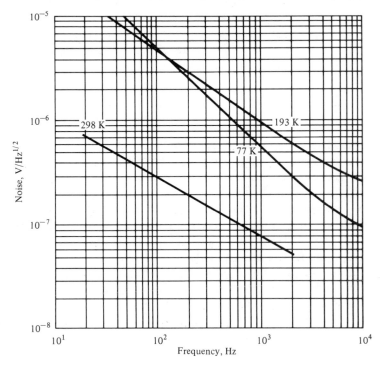

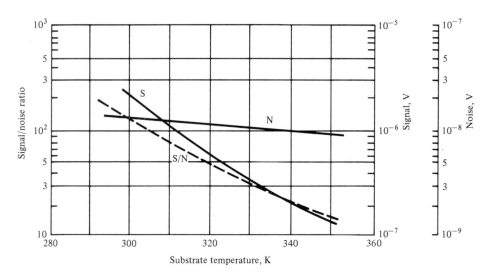

Figure 7.8 Variation of signal, noise, and signal-to-noise ratio with substrate temperature for room-temperature, photoconductive InSb. Test conditions: current, 10 mA; incident energy, 68 μW/cm^2 at 4.4 μm; modulation frequency, 800 Hz; amplifier bandwidth, 50 Hz. (Mullard Bull. TP 1080.)

where k is Boltzmann's constant, T is the operating temperature, R is the source resistance, and Δf is the bandwidth.

Typical D^* characteristics for a type I detector are shown in Fig. 7.5. The type III detector with a guard ring has a lower surface leakage current, the main contributor to dark current, and hence has a lower shot noise. Also, the guard ring reduces thermal noise because reduction in leakage current effectively increases the detector impedence. The noise depends on frequency and bias voltage (Fig. 7.6).

Lead sulfide detectors have the noise characteristics shown typically in Fig. 7.7, and those for indium antimonide, in Fig. 7.8.

Electronic Circuit Noise

This noise arises mainly from resistor and transistor contributions, although mutual coupling (electromagnetic or electrostatic) between closely spaced, unshielded wires can contribute. Noise sometimes can have a common source, such as that from an imperfect isolation between circuits or through a common power supply. Mutual coupling can occur from the use of separate grounds to different points on a spacecraft or aircraft, rather than to a common ground point. Where radio frequency (rf) transmitters are close to low level signal circuits, the rf can couple into the input circuits and cause interference and instability.

In the design of the electronic system, the electrical bandwidth must be made wide enough to handle the very high rate digital data stream (typically 85 Mbps or higher). With the wide bandwidths required, the relative noise level is increased. Techniques for optimizing bandwidths have been extensively described in the literature.

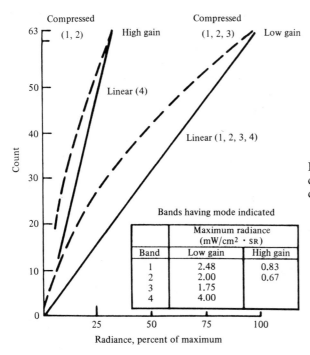

Figure 7.9. MSS digital output count as a function of radiance, compressed and linear modes.

Band	Maximum radiance ($mW/cm^2 \cdot SR$)	
	Low gain	High gain
1	2.48	0.83
2	2.00	0.67
3	1.75	
4	4.00	

A technique used in LANDSAT 4 is to do all processing (including amplification, track and hold, and dc restoration) before the A/D conversion. Also, provision is made for linear or nonlinear amplification via four-segment, quasilogarithmic amplifiers to improve the signal-to-noise ratios in bands 1, 2, and 3 (see Fig. 7.9). By compressing high radiance level signals, the quantization noise more closely matches the photomultiplier noise. The band 4 signals from silicon photodiodes are not compressed because the equivalent load resistor noise is best matched by linear quantization. In the high gain mode, the gains of bands 1 and 2 are increased by a factor of 3 to allow use of a system dynamic range for scenes of lower radiance.

If a polarimetric capability were to be added to a given system (i.e., LANDSAT 4), the TM data handling requirement would be approximately doubled [i.e., 170 megabits per second (Mbps)]; the MSS would require 30 Mbps, making the system requirement 230 Mbps for polarization on both sensors.

Mechanically Induced Noise

Multispectral scanners produce imagery sequentially. The ground target is imaged and scanned a line at a time with an optical–mechanical system. The radiation is focused onto one detector at a time to establish an instantaneous field of view (IFOV), the total scanned raster being the total field of view (TFOV). The motion of the vehicle along its orbital track provides the along-the-track scan motion of the sensor system while the line scanner provides the across-track scan. The across-track scan has been modified in the LANDSAT 4 with a scan line corrector (Fig. 7.10) that rotates the TM line of sight slightly backward along the ground track during the active scan time of the scan mirror to compensate for the orbital motion of the spacecraft. This action generates scan swaths that are then perpendicular to the ground track and perpendicular to each other (Fig. 7.11). Where a mechanical scanning system is involved, there is a tolerance on the scan parameters as well as on the stability and pointing accuracy of the spacecraft (or aircraft). No system has a

Figure 7.10 Optical system and detector projection on ground track.

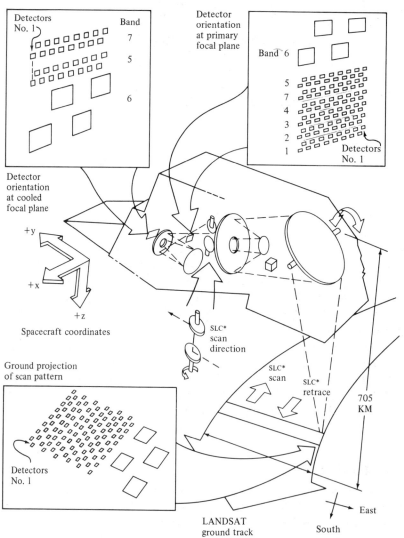

*Scan line corrector

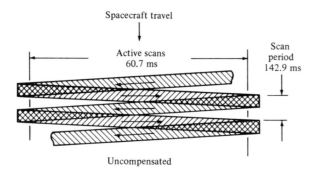

Spacecraft travel

Active scans
60.7 ms

Scan
period
142.9 ms

Uncompensated

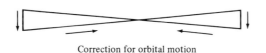

Correction for orbital motion

Figure 7.11 TM scan line
correction.

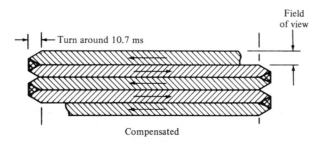

Field
of view

Turn around 10.7 ms

Compensated

zero tolerance on these parameters and the cost of reducing the tolerances rises exponentially.

For instance, the LANDSAT 4 has a high-precision zero-momentum altitude control subsystem with a positioning accuracy of 0.01° and a stability of 10^{-6} °/sec. In the worst case, for an orbit of 705.3 km, the pointing could be off on the ground by 123 m; also, as an extreme case—and an improbable one—during one scan (6.9967 ± 0.02 scans per second), the scan has the capability of being changed by 1.3% of the ground swath length of 185 km (= 2.4 km). This constitutes a source of possible noise that would require cleanup by subsequent data processing. The scan mirror has certain operational parameters (Fig. 7.12); the mirror itself is a lightweight low-inertia beryllium eggcrate structure that floats nearly inertially free during the active scan through the use of magnets that compensate for the flex spring pivot forces, and torque is applied to the mirror only during the turnaround times. Exceptional care is thus necessary to minimize mechanical effects. A scan angle monitor gives the angular position of the scan mirror. In the data sequence a preamble maintains bit synchronization from scan to scan. As the mirror arrives at

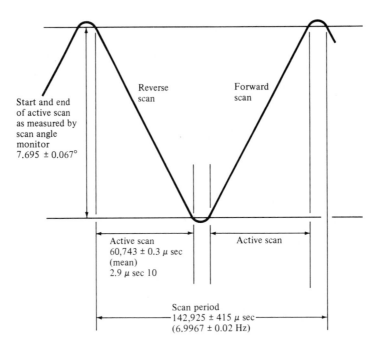

Figure 7.12. Scan mirror operational parameters.

the western edge of the area to be imaged, a line-start code is produced. This code interrupts the detector sampling sequence and causes detector A, band 1 to be sampled. A mirror-frame synchronizing signal (MNFS) is also produced (on channel 25), to indicate a new sequence of detector video, starting with detector A, band 1, is being produced; thus the data are tagged throughout each line scan.

Thus, although every reasonable attempt is made to minimize or compensate for mechanical noise, there is still a need to recognize the possibility of its existence.

Downlink Noise

The downlink to the domestic or foreign ground stations from a satellite is by microwaves (Fig. 7.13). The LANDSAT 4 digitized TM and MSS data are transmitted at X band. A separate S-band link, compatible with that used for LANDSATS 1, 2, and 3, is also provided to transmit MSS data to those stations so equipped. This S-band link served as the primary communication path prior to the availability of a Tracking and Data Relay Satellite System (TDRSS). Spacecraft telemetry and command communication paths are via TDRSS and through S-band NASA ground stations.

The communication links used by LANDSAT 4 are greater in number and sophistication than on preceding spacecraft because the TM data is transmitted at 85

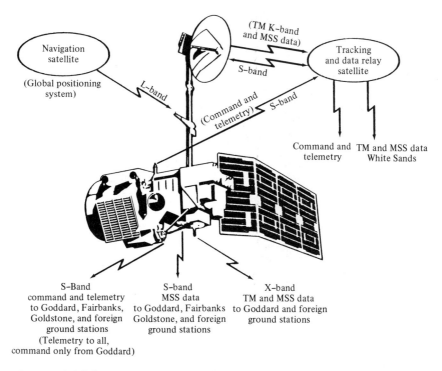

Figure 7.13 Flight segment communication links. Image data transmission characteristics.

Mbps rather than 15 Mbps. The frequency of X band (8.2125 GHz) thus uses approximately 100 Hz at 10 GHz to transmit one bit of information. At the S-band frequency (2.2655 GHz), there would only be about 25 Hz to carry a bit of information, degrading the signal-to-noise ratio. There is an additional capability to transmit a multiplexed X-band signal containing both TM and MSS data.

A few of the characteristics of the X- and S-band telemetry and data are shown in Table 7.1. Use of a high gain antenna on the spacecraft as well as at the ground station, unbalanced quadrature phase-shift keyed modulation, sufficient power at the transmitter, and a sufficiently high telemetry carrier frequency provide excellent noise rejection. Any interfering spurious signals, whether man-made or natural, would have little effect except if introduced along the line of sight between the spacecraft and ground antennae. Refraction through the ionosphere would produce a minor effect in introducing amplitude or phase distortion.

For a polarimetric/photometric system, based on the LANDSAT 4 sensors, the data link would require about 2.5 times the frequency (i.e., K or Ku band at ~ 20 GHz).

Tape Noise

Satellites are occasionally equipped with tape recorders to record remotely sensed data when the satellite is out of range of a ground tracking station. Further, tape recorders are used at ground stations to produce computer compatible tapes (CCT)

Table 7.1 X-band Image Data Transmission Characteristics

Frequency: 8.2125 GHz (X-band)

Transmitter power: 44 W

Shaped-beam antenna

Unbalanced quadrature phase-shift keyed (UQPSK) modulation

The TM data are normally on the "I" carrier channel, and the MSS data
are modulated on the "Q" carrier channel with a 4-to-1 power split.
There are three operational modes as follows:

Mode	I-Channel Data	Q-Channel Data	Modulation
1	PN (84.903 Mbps)	MSS (15.0626 Mbps)	UQPSK
2	TM (84.903 Mbps)	TM (84.903 Mbps)	BPSK
3	TM (84.903 Mbps)	MSS (15.0626 Mbps)	UQPSK

The TM data are replaced with pseudonoise (PN) code for mode 1, in which only the MSS is operating. When only the TM is operating, the MSS data may be replaced with TM data. The TM data are PN-encoded within the instrument electronics. The MSS and TM are differentially encoded by converting from NRZ-L to NRZ-M for downlink transmission.

X- and S-Band Communications To Foreign Ground Stations

Foreign ground stations can acquire TM image data by the X-band link only. TM payload correction data can be acquired either from the X-band image data stream or the S-band 32-kbps telemetry data link. MSS image data can be acquired from the X-band link in addition to the S-band link. MSS telemetry data can be acquired from the S-band 8-kbps link. If required, S- and X-band communications links can be operated simultaneously to satisfy foreign ground station coverage requirements for common areas. Simultaneous S- and X-band image data transmissions to one station will not be supported, however.

S-Band Image Data Transmission Characteristics

Carrier frequency: 2265.5 MHz

Transmitter power: 10 W

MSS data rate: 15.0626 Mbps

Shaped-beam antenna

NRZ-L PCM/FM modulation

Deviation ± 5.6 MHz $\pm 5\%$

S-Band Telemetry Data Transmission Characteristics

Real-time spacecraft telemetry (housekeeping and GPS data), narrow band tape recorder telemetry, the payload correction data, and onboard computer data are downlinked by the S-band transponder. S-band telemetry will be commanded on in response to a foreign station's request for telemetry data to support their MSS image data reception (by either S or X band). Foreign ground stations can use the real-time spacecraft telemetry or the payload correction data only, or they may use both.

Frequency: 2287.5 MHz

Effective isotropic radiation power: $+3.2$ dBW

PCM/PSK/PM modulation 8-kbps telemetry data on 1.024-MHz subcarrier

PCM/PM modulation 32-kbps PCD (payload correction data) phase-modulated on carrier for transmission to the user community. Magnetic tapes are also used to change data rates; for instance, the LANDSAT 4 TM produces data at 85 Mbps, and the TDRS can relay this data to a communications satellite at a maximum of 50 Mbps; thus the data must first be recorded at the TDRS and played back at a slower rate.

There has been much material in the literature on "tape noise," which not only includes tape, but the cleanliness of the heads, storage and shipment of tapes, coatings used for the magnetic recording on tapes, and space qualification of tape units. The early LANDSATs have had difficulties with tape recorders in space; LANDSAT 4 does not use one, but makes data available in real time.

Tape shipment is generally by air freight and is reasonably reliable, and storage of tape is generally in an air-conditioned vault. The general remedy for a bad tape is to rerecord the data.

Scanning Noise

Noise may be random or may have a definite spectrum (temporal or frequency). When noise is random, and a set of noise distribution functions are superimposed, then

1. the direct-current components of the sets are additive, and
2. the alternating-current quadratic contents are additive for any frequency range.

Random noise is any stochastic function of time of specified duration whose Fourier series components each have a two dimensional normal distribution and random phase, providing that the quadratic content of no single component is a significant percentage of the total.

Where noise is not random, and has a periodicity, the particular Fourier series components involved must be summed appropriately.

As an example, the LANDSAT 4 TM noise specification in the various spectral bands (Table 7.2) is based on the sensor system performance and is specified in terms of a "noise equivalent reflectance" (Table 7.2). The usable range of scene radiances is specified together with the corresponding specified and measured SNR. The measured SNR ratios are seen to lie between 21 and 342.

Table 7.2 Radiometric Sensitivity Bands 1–5 and 7

Band	Noise equivalent reflectance	Scene radiance (mW/cm² sr) Minimum specified	Maximum specified	SNR Minimum scene Measured	Specified	Maximum scene Measured	Specified
1	0.8%	0.28	1.00	52	32	143	85
2	0.5%	0.24	2.33	60	35	279	170
3	0.5%	0.13	1.35	48	26	248	143
4	0.5%	0.19	3.00	35	32	342	240
5	1.0%	0.08	0.6	40	13	194	75
7	2.4%	0.046	0.43	21	5	164	45

Table 7.3 Square Wave Response (SWR) (Band Average)

Band	30 m Bar		45 m Bar		60 m Bar		500 m Bar	
	SWR	σ	SWR	σ	SWR	σ	SWR	σ
1	0.46	0.01	0.76	0.03	0.94	0.02	1.0	0.0
2	0.44	0.02	0.72	0.04	0.96	0.03	1.0	0.0
3	0.41	0.01	0.72	0.02	0.91	0.02	1.0	0.0
4	0.43	0.01	0.76	0.03	0.95	0.03	1.0	0.0
5	0.42	0.02	0.78	0.03	0.89	0.03	1.0	0.0
7	0.44	0.02	0.76	0.02	0.92	0.02	1.0	0.0
Spec	0.35		0.70		0.85		1.0	

Band	120 m Bar		180 m Bar		240 m Bar		2000 m Bar	
	SWR	σ	SWR	σ	SWR	σ	SWR	σ
6	0.44	0.04	0.78	0.01	0.94	0.00	1.0	0.0

The sensor system square wave response (SWR) is presented in Table 7.3, as well as the standard deviations (σ) of the channel-to-channel response variation for the 16 channels in each of bands 1–5 and 7 as well as the four channels in band 6.

The SWR is the equivalent of the OTF. It is seen that as the bar size is decreased from 2000 to 30 m, the response drops from 1.0 to 0.4; thus the 30 m is defined as the resolution of the sensor system.

System Optimization

In the design of a remote sensing system, a series of specifications are delineated by the users. Sometimes the users may extrapolate from the known capabilities of a remote sensing system to a more sophisticated version. Studies are made of hypothesized configurations, with parametric variations of spectral, spatial, and temporal coverage. Data handling requirements are characterized, and user or applications of the data are envisioned. But a dominating, underlying factor is economics; what will it cost? Thus we start with the trade-offs: what can we achieve with a given system; at what expense; and above all, can it be done at all?

Usually a "straw man" system concept is set up, the idea being to start somewhere with something. Then one adds or subtracts pieces from the "straw man" system concept, keeping in mind whether existing technology can bring the concept into reality. Progress, in the material sense, involves extending technology a bit further than the present.

The goal of the general theory of system design is threefold:

1. Obtain the explicit structure of an optimum or near optimum system to fit the requirements, including reducing the system to an ordered sequence of physically realizable elements.
2. Quantitatively evaluate the performance of such systems.
3. Compare the theoretical optimum with the actual suboptimum systems.

Following digitizing of the remotely sensed signals in LANDSAT 4 a binary detection system is involved wherein there are two decisions corresponding to two hypotheses, the presence or absence of a bit. Prior to digitization, there is a multiple alternative detection system in operation that makes more than two decisions corresponding to the various sensor photometric levels. Again, following data reduction, we have a multiple alternative detection system to resolve ground targets through the atmosphere. Needless to say, a complete detailed system analysis would be a continuing program. In practice, the sensor systems (from the optical system through the data phase) are subject to a rigorous optimization analysis during the design phases, and the user is relegated to performing the optimization of the data for his application. Polarimetric remote sensing adds another factor in a remote sensing system optimization, as well as in the data analysis.

Orbit Time of Day: Solar Angle

In general, remote sensing of the earth is best at high solar angles (near zenith) and becomes increasingly difficult as the sun goes much below 30° from the horizon. The exceptions are those applications for which shadowing is beneficial (such as in some geological and cartographical surveys). It may be that many more user applications will be satisfied by higher rather than lower sun angles. Near-noon orbits yield best photometric information (maximum brightness); water areas, however, are affected adversely by sun glint within approximately a 10° cone, while recognition of some types of vegetation is facilitated at or near solar opposition.

The most important factor that limits the range of useful sun angles is the increased shadowing effects that become apparent as the sun goes below about 30°. Target reflectivities change with phase angle because of shadowing, as does the quality of the image seen from space, owing to increased atmospheric scattering and absorption.

Figure 7.14 Solar angle vs months for 0930 and 1200 time of day, 30° and 40° latitude.

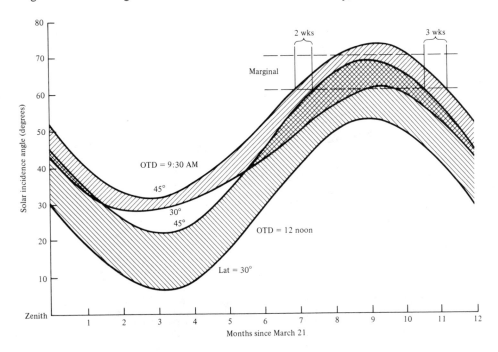

With this as a guideline we may look at the available range of sun angles as determined by the orbit time of day from the point of view of the user. Apart from considerations of local weather and visibility, cloud cover, etc., we may say that the optimal launch time for meeting the greatest number of user requirements will be the one that allows the highest solar elevations. Figure 7.14 shows the seasonal variation of solar angle ranges seen by a satellite over CONUS for a 9:30 a.m. and 12 noon orbit. The latitudes between 30° and 45° cover most of the CONUS. It is apparent that a near-noon orbit provides good viewing conditions throughout most of the year for much of the U.S., and that a good portion of the U.S. has a good range of sun

Figure 7.15 Sun angle vs time of day for different latitudes.

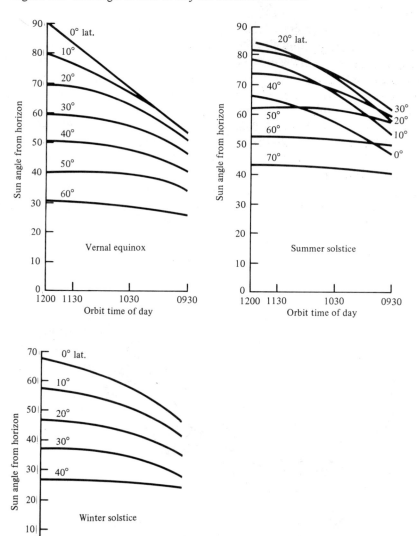

angle over the entire year. Naturally those users requiring some amount of shadowing find that some regions of the southern U.S. are less than optimal for a good portion of the year. On the other hand, a 0930 orbit provides good viewing conditions for all users over the entire U.S. for about 10 months of the year, while for a period of about 2 or 3 months in winter some sections of the U.S. have rather poor viewing conditions for those users for whom shadowing is not an advantage. If a 30° sun elevation is considered minimal, then a 12 noon orbit gives five weeks more of usable coverage for the northernmost CONUS than the 0930. Another plot (Fig. 7.15) showing sun angle versus time of day for different latitudes was prepared to show the impact of time of day on the shape of the sun angle curve at the higher latitudes. Sun angle versus orbit time of day does not change rapidly for low sun angles at high latitudes; however, at lower latitudes nearer-noon orbits give significantly higher average sun angles.

Viewing conditions over Alaska (or Antarctica) are relatively insensitive to orbit time of day, and in general there will be low sun angles corresponding to the Alaskan latitude range of 55°–70°. For example, on June 21 the sun reaches a maximum elevation angle of 33°–57° depending on latitude, and on December 21 some regions are in total darkness throughout the day. In general, therefore, one may expect less than optimal viewing conditions for a good part of the year over much of Alaska for applications not involving shadowing effects. However, the more southerly (northerly) latitudes are available for remote sensing for a good part of the year during clear conditions.

Circular Polarization and Contrast

The range at which a target can be perceived through a turbid medium is affected by the scattering and absorption in that medium. For media where absorption dominates, the visibility range is determined by the brightness of the light source and the sensitivity of the sensor. When scattering occurs, the effect is reduction of the apparent contrast between the target and background.

The standard definition of contrast level as seen at a distance R of target versus background is

$$C_R = \frac{B_T(R) - B_B(R)}{B_B(R)},$$

where B is the brightness in watts/cm^2 steradian and subscripts T and B are target and background, respectively.

Contrast transmittance is defined for a specific set of viewing conditions (look angle, distances from target, etc.) and is the ratio of the apparent contrast seen by the instrument to the contrast at ground level.

The background radiance is directly proportional to scattering. When the scattering is large, the contrast transmittance approaches zero, and no increase can be brought about by an increase in source luminance or sensor capability. However, through the use of circular polarization, contrast improvements of a factor of 5–20 can be achieved.

The improvement is based on the fact that the turbid medium scatterers are widely separated, and that the return reflection is the result of single scattering.

However, a surface has many facets, and multiple scattering occurs at and within the surface. But the handedness of circularly polarized light changes with each reflection (Shurcliff and Ballard, 1964). Thus singly scattered, circularly polarized radiation will have the opposite handedness of the incident rays, while that from the target will have both types of handedness.

As a result, by using a circular polarization analyzer, the medium scatter is reduced much more than that from the target, and contrast is improved. This effect has been demonstrated to improve underwater visibility (Gilbert and Pernicka, 1967).

8

Calibration

The subject of calibration, as described in our context, is sensor system photometric and polarimetric calibration, applying to laboratory, aircraft, and satellite systems. Sensor calibration alone is but one aspect of overall system calibration that relates the system response to absolute photometric and polarimetric standards. The constancy of the calibrator unit also must be assured. Three calibration techniques will be described: one for true color and false color films, one for a nonimaging polarimeter/photometer, and one for the LANDSAT calibrator as modified for polarization. Use of large ground targets as calibrators will also be discussed.

Color Film Calibration

An incentive to investigate calibration techniques for Type 8443[1] and Ektachrome color reversal films was motivated by a desire to make photographic polarization measurements of forest and agricultural areas to permit better differentiation. Polarization measurements require accurate control of photographic exposures, comparable to that of precision astronomical photometry, with a properly oriented polarization analyzer inserted in the path of the light entering the camera. The small differences in recorded brightness which are indicative of polarization require careful calibration and control of the over-all photographic recording process.

Colorimetry with color films is at a very early stage; without exaggeration, it may be stated that the use of three color emulsion layers, instead of one black and white emulsion layer, makes the problems at least 9 times as great. There is no effort at present directed toward precision color control by users of Type 8443 and Ektachrome films. Users of single emulsion black and white film, with an associated color filter for color reproduction, generally recognize the need for careful color control,

---

[1]At the time of the work described in this section, the false color infrared film designation was type 8443; subsequently the designation was changed to type 2443.

and do maintain an evaluation program of the photographic emulsion; admittedly, the task is much simpler with a single emulsion layer.

The time for standardization of calibration and control of Type 8443 and Ektachrome films is now, when the field is young. The work in calibration and control that will be described is not to be regarded as the final solution. These efforts are only a first step, and as techniques are improved, refinements are most surely possible.

The primary use of color photography in earth resources application is the demonstration of color contrasts. These can be easily seen. There has been some effort in calibration with the Munsell color standards, with the involvement of human judgment, with its recognized inherent limitations.

The philosophy of remote sensing is the acquisition of information and representing it in a way that permits the most expeditious utilization. Photographic recording is convenient in that much information is obtained through the photographic medium, and a representation—the photograph—presents the data in a convenient way amenable to analysis by photointerpreters.

However, the amount of data obtained in remote sensing by satellites, or even by aircraft, is astronomical in magnitude. Therefore, the use of computer analysis techniques on photographs would permit the ultimate in speed of data analysis without necessitating the use of a battery of highly trained photointerpreters. This requires accurate control of the information transfer medium (the color photograph) so that maximum information can be obtained.

Over the past 50 years, photography and cinematography have reached a high level of standardization in the United States (McCamy, 1968). Motion picture standardization had been accomplished through the Society of Motion Picture Engineers and the Society of Motion Picture and Television Engineers, fully a generation before still photography. The impetus was given by the fast growth of the motion picture industry. Still photography standards had their roots in the International Federation of National Standardizing Associations (ISA) and the American Standards Association (ASA), spearheaded by L. A. Jones of Eastman Kodak with committees sponsored by the Optical Society of America. The International Organization of Standardization (ISO) replaced ISA in 1946. At present, the United States of America Standards Institute (USASI), organized in 1966, implements standardization; however, none exists for color film or Infrared Aero Film, Type 8443.

The need for calibration and control techniques is essential when color photography (particularly the use of 8443 film) is to be used quantitatively, rather than qualitatively. Some present qualitative cases of Type 8443 false color infrared film depict gross foliage differences between conifers and deciduous trees, and camouflage or moisture content of the soil (Fritz, 1967). In actual use, the colors obtained with 8443 film have been found to have enormous range depending on the following four factors of the photographic process:

1. film manufacture and storage,
2. exposure,
3. development, and
4. evaluation—densitometry.

These factors will be discussed sequentially in the following sections. The use of color over black and white photography has an advantage: it has been cited that a trained observer can distinguish as many as 10 million colors in daylight on the basis of hue, value, and chroma (Judd and Kelley, 1939), whereas only 100–300 black, gray, and white tones can be perceived (Meyers and Allen, 1968). Concurrently, a control of these colors recorded by the film to yield reproducible results is implied. This color evaluation by the human eye has a considerable degree of subjectivity, which would be eliminated by automatic densitometry.

Film Manufacture and Storage

Color photography has been considered an art form and a scientific tool and is used widely in both applications. As an art form, color fidelity is of lesser importance than an aesthetically pleasing representation. For scientific applications, the exactness of the reproduced color (even though not the true color) and the consistency with which this color is subsequently duplicated are of utmost importance. Our interest is in the scientific application of color film, specifically Type 8443 infrared Ektachrome and normal Ektachrome color reversal film.

Color reversal film is composed of three layers, which have representative sensitivities (Neblette, 1962) as illustrated in Fig. 8.1: one layer is sensitive in the blue spectral region, one to the green, and one to the red. This figure is representative, because there are deviations in the exact shape of the curves. Also, in furnishing sensitometric curves, the Eastman Kodak Company requires that a disclaimer accompany the curves, indicating that "these data should not be construed as representing a standard or specification which must be met by the Eastman Kodak Company. The company reserves the right to change and improve product characteristics at any time. Further, the conditions of exposure and processing of the films from which these data were taken may not duplicate those encountered in your establishment. While Eastman Kodak Company has no objection to your using the

Figure 8.1 Typical sensitivity curves for the three emulsion layers of a color reversal film.

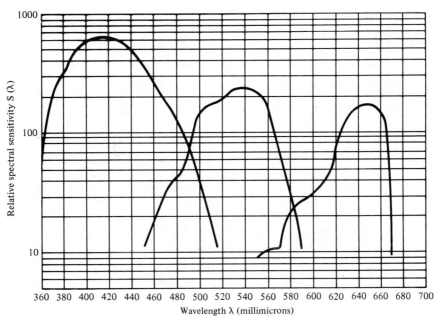

information, it feels that this use should be a matter for your own investigation and decision and entirely without obligation, responsibility, or liability on its part."

Referring to Fig. 8.1, one can see that the spectral sensitivities are not monochromatic, i.e., the sensitivity of each emulsion extends over a range of wavelengths. Variations can be expected in terms of emulsion types and under manufacturing processing for a specific emulsion. In general, the sensitivity is not a symmetrical or simple function about the mean wavelength, nor does it have a smooth variation with wavelength; all of these characteristics are not amenable to simple mathematical analysis to account for the spectral bandwidth.

If the color sensitivities were monochromatic, the resulting visual color arising from an additive color combination could be predicted from the periphery of a chromaticity diagram (Fig. 8.2). (We will disregard for the moment the fact that

Figure 8.2. Chromaticity diagram for standard source "C" (artificial daylight) with locus of extreme purities of pigments, dyes, and inks (x and y are the normalized contributions of the CIE nonphysical primaries composed of 700.0-, 546.1-, and 435.8-mμ light).

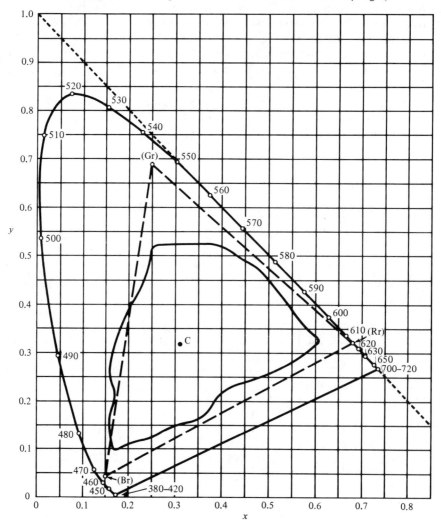

reversal color photography is a subtractive color process.) The chromaticity diagram is the projection of a tristimulus representation of color on a two dimensional plane. An infinite number of projections is possible. Transformations are generally possible between various color representations.

The periphery of the color diagram is the locus of pure spectral colors ranging from 380 to 780 mμ, and is shown by the solid line with the spectral wavelengths as circles. If three primary colors were selected and joined by straight lines to form a triangle, any color lying within the triangle could be reproduced; those colors, particularly spectral colors other than the selected three primary ones, could not be reproduced in a three-color additive system. The color triangle shown is for those existing in a color television system receiving tube, where the three colors are based on Illuminant C (average daylight, having a color temperature of 6750 K) and Wratten Nos. 25, 58, and 47 filters for red (Rr), green (Gr), and blue (Br). The positions of these three colors lie within those for the pure spectral hues, and the range of colors lying within the triangle are even fewer. The solid inner envelope is

Figure 8.3 Chromaticity diagram for standard source "C" (artificial daylight), showing the colors of various Kodak Wratten filters when illuminated by the standard source "C" (x and y are the normalized contributions of the CIE nonphysical primaries composed of 700.0-, 546.1-, and 435-mμ light).

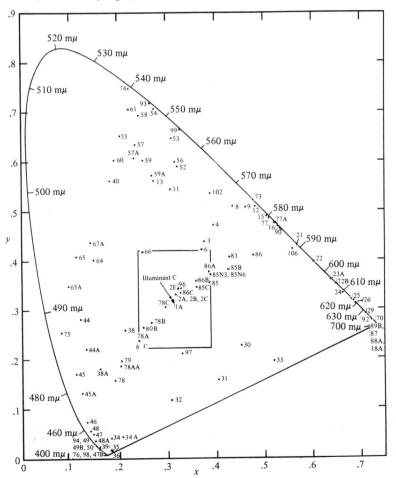

the locus of the maximum purities of colors covered by the Munsell and Ostwald samples, the Textile Color Card Association, and the gamut of printing inks (Wintringham, 1951).

Thus, we may represent an effective color on the chromaticity diagram considering a specific illuminant and a filter. This is shown for the Kodak Wratten filters in Fig. 8.3, again for Illuminant C (Eastman Kodak, 1968). If we choose any three filters, and form a triangle, the additive colors that can be produced lie within the triangle. The filter colors may be slightly varied by the use of color correction filters in the fashion indicated by Fig. 8.4, also for Illuminant C (Kodak, B-3, 1968).

However, the reversal color film process involves subtractive colorimetry. Thus, for instance, the positive image should be dense where there was no red and thin where the subject contained large amounts of red; the color of the transmitted light

Figure 8.4 Chromaticity diagram for standard source "C" (enlarged section) showing colors of Kodak Color Compensating and Kodak Light Balancing filters (x and y are normalized contributions of the CIE nonphysical primaries composed of 700.0-, 546.1-, and 435-mμ light).

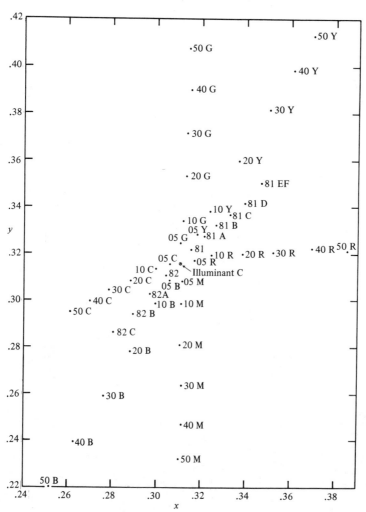

would be minus red, or cyan (a mixture of blue and green). A representative set of curves for yellow (minus blue), magenta (minus green), and cyan (minus red) is shown in Fig. 8.5 (Neblette, 1962) from left to right. The uppermost curve represents the total absorption which should be a straight line to represent a true neutral. Thus, the dyes are an approximation. Further, the dyes used vary between film types and are subject to production and development variations. The exact positions representing the three transmitted colors blue, green, and red thus depend on the film, and form a triangle within the spectral hue curve on the chromaticity diagram. Thus, there are colors that cannot be reproduced by the dyes in the emulsion, in addition to those that cannot be reproduced by the emulsion sensitivities.

If we consider 8443 infrared Ektachrome film, the above discussion applies in general, except that the chromaticity diagram applies basically to visual sensitivity. When a film, such as 8443, has infrared sensitivity, we speak of a false "color," representing the radiation extending from infrared to green, sensed by the film (with a No. 15 Wratten filter eliminating the blue sensitivity). A representative curve is shown in Fig. 8.6 (Fritz, 1967), subject to production variations.

But the infrared sensitivity of the 8443 is also seriously subject to storage effects; the film must be kept in a deep freeze (0°F), not simply cooled, or the infrared sensitivity is seriously decreased. The 35-mm, Type 8443 films commercially available in photographic stores have been found to vary considerably in infrared sensitivity. Photographic stores generally have a refrigerator for film, not a deep

Figure 8.5 Spectral absorption curves for the three dyes of a typical reversal color film.

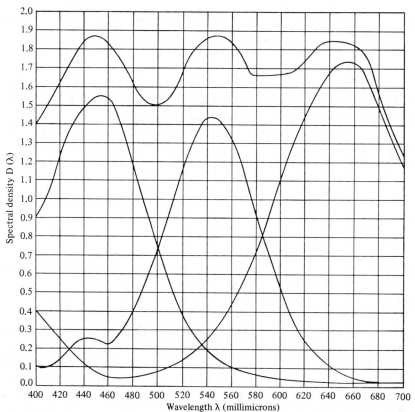

Spectral density $D(\lambda)$

Wavelength λ (millimicrons)

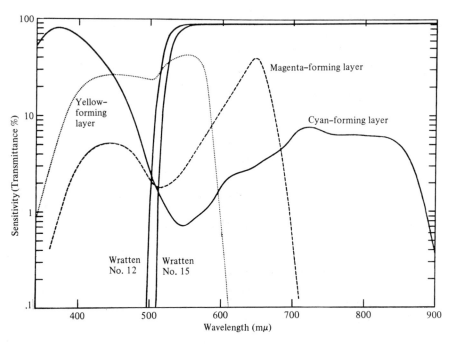

Figure 8.6 Spectral sensitivities of three layer type 8443 false color Ektachrome film; spectral transmittance of Kodak Wratten filters as indicated.

freeze. Type 8443 film is best obtained on special order direct from Eastman Kodak Company, with a minimum amount of handling or room temperature storage. It then must be checked for infrared sensitivity by procedures described in the following sections.

Rather than put the burden of stabilization of the Type 8443 film on Kodak, and wait for a solution, the user is necessarily obligated to check the properties of the film he purchases. Film manufacture dating is not the answer.

The author has been told of an instance when a film roll was stored in a refrigerator, with the film roll lying on its side. Because the film roll edge in contact with the refrigerator compartment was cooler, the characteristics were different and a banding occurred on the developed film.

Exposure

For scientific applications, we expose the film to record scientific data, and we concurrently record for a density calibration. In recording scientific data the system transfer function is involved, which includes the optical system, generally with a lens and shutter. The transfer function is affected by the lens f/number setting, the lens aberrations as a function of f/number setting, the focusing accuracy, the shutter, and internal scattering from the light-tight housing of the camera. A focal plane shutter was found to be particularly critical in exposing Type 8443 film; the focal plane shutter which is near the image plane in 35-mm cameras may have at best production inaccuracies of 15% in closure time and has been found to be a problem with the 8443 film; a nonuniform exposure across the photograph was produced that was not apparent with Ektachrome; this is apparently because of diminished film exposure latitude (Fritz, 1967). This effect is not visually apparent with ordinary Ektachrome film. Thus, it is advisable to use a compur shutter (multiple leaf type) or extremely accurate focal plane shutter with 8443 film. In aerial exposures, image motion can affect the system transfer function and exposures should be as short as possible, unless image motion compensation is incorporated in the camera.

Another effect, not generally recognized, is that the integrated color of a scene usually depends on viewing angle. Figure 14.4a (Chapter 14) is a set of laboratory photometric curves for alfalfa taken at the wavelengths of 0.350, 0.566, and 1.0 μm for incident radiation at an angle of 30° to the average surface normal; the leaves were measured in two orientations, parallel and perpendicular to the plane of incidence. It can be seen that the color photometry is a pronounced function of wavelength. This effect has been observed in the laboratory for other farm crops and foliage (Chapter 14). Also implied is that the color will vary with illuminating angle of the sun for outdoor photography, not only from Rayleigh scattering by the atmosphere but from geometry effects.

The second reason for exposure in scientific work is for calibration of a scientific observation. The calibration exposure may be made with the same camera configuration as for the scientific observation, but this involves the system transfer function. The best way to check the sensitivity of a film is with the simplest system possible since it reduces the number of possible variables. The use of field aerial color panels does not do this, even though the paints used for the panels may have well known characteristics. However, the use of aerial color panels in conjunction with a film calibration and control program is quite useful. One simple approach involves a sensitivity exposure from a density scale (calibrated step tablet). Many techniques have been described in the literature for this purpose, and for color film calibrations, the color temperature of the illuminating source becomes important.

One approach that has been suggested is the use of the Model 101 Kodak Process Control Sensitometer. This sensitometer provides a $\frac{1}{5}$-sec exposure produced by a shutter disk connected directly to a synchronous motor. Various step wedges may be used, as well as color compensating filters for the exposure of color films and papers. The unit has a voltage regulator and means for precisely adjusting the current to a projection lamp used for exposures. The sensitometer is not presently available as a Kodak catalog item.

A second approach used at Grumman Aerospace Corporation, based on a technique used in the Kodak Research Laboratories, uses a 500-W DMS 3200 K projection lamp. The lamp has been calibrated by a NBS secondary standard optical pyrometer to a color temperature of 2850 K. Two glass filters are used, one obtained from Kodak Special Product Sales to convert the 2850 K to 6100 K (Corning No. 5900), and another (1 mm No. 2043) from Pittsburgh Plate Glass Company to optimize the infrared color balance (Fritz, 1967). A leaf type shutter is interposed to control the exposure time to a calibrated step tablet. The light path length is about 1 m, and the measurement geometry is fixed.

The author is aware of at least one laboratory that uses a simple photographic strobe light as a calibration source; this technique is inaccurate. A pulsed gas discharge lamp has been used for photometric calibration where flashlamps are used for exposures of the order of microseconds; conventional exposure techniques do not take into account reciprocity failure in this application. Problems in repeatability of the flash limit the present accuracy to about 1%.

The step wedge that is used in the Grumman densitometer is a photographic silver density type, which cannot be used for an image-forming beam. The Collier

"Q" factor is given as follows (Eastman Kodak, P-114, 1967):

$$\lambda = 0.420 \; \mu\text{m}: \quad Q = 1.40;$$

$$\lambda = 0.550 \; \mu\text{m}: \quad Q = 1.41;$$

$$\lambda = 0.680 \; \mu\text{m}: \quad Q = 1.40.$$

It has 21 steps, with a diffuse density range from 0.05 to 3.10. However, any suitable calibrated density standard may be used. In exposing Type 8443 with a sensitometer, a No. 12 Wratten filter must also be interposed in the light path.

The exposure technique just described will yield integral densities, since no attempt is made to isolate the separate contributions of the three dye layers. However, analytical densities, which measure the individual effects of the dyes, can be misleading because of interlayer effects during film development. Analytical densities would be obtained experimentally if each of the three film emulsion layers could be isolated and exposures made on each. The use of step wedges in three colors does not isolate the contributions of each layer.

Another highly desirable approach to film calibration, which requires camera design changes, is the printing of a step wedge on a concurrently exposed aerial frame by a precision photometric illuminator located in the aerial camera. It is not of any use to put the step wedge on the film at Kodak, before shipment of the film to the user, because of changes caused by the fading of the latent image of the step wedge. Some short focal length cameras use antivignetting filters to compensate the overall exposure in the film plane. The results obtained with color films must be carefully evaluated to eliminate the resultant possible nonuniform exposure effect on color.

Development

The recommended development processes have been described in detail by Fritz (1967) for Type 8443 Ektachrome film. However, when accurate control of film development is necessary in order to minimize development effects, some form of automated technique is essential. A Nikor development tank is hardly the answer. If a deep tank with nitrogen agitation is used, a frame may be made to hold the film (or a Nikor reel may be used to hold the film, with this then inserted into the nitrogen agitation tanks); such a procedure has been followed with 35-mm films with reasonable results. The use of a Kodak Ektachrome RT Processor, Model 1811 M (formerly Model 1411M), with calibration exposures interspersed at intervals on the film, offers a good approach. Complete automation, as is possible for black and white films with the Versamat processor, appears best.

However, if there is no precise control of development, the use of a step wedge exposure on a test strip, simultaneously developed with the conventionally exposed film, offers the possibility of correcting in the densitometry measurement for varying development procedures. This assumes no variation in the uniformity of development of the processed batch. The color correction procedure is involved, since it is a function of three emulsions. Also, it is almost impossible to correct for nonuniformities in the development process. In any case, Type 8443 film should be developed as soon after exposure as possible to eliminate effects due to fading of the latent image. If the film cannot be processed immediately, it may be placed into cold storage; however, the effect of fading of the latent image has not yet been accurately evaluated.

Evaluation — Densitometry

In the evaluation of color densities, the densitometers must have higher sensitivities because fairly narrow spectral bands are measured. Three wavelengths are used; a typical set is 630, 530, and 440 mμ, corresponding to red, green, and blue. Narrow band interference filters with half-widths of the order of 50 mμ are available to isolate these wavelengths conveniently for density measurement of transparencies; another alternative is the use of spectral lines from a mercury–cadmium arc.

It has been noted above that the integral densities will be obtained when all three color emulsions are exposed simultaneously. The densitometry that will be subsequently described will determine this. There are techniques for determining the analytic densities by measuring the equivalent neutral densities, and also methods for transforming integral densities to analytic densities are described in the literature (Neblette, 1962); these involve solving three simultaneous equations, providing the characteristics of the positive dyes are known as a function of wavelength.

A representative plot of analytic density versus the log exposure is shown in Fig. 8.7 for Type 8443 film (Fritz, 1967). Similar curves exist for Ektachrome film, but the ordinate may be plotted as equivalent neutral density. It is necessary to have the

Figure 8.7 Representative sensitometric curves for type 8443 infrared Ektachrome film.

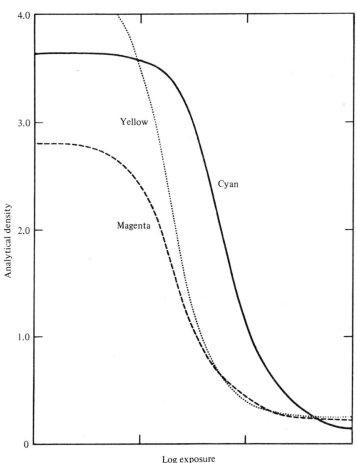

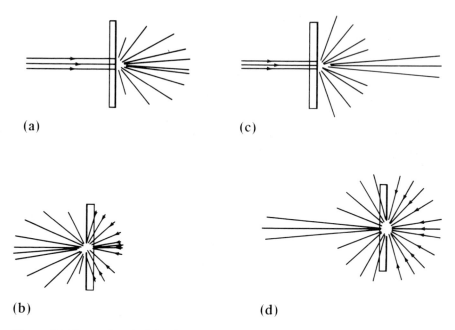

Figure 8.8 Types of optical density.

analytic density curves for color correction work, but for color monitoring, the integral density curves versus exposure are sufficient.

In this work, we have used a Joyce–Loebl microdensitometer with a range of 0–6 D. The unit essentially determines the specular density of an area of a transparency (Fig. 8.8a), as contrasted with the Macbeth densitometer, which determines the diffuse density (Fig. 8.8b or 8.8c) (Meyers and Allen, 1968). The Joyce–Loebl unit permits areas as small as 1 μm^2 to be measured, whereas the smallest area possible with the Macbeth is 1 mm^2 (10^6 μm^2). It is not to be inferred that measurements will be made over a 1 μm^2 area, because this approximates the size of grains. The use of a microdensitometer is necessary to evaluate the small color images resulting from aerial photography, and any other subtle effects. (A punched tape output is useful for computer analyses.) Thin film interference filters nominally centered at 0.633, 0.533, and 0.433 μm may be inserted into the measuring beam to isolate the desired wavelength bands. Narrow band filters permit better band isolation than the use of broadband Wratten filters.

Integral density curves obtained on the Joyce–Loebl microdensitometer for Type 8443 and Ektachrome films are shown in Figs. 8.9 and 8.10, respectively. The curve for Type 8443 (Fig. 8.10) shows reasonable agreement with the analytic density trends shown in Fig. 8.7.

Discussion

The calibration and control technique described is essentially an in-house standardization program to permit an evaluation of color film processing, which is used in the assessment of color differences appearing in aerial color transparencies made with Type 8443 and Ektachrome films. There is a relation between our technique and that

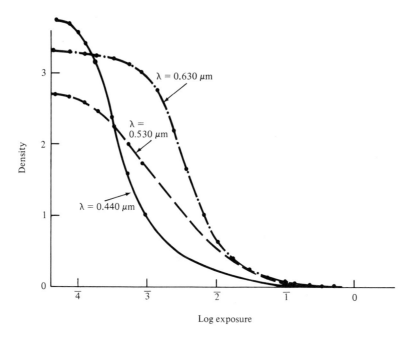

Figure 8.9 Integral density curves for type 8443 infrared Ektachrome film obtained on Joyce–Loebl microdensitometer.

used by the Kodak Research Laboratories, but there exists no "standard." In other words, if we measure a color with our microdensitometer on our transparencies, how can we communicate this color to other laboratories, or to field observers in terms of coordinates on a chromaticity diagram or tristimulus coefficients? The Munsell Book of Color involves reflectance standards and could serve as a bridge between laboratories and field observers. Reflectance measurements must be related to transmission measurements, but it is undesirable to use subjective human estimates of color.

Then, if we measure a "color" in terms of chromaticity coordinates, what does it mean? It has a certain significance in the laboratory where it was measured, but depending upon color controls and standardization in other laboratories, it will have more, less, or no meaning.

There is no question that qualitative color contrasts can be readily seen with color film, and these contrasts can be influenced through the use of filtering and emulsions. In an autumn Ektachrome aerial photograph of a group of red pine stands (in Connecticut) surrounded and separated by mixed hardwood trees, we can identify red pine trees that are dying from a certain pathologic cause; they appear lighter green in color at the centers of the crowns. The crown damage is more readily recognized on aerial photographs than at ground level. In an Ektachrome infrared photograph of the same area, one can also pick out the lighter crowns of the red pine trees that are suffering pathological attack or physiological change. A point should be made here: crown shapes of trees have been used to differentiate tree species (Heller et al., 1964); but in the closely spaced hardwood separation rows between the red pine stands, the crowns have been deformed. It is impossible to identify the

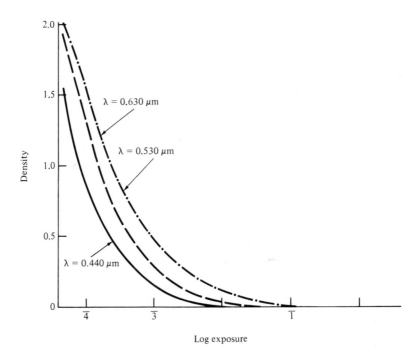

Figure 8.10 Integral density curves for type EH high speed Ektachrome film obtained on Joyce–Loebl microdensitometer.

species by crown shape. Color (Ektachrome and Type 8443) appears to be of considerably more use in species identification. This requires careful control of color.

It has been suggested that color differences be used, for instance, between a tree taken as "standard" and other trees that are under examination. This approach is of limited utility because only relative information is obtained; first, the tree taken as "standard" will vary, but its variation will not be measured; second, because film batches will vary, and if no control technique is employed, the comparisons against the "standard" will vary, and erroneous information will be obtained.

While the field of color and infrared false color aerial photography is still young, standards of color control and calibration must be thoughtfully considered and evaluated. Interlaboratory and field controls must be established. If the color control and calibration standards are delayed, this new science will suffer and each worker in the field will develop his own techniques, and possibly have his own problems in reproducibility. Further, as a corollary, each laboratory working in color aerial photography requires a color standardization control that permits exact evaluation of the color photographs produced by the laboratory.

Another critical factor is the duplication of color photographs; although the original producer of the color transparencies has them at his disposal, if he wishes to present color information to other workers in the field, extreme care is necessary in the production of the duplicates. A numerical specification of the colors would be a useful adjunct, not only for conveying color information from one laboratory to another, but to facilitate data analysis.

Nonimaging Sensor Calibration

A spectro-polarimeter/photometer that has been used in a light aircraft for observations of forest and sea areas will be described as representative of nonimaging sensor calibration. For these remote sensing observations, an optical, polarimetric/photometric sensing head was designed to permit adaptation for laboratory, field, surface vehicle, and aircraft observations in the spectral region from 0.3 to 3 μm. The unit specifically measures the Stokes parameters, I and Q, with the observational direction chosen to make $U = 0$ (i.e., observations with the average surface normal in the plane of vision).

The sensor unit consists of the appropriate sensor for the spectral region being investigated, appropriate polarization analyzers, spectral filters, and optics; these are all enclosed in a portable light-tight housing. The sensor unit may be mounted on a tripod, or on a truck for terrestrial field use, or in an aircraft for airborne observations.

The system schematic is shown in Fig. 8.11. The incoming radiation enters two parallel optical systems: one is the polarimetric/photometric sensor unit and the other is an adjacent, but separate, 16-mm framing camera. (The framing camera allows a 7° angular field of view to be recorded simultaneously with the 1° FOV polarimetric/photometric observations for later correlation.) The optics of the sensor consist of a 5-in.-diameter precision quartz lens of 12-in. focal length, followed by an aperature stop to define a 1° angular field of view. This is followed by a field lens and field stop, filters, polarization analyzer, and sensor. The polarization analyzer is a precisely mounted HNP′B, HN 22, or HR polaroid (made by the

Figure 8.11 Block diagram of aircraft polarimeter/photometer system.

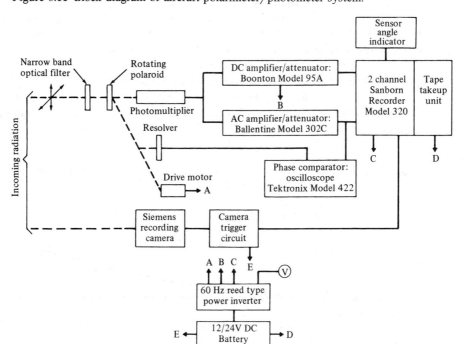

Polaroid Corporation) for the ultraviolet, visible, and infrared spectral regions. The sensors for the corresponding regions are either 6199 or 7102 end window photomultipliers or a balanced PbS detector. Narrow pass interference filters or broadband filters can be used in the system.

In operation, the spectral region is chosen by the appropriate filter–sensor combination. The polarization analyzer is rotated by an electric motor at 6.5 Hz, producing a fluctuating voltage, and for a polarized incoming light it is 13 Hz. Simultaneously, a sine/cosine resolver electrically indicates the angular position of the polaroid on the oscilloscope. The voltage from the output anode of the photomultiplier is simultaneously fed to both an ac and dc amplifier. The amplifiers serve mainly to amplify or deamplify signals as required for input to the Sanborn recorder. Provision is made for elimination of zero current (dark current) from the photomultipliers or PbS sensor. The Sanborn records, on a moving chart, voltages proportional to the Stokes parameters I and Q from the dc and ac amplifiers; the Sanborn also records the exposures made by the Siemens camera, as well as angular marking pips indicating the sensor viewing angle.

For laboratory measurements, percent polarization is automatically computed directly from I and Q; however, for airborne operations from a light aircraft, the computer power consumption in this application was excessive, and hence the polarization was computed subsequently from the data. It is also possible to record the I and Q data on magnetic tape and subsequently analyze it on the ground.

The calibration technique involves photometry and polarization. First, the photometric uniformity of the field of view (FOV) must be checked, and then the residual polarization of the system must be determined to be at a negligible level for the measurements to be made.

The photometric calibrations involve checking the uniformity of the 1° FOV sensing region. To accomplish this objective, the end window photomultiplier is replaced by a double ground glass diffuser illuminated by a frosted glass lamp; in effect, the sensor system is inverted to function as a source. The uniformity of the FOV is checked using a 1-mm aperture sensor that can be moved about in the projected beam. By adjusting the positions of the field lens, and field stop in relation to the aperture stop and objective lens, the residual field nonuniformity was capable of reduction to less than 1%. The introduction of interference filters into the system did not change the uniformity because the location was in the collimated beam location.

Figure 8.12 Polarimeter calibration curve (standard surface: freshly scraped magnesium carbonate, perpendicular incident radiation).

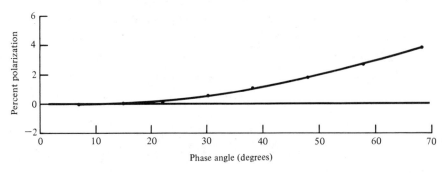

Subsequent to the initial alignment, the photomultiplier is replaced and the appropriate rotating polarization analyzer is placed in the collimated beam region. The sensor is then set to view a suitable nonpolarizing white photometric standard such as a freshly scraped $MgCO_3$ block, or where this is inconvenient, a white Nextel painted panel is used. The photometric properties of $MgCO_3$ and white Nextel paint are close to Lambertian and the polarimetric properties of $MgCO_3$ are shown in Fig. 8.12. At perpendicular viewing for these standards, the polarization is negligible (0.1%) when illuminated by an unpolarized light source.

The unpolarized source for these calibration measurements was an incandescent lamp, a GE type 2331 followed by a Lyot depolarizer; the lamp was located at the focus of a 5-in.-diameter quartz lens; this source was also checked for uniformity and found to be better than 1%; the residual polarization was < 0.1%. These checks were made with a 1-mm aperture sensor both without and with a polarization analyzer.

The polarization sensor system was found to have a residual polarization of < 0.1% for polarization configurations using either the HNP′B, HN22, or HR analyzers.

For higher accuracies, birefringent prisms must be used for the polarization analyzers, necessitating higher cost and more involved mounting configurations.

Field calibration is accomplished by placing a white Nextel plate in the FOV of the sensor, noting the response, and concurrently measuring the plate brightness with a calibrated photometer. The polarimetric calibration consists of placing a 5-in. diameter polarization analyzer in front of the objective lens of the polarimeter and noting the response for two mutually perpendicular positions of the analyzer. The polarization calibrations are accomplished with the sensor internal polarization analyzer rotating.

Another technique used to check the response of the polarimeter at intermediate viewing angles consists of measuring the ratio of the parallel and perpendicular components (relative to the plane of incidence) of the light reflected off a glass plate specularly at a variety of angles; the polarization may be computed using Fresnel's equations, taking into account the effect of light reflected off the back surface of the plate.

The interference filters used in the collimated region of the sensor beam were found to have an effect of < 0.1% on the residual polarization of the sensor.

Imaging Sensor Calibration (TM of LANDSAT 4)

The on-board calibrator optical system for the TM is shown in Fig. 8.13. Three tungsten filament lamps furnish the source for bands 1–5 and 7, and a blackbody for band 6, all through a flex-pivot-mounted resonant shutter oscillating at the same frequency (7 Hz) as the mirror assembly. The shutter introduces a black dc restoration surface as well as the calibration sources into the detector fields of view during the scan mirror turnaround time. The six rectangular integrated light pipes with prisms and cylindrical lenses flood the detectors with light as uniformly as possible.

If the TM detectors were modified with beam splitters to permit polarization measurements, a dual light pipe system would be necessary to sequentially illuminate the detectors with polarized radiation in two mutually perpendicular directions. The

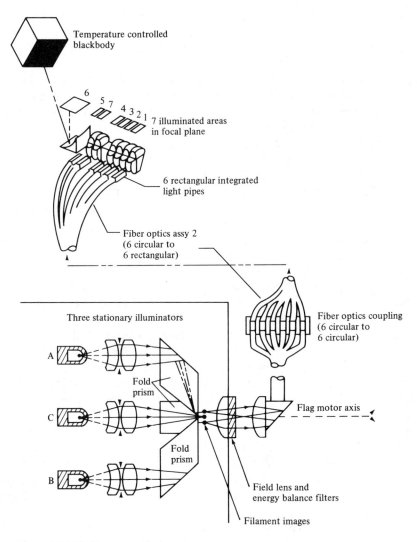

Figure 8.13 Calibrator optical system.

simple polarizers could be those made by the Polaroid Corporation for the visible and near IR ranges, and wire grid polarizers for the longer wavelength IR.

Ground Calibrations

Underlying the entire textual material is an unrelenting emphasis on quantitative calibration. This must not be taken to impugn the quality of much good remote sensing work that has been accomplished to date, but is meant to emphasize that training areas are not conveniently available to do comparisons in sea areas as on land. Local wind conditions, which may vary near land areas, affect the sea surface roughness, which in turn changes the photometric properties as a function of time. Further, for remote sensing of water quality, large areas with similar solar incident radiation angles, with controlled and defined turbidity, chlorophyll, carotenoids, or effluent wastes, do not generally occur.

Also, in order for intercomparisons between observational data to be made, a careful calibration and control program must be maintained, as painful as this requirement is. For instance, the quantitative contribution of a varying atmospheric absorption and scattering must be deduced for each remote sensing observation. This is particularly true for the higher resolution photometry of LANDSAT 4. The LANDSAT 4 TM has 256 photometric quantization levels, compared with 64 for the MSS on LANDSATS 1, 2, and 3. Also, the instantaneous field of view (IFOV) of the TM is 30 m, compared with 79 m for the multispectral scanner (MSS) on LANDSATS 1, 2, and 3 (82 m on LANDSAT 4), resulting in more possible scene noise.

An example of satellite ground calibration is the use of a large coral sand beach area to calibrate the LANDSAT 1 in a water quality experiment in the U.S. Virgin Islands. Concurrently with the overpassage of LANDSAT 1, four band aerial photographs were taken of the area. The aerial photographic imaging could have been accomplished just as readily with a scanning sensor located in the aircraft, without the inherent calibration difficulties associated with the photographic process. Test panels were used to calibrate the low altitude aircraft imagery.

The spectral reflectance of the test panels was determined in the laboratory (Fig. 8.14) and the associated photometric properties similarly determined (Fig. 8.15); from these data the apparent spectral reflectance for the given incident solar and sky radiation was determined. Also, the coral beach sand (used as a calibration standard, for the LANDSAT MSS) photometric properties were similarly determined spectrometrically (Fig. 8.16) and photometrically (Fig. 8.17). (Note that the phase angle is the angle between the incident radiation and the scattered ray.)

Figure 8.14 Spectral reflectance (relative to $MgCO_3$) of test panels at an incident angle of 40° and a phase angle of 3° (W, white; G_1, G_2, G_3, gray shades; R, red; G, green; B, blue).

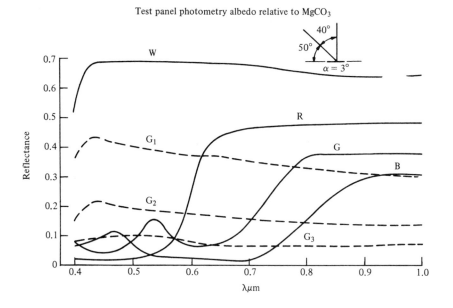

Test panel photometry albedo relative to $MgCO_3$

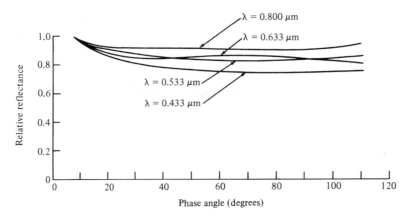

Figure 8.15 Spectrophotometric properties of 3-M Nextel White 110-A-10 paint as a function of phase angle and wavelength for an incident illumination angle of 40°.

The test panels necessary for the implementation of the optical ground truthing and aerial photography were placed on an open coral sand area on Brewers Bay Beach. The ground color panels were then used for ground irradiance measurements, to monitor color bias by the atmosphere, to make *in situ* relative measurements of the beach sand reflectance, and to monitor the response of the aerial color films. During the aerial photography, the 6-ft color panels and adjacent beach areas were

Figure 8.16 Spectral reflectance (relative to $MgCO_3$) of Brewers Bay beach sand and Cape Hatteras beach sand at an incident angle of 40° and a phase angle of 3° (—, central area Brewers Bay beach at test panel location; - - -, area on Brewers Bay Beach adjacent to the water; - · -, Cape Hatteras beach sand.

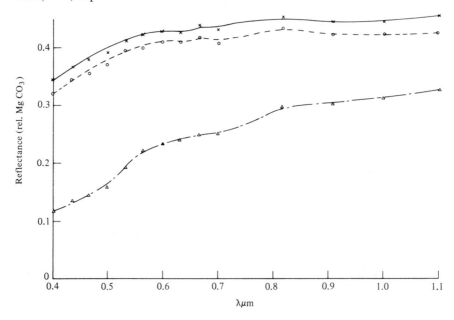

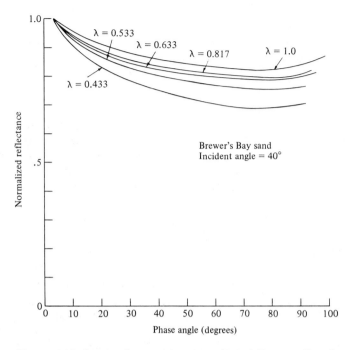

Figure 8.17 Spectrophotometric properties of Brewers Bay beach sand from test panel location at an incident angle of 40° and a phase angle of 3°.

photographed on at least 1 frame in 12. The beach area on which the panels were placed has a size of 350 by 900 ft. Although some trees and a road exist, 200 by 900 feet are relatively clear and sufficiently resolvable by the LANDSAT 1, in at least one line element, to serve as a ground calibration to permit atmospheric effects to be subtracted. It is to be noted that the placement of the panels on a white sand area is detrimental as far as using the panels for aerial photography; a dark background would be better because the solar radiation scattered by the sand into the atmosphere above the panels adds to that scattered directly by the atmosphere above the panels.

Depolarization

As seen from Fig. 8.12, even an accepted photometric standard such as $MgCO_3$ has some residual polarization at nonnormal viewing. Further, as will be discussed in Chapter 20, in many cases lasers may be appropriate as radiation sources for remote sensing. Lasers normally produce 100% plane polarized output radiation, and hence the depolarization properties of two naturally occurring materials (silica beach sand and Haleakala volcanic ash) as well as $MgCO_3$ will be presented. The silica beach sand would be analogous to the previously discussed beach sand photometric reference, and the volcanic ash somewhat analogous to that existing on the lunar surface.

Measurements

In the experimental arrangement, the apparatus was large scale and was designed to measure the polarimetric and photometric scattering properties of samples up to 9 cm in diameter. The axes of both the source beam and the photometer/polarimeter defined a vertical scattering plane containing the normal to the average surface of the sample. The scattered light from the sample was measured by a photometer/polarimeter located at a 60° viewing angle. The incoherent unpolarized source was a 1000 W tungsten iodine lamp filtered to produce a limited bandwidth, with an effective wavelength centered in the passband, with light collimated to within 0.5° to 2° degrees. The passbands are 0.48 μm (blue), 0.54 μm (green), and 1.0 μm (IR), with 3 dB half-bandwidths of 0.039, 0.036, and 0.17 μm, respectively.

The high coherence source was a Spectra-Physics Model 125 He–Ne laser operating at a wavelength of 0.6328 μm; its radiation was 100% plane polarized in a direction determined by the orientation of the Brewster angle windows at the ends of the gas discharge tube. The orientation of the plane of polarization of the laser was set within a fraction of a degree, with the electric vector either parallel ($E_{\parallel}$) or perpendicular ($E_{\perp}$) to the scattering plane. The laser had an output power of about 80 mW. The beam was expanded by a telescope to produce a 10-cm-diameter beam incident upon the sample about 3 m away. The divergence of the laser beam was less than 1 minute of arc.

A type 6199 end window photomultiplier (S-11) was used for blue ($\lambda_{eff} = 0.48$ μm), green ($\lambda_{eff} = 0.54$ μm), and laser (0.6328 μm) observations, and a type 7102 end window photomultiplier (S-1) was used for infrared ($\lambda_{eff} = 1.0$ μm). Rotating polarization analyzers were located in the collimated beam in front of the photomultipliers to measure polarization of the light scattered by the samples. The residual polarization in the incoherent source and polarimeters was of the order of a tenth of a percent. The polarization analyzers were sheet polaroid types HN-22 for blue, green, and laser illumination, and HR for infrared. To polarize the low coherence incandescent source, the corresponding sheet polaroids were used in the collimated source beam.

Thus, the scattered polarization (including the depolarized components) is a function of the incident polarization (Beckmann, 1963), and therefore coherent and incoherent radiation must be expected to be depolarized in different ways.

Samples

The laser has been suggested for lunar and terrestrial remote sensing investigations. Laser and incoherent light measurements were made, therefore, on materials that were reasonably representative of simple particulate surfaces. One of these materials was Haleakala volcanic ash composed of dull vesicular gray–brown particles measuring between 37 and 88 μm. Another was a white silica beach sand from Rockaway Beach, New York, that had predominating silica particles of about 0.1 to 0.5 mm. These samples were lightly sieved on a platen that filled the field of view of the photometer/polarimeter. The photometric properties of these samples were determined by optically referencing them to a block of $MgCO_3$.

Although it would have been highly desirable to determine the reflection coefficients of the samples when smoothly polished, this was not possible for the present samples. The Haleakala volcanic ash was a vesicular material crushed to provide the 37–88-μm particles. The average index of refraction, determined visually by the

Becke line method, was about 1.65. The Rockaway Beach sand, arising from the continental shelf, is essentially an amorphous SiO_2, being birefringent with indices of 1.544 and 1.553 (*Handbook of Chemistry and Physics*, 1983) checked by the Becke line method. The $MgCO_3$ was U.S.P., made in France by a precipitation process and pressed into blocks, and was imported by Chas. L. Huisking and Co., Inc., New York, N.Y. The particles were micron size, and the indices of refraction were 1.717 and 1.515 (*Handbook of Chemistry and Physics*, 1983).

Depolarization Results

The depolarization properties of the silica beach sand for a 60° viewing angle under conditions of varying wavelength are shown in Fig. 8.18a for an incoherent source, and in Fig. 8.18b for highly coherent laser radiation.

Figure 8.18 (a) Depolarization characteristics of silica beach sand [blue, ($\lambda_{eff} = 0.48 \ \mu$m), green ($\lambda_{eff} = 0.54 \ \mu$m), and infrared ($\lambda_{eff} = 1.0 \ \mu$m)]. (b) Depolarization characteristics of silica beach sand (6328-Å laser).

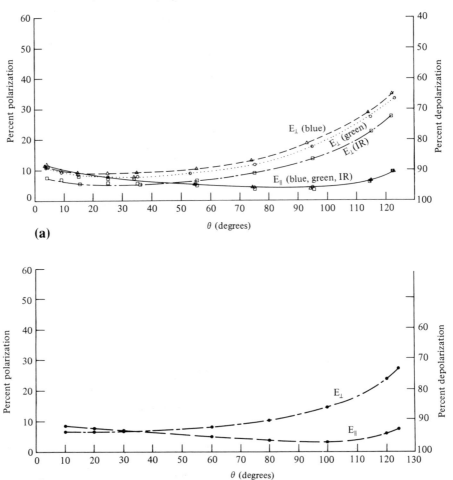

It can be seen (Fig. 8.18a) that the radiation parallel to the plane of incidence is most depolarized at higher θ for all wavelengths, whereas the perpendicular incident component is least depolarized in the blue. The parallel component could be the result of much multiple scattering and may be the most depolarized, while the perpendicular component is dominated by a single facet scattering with less de-

Figure 8.19 (a) Depolarization characteristics of Haleakala volcanic ash, 37–88-μm sizes [blue ($\lambda_{eff} = 0.48$ μm), green ($\lambda_{eff} = 0.54$ μm), and infrared ($\lambda_{eff} = 1.0$ μm)]. (b) Depolarization characteristics of Haleakala volcanic ash, 37–88-μm sizes (6328-Å laser).

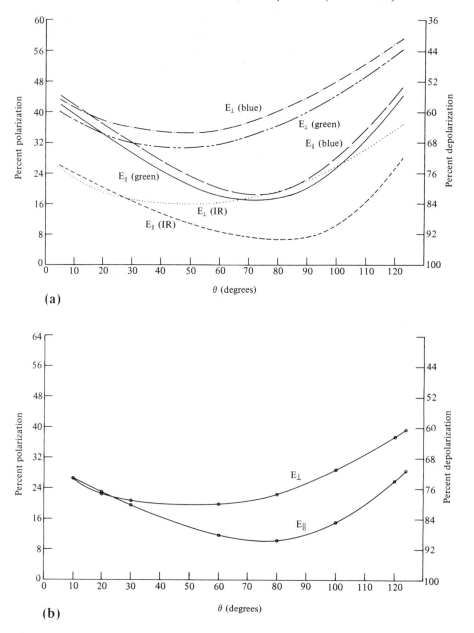

(a)

(b)

polarization, with the blue being best reflected by facets that are small compared to the wavelengths of green and infrared.

Comparing this to the laser depolarization in Fig. 8.18b, we observe that there is generally less overall depolarization for the perpendicular incident radiation polarization plane direction. This might be the result of the fact that the source is monochromatic, with consequent reduced mixing of various scattered wavelengths, which could result in reduced depolarization.

When we consider the depolarization properties of the Haleakala volcanic ash under incoherent illuminating radiation of varying wavelengths (Fig. 8.19a), we observe decreased depolarization compared to the silica sand, probably the result of a decreased refractive component contribution to the scattered light by the darker

Figure 8.20 (a) Photometric characteristics of silica beach sand [blue ($\lambda_{eff} = 0.48$ μm), infrared ($\lambda_{eff} = 1.0$ μm)]. (b) Photometric characteristics of silica beach sand (6328-Å laser).

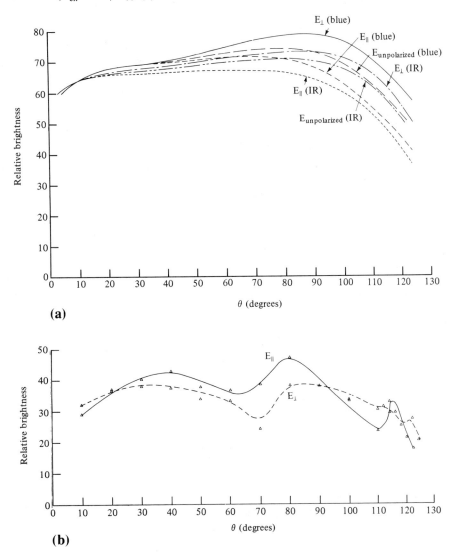

volcanic ash. Again, the corresponding blue radiation components are depolarized least and the infrared most. For the parallel incident light, however, there is a wide difference in depolarization, from which could be inferred multiple surface scattering with a very small refractive light contribution. Comparing these to laser depolarization (Fig. 8.19b), we again see generally less depolarization for the perpendicular component, comparable to that for the incoherent IR radiation.

Photometry Results

The relative photometric properties of the silica beach sand and Haleakala volcanic ash are shown in Figs. 8.20a and 8.21a, respectively, for blue and infrared low coherence light; for laser illumination, referenced to $MgCO_3$, the results are shown in Figs. 8.20b and 8.21b.

The sand (Fig. 8.20a) tends to a higher reflectance in the blue and to be specular, with a maximum at about $\theta = 90°$. The perpendicular component is scattered more

Figure 8.21 (a) Photometric characteristics of Haleakala volcanic ash, 37–88 μm sizes [blue ($\lambda_{eff} = 0.48$ μm), infrared ($\lambda_{eff} = 1.0$ μm)]. (b) Photometric characteristics of Haleakala volcanic ash, 37–88 μm sizes (6328-Å laser).

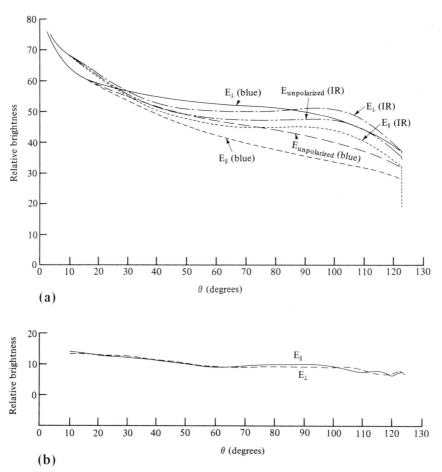

strongly, again evidence of a large single scatter contribution. The laser curves (Fig. 8.20b) present similar trends except for minima at normal incidence, analogous to the $MgCO_3$, discussed below.

In the volcanic ash incoherent light photometric curve (Fig. 8.21a), the relative effect of wavelength is different for small and large θ, probably the result of shadowing effects with the relatively nontransparent volcanic ash. The laser photometric curve (Fig. 8.21b) is relatively flat, possibly indicating extensive diffraction effects.

The photometric properties of $MgCO_3$ were checked because of the variations, mentioned above, in the photometric properties of the silica beach sand and Haleakala volcanic ash for varying coherences of source illumination.

The results were quite interesting (see Fig. 8.22). The photometric curves for sources nonpolarized and polarized perpendicular and parallel to the scattering plane approximated the superimposed cosine curve. The perpendicular source polarization produced the highest relative brightness, presumably the result of a dominant single-scatter component. The parallel source polarization produced lower relative brightness, possibly because of the greater effect of multiple scattering. The unpolarized source curve is the average of the two component curves.

For laser illumination, however, the photometric properties indicated a pronounced dip around normal incidence, with a fairly large difference between the curves for the laser radiation perpendicular and parallel to the scattering plane.

The existence of this unusual photometric minimum was checked and verified on another photometer, a Spectra Pritchard model. The same kind of dip was noted at a 0° viewing angle. To determine the cause of this minimum at normal incidence, we made a series of high magnification scanning electron microscope studies of the

Figure 8.22 Photometric characteristics of $MgCO_3$ photometric standard [green ($\lambda_{eff} = 0.54$ μm); (6328-Å laser)].

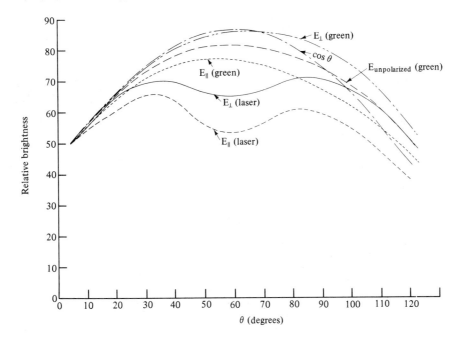

138

Figure 8.23 Scanning electron microscope views of magnesium carbonate photometric standard. (a) magnification 1000 X; (b) magnification 10,000 X.

Figure 8.24 Polarization characteristics of silica beach sand (effect of source coherence).

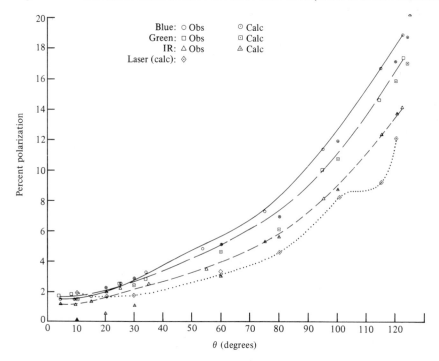

MgCO$_3$ surface with a Joel JSM-2. The images obtained are shown in Fig. 8.23 for magnifications of 1000 and 10 000 ×.

It can be seen (Fig. 8.23) that the MgCO$_3$ is composed of platelets of the order of 0.1 μm thick in a very porous matrix. The cause of the dip in the photometric curves for coherent light at near normal incidence may be the result of diffraction effects of the large number of platelet edges oriented perpendicular to the sample plane. These edges could scatter light into a 2π solid angle more effectively at normal incidence because their projected area is smallest to the incident beam.

Polarization Comparisons

After computing the polarization characteristics of the samples, as would be expected in blue, green, and infrared incoherent light, and comparing them to the measured values, we obtain the curves shown in Figs. 8.24 and 8.25, which agree within experimental error. The curves are computed for a 60° viewing angle in all cases. Also shown are the computed curves that we should obtain if an unpolarized laser source were available.

We see that the positive polarization is greatest for blue incident incoherent radiation and least for the infrared, with the 0.6328-Å laser curve peak falling below the other three. The probable explanation is that coherence effects had not been taken into account.

There are effects, which are probably the result of the degree of coherence of the incident radiation, that can affect the depolarization characteristics observed from surfaces. These effects appear to be the result of the fixed geometrical relationships

Figure 8.25 Polarization characteristics of Haleakala volcanic ash (effect of source coherence).

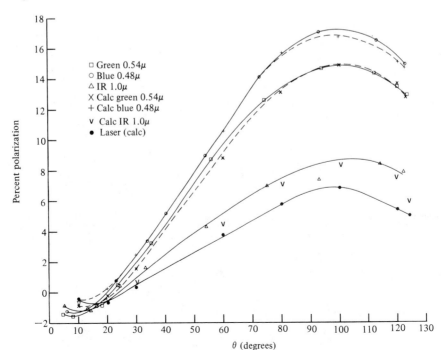

between the surface particles and are in contrast to aerosol multiple scattering, where the particles are in constant motion and an average effect is observed.

Consequently, caution is advised in interpreting data when coherent radiation is used in remote sensing, since the effects observed could be appreciably influenced by coherence.

A final point: the photometric investigation of a widely used diffuse reflectivity standard, $MgCO_3$, revealed unusual photometric properties with coherent radiation. One therefore concludes that acceptable diffuse reflectivity standards for incoherent light sources may no longer be acceptable for highly coherent sources.

9

Atmospheric Effects

In order to predict the solar radiation incident on the earth's surface (both specular and diffuse) with a realistic atmosphere and particulate loading, the flux divergence in a nonhomogeneous curved atmosphere must be computed for a spectral region between 0.3 and 100 μm. Scattering, absorption, and emission by particles and gas molecules must be included (as well as multiple scattering when appropriate), and the effects of clouds and the underlying terrestrial surface. The polarimetric properties of the atmosphere and underlying surface should also be in the formulation. As yet no such complete program has been undertaken.

There have been a number of simple photometric models applicable over a limited wavelength range that account for aerosols. The first investigators to include multiple scattering in a plane parallel atmosphere with an aerosol layer just above the ground were Rasool and Schneider (1971). They also included 50% cloud cover with an effective cloud top height of 5.5 km, and two stream calculations were made at a mean wavelength of 0.55 μm. The effect of outgoing terrestrial radiation was also estimated to cause a decrease of between 3.5 and 4.9°C in the earth's surface temperature depending upon the ratio of scattering to absorption of aerosols, if the optical depth of the atmosphere is increased from 0.1 to 0.4.

In principle, the equation of photometric radiative transfer may be solved using one of several numerical techniques. The problem arises when one considers the huge amount of computation time necessary with variation of a large number of parameters to explore a radiation field reliably and completely. The technique most acceptable is the successive-scattering iterative technique of Dave and Gazdag (1970) in which the phase function for aerosols is expressed in a Fourier series (Dave, 1970; Braslau and Dave, 1972). The model developed (Dave, 1972) presents the effects of aerosols on the incident solar energy absorbed, reflected, and transmitted through the earth's atmosphere between wavelengths of 0.285 and 2.50 μm. Adequate spectral resolution is provided to account for gaseous band absorption, and sufficient vertical spatial resolution, to account for inhomogeneities. Neither horizontal inhomogeneity nor earth curvature are included.

An extension of the original scalar Dave (1972) model includes polarization in a vector version. The amount of computing time is formidable.

In the model, a cloudless plane parallel semi-infinite atmosphere is assumed to have unidirectional, unpolarized solar flux incident at the top. The vertical concentrations of oxygen, ozone, water vapor, and carbon dioxide are specified as well as that of the aerosol. As the solar flux penetrates the atmosphere, it undergoes absorption and scattering at each level to produce an attenuated direct flux and diffuse upward and downward components. At the bottom of the atmosphere there is either (1) an idealized Lambert ground reflecting surface (producing unpolarized, isotropic scattered radiation, independent of the polarization or angular distribution of the incident radiation), (2) a Fresnel reflecting ground surface or (3) a realistic ground surface represented by the angular Stokes parameters determined from actual measurement of a surface or from theory. The Lambert ground reflectivity is defined as the ratio of upward scattered flux to downward flux on to the surface. Contributions from thermal emission of the atmosphere or surface are neglected because the temperatures are too low to cause a significant contribution in the spectral range. The aerosols in the original program are considered to be spherical; however Egan (1982) has proposed an empirical size modification to account for aerosol nonsphericity, edges, and asperities.

The vector program consists of five separate sections, each involving a separate computer run:

Part I (VPA) computes the Legendre series coefficients for four different scattering functions that are used to determine the intensity, amount and direction of plane polarization, and circular polarization, of a spherical particle as a function of size and optical complex index of refraction.

Part II (VPB) computes the Legendre series coefficients representing four scattering functions of a unit volume scattering phase matrix containing a given size distribution of spherical particles of the same material when illuminated by a polarized monochromatic, unidirectional beam of radiation.

Part III (VPC) computes the Fourier series coefficients $M_{ij}(\theta', \theta)$ of a normalized 4×4 scattering phase matrix of a unit volume with an arbitrary distribution of spherical particles, where θ' and θ are the directions of the scattering and incident directions with the zenith. The Fourier series argument is the difference between azimuthal angles of the meridian planes containing the incident and scattered radiation.

Part IV (VPD) computes the intensity, degree of polarization, direction of polarization, and the ellipticity of polarization of the scattered radiation emerging from selected levels of a plane parallel nonhomogeneous atmosphere. The model allows the insertion of an arbitrary aerosol number density, and the addition of molecular absorbers such as ozone. The lower bound is assumed to be a Lambert type reflecting ground of a given reflectivity.

Part V (VPDF) is similar to Part IV, except that a Fresnel reflecting ground, with a given optical complex index of refraction, is substituted for the Lambert reflecting surface; or the Stokes parameters of an actual surface may be inserted into the program. The input parameters for the Dave (1972a, b) models are

1. the optical complex index of refraction of the aerosol,
2. the size range of the aerosols,
3. wavelength of radiation,

4. number density distribution of the aerosols,
5. modification of number density to account for particle shape,
6. vertical distribution profile for the aerosols,
7. Raleigh atomic and molecular scattering,
8. ozone absorption,
9. other molecular absorption as necessary,
10. solar zenith angle θ_0,
11. sensing zenith angle θ,
12. sensing azimuth angle ϕ,
13. sensing altitudes, and
14. either the diffuse surface albedo or the photometric/polarimetric Stokes parameters of surface.

Thus, a considerable number of factors are adjustable and experiment yields the appropriate values where available.

In order to clarify various specific aspects in the application of the Dave (1972a, b) vector programs, it is important to consider the input quantities for the five programs.

The first program, Vector Program A (VPA) contains as input on one card n_1, the real part of the refractive index of the aerosol; n_2, the imaginary part of the refractive index of the aerosol; x_1, the first value of x ($= 2\pi a/\lambda$) for which output is required; Δx, the increment on x; and the option

$$\text{NOPT} = \begin{cases} 1 & \text{for magnetic tape output for VPB,} \\ 2 & \text{for tabular output.} \end{cases}$$

The index of refraction (n_1, n_2) may be taken from Egan and Hilgeman (1979) or from Appendix A of this book for the appropriate aerosol material. A multitude of materials are listed, leaving one to wonder where to start. In general, silica sand or farm soil would be a good start for terrestrial modeling over land, whereas water would dominate over sea areas (perhaps with NaCl) or during humid conditions (the index of water is available in the literature, e.g., Irvine and Pollack, 1968; Hale and Querry, 1973). If ice exists, this same reference contains the index (also Bertie et al., 1969); in the stratosphere, as a result of volcanic eruptions, andesitic glass or sulfuric acid (index in Palmer and Williams, 1975; Majkowski, 1977) may be appropriate for aerosols. Where combinations of various aerosol materials exists, an average value must be used. The scattering in the final radiative transfer model is determined mainly by the real portion of the index of refraction, and the absorption by the imaginary part. At present, our spectral polarimetric/photometric observational capability accuracy limits the complete application of the detailed spectral variation of the complex index of refraction; but as instrumentation improves, the detailed variations will become important in analyses. A VPA program modification would permit the phase functions of two or more materials to be calculated for input to VPB.

The quantity x ($= 2\pi a/\lambda$), where a is the particle radius and λ is the wavelength considered) is set empirically by the smallest aerosol actually found under the model conditions. The minimum size a that is effective in scattering is of the order of 0.1λ to 0.01λ. The largest size is of the order of 10λ to 100λ. The quantity x is then chosen to yield about 100 increments. For tape input to VPB, NOPT = 1, and for tabular output of phase functions, NOPT = 2.

It may be remarked that one could combine all four or five programs into one program, to run faster and more efficiently, rather than have tape (and card) outputs for feeding into the next succeeding program. This would definitely be more efficient; however, the Dave program is a research type program where many intermediate outputs are available, for tweaking parameters or optimizing variables, and determining their effects rather than burying everything in a program, and having intermediate values listed. It frequently happens, as will be mentioned for VPB, that the actual measured volume extinction coefficients drastically disagree with the values calculated from the program. To continue a run under these conditions would be nonsense, although one could insert a statement to stop the execution of the program if such a disagreement occurs.

The second program (VPB) allows the selection of three different analytical size distribution functions for the aerosols (discontinuous power law, modified gamma, and lognormal) in the calculation of the normalized scattering phase matrix of a unit volume. As pointed out by Egan (1982), the selection of the break point on the discontinuous power law distribution can have a drastic effect on the volume extinction coefficient, sometimes as much as a few orders of magnitude. The modified gamma function appears to be more realistic (Deirmendjian, 1969) in terms of curve shape. Also the limits on particle radii are easily related to those actually observed (Table 9.1). The aerosol particle size limits are immersed in an exponential expression involving a standard deviation.

For the modified gamma distribution, the values of the parameters may be those suggested by Deirmendjian (1969) or Egan (1982). In order to account for particle edges or asperities, the small particle size range and number must be increased (see Egan, 1982).

The output of program VPB is a set of cards that are used as inputs to VPC. There is no further input data necessary to run VPC to produce one or more magnetic tapes for input to VPD. Also VPD requires input data that describe the variation of atmospheric pressure, ozone (or other gas), and particle number density with altitude (Fig. 9.1).

The first line below the alphanumeric information contains two scaling factors that modify the input ozone and particle size distribution by a fixed amount. The next line contains the wavelength, Rayleigh scattering for 1 atm, and ozone absorption for 1 atm cm. The next line gives the number of layers (i.e., 160/1, 2, 4, 5, 8, or 10), the angular increments on viewing angle (i.e., $2°$, $6°$ or $10°$ for 1, 3, or 5), and

Table 9.1 Distribution Parameters[a]

In VPB	Discontinuous power law	Modified gamma	Log normal
DISCON (1)	r_{min}	a	σ
DISCON (2)	r_m	α	r_m
DISCON (3)	r_{max}	b	—
DISCON (4)	C	γ	—
I POWER	ν	—	—
JDIS	1	2	3
RAD1	—	r_{min}	—
RAD2	—	r_{max}	—

[a]for log normal distributions $r_{min} = \exp(\ln r_m - 4\sigma)$, $r_{max} = \exp(\ln r_m + 4\sigma)$.

```
      HEIGHT - PRESSURE AFCRL MIDLATITUDE, SUMMER
      HEIGHT - OZONE : AFCRL MIDLATITUDE SUMMER
      HEIGHT - PARTICLE NUMBER DENSITY - DAVE AND BRASLAU
        0.0 1.013E+03 6.000E-05 3.780E+03
        1.0 9.020E+02 6.000E-05 1.250E+03
        2.0 8.020E+02 6.000E-05 3.100E+02
        3.0 7.100E+02 6.200E-05 1.300E+02
        4.0 6.280E+02 6.400E-05 4.300E+01
        5.0 5.540E+02 6.600E-05 1.400E+00
        6.0 4.870E+02 6.900E-05 1.090E+00
        7.0 4.260E+02 7.500E-05 8.400E-01
        8.0 3.720E+02 7.900E-05 6.470E-01
        9.0 3.240E+02 8.600E-05 6.500E-01
       10.0 2.810E+02 9.000E-05 6.550E-01
       11.0 2.430E+02 1.100E-04 6.600E-01
       12.0 2.090E+02 1.200E-04 7.350E-01
       13.0 1.790E+02 1.500E-04 8.000E-01
       14.0 1.530E+02 1.800E-04 8.900E-01
       15.0 1.300E+02 1.900E-04 9.600E-01
       16.0 1.110E+02 2.100E-04 1.050E+00
       17.0 9.500E+01 2.400E-04 1.170E+00
       18.0 8.120E+01 2.800E-04 1.290E+00
       19.0 6.950E+01 3.200E-04 1.400E+00
       20.0 5.950E+01 3.400E-04 1.550E+00
       21.0 5.100E+01 3.600E-04 1.390E+00
       22.0 4.370E+01 3.600E-04 1.190E+00
       23.0 3.760E+01 3.400E-04 5.800E-01
       24.0 3.220E+01 3.200E-04 2.700E-01
       25.0 2.770E+01 3.000E-04 1.230E-01
       30.0 1.320E+01 2.000E-04 4.700E-02
       35.0 6.520E+00 9.200E-05 1.700E-02
       40.0 3.330E+00 4.100E-05 6.200E-03
       45.0 1.760E+00 1.300E-05 1.800E-03
       50.0 9.510E-01 4.300E-06 5.000E-04
       55.0 4.850E-01 1.600E-06 1.450E-04
       60.0 2.500E-01 6.000E-07 4.100E-05
       65.0 1.300E-01 2.250E-07 1.150E-05
       70.0 6.710E-02 8.600E-08 3.300E-06
        1.0          .56811
  0.50000        0.15080       0.02460
   10      5      20
    2      9      13
  0.00000        0.40000
WAVELENGTH = 0.50000    R SUB MIN =  0.0050      R SUB MAX =  4.0000
       DELTA X =   0.5027     RFR =   1.33600      RFI =   0.00609
       WAVELENGTH AND RADII ARE IN MICRON UNITS.   0.50000
MODIFIED GAMMA DISTRIBUTION:  A =   5.3330D+04    ALPHA =  1.0000D+00
       B =    8.9443D+00      GAMMA =   5.0000D-01
VOL.EXT. COEF. PER PARTICLE PER CM.=   1.08268D-08
VOL.SCAT.COEF. PER PARTICLE PER CM.=   9.78325D-09
VOL.ABS. COEF. PER PARTICLE PER CM.=   1.04356D-09
   3   3   3   3   3   3   3   3   3   3   3   3   3   3   3   3   3   3   3   3   3   1
   3   3   3   3   3   3   3   3   3   3   3   3   3   3   3   3   3   3   3   3   3   2
   3   3   3   3   3   3   3   3   3   3   3   3   3   3   3   3   3   3   3   3   3   3
   3   3   3   3   3   3   3   3   3   3   3   3   3   3   3   3   3   3   3   3   3   4
   8   9  10  10   9   8   8   8   8   7   7   7   7   7   7   6   7   6   7   6   7   6   6   5



  53  51  49  46  44  42  41  39  39  37  35  33  31  29  28  26  24  22  20  17  15  12  10170
   8   6   3   3   5   7   9  11  14  18  20  24  26  30  33  36  38  40  43  40  42  44  46171
  48  50  52  52  54  56  56  55  53  55  58  56  55  57  56  55  57  54  60  75  97  98  99172
  99  99  99  97  73  81  79  73  64  61  58  56  58  56  57  55  57  59  57  57  58  56  54173
  52  50  48  46  43  41  41  41  37  35  35  30  27  29  26  21  21  19  16  14  12   9   7174
   6   3   3   6   7   9  11  11  15  18  23  26  28  31  34  36  39  41  38  42  43  45  47175
  49  51  53  53  55  57  57  56  55  56  58  57  55  57  60  55  53  55  60  76  97  99  99176
  99  99  99  97  77  68  61  58  56  57  55  56  58  59  57  55  57  55  56  56  55  55  53177
  51  49  47  45  43  40  38  42  40  37  34  32  29  24  22  18  16  14  14  11   9   7   5178
   3   3   6   7   9  11  12  13  15  19  23  28  31  34  36  37  39  43  39  42  44  46  48179
  50  52  54  54  56  55  54  56  54  56  59  57  56  58  57  56  58  55  60  76  97  99  99180
  99  99  99  97  76  60  55  58  56  57  58  56  57  59  56  54  56  54  55  56  54  54  52181
  50  48  46  44  42  39  43  39  37  36  34  31  28  23  19  15  13  12  11   9   7   6   3182
       26.0        31        7
          0                30              60              90             120
        150             180
```

Figure 9.1 Input to VPD.

the number of iterations allowed for a preset convergency criterion. The next line gives the number of ground reflectivities desired, and selected levels for intermediate output; the next line gives the desired Lambert ground reflectivities.

The next 190 lines give the aerosol data from VPC.

Following this is the value of the solar zenith angle θ_0, $(\theta_0/2)+1$, and the total number of azimuth angles for which output is desired. The angles follow in the next two lines.

```
      50.0 9.510E-01 4.300E-06 5.000E-04
      55.0 4.850E-01 1.600E-06 1.450E-04
      60.0 2.500E-01 6.000E-07 4.100E-05
      65.0 1.300E-01 2.250E-07 1.150E-05
      70.0 6.710E-02 8.600E-08 3.300E-06
       1.0      .56811
   0.40000   0.38250    0.00000
   10    5   10
   0.00000   0.00000     9   13
WAVELENGTH = 0.40000   R SUB MIN =   0.0050        R SUB MAX =  4.0000
   DELTA X =   0.6283        RFR =    1.50100        RFI =    0.00211
        WAVELENGTH AND RADII ARE IN MICRON UNITS.   0.40000
MODIFIED GAMMA DISTRIBUTION:  A =    1.1800D+07    ALPHA =  2.0000D+00
    B =    2.0000D+01             GAMMA =    1.0000D+00
VOL.EXT. COEF. PER PARTICLE PER CM.=    2.89144D-09
VOL.SCAT.COEF. PER PARTICLE PER CM.=    2.85133D-09
VOL.ABS. COEF. PER PARTICLE PER CM.=    4.01100D-11
 3  3  3  3  3  3  3  3  3  3  3  3  3  3  3  3  3  3  3  3  3  3  3
 3  3  3  3  3  3  3  3  3  3  3  3  3  3  3  3  3  3  3  3  3  3  3
 3  3  3  3  3  3  3  3  3  3  3  3  3  3  3  3  3  3  3  3  3  3  3
 3  3  3  3  3  3  3  3  3  3  3  3  3  3  3  3  3  3  3  3  3  3  1
 5  5  6  6  6  6  6  6  6  6  6  6  6  6  6  6  6  6  6  6  6  6  6
 6  8  5  5  6  8  6  6  6  6  6  6  6  6  6  6  6  5  5  5  5  5  5



 6  3  3  5  6  7  9  9  9 11 11 13 13 13 15 15 16 16 17 18 18 19 20175
20 20 20 22 22 22 22 22 24 23 23 24 24 26 27 27 27 28 29 29 29 29 29176
29 29 29 29 29 29 27 27 27 27 27 27 27 26 26 25 23 23 23 23 23 22 21177
21 21 19 19 19 17 17 17 17 16 15 15 14 13 12 12 11 10 10  8  8  7  6178
 3  3  5  6  8  8  9 10 11 11 12 13 13 15 15 15 17 17 18 18 19 20 20179
20 20 22 22 22 22 22 24 23 23 24 24 26 27 27 27 27 27 29 29 29 29 29180
29 29 29 29 29 29 27 27 27 27 27 26 24 24 23 23 24 22 22 22 22 22 20181
20 20 20 19 18 18 17 17 15 15 15 13 13 12 11 11 10  9  8  8  6  5  3182
   00.00         0.4180E+00       0.4120E+00     -0.4150E+00          0.0
   10.00         0.4290E+00       0.4270E+00     -0.4280E+00          0.0
   20.00         0.4000E+00       0.4000E+00      0.4000E+00          0.0
   30.00         0.4410E+00       0.4290E+00      0.4350E+00          0.0
   40.00         0.2930E+00       0.2770E+00      0.2850E+00          0.0
   50.00         0.2510E+00       0.2390E+00      0.2450E+00          0.0
   60.00         0.2340E+00       0.2060E+00      0.2200E+00          0.0
   70.00         0.2110E+00       0.1790E+00      0.1940E+00          0.0
   80.00         0.2030E+00       0.1670E+00      0.1840E+00          0.0
   90.00         0.2040E+00       0.1660E+00      0.1840E+00          0.0
      36.0    31     7
         0          30        60          90         120
       150         180
  1
```

Figure 9.2 Input to VPDF.

The program VPDF input data are similar to those of VPD except that the Stokes parameters or Fresnel reflection coefficients are inserted into the data statement following the 190 lines of aerosol data (Fig. 9.2). The Stokes matrix may be computed in VPDF for an input of surface complex index of refraction in the data statement (i.e., set to zero for default in Fig. 9.2) or determined from the spectral photometric/polarization curves for the surface, empirically or theoretically obtained.

The Dave programs may be used in line-by-line modeling, except that the molecular line absorption (and scattering) for each layer must be separately calculated and then inserted into the program (see Section III, Chapter 21, and Egan, Fischbein, Smith, and Hilgeman, 1980).

10

Data Handling and Analysis

Remote sensing data take many forms. They may be digital, representing photometric (or polarimetric) levels, or be some form of imagery obtained directly or through conversion of digital data to single or multicolor imagery. Digital data are most easily manipulated in terms of enhancement or color comparisons, and statistical analyses. What we do with the data depends upon the information that we expect the data to yield, and the form in which the information is to be presented (i.e., graphical, tabular, imagery, etc.).

The analysis of photometric data would form the basis for polarimetric analyses, except that the database would be polarimetry, or polarimetry in combination with photometry. Various interpretation factors will be discussed for minerals, particle sizes and combinations, porosity, orientation, chemistry, foliage, atmosphere and aerosols. Then some specific techniques will be mentioned for computer recognition of wetlands, optical data processing including the application of the Fourier transform, the use of matched filters, and correlation matrices. The interpretation factors are described in the following paragraphs.

Minerals

The effect of surface minerals and associated structure on the polarimetric and photometric curves may be seen by comparing Figs. 10.1–10.6 (from Egan and Hallock, 1966). Ilmenite (curves of Figs. 10.1 and 10.2) is a blue–gray black, opaque, hexagonally crystalline mineral (Hurlbut, 1966). The sample consisted of 1–10 cm chunks of ore (from N.Y. state) unmixed with foreign matter. It is seen that the infrared polarization is highest (Fig. 10.1) and the blue is lowest for phase angles above 60°. The photometry (Fig. 10.2) indicates a fairly diffusing surface for the ore because of the lack of a specular photometric peak at a phase angle equal to twice the viewing angle. A niccolite sample (from Cobalt, Ontario), consisting of chunks 1–5 cm in size, is largely composed of niccolite (NiAs) with traces of cobaltite (CoAsS) and calcite (CaCO$_3$). Niccolite is pale copper red and has a hexagonal

148

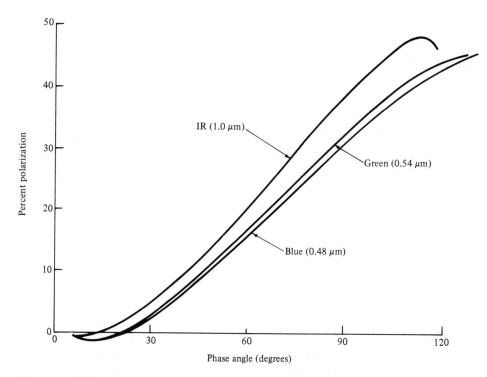

Figure 10.1 Polarization curves of ilmenite: FeTiO$_3$ (New York) $\epsilon = 60°$.

Figure 10.2 Photometric curve of ilmenite. $\epsilon = 60°$ at 0.54 μm.

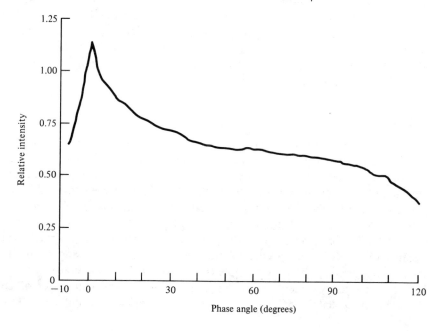

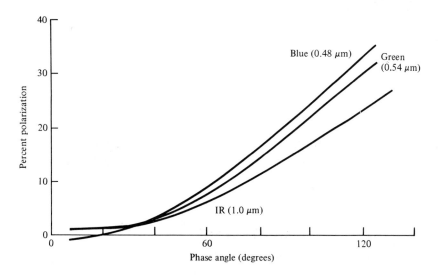

Figure 10.3. Polarization curves of niccolite–NiAs (Cobalt, Ontario). $\epsilon = 60°$.

crystal lattice (Hurlbut, 1966). The polarimetric curves (Fig. 10.3) do not reach a peak and have an inverted color order, with the highest polarization in the blue and the lowest in the infrared. Photometrically, however, the sample produces a peak at about 100° (Fig. 10.4), and less of a backscatter at 0° than the ilmenite.

A sample 10-cm block of rose quartz (low quartz) with a hexagonal crystal lattice, translucent and birefringent, produced polarization curves that remained slightly negative up to about a 70° phase angle, where they rose steeply (Fig. 10.5). The polarization in blue light was higher than that in infrared at a phase angle of about

Figure 10.4 Photometric curve of niccolite. $\epsilon = 60°$ at 0.54 μm.

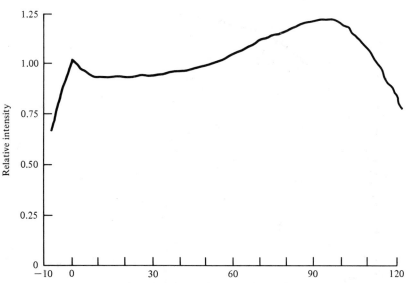

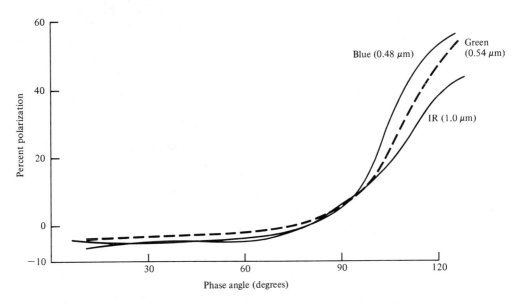

Figure 10.5 Polarization curves of quartz (rose) −SiO$_2$ (Maine). ε = 60°.

110°. Photometrically (Fig. 10.6), the quartz produced a specular photometric peak near a 120° phase angle, and negligible backscatter at 0°. A small amount of titanium, as an impurity, appears to cause the rose color.

It should be pointed out that surface structure has been found to affect polarimetric curves (see the following), and the curves representing effects of mineral composition could be expected to be affected somewhat by surface structure. Also, contaminants existing in the niccolite sample can affect the polarimetry.

Figure 10.6 Photometric curves of quartz (rose). ε = 60° at 0.54 μm.

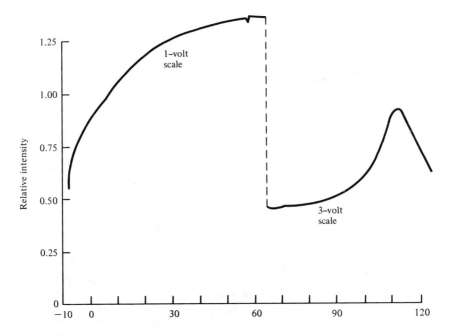

The photometry has been presented only in green light because it is largely unaffected by the observational light wavelength (Halajian and Spagnolo, 1966).

Particle Sizes and Combinations

Although implicit in the investigations of Dollfus (1957) and others, the explicit effect of particle size on the polarimetric curves was shown for a sample of Haleakala (Hawaii) volcanic ash (Egan, 1967). Representative curves are shown in Figs. 10.7 and 10.8 for a particle size ranging from 6.35 mm to below 1 μm. The investigation was directed toward lunar surface simulation, and hence comparison curves for a maria (Crisium) and a highland (Clavius) are shown, derived from data of Gehrels, Coffeen, and Owings (1964). The greatest effect on polarization is for the smaller particle sizes (Fig. 10.7), where it is seen that the smaller, brighter (higher albedo) particles produce the lowest polarization. There is a smaller effect on polarization for the larger particles (Fig. 10.8). These results would be expected to vary somewhat depending upon how the crystalline structure of a material would affect its fracture into smaller particles. This could be influenced by the technique of pulverization.

The photometric curves also depend upon particle size (Halajian, 1965), differing most at the highest phase angles where the smaller particles have the higher photometric curves.

An effect of combinations of polarimetric and photometric models can be seen in Fig. 10.9. Here, a coarse Haleakala volcanic ash has been lightly coated with the

Figure 10.7 Haleakala volcanic ash: percent polarization as a function of particle size; integrated visual light ($\lambda_{eff} = 0.5 \ \mu$m); 5° albedos.

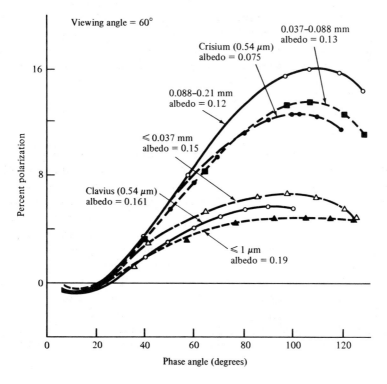

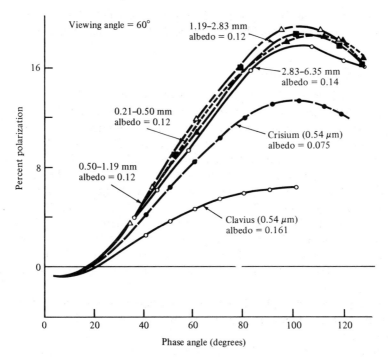

Figure 10.8 Haleakala volcanic ash: percent polarization as a function of particle size; integrated visual light ($\lambda_{eff} = 0.55$ μm); 5° albedos.

same powder smaller than 1 μm. This reduces the maximum polarization (here intended to achieve a better match to Mare Crisium), but the photometry is relatively unaffected. The composite photometric curve is shown in Fig. 10.10. The photometry is basically produced by shadowing by the larger sample scale, and relatively unaffected by the powder (see Halajian and Spagnolo, 1966, for instance). The polarization, on the other hand, is primarily the result of the particulate surface scattering.

Porosity

The term porosity may be used to express the looseness or denseness of a granular material such as soil. Its numerical value is given by the ratio of the void volume to the total volume of the soil mass. Because many soil properties (strength, permeability, etc.) are dependent to some degree upon porosity, it would be useful to establish a porosity–photometry or porosity–polarization relationship. The use of surface porosity, in lieu of bulk porosity, is necessary for this type of relationship because the polarization characteristics of a powder are basically a function of surface or near-surface effects in the visible and near infrared spectral regions. (The surface porosity of a material may be defined as the ratio of the apparent surface voids to the total projected surface area.) Egan and Nowatzki (1966) have shown that a surface porosity–polarization relation does exist for basalt powder consisting of particles less than 38 μm when it constitutes a surface having Z-axis symmetry. Three samples were investigated, all having the same particle size, but each having a different surface porosity. The surface roughness of each sample was defined in

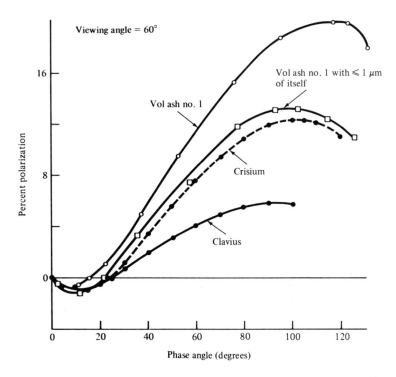

Figure 10.9 Effect of combination of models; a Haleakala volcanic ash is compared to itself when coated with a light powder.

Figure 10.10 Photometry of composite model of a Haleakala volcanic ash topped with particles of itself $\leq 1\ \mu m$ (integrated visual light, $\lambda_{eff} = 0.54\ \mu m$). For lunar crater floors and continents, see Egan (1967).

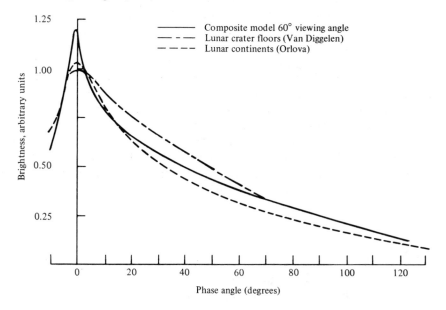

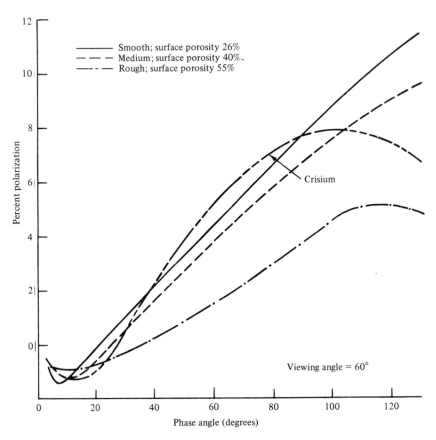

Figure 10.11 Effect of surface porosity on polarization curves of basalt powder $< 38\ \mu\mathrm{m}$.

terms of the distribution of slope lengths and was determined semiphotographically by a technique described elsewhere (Egan and Nowatzki, 1967). It was shown that the less porous a surface was, the greater was its total length of near horizontal slopes. The curves in Fig. 10.11 clearly show the trend of increasing polarization for decreasing porosity in the infrared $(1.0\ \mu\mathrm{m})$ region at a $60°$ viewing angle. This effect may be the result of the greater length of near horizontal slopes producing a specular type reflection and a greater positive polarization maximum at high phase angles. The results of photometry measurements on the same samples are shown in Fig. 10.12 and indicate that the surface with the lowest porosity has the highest specular peak and highest backscatter. The albedo tends to decrease with increasing porosity.

Before a universally applicable authoritative relationship between porosity and polarization or photometry can be established, it will be necessary to obtain much more experimental evidence. Analytically, the subject is very difficult and at this time lends itself best to an empirical approach.

Orientation

Following the results of the investigation into the porosity–polarization relationship of a powdered material, the question arose as to whether the same concepts applied to an intact vesicular material. Prompted by the anomalous polarimetric results at

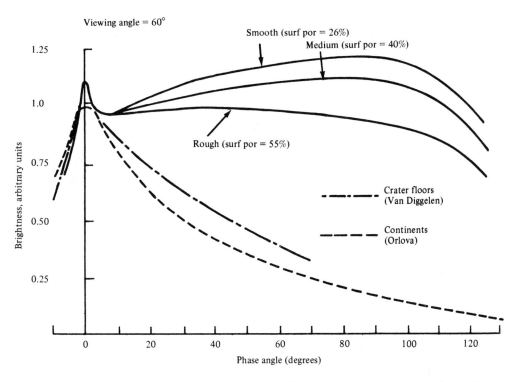

Figure 10.12 Effect of surface porosity on photometric curves of basalt powder $< 38~\mu m$ (integrated visual light). For crater floors and continents, see Egan (1967).

low phase angles obtained by Egan and Hallock (1966) from experiments on rhyolytic pumice, a study was initiated to investigate both the porosity question and the possibility of an orientation effect for this type material. As could be deduced from the fact that a given pumice specimen has the same surface porosity regardless of the viewing direction, a porosity–polarization relationship could not be established. However, it was observed that the polarimetric characteristics of a given pumice specimen were very much dependent upon the plane of vision. Figure 10.13 shows the results of measurements made on a typical pumice specimen. When the axes of the elongated pores on the surface of the specimen are oriented parallel to the plane of vision, the maximum value of the percent polarization is larger and the minimum value less than when the pore lineation of the same specimen is oriented perpendicular to the plane of vision. An inversion occurs in the parallel orientation which does not occur in the perpendicular orientation. From this, it was inferred that the orientation of surface pore lineation appears to control the inversion angle and the minimum value of the percent polarization at low phase angles. The effect of surface structure, defined in terms of a distribution of slope lengths, was also investigated. Figure 10.14 shows that a greater total length of near horizontal surface slopes occurs when the pore lineation of the sample is oriented parallel to the plane of vision. The fact that the greater maximum polarization occurs in this orientation seems to confirm the results of the porosity–polarization study performed on powders, namely, that the greater length of near horizontal slopes produces a greater positive polarization maximum. Work is currently in progress to determine a parameter that could be used as an orientation index. Preliminary results suggest

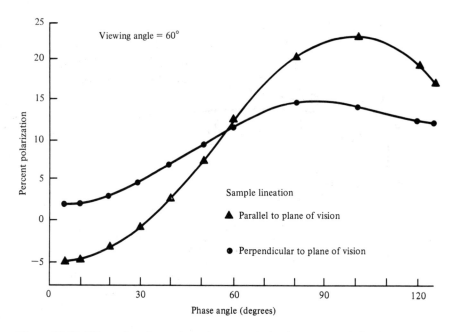

Figure 10.13 Effect of surface orientation on polarization curves of rhyolytic pumice (Mono Crater, California Region).

Figure 10.14 Effect of surface orientation on surface slope distribution at 90° phase angle (rhyolytic pumice).

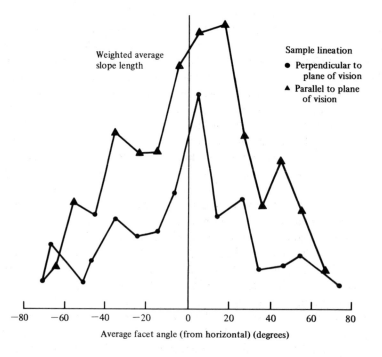

that the difference in residual polarization between two orientations be used. The residual polarization is defined as the percent polarization (positive or negative) when the curve is extrapolated to zero phase angle.

The research reported here is admittedly preliminary, and it would be naive to say that the effect of a specific parameter has been isolated. For instance, neither the powder study nor the pumice study has specified the effects of diffraction, microstructure, or multiple scattering. However contrived laboratory models have been used in which the effect of some of these other influences are minimized; the general effects of porosity and orientation on the polarization have been verified.

Chemistry

Attempts have been made to explain the seasonal color changes for certain regions of Mars by chemical reactions (Cohen, 1966). Investigations (Mumford and Thompson, 1966) have shown that iron-rich brucite of the New Idria serpentinite formation of California, when exposed to various controlled environments of oxygen, carbon dioxide, and water, reacts chemically with these agents, and may form products interesting from the viewpoint of color transformation. Table 10.1 shows the color change results of laboratory experiments starting with bluish-gray reactant minerals. An experiment was conducted at Grumman on a serpentine–brucit slurry in $CO_2 + O_2$ atmosphere at reduced pressure, to see if any of the reactions reported in Mumford and Thompson (1966) could be duplicated. Although the anticipated color change was achieved, the reverse reaction could not be effected. The photometric and polarimetric characteristics of the material were obtained before and after the chemical reaction. The results show that although neither match

Table 10.1 Laboratory Reactions of Iron-Rich Brucite from the New Idria Serpentinite

Experiment	Starting material	Sample preparation	Gas environment	Time period	Products[d]	Color
I	Serpentine + brucite[a]	Dry mount[b]	Lab. atmosphere	6 months	Serpentine + s. coalingite + s. brucite	Tan
II	Serpentine + brucite	Slurry	$O_2 + CO_2$[c]	1 hr	Serpentine (+ hydromagnesite precipitate from solution)[e]	Brown
III	Serpentine + brucite	Dry mount	$O_2 + CO_2 + H_2O$	5 days / 16 days	Serpentine + s. coalingite / Serpentine + nesquehonite	Brown / Brown
IV	Reagent grade brucite	Dry mount	$O_2 + CO_2 + H_2O$	3 days	Nesquehonite	Gray
V	Serpentine + brucite	Slurry	O_2	50 days	Serpentine + tr. coalingite	Tan
VI	Serpentine + brucite	Dry mount	$O_2 + H_2O$	50 days	Serpentine + tr. coalingite	Tan
VII	Serpentine + brucite	Slurry	CO_2	1 hr	Serpentine	Olive green
VIII	Coalingite concentrate	Slurry	CO_2	18 hr	Amorphous (+ hydromagnesite precipitate from solution)[e]	Red–brown

[a]Sample 487–52–7, about 25% brucite, powdered, magnetite removed.
[b]Packed into an x-ray holder.
[c]Gases bubbled through a water trap and then onto the dry sample or into the water slurry.
[d]Identified by x-ray diffraction technique; s = some; tr. = trace.
[e]The clear filtrates were heated for about 1 hr; a white fluffy precipitate of hydromagnesite then formed.

the observed Martian surface characteristics, there is a distinct difference between the two. A chemical change has occurred between the initial and final states of this material. This chemical change affected the albedo and probably the physical structure. Thus, chemistry is not a readily isolated variable, but produces a complex change involving other variables affecting polarimetry.

The chemistry of a mineral—such as limonite, which has been hypothesized as the major constituent of the visible Martian desert surface (Dollfus, 1957)—could fundamentally affect its polarization and photometry. Limonite is largely $FeO(OH) \cdot nH_2O$ with some $Fe_2O_3 \cdot nH_2O$, often containing small amounts of hematite, clay minerals, and manganese oxides, with widely varying water content (Egan, 1967a). Individual samples of limonite thus may vary. Some of the discrepancies in the observed polarimetry (Egan and Foreman, 1967) of limonite may arise because of the chemistry of the specific type of limonite used.

Foliage

The polarimetry of lichens has been investigated by Dollfus (1957); a grass turf was observed by Coulson (1966); and the author measured the polarimetric properties of *Rhododendron maximum* (Rosebay) leaves. The curves of polarimetry and photometry are shown in Figs. 10.15 and 10.16, as a case in point and not as representative of all foliage. The polarization curves are widely divergent between the green and infrared, and a peak occurs at about 120° in infrared light, whereas the maximum is only approached in green light. The leaves were placed in the beam of the large scale photometer on a platen; they would not normally be arranged in such a manner on a

Figure 10.15. Polarization curves of *Rhododendron maximum* (Rosebay) leaves. $\epsilon = 60°$ at 0.54 μm.

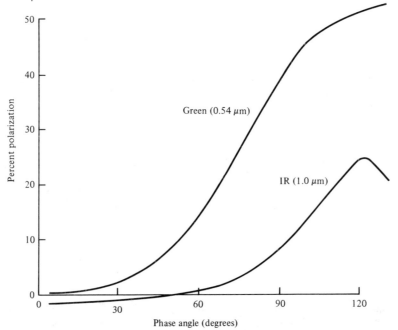

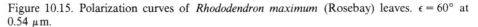

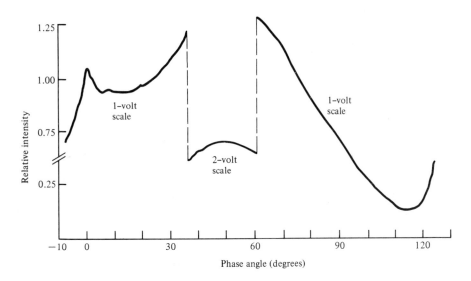

Figure 10.16. Photometric curves of *Rhododendron maximum* (Rosebay) leaves. $\epsilon = 60°$ at 0.54 μm.

bush. It is expected that a polarimetric orientation effect would exist (see above) for leaves occurring in nature.

The photometric curves are presented in Fig. 10.16; it can be seen that there is a peak at about 45°. This could be the result of the physical placement of the leaves. Also, another photometric effect has been observed by Chia (1967) whereby the albedo of Barbados sugar cane increases with crop height until the leaves reach complete cover. The albedo of the vegetation, but not the bare soil, depended upon the zenith angle of the sun.

Atmosphere and Aerosols

The effect on polarimetry of an atmosphere above a surface has been noted by Lyot (1929), Dollfus (1957), Coulson (1966), Egan and Foreman (1967), and Rea and O'Leary (1965). Essentially, the effect depends upon the optical thickness and scattering properties of the atmosphere. The contributions of the molecules of the air are analyzed in terms of Rayleigh scattering, but aerosol polarimetric effects must be deduced in terms of the Mie theory.

A typical set of theoretical curves depicting the comparative effects of a molecular atmosphere hypothesized for Mars, containing aerosols, is shown in Fig. 10.17 [from Rea and O'Leary (1965)]. Curve a depicts Rayleigh scattering for a pure molecular mixture, while curve b depicts Mie scattering for aerosols of 0.58-μm diameter at $\lambda = 6100$ Å; a mixture of the molecules a and aerosols b in a relative brightness ratio of 25 to 65 at 0 phase angle is shown by c. These curves are illustrative and serve to exemplify the effect of aerosols on the polarization curves at small phase angle. Experimental laboratory investigations of the effect of aerosols, such as ice crystals, have been initiated by Lyot (1929), Dollfus (1957), Zander (1966), and Egan and Foreman (1967). Problems exist in the control of the aerosol particle sizes, and containment for polarimetric observations.

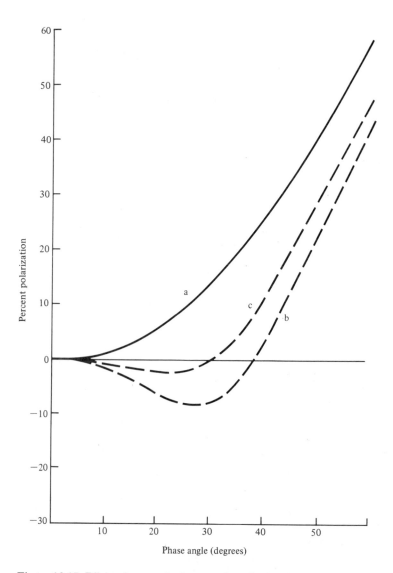

Figure 10.17 Effect of atmospheric aerosols (calculated) (from Rea and O'Leary, 1965).

Significance of photometric effects of aerosols have been indicated by most of the aforementioned polarimetric investigators. It appears that the separation of the various parameters relating to the atmosphere (with and without aerosols) may be accomplished by narrow spectral band polarimetric and photometric observations over a wide spectral range for many different phase and viewing angles.

It should be mentioned that in regard to the atmosphere Rayleigh polarization of the incoming sunlight occurs strongly at sunrise and sunset. At noon, and for 3 or 4 hours each side of noon, the sunlight polarization is small and has been observed by Egan (1967) to be under 1% in the Long Island area on a cold winter day. Toward sunset, the polarization of the sunlight increased to about 5% on the same day. This polarization of the illuminating light, as well as sky light polarization, must be taken into account when using the observed or inferred depolarization characteristics of the surfaces observed polarimetrically.

11

Interpretation and Information

Satellite remote sensing systems produce a prodigious amount of data; from this data, following suitable processing, an interpretation is made and information extracted. This final step of transforming remote sensing raw data into information that is intelligible and useful to resource managers involves the interpretation and analysis of imagery or digital data.

Interpretation Procedures

Various approaches are utilized to extract information from remote sensing data (National Research Council, 1977). In the order of increasing cost and sophistication, they are

1. manual interpretation of standard photographic imagery using simple, inexpensive instruments;
2. manual interpretation aided by photographic image enhancement, using more costly optical equipment;
3. manual interpretation of special digitally enhanced photographic imagery using the same more costly optical equipment;
4. digital analysis of the computer-compatible tapes in man–machine interaction to produce the desired computer output, which is subject to subsequent human interpretation and analysis (i.e., interactive); and
5. digital analysis of computer-compatible tapes without man–machine interaction, the computer determining the classification, based on previous algorithms, and outputting the desired information in graphical or digital form, i.e., (noninteractive).

The first of these procedures, the manual interpretation of standard photographic imagery, utilizes the skills of photointerpretation very familiar to geologists, geographers, foresters, urban and agricultural analysts, and other resource scientists. This interpretation is usually aided by the use of enlargement equipment, from the most simple form, a microscope or binocular magnifier, to more advanced photographic enlargement. The various imagery bands may be represented in true or false color, and combinations of imagery may be made using light tables or simple projectors. It goes without saying that the wider the background of the interpreter, the more experienced he is as a resource scientist and interpreter, and the more familiar with the area under study, the better will be the interpretation. Besides the usual black and white transparencies of imagery, and subsequent color projection, one may make color prints, either of one color or combined. The relatively simple diazo process has been used to advantage to produce false color imagery at the EROS Data Center. It is an extension of the standard blueprint process, and utilizes ammonia vapor to produce vivid color transparencies from black and white transparencies.

By using three photographic MSS or TM band black and white film products printed into diazo transparent bases of appropriate color, a standard color transparency results in which the gray tones appear as color tones. Experimentation is easily possible with a variety of colors to enhance specific features.

With polarization as an additional parameter, the number of dimensions of combination possibilities is more than doubled. However, the highest positive polarization occurs at the higher phase angles and is highest for dark surfaces; for a satellite passing over a ground area at near local noon time, the solar phase angle depends upon the time of the year and is least during the local summer, and near the equator.

This has the advantage, though, of producing small phase angles where negative polarization may occur, as well as the inversion angle where the polarization changes from negative to positive. A sensor that permits a choice of pointing direction, such as the proposed High Resolution Pointable Imager (HRPI), would provide a range of phase angles, with corresponding photometry and polarization, adding still another variable—the phase angle in bidirectional reflectance. Going a step further, the incident and scattering angles (relative to the surface normal) may be used in specifying bidirectional reflectance, instead of the phase angle between the incident and scattered rays lying in the plane of vision. The level of analysis is then such that a computer program is generally necessary to extract the pertinent information, or at least to determine the significant information, or to determine the significant data that should be used in the process.

Returning to data analysis techniques, we note that many universities and government departments of developing countries are now modestly equipped with color photography laboratories. In these laboratories special equipment such as zoom transfer scopes, color additive viewers, and microdensitometers may be used to assist the interpreter. The cost levels of these three approaches are $8,000–14,000, $15,000–25,000 and $12,000–15,000. The zoom transfer scope permits data from enhanced images to be transferred to resource maps of various scales. A color additive viewer permits viewing different spectral (or polarization) wavelength bands through separate color filters with differing illumination intensities. By this technique, specific surface characteristics can then be emphasized for the data user. A microdensitometer measures the color density of a composite color reproduction or color transparency, and may be used to quantify certain aspects of image analysis. But a word of caution: the validity of the microdensitometer approach depends upon the characteristics of the photoreproduction process, which are linear only in the intermediate range; a careful calibration process involving the use of a density step wedge and controlled film development is necessary for quantitative analysis. For photometrically noncalibrated analyses, where only contrasts are used in relation to ground truth, the calibration process is unnecessary, but has the ultimate fallacy that under different solar illumination angles, the contrasts will be different, requiring ground truth for interpretation. In comparison, absolute photometry allows the use of a general reference library, not requiring ground truthing, to interpret data. Further, the use of photographic imagery produced from digital satellite data has the problem of a line structure (from each scan line), which is seen by the microdensitometer, to confuse analysis.

In the use of photographic techniques, usual requirements are for emphasizing boundaries or edges of surface features, superimposing incongruent registration of several images that may have been taken on different dates; also, different colors may be assigned to different image density ranges or "slices," to more sharply distinguish different densities.

The third level of information acquisition may use all of the preceding methods, but in addition uses special digital enhancement to produce higher contrast photographic imagery. These products are obtainable from commercial organizations, government, or university research laboratories. A computer is required to read the digital tapes, and enhance the differences in photometric levels on the tape; however, the absolute photometry is generally lost in such a process, the actual enhancement technique being known only to the group doing the enhancement. Another technique, termed, "band ratioing," uses the ratio between signal intensities (of the same ground area) sensed by individual spectral bands to provide another factor in image analysis. Again, the absolute photometry is lost, as well as possible significant atmospheric information; the ratio between close bands tends to minimize effects of atmospheric absorption and scattering. Mining, petroleum, and timber companies have utilized this technique at a scale of 1:250,000 at a cost of about $1,000/scene (October 1976). The approach is cost effective where major exploration costs are involved.

The next level of analysis is more complex, but most reliable; above all, it preserves the photometric (and polarimetric) fidelity of the imaging process. A computer-compatible tape is necessary and the output may be in a variety of forms: graphs or digital maps, in which landscape features are enhanced or appropriately colored to define location or extent. Tabular or statistical tables are frequently the initial output, but they have less of an impact conceptually than some graphical display forms.

In this level of analysis, time is not an important factor; in other words, the analysis of the data may be done in days, weeks, months, or years. However, there is an increasing need for information that is of interest at the moment the data are acquired. Such information could be air traffic or space traffic, or where resolution permits, city and highway traffic. In such situations, a computer program could be devised to use pattern recognition to sense the object of interest and rapidly yield the desired information.

The latter two techniques are amenable to the facilities of major universities and government departments in the United States, England, Japan, France, Germany, and other developed countries. The investment in digital processing and display equipment begins at $1,000,000 and runs upward depending upon the peripheral display and color recording devices. Because this computer equipment is a major capital investment, careful exploration of the cost effectiveness is necessary; sometimes a multipurpose computer is desirable, because it can be used for other types of research or simply accounting.

The application of photometric remote sensing to resource information is well documented. The use of polarization to augment photometry and the establishment of absolute reference standards are only beginning. Polarization will furnish microstructural information as contrasted to photometry, yielding macrostructural information (Egan et al, 1967). Thus, in agriculture, spectropolarization in combination with spectophotometry could permit crop yield forecasting. When the ground is visible, the scattering and absorption effect of the atmosphere could be factored out, and the growth stage of a particular crop determined on an absolute basis. When cloud cover exists over an area that causes precipitation or simply high humidity, the polarimetric properties of the scattered radiation would indicate moisture droplet size; then in association with other meteorological data, the effect of moisture on growth could be extrapolated to give an anticipated growth and the yield. But in all phases of remote sensing, as presently constituted or in the future, a gap will always remain between the technical experimenters and the ultimate users of earth resource information. It is anticipated that bridging the gap involves close liaison between remote sensing technologists, planners, and decision makers.

Advanced Pattern Recognition

For the automatic recognition of areas in imaging remote sensing displays, a good machine vision system is necessary. The system should provide rapid, automatic shape recognition and have wide adaptability. Image variations such as position, orientation, illumination, and background noise should have negligible effect. One such nonmilitary processor, the Cytocomputer, has been developed by the Environmental Research Institute of Michigan. It permits discrimination, in real time, of two and three dimensional (gray level) shapes. The system is based on a functionally complete algebra of images in association with a computationally complete architecture of cellular neighborhood processing hardware (Becher and Peace, 1982).

The formal image algebra formulates the cellular image processing alogrithms into algebraic expressions, the variables of which are images. This image algebra (Sternberg, 1980) is based on the concept that binary images are sets in two dimensional space. The binary images are then silhouettes of black (foreground) on white (background) regions delineated by partition of the original image plane.

Then, set geometric and theoretic operators can be functionally applied to these binary images. The common operators are single and multiple. Single image operators are (1) complementation, the interchange of black and white regions; (2) reflection, producing an image symmetric with the original image through the origin; (3) translation, the shift of an image from the original position. Multiple operators are (1) two image union, the formation of a new image that is black at all points where either of the two images are black; and (2) two image intersection, the production of a new image that is black only where both images are black.

It is convenient to include two more image operators, dilation and erosion. Image A (Fig. 11.1) when dilated with image B yields image C, formed by the union of all translations of the origin of A by each of the points within B. In nearly inverse operation, image A is eroded by image B forming image D as the union of all points

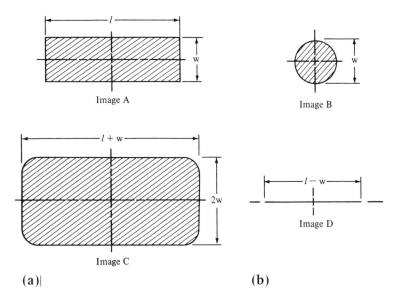

Figure 11.1. Two image operations. (a) Dilation: $A \oplus B = C$. (b) Erosion: $A \ominus B = D$.

of the origin of image B where all of B is completely contained with A. Symbolically,

$$A \oplus B = \overline{\overline{A} \ominus B'} \quad \text{and} \quad \begin{array}{l} \text{Dilation is expressed as } A \oplus B = C \\ \text{Erosion is expressed as } A \ominus B = D \end{array}$$

where the prime denotes reflection and the overbar represents completion.

These operators together with the properties of inclusion and equality, and null and identity images yield a complete image algebra.

As an example of the application of this image algebra, the template matching algorithm (Sternberg, 1979) illustrated in Fig. 11.2 is considered. There are three points where triangle B is located in complex pattern A, which are identified simply by eroding A with a template of image B. The actual triangles (image C) are then produced by dilating the three points with the template of image B. This simple example, in a noise free environment, does not represent situations where angular, size, or shape variations occur.

The Cytocomputer developed from the cellular theory of Von Neumann (1961–1963). In Fig. 11.3 a two dimensional parallel array implementation is shown as proposed by Unger (1959). Assume that the image is produced by a raster sensor

Figure 11.2 Template matching.

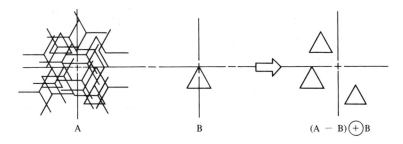

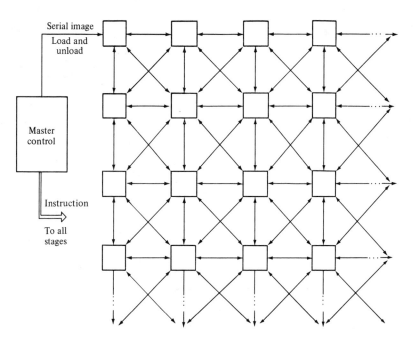

Figure 11.3 Parallel image processor.

scan with the lines of picture elements stored consecutively in the rows (squares) of the array shown. In the array, each square is a processing element connected to the set of eight nearest neighbor elements.

The picture processing is produced by programming all of the array processors with a transition instruction to produce a new pixel state based on the original pixel state, and those of the eight nearest neighbors. The instruction affects all pixels in parallel, resulting in a new image from a single image algebra operation. The complete image transformation follows from a sequence of individual instruction steps with a series of instruction loads followed by pixel state neighborhood transformations.

The problems with this parallel array processor are the need to continually reprogram the processor after each state transition and the need to have an array with as many processor elements as picture pixels. High resolution sensor systems may have as many as 1024×1024 pixels, necessitating over a million processors in the array.

The Cytocomputer (schematically shown in Fig. 11.4) stage alleviates these problems through the use of shift registers. The image array is at the left with pixels $A_{1,1}$, $A_{1,2}$, etc., corresponding to the sensor acquisition sequence that composes the image. To the right is a typical processing stage composed of shift registers. The number of shift registers corresponds to the N pixels on one line of the image and are stored in each of the two rows of the register elements.

Thus, the pixel elements lying in the 3×3 array (window) to the left of the stage corresponds to a pixel with its eight nearest neighbor pixels. The figure assumes that pixel $A_{6,6}$ has just been sensed and stored in the first stage register. Previously collected pixels are shifted through the register and are shown as stored. With the

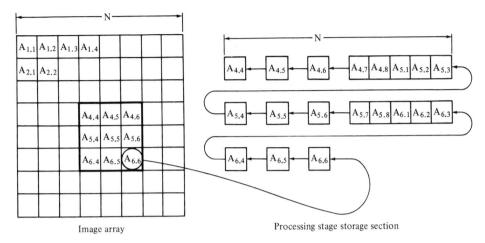

Image array   Processing stage storage section

Figure 11.4 Cytocomputer moving window stage section.

3×3 array connected to a nearest neighbor processor, the processor output will be the state determined by the instruction (Lougheed et al., 1980).

For real time operation, the Cytocomputer stages are connected in cascade.

The system described is applicable only to binary images. Most remote sensors which collect image data are based on gray-scale levels, and it is not a simple matter to convert these levels to binary images because of scene shading (phase function) and noise.

A representative waveform for a single line raster scan is shown in Fig. 11.5a, with image intensity as the ordinate. The rectangles represent objects on this varying intensity background. Simple thresholding will not distinguish the objects. However, if a third dimensional plane is added, considering the area above the waveform as background (white) and the area below as a silhouette (black), the image can then be processed as a two dimensional one.

To illustrate, in order to remove background variation, the image of Fig. 11.5a is eroded with a disk of slightly larger diameter than the object width. This will result in a black image below the locus of points connecting the disk centers (Fig. 11.5b). Dilation of this image produces a black image below the locus of disk tangents.

Figure 11.5. Rolling ball three dimensional processing. (a) sensor waveform; (b) rolling ball processing; (c) locus of tangents subtracted.

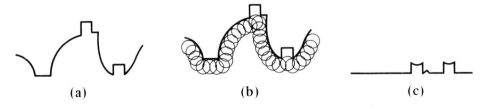

(a)   (b)   (c)

Subtracting this locus from the original image yields Fig. 11.5c, permitting constant level thresholding to be applied.

Further, by filtering with a ball instead of a disk, the image direction orthogonal to the scan line can be analyzed, producing a two dimensional image analysis.

The technique described is a representative system that may ultimately be adapted to semi-real time as well as real time remote sensing.

III

APPLICATIONS

Photometry has been recognized as a technique for optical sensing ever since man first looked upon the horizon. Interpretation of photometric data is of value in obtaining geological or geophysical information from aircraft, earth meteorological or planetary satellites, or planetary landers. However, extensive spectral reflectance measurements are necessary to reduce ambiguity in interpretation. Vast amounts of photometric data are readily obtained from observations of planetary surfaces. Such measurements were made by Krinov (1953) on various soils, rocks, sands, and some vegetation. Additional measurements were made by Bauer and Dutton (1962) on farmlands and wooded hills of Wisconsin, and by Romanova (1964) on sand deposits. The data ultimately may be presented graphically, or as a picture.

However, reflected light in itself is a function of the angle of incidence i of radiation upon the surface being investigated as well as the viewing angle ϵ. Also, the reflected light may be depicted in terms of a phase angle α, and measured in the "plane of vision" containing the lines of incidence and emergence (e.g., the source, planetary surface, and sensor). The effect of incident angle has been shown by Kondratiev and Manolova (1955), and the effect of viewing angle by Ashburn and Weldon (1956) and Coulson (1966). In lunar surface simulation, Orlova (1952), Hapke and Van Horn (1963), and Halajian and Spagnolo (1966) have sought to depict the directional reflectance of the lunar surface.

The data are frequently ambiguous, but the ambiguity may be reduced or the data rendered unique by the use of polarimetry. Polarization information already exists in the sensor photometric data, and usually may be obtained by a simple modification of the sensor optical path for polarimetry during data acquisition.

Polarimetry, too, is not a newcomer, having been used by Lyot (1929), Dollfus (1957, 1961), Gehrels, Coffeen, and Owings (1964), and Coffeen (1965) for lunar analysis and simulation. Coulson (1966) and Ashburn and Weldon (1956) have studied the polarimetric properties of a desert surface. Further work has been done on sands, loam, and grass by Coulson, Gray, and Bouricius (1966) and Coulson (1966). Also, Egan and Hallock (1966) investigated the polarimetric properties of various minerals and some vegetation, and Egan (1967) has investigated simulated lunar and planetary surface materials.

However, the effect of the atmosphere, through which the polarimetric and photometric observations are made, must not be neglected. The effect can be photometric (affecting the intensity or spectral distribution of the illuminating light) or polarimetric. The intensity effect has been studied by Gordon and Church (1966) and Coulson (1966), and the specular effect, by Henderson and Hodgkiss (1964). Atmospheric polarization has also been considered by Coulson (1966), Lyot (1929), Dollfus (1957) and Egan and Foreman (1967).

Because of all the factors involved, the interpretation of high spectral resolution photometry/polarimetry from surfaces beneath an atmosphere is complicated, but not beyond the realm of analysis. In other words, the spectral photometry and polarimetry are functions of the surface and the angles of illumination and viewing, as well as the effect of the atmosphere. The factors involved may be readily separated by sequential optical measurements, as from a satellite, and by subsequent analysis in terms of "ground truth" data obtained by previous observations and measurements. For unknown planetary surfaces, the interpretation is dependent upon terrestrial simulation until the first *in situ* verifying measurements are made.

From the details of Parts I and II, the mathematical representation of polarization and the remote measurement of it are evident. Beyond being just something else to measure, there must be justification for the use of polarization—rooted in what *it* can yield that is not obtainable or difficult to obtain from photometric observations.

Basically, high polarization is obtained from single reflection off shiny (specular) surfaces. Diffuse surfaces have lower polarization, as well as macroscopically irregular (multiple scattering) surfaces, or groups of objects such as plant or tree leaves. However, macroscopically irregular surfaces (pine trees) may have enough gross average orientation (the needles of the trees) to produce significant polarization. The state of compaction of a surface affects the polarization as well as the optical complex index of refraction of the surface particles (particularly absorption). Darker photometric surfaces typically have the higher polarization, and are thus more readily distinguishable using polarization rather than photometry. Where a combination of large scale macrostructure and a microstructure exists (agricultural or disturbed sea surface areas for instance), polarization complements photometry in describing the surface remotely.

In this part, various applications of polarization will be discussed. Being a relatively new field in terms of remote sensing, the depth of coverage will vary; in astronomy, polarization has been used for over half a century in planetary and stellar observations. Terrestrial observations of polarization approximately coincide with the astronomical observations, and detailed terrestrial atmospheric work has extended over about half this period.

12

Hydrology

Examples will be presented demonstrating the application of calibrated photographic remote sensing to optical remote sensing of shallow water bioresources and hydrology. It is acknowledged that the techniques described are photometric; to add polarization would require the use of four (instead of two) cameras, each with appropriate polarization filters in the optical systems. The same techniques described would apply to the polarization/photometric data obtained.

Introduction

Wetlands are land–water edge areas that exist in every state of the United States as well as all areas of the world. These areas are characterized by unique plant communities as a result of temporary, cyclic, or permanent submersion by rain, tides, or storms. Wetlands may be known locally as a salt marsh, tidal marsh, marshland, tideland, submerged land, swamp, swampland, gut, slough, pothole, bog, mud flats, wet meadow, overflow land, or flood plain. Such areas have evolved over a period of many years, and once disturbed or destroyed cannot be replaced by engineering. They are of supreme importance for aquatic flora and fauna, in that many essential nutrients have their origin in wetlands. Further, many aquatic fauna during certain stages of their life cycles spend time in wetland areas for reproduction.

Due to their worldwide distribution and ecological importance, a current need exists for the detailed mapping of wetlands. This need has been further emphasized by studies (see, for instance, Maryland, 1970; USDI, 1970), and by wetland preservation laws enacted by various states, notably Maryland and New Jersey. The true impact of alterations of these areas by dredging and fill, bulkheading, and establishment of recreational sites cannot be adequately assessed without a basic inventory in regard to spatial distribution, primary production, and their role in the marine food webs.

Physical mapping by ground surveys is slow, expensive, and at times practically impossible because of the physical topography of wetlands. Thus, a very practical solution to the problem of the mapping of wetlands is the use of remote sensing. Aerial photography, as a remote sensing technique, allows large scale rapid mapping, provides a baseline for comparison with future development, and, since several surveys can be completed in one year, allows time-dependent phenomena to be assessed.

In response to the widespread need for wetland mapping and to the desire for standardizing techniques in a way that would allow the automated analysis of remote sensing data, an optical remote sensing program was performed on selected wetland areas on Long Island (Egan, 1970). Also, a test site in Calvert County (Md.) has been the object of a detailed multialtitude operational aerial photographic survey with simultaneous acquisition of ground truth, followed by a computer-oriented data analysis.

In contrast to present photographic remote sensing, whereby visually observed color contrasts on transparencies or prints are cues to physiological stress or disease, the approach involves detailed color calibration and control procedures on the imagery obtained with true color and false color Ektachrome films, with evaluation by microdensitometry. The calibration procedures were described in Chapter 8 (Egan, 1969) and the techniques were applied to the aerial photography of the Long Island and Maryland wetlands (Egan, 1970). Both programs involved the use of true color and false color infrared imagery.

Optical techniques (involving the visual and near visual spectral ranges) have been highly developed as remote sensing tools. Predominantly, two classes of sensors have been used: photographic and electronic.

Photography has been extensively used in remote sensing, but has been limited to applications where color contrasts are evident or may be effected by display techniques. The accurate objective specification of color is limited by control problems associated with film: (1) manufacture and storage, (2) exposure, (3) development, and (4) evaluation. Photometric accuracies of the order of $\pm 2\%$ or better are indicated when precision calibration and control techniques are used by astronomers for photographic spectrophotometry. Without a calibration and control program on the film, attempts at photographic photometry are practically useless.

Infrared film (single emulsion and type 8443 false color Ektachrome) have been used widely for photographic remote sensing of earth resources. Such applications range from hydrology and bioresources to forestry and agriculture. Interpretation of colors may be made subjectively by using Munsell color standards, or objectively by spectrophotometry. However, colors are meaningless unless a careful calibration program supplements the photography.

Present electronic techniques—similarly—basically sense the intensity of the scattered light from earth surface areas. The intensity of the light usually varies with the wavelength of the scattered radiation. The color of the surfaces depends on the color of the incident illumination. However, it is not generally recognized that the color of almost all surfaces also depends on the illumination angle and viewing angle. Further, electronic devices may not always be faithful reproducers of the light that they sense, and they too must be calibrated for accurate results. Scanning sensors such as vidicons, for instance, may have shading errors as high as 20%; image dissectors are limited to errors of a few percent by the uniformity of their photosensitive surfaces; photomultipliers, used basically in nonscanning applications, are limited by noise at high sensitivities and by saturation at high light levels.

A more general approach to remote sensing that makes use of all the optical information contained in the scattered light appears to offer increased accuracy in characterization of earth resources. This information is embodied in a set of parameters, termed the "Stokes parameters," which completely describe the scattered radiation. The Stokes parameters are variables dependent on the wavelength and geometry of viewing. There are four parameters: the first describes the brightness; the second, the relative polarization; the third, the plane of polarization; and the fourth, the circular polarization.

Techniques have been evolved, based on instrumentation used for astronomy, to sense remotely the Stokes parameters of water and land environments. The Stokes parameter approach appears to permit the identification of bioresources which would be below the resolution limit for a given set of hardware. Also, it has been noted that orientation effects of surfaces appear to produce more distinctive effects on the second Stokes parameter than on the first.

Visual impressions furnish man with perhaps the greatest information input, above that obtained from all the other human senses. It is no wonder then that optical remote sensing should be looked on as an extension of sight; people prefer to "see" the results of remote sensing surveys in some form of visual display. However, the unaided eye has its limitations; it is far from being a linear device, and responds best to contrasts, either color or brightness. Therefore, when we pursue extensions of optical sensing, we attempt to implement the desire for contrasting visual displays of information, and supplement these displays by accurate evaluation of the data content thus presented.

In terms of the bioresources of shallow water environments, we are able to employ optical remote sensing to advantage when the optical absorption of the scattered radiation by the atmosphere is not too great. Thus, haze, mist, rain, snow, smog, and fog interfere with remote sensing of bioresources which generally lie on the surface of the earth or beneath it. However, these interfering influences may, and generally do, exert a profound effect on the earth's surface. They thus may form part of the ecological system which should be sensed in connection with the bioresources.

Optical Properties of Bioresources

Perhaps the easiest way to approach the general determination of the optical properties of bioresources is to start with an example that embodies a broad range of optical properties, analyze the optical properties in terms of the actual biological properties, and then extend the analysis to other biological environments. This is the approach we will use.

Long Island Area

The Connetquot River, on the south shore of Long Island, empties into Nicoll Bay, which in turn is part of Great South Bay (see Fig. 12.1, upper right corner). At the mouth of the Connetquot River is Timber Island. This island is essentially a base of sand covered by salt marsh vegetation. It is subject to tidal effects, and may be almost completely covered with water during extremely high tides. An aerial infrared photograph (taken in August 1969) of the island is shown in Fig. 12.2. There are drainage ditches visible in the photograph, and higher land exists at the eastern end of the island. The aerial infrared photograph was originally made on type 8443 infrared Ektachrome film, and subsequently printed in black and white. The colors

174

Figure 12.1 Great South Bay Area on Long Island (from Nautical Chart 120-SC U. S. Department of Commerce, E.S.S.A., Coast and Geodetic Survey, Washington, D.C.; Sixth Edition, December 1968).

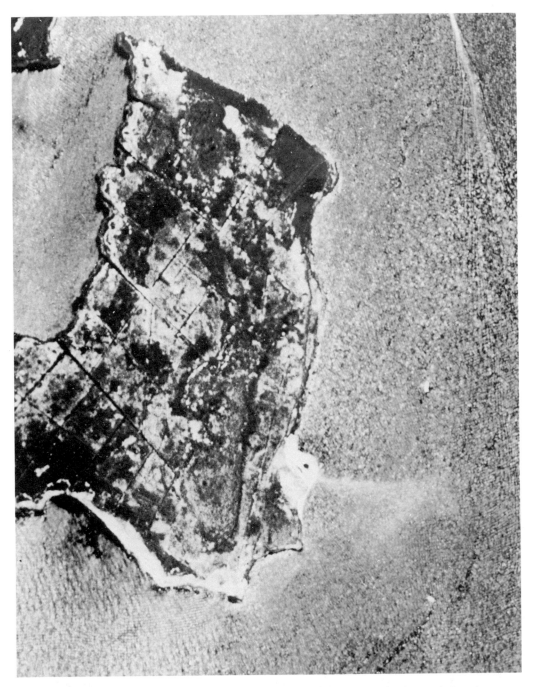

Figure 12.2 Aerial vertical infrared photograph of Timber Island (produced from infrared Ektachrome type 8443 film, 21 August, 1969, altitude 8000 feet, color controlled).

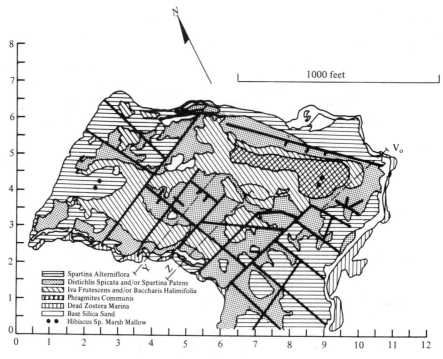

Figure 12.3 Timber Island: general flora areas (1969).

of various areas on the original photograph were different, and ground surveys ("ground truth") indicated zonation in the vegetation. Figure 12.3 is a tracing of Fig. 12.2, wherein various areas of color contrasts are outlined. Also shown are three ground truth transects that were made by Dr. Malcolm Hair of Adelphi Institute of Marine Science for the purpose of ultimately evaluating the photographic representation of the island. The area surrounding the island is estuarine (salinity ~ 26 ppt) and affected by tidal inflow.

The following varieties of ground cover exist:

Salicornea europaea (saltwort),   *Hibiscus palustris* (marsh mallow)

Baccharis halimifolia (groundsel tree),   *Sassafras albidum*,

Spartina patens (salt hay),   *Phragmites communis* (marsh reed),

Iva frutescens (marsh elder),   *Zostera marina* (eel grass),

Distichlis spicata (spike grass),   *Spartina alterniflora*, and

Amelanchier laevis (shadbush),   decomposing vegetation.

The reflectances vary for each variety of flora as a function of wavelength of the scattered light, the angle of incidence of illumination, the angle of scatter of the emergent light, the amount of ground cover, the type of soil, the state of vigor of the flora, and the physical arrangement of the foliage. There is another optical property that varies with the aforementioned factors in a subtle way—the polarization (this will be discussed in the following section).

Also seen in the photograph are water flow lines around the island, as well as shallow water areas where the white sand bottom lightens the color of the water. There are also patches of zostera in the open water some distance from the island. The zostera usually forms wind rows along the shore, but has occasionally become an obstacle to navigation.

The infrared photograph is used as an illustration here because flora have a high infrared reflectance, and differences may more easily be noted in the infrared as compared to ordinary color film representations.

The color of the water can be indicative of bottom flora, as well as phytoplankton and cladophora blooms, wind rows of zostera, and bottom topography.

Maryland Area

A second example is a more southern tidal wetland area. The wetland area covered by aerial photography for this example was along the Patuxent River in Maryland, between Hunting Creek on the south and Chew Creek on the north (Fig. 12.4). The Patuxent River is an estuary area that is amenable to remote sensing from aircraft and satellites (Egan et al., 1969). The area is undeveloped industrially and is mainly farmland, interspersed with wetlands and forested areas. The Chalk Point power plant is located south of the survey area, with the condenser outfall near the south boundary of the survey area.

Imagery of the area (Fig. 12.4) was obtained by Grumman Ecosystems Corporation Gulfstream I aircraft on 25 September 1970. The entire area was covered at 3000, 6000, 9000, and 12000 ft using Wild RC-8 cameras and Kodak Aero Neg type 2445 and Kodak false color infrared Ektachrome type 2443 film. All imagery was obtained with 60% forward overlap for stereo pairs. The camera lenses were 6-in. $f/5.6$ Universal Aviogons with $+1.4$ antivignetting filters, combined with a minus blue filter for the camera containing the type 2443 film.

A ground survey team visited the test site from 24 September to 27 September 1970. This team, consisting of an ecologist, a botanist, an aquatic microbiologist, and an engineer with forestry and photointerpretation experience, visited each marsh area along the eastern shore of the river. Specific areas for study were initially located using standard USGS quadrant maps: Benedict, Md. SW/4 Prince Frederick 15′ Quadrangle, and Lower Marlboro, Md. NW/4 Prince Frederick 15′ Quadrangle. Each side was visited by land, and notes were made of the general utilization and development of the wetlands and surrounding areas. Due to inaccessibility of some areas, and land-owner restrictions, these areas were also visited by boat, and each creek followed to its farthest upstream navigable point.

In each marsh area, field notes were made of the dominant marsh vegetation and a photographic record kept of each representative habitat type. Representative examples of each dominant species were taken for later checks on identification (Fernald, 1950). In addition to vegetative analysis, temperature and salinity records were kept for intervals along the creeks and for the river proper.

Where topography allowed, white plastic strips (18 in. wide) were stretched across certain areas showing marked vegetative zonation or relatively homogeneous stands of a dominant species. These plastic strips were also used to denote the mean high water line along the shore at Deep Landing at the time the imagery was obtained.

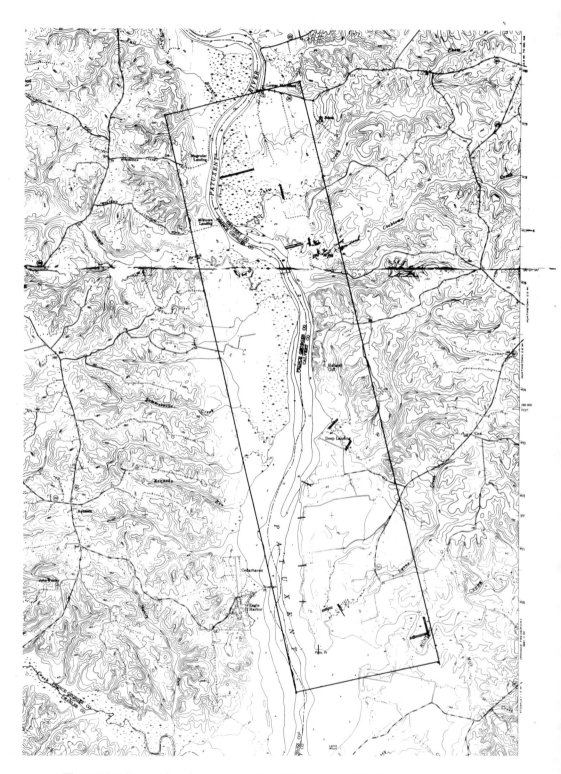

Figure 12.4 Survey site, showing ground truthing areas and flight coverage.

Of all areas visited, only Chew Creek, Cocktown Creek, Deep Landing, Little Lyons Creek, and Hunting Creek showed significant marsh development. All areas were within the range of tidal influence as shown by their salinity values. Anderson et al. (1968) showed the salt front to extend at least 13 miles north of Benedict.

Dominant marsh vegetation in all areas consisted of *Spartina alterniflora* as a narrow band along the intertidal shore; *Spartina cynosuroides* on the dryer areas bordering on the creek beds; *Typha angustifolia* as a band in the center of the marsh proper; *Peltandra virginica* and *Pontederia cordata* as narrow bands along the upper fresher reaches of creek borders. Mosaics of these species occurred in each area depending on local topography and distance from the main creek channels. *Scirpus americanus* and *Iva frutescens* were also found as mixtures with *Sp. cynosuroides* and *Typha sp*. Ground surveys of the marsh proper are almost impossible in all cases, owing to the unconsolidated hummocks and soft muck subsurface.

The river border is characterized by sharply sloping bluffs dominated by *Quercus* species (*Quercus ilicifolia*, *Quercus rubra*, *Quercus velutina*, and *Quercus stellata*) in almost all areas except where creeks empty into the river.

Calibration was accomplished with seven 6-ft-square panel arrays located at Deep Landing. For ease of transport, each 6-ft-square array was made up of four anodized aluminum panels painted with 3-M Company Nextel Brand Velvet Coating 110 Series. The gray shades were obtained by combining appropriate parts of white 110-A-10 with black 110-C-10. Red 110-D-4, green 110-G-10, and blue 110-H-10 were also undiluted.

The spectral reflectances of the paints, on the anodized aluminum surfaces, were measured on a large scale laboratory photometer in the wavelength range between 0.400 and 1.0 μm (Part III, Chapt. 15). The geometry was that for an incident sun

Figure 12.5 Spectral reflectance (relative to MgCO₃) of test panels at an incident illumination angle of 40°, and a phase angle of 3° (W = white; G_1, G_2, G_3 = gray shades; R = red; G = green; B = blue).

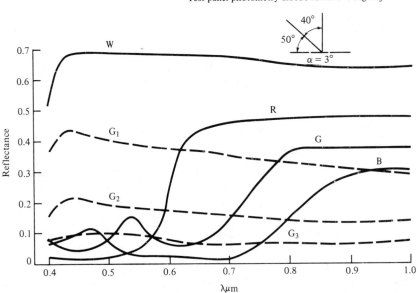

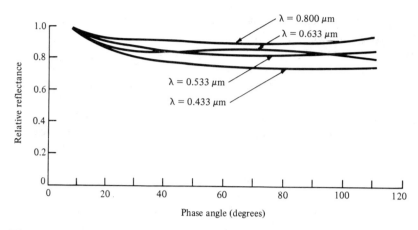

Figure 12.6 Spectrophotometric properties of white 110-A-10 as a function of phase angle and wavelength for an incident illumination angle of 40°.

angle 40° above the horizon (that existed during the aerial survey), at a phase angle of 3°, and the results are plotted in Fig. 12.5 (the phase angle is the angle between the incident and scattered rays). The red is more saturated than the blue and green, and all three colors have appreciable infrared reflectance, with the red being the strongest. The white and three gray shades vary somewhat in constancy of reflectance as a function of wavelength.

The flatness (nonspecularity) of the paints was also measured for the seven painted panels at 0.433, 0.533, 0.633, and 0.8 μm (corresponding to emulsion sensitivity peaks), and a typical set of photometric curves for the white panel are shown in Fig. 12.6. As the phase angle becomes smaller, there is a distinct increase in backscatter.

These spectral reflectance and photometric curves are used to calculate the relative brightness of the test panels when placed on the ground and photographed in the course of the aerial survey.

The application of the two general classes of optical remote sensing techniques, photographic and electronic, will now be discussed in detail in this and the next section. While these classifications are broad, there generally is overlap because electronic analysis is generally involved in the evaluation of photographs.

Imagery Analysis: Long Island Area

Photographic Techniques

A camera is a small and convenient optical instrument for recording data, with perhaps a theoretical limit of 10^8 data points possible on a 35-mm transparency with a resolution of a few hundred lines per millimeter. But let us not be misled; black and white film can be used, and has been used by astronomers, to record intensity data accurate to ±2% (Hiltner, 1962); but this is possible only if a careful program of control and calibration is associated with the use of the film. Further, when three color film is used to record data, the problems are magnified perhaps ninefold; these difficulties have been discussed previously in Chapter 8, and a practical calibration technique suggested. The same considerations apply to "false color" infrared Ektachrome film, type 8443.

Essentially, color photographic film records the first Stokes parameter by means of three emulsions; if narrow band filters are used with black and white film, many more bands may be recorded; however, the simultaneity or registration of the transparencies may be sacrificed. It is at present difficult to accurately record the other Stokes parameters photographically.

Positive transparencies are emphasized here as a convenient way to record data; they are amenable to densitometric analyses (Fig. 12.7) because the negative to positive conversion process is carried out in the film. Further, the photographs

Figure 12.7 Microdensitometer trace of transect V on Fig. 12.3 taken at a wavelength of 0.633 μm (corresponding to the infrared emulsion).

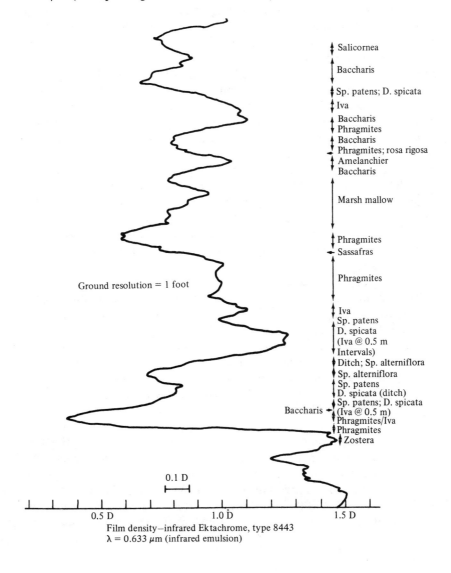

Salicornea

Baccharis

Sp. patens; D. spicata
Iva
Baccharis
Phragmites
Baccharis
Phragmites; rosa rigosa
Amelanchier
Baccharis

Marsh mallow

Phragmites
Sassafras

Phragmites

Ground resolution = 1 foot

Iva
Sp. patens
D. spicata
(Iva @ 0.5 m
Intervals)
Ditch; Sp. alterniflora
Sp. alterniflora
Sp. patens
D. spicata (ditch)
Sp. patens; D. spicata
Baccharis (Iva @ 0.5 m)
Phragmites/Iva
Phragmites
Zostera

0.1 D

Baccharis

0.5 D   1.0 D   1.5 D

Film density—infrared Ektachrome, type 8443
$\lambda = 0.633$ μm (infrared emulsion)

should be vertical views, rather than oblique, to minimize extreme viewing angle dependence of the scattered light in the oblique views. Whereas oblique views on one hand give perspective information, they are not very useful colorimetrically or metrically.

Electronic Techniques

The present electronic techniques used to sense bioresources remotely are dominantly photometric; that is, they sense the first Stokes parameter, generally with a high resolution scanner. The scanning may be electronic as with a vidicon or an image dissector, or may be mechanical as with the TRW Systems Group Wide Range Image Spectrophotometer, WISP (White, 1969). The RCA return beam vidicons have been discussed for remote sensing applications (Weinstein, 1969); although shading errors in vidicons may be as great as 20% (thus affecting the photometric response by the same amount), it has been suggested that the ultimate interpretation error may be reduced to the order of a few percent by built-in photometric calibration. The ITT image dissector, which is simply a scanning photomultiplier, has a good linearity and is basically limited to a few percent photometric accuracy by the uniformity of the photosensitive screen; however, with a photometric calibration system, it has been suggested that the ultimate interpretation error may be reduced to a few tenths of a percent (Eberhardt, 1969).

A more recent pioneering effort by Grumman has been to make use of all the information embodied in the scattered radiation, that is, measure the four Stokes parameters. An experimental spectopolarimeter/photometer has been flown over the Great South Bay area of Long Island and Long Island Sound. Results are described in the second section of Chapter 8. The initial determinations were made for the first three Stokes parameters of land and water areas, with a 20-ft ground resolution. The ground resolution may be easily improved to 2 ft by an aperture change, and another modification permits measurement of the circular polarization (i.e., fourth Stokes parameter). The magnitude of the circular polarization is small except for metallic reflection, which produces an appreciable amount, and is thus not of importance for sensing bioresources.

Another electronic technique is the use of fluorescence to detect chlorophyll. When chlorophyll is irradiated with broadband ultraviolet radiation at 0.400 μm wavelength, it fluoresces at about 0.670 μm. This technique has been used in aircraft remote sensing (Hemphill, 1969) and has been employed aboard the Grumman submersible, *Ben Franklin*, in the Gulf Stream Drift Mission from West Palm Beach (Egan, 1969). While the remote sensing system in the *Ben Franklin* sensed areas only tens of feet away from the vessel, determinations were made aboard immediately, ("in situ") at many depths up to 2000 ft. Data correlations were made with "gelbstoff"/dissolved fluorescent minerals, sea properties, downwelling, and scattered light.

The same fluorescence sensing system is now being used in the Long Island shallow water environments in an augmented form; it now measures turbidity, pH, and dissolved oxygen as well as the fundamental water properties of salinity/chlorinity/conductivity, temperature, speed of flow, and direction. Other chemical probes will be added in the near future. The remote sensing platform permits large scale surveys of coastal waters at varying depths with immediate availability of reduced data. Calibrations of the system were accomplished through the cooperation of the Adelphi Institute of Marine Sciences on Long Island.

Imagery Analysis: Maryland Area

Photographic Techniques

The aerial photography of the Maryland wetlands produced 350 9×9-in. photographs, half being true color and the other half in false color Ektachrome. Of these, 188 were at 3000-ft altitude. At the onset of the program, it was not clearly evident that recognition of wetlands could be accomplished with a computer-oriented approach. To check the validity of an automated delineation of wetlands from photographic imagery, a manual photographic analysis was made concurrently (Hair, 1970).

Photo interpretation was performed on 9×9-in. prints of stereo pairs of both color and color IR imagery. Sets of prints from each altitude were examined for gross topographic and vegetative features. To prevent duplication, acetate overlays were made only on those frames covering the east shore of the river from the mouth of Hunting Creek to the northern end of the Chew Creek marsh, and by eliminating stereo coverage. However, each area was examined in stereo.

Changes in the degree and frequency of submergence either by stream flow or tides are characterized by variations in the dominant species composition within these areas. Vegetative patterns were delineated on each overlay and correlated with ground truth observations and measurements of temperature and salinity. A dark line was used to delineate the maximum upland border of the designated marsh areas. Where no marsh vegetation was apparent, this line was drawn along the shore to indicate that there is an intertidal zone present but its extent is not visible in the imagery.

In each marsh area, certain features were noted for each frame and altitude. The upland–marsh border was delineated mainly on the existence of terrestrial tree species. In most cases, this was rather simple due to the abrupt rise in topography along the river and the resulting sharply defined tree line. However, in those areas where the topography showed only slight slopes toward the upland, pioneer tree species could often be found invading the marsh. In these cases, this area was included as part of the marsh but shown as "PIONEER" on the overlays.

Each overlay included the river–marsh shore, the upland forest border, farm or field, watercourse delineation, and gross vegetative patterns of marsh species. Letters were used on the overlays to denote certain recurring features. Cross vegetative patterns were delineated based on variations in color contrast and height as determined by stereo inspection.

Not all features were marked in each area, to avoid needless duplication. Cross vegetative patterns, marsh, upland, and river borders were marked on all 3000-ft imagery.

The imagery at 3000 ft was analyzed and overlaid first, and this may be cause for subjective bias as to presence or absence of some feature in the later 6000-, 9000-, and 12 000-ft imagery. However, since this altitude gave the most complete coverage and best resolution, it was thought better to start with 3000 ft first for identification and correlation purposes.

Electronic Techniques

The automated analysis of the imagery is accomplished by the procedure indicated in Fig. 12.8. The 9×9-in. positive transparencies, either true color or false color infrared, are density scanned automatically at three wavelengths (0.433, 0.533, and 0.633 μm), corresponding to peaks in the transmissions of the three emulsions. The parameters of the scan, such as lines per film frame, scanning aperture size, scanning speed, and optimum wavelength must be determined for the particular image under analysis. The output of the density scans are fed to a minicomputer programmed with appropriate recognition criteria. The recognition criteria are based on the film sensitometry and the spectrophotometric functions of the surfaces being photographed. If the recognition criteria are simple, and one color densitometry is adequate, a map overlay printout is obtained from the minicomputer directly. If correlation between colors and/or thermal data is necessary for recognition, the facilities of a larger computer, such as a time-shared IBM 360/67, are necessary. The time-sharing feature is essential because of the variability of the spectrophotometric

Figure 12.8 Computer data analysis flow chart.

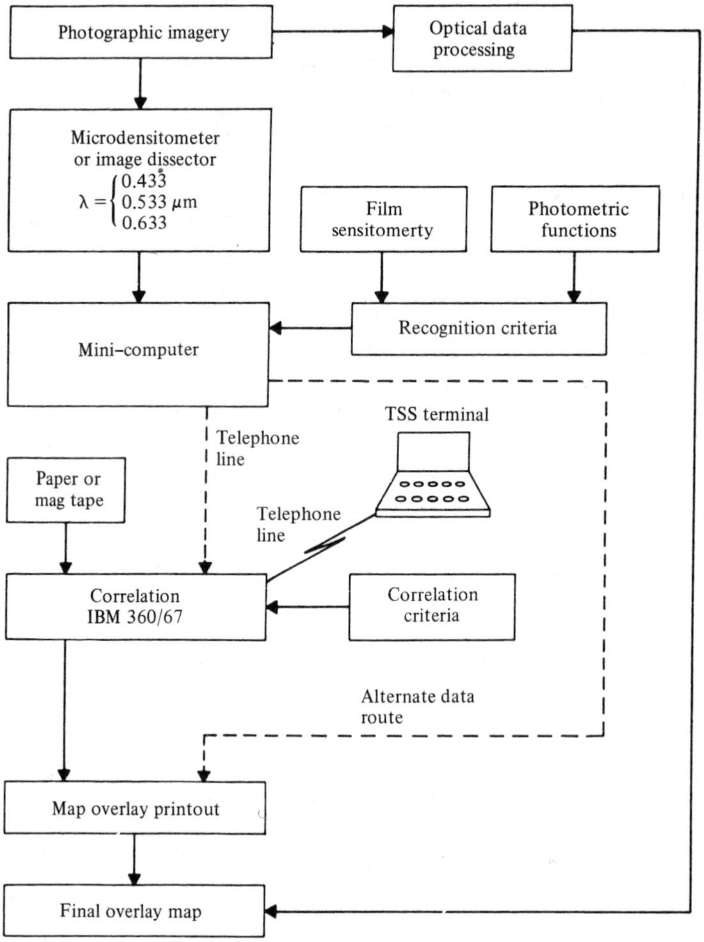

properties of a biological environment and the necessity of making minor program corrections during a computer run.

A refinement of the recognition technique may be included whereby optical data processing can be used in certain instances for recognition of regularly recurring features, such as crop furrows.

Results

To illustrate the application of an automated delineation of wetlands, a representative 3000-ft-altitude frame, No. 5592 (Fig. 12.9), will be used to illustrate both manual and automated analysis. The manual analysis was made on a true color print obtained from the Aero Neg type 2445 format, and the automated analysis performed on a true color transparency.

The manually produced overlay of Fig. 12.10 delineates the wetlands at the area of Little Lyons Creek and Hunting Creek. The code for the notations in the figure is listed in Table 12.1.

To produce an equivalent computer overlay, the red emulsion on the true color transparency was selected. This selection was based on the calibration curve for the true color transparencies resulting from the photometric properties of the test panels (Fig. 12.11). It is to be noted that the test panel reflectances must be corrected for their location in the field of view of the camera; the overall transfer function of the camera is a strong function of the location in the plane of the negative and not completely compensated by the antivignetting filter (Duddek, 1967).

By referring to Fig. 12.11, it is observed that the blue ($\lambda = 0.433$ μm) emulsion density does not increase as the standard panels decrease in reflectance because atmospheric scattering produces a residual brightness level. The red emulsion is unaffected by atmospheric scattering and shows a high contrast for green tree foliage. This is also verified by measurements of the density ranges obtained on the true color transparencies.

The false color infrared transparencies were not used because the infrared emulsion was less differentiated between wetlands and forest areas surrounding the wetlands. The infrared emulsion normally would have had the highest contrast of the three emulsions on false color infrared film (see Fig. 12.12 ($\lambda = 0.633$ μm) and include the least atmospheric scattering effects).

A typical microdensitometer scan, based on a 200-line raster for the 9-in. format, is shown in Fig. 12.13. The size and shape of the scanning aperture crucially affects the scan results. A 1-mm-square effective aperture size was optimum for the present analysis, but circular or rectangular apertures with dimensions on the order of microns can be more suitable for data analysis in other instances. The smaller apertures concurrently produce more information. The scan raster starts with line number 1 at the lower left (southwest) corner of Fig. 12.9 and ends with line number 200 at the upper right (northeast) corner of Fig. 12.9. Farmlands, trees, wetlands, and the Patuxent River are designated in the margin. It is seen that tree regions are darker as a result of the reflectance properties of oak leaves (Egan, 1970) and shadowing effects. The reflectance of black oak leaves is known to vary with phase angle (Egan, 1970), and also the angle relative to the plane of incidence (plane including the normal to the ground surface and the incident sun rays).

There will be some differences in scattering between oak leaf varieties, but the general properties of a shiny leaf, as exhibited by the black oak leaves, will be

Figure 12.9 Black and white rendition of true color transparency (Frame No. 5592); Little Lyons Creek–Hunting Creek area, Patuxent River, Maryland.

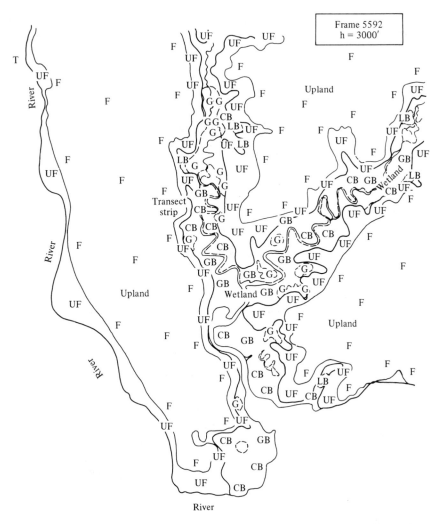

Figure 12.10 Reproduction of manually produced delineation of wetland areas for Frame No. 5592 (see Fig. 12.4). See Table 12.1 for coding.

followed. The computer program must include these effects (Figs. 12.14 and 12.15). The minimum photometric threshold of the trees varies from the south to north of the transparency (as a function of raster scan line number) (Fig. 12.14), and very greatly across the raster for a particular raster scan line number (Fig. 12.15) because of shadowing effects and camera lens vignetting. The tree minimum photometric threshold (Fig. 12.14) indexes the left edge of the convergence of the three photometric calibration lines of Fig. 12.15. This variable index serves as the discrimination threshold to delineate the trees, and the photometric function (Fig. 12.15) varies with raster trace number. The wetlands, however, because of the diverse nature of the vegetation, do not have a well-defined photometric range. Because of the lack of shadowing, there is a negligible east–west variation. This can be seen by referring to photographs made of the wetland areas. In the Hunting Creek area, a typical

Table 12.1 Letter Codes for Figure 12.10

BW	Backwash pattern in river
CB	Creek bed in uplands
D	Discontinuity in river of stream flow patterns
DP	Drainage pattern in marsh or upland
F	Farm or field
GY	Gray area on marsh
G	Green area on marsh
GB	Green-brown area on marsh
LB	Light-brown area on marsh
OF	Old field
OSB	Old stream bed
PA	Path
P	Pond
S1 through S11	Station number for ground site
SG	Sun glitter
UF	Upland forest

overview of the marsh shows the sharp delineation between the *Typha spartina* communities and the upland oak forest. The marsh vegetation is of almost uniform height. The only shadow effects apparent are immediately adjacent to open water areas. The typical shoreline development along the streams in the study area produces an almost fencelike structure of the *Spartina cynosuroides* with uniform height of the vegetation. Plastic strips were used to delineate differences in dominant species types for better photointerpretation. A strip was placed in an east–west direction in Little Lyons marsh. The sharp change in height between the *Typha sp.* on the north of the strip and *Spartina patens* and *Scirpus* on the south is easily seen in the aerial photographs.

Figure 12.11 Color test panel calibration on true color transparencies.

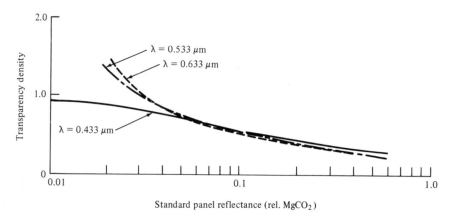

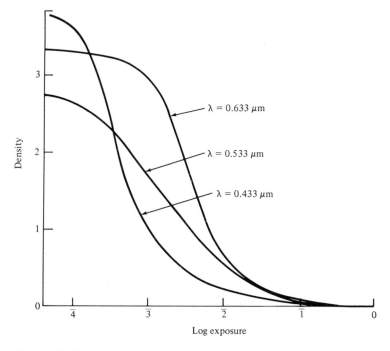

Figure 12.12 Integral density curves for type 8443 infrared Ektachrome Film obtained on Joyce Loebl microdensitometer.

Figure 12.13 Microdensitometer scan of raster line no. 80 showing ground truth results.

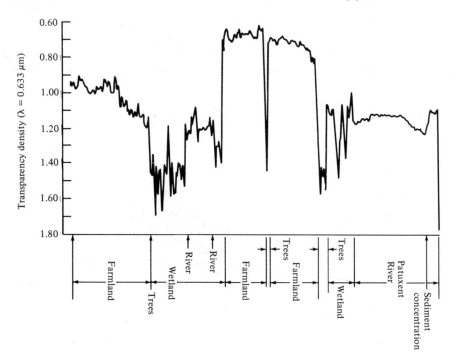

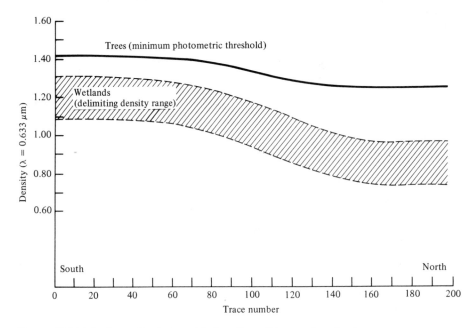

Figure 12.14 North–south photometric function of tree and wetland areas.

By applying the decision criteria, a computer-printed map is obtained delineating the wetlands in terms of the forest areas bounding them (Fig. 12.16). The areas delineated by Fig. 12.16 agree with those manually deduced (Fig. 12.10). The accuracy of location of the automated delineation is of the order of a few feet when a sharp line of demarcation exists between trees and wetland. If there is a pioneer tree area, the edge of the wetland area is less well defined.

As an alternative, the wetland density criteria may be used to obtain a printout that requires further analysis. This is accomplished through optical data processing, whereby farmlands, which may have the same reflectance range as the wetlands, are

Figure 12.15 East–west photometric function of tree areas.

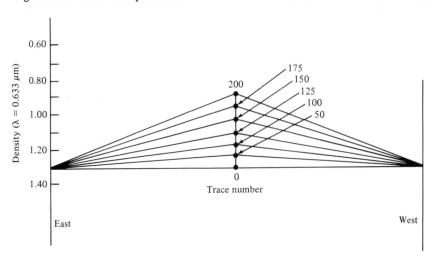

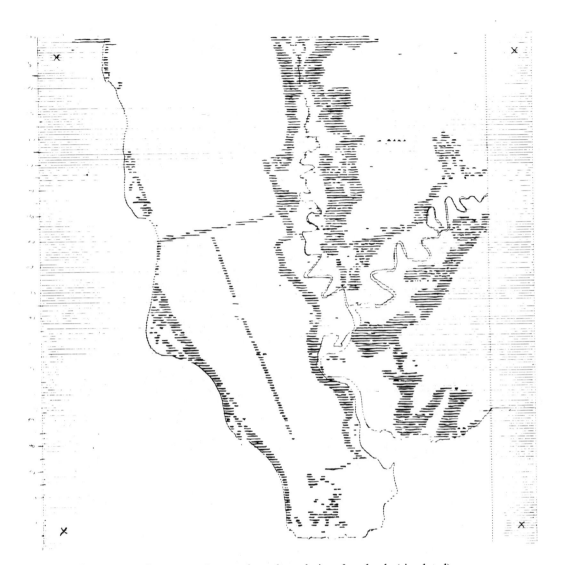

Figure 12.16 Computer printout of tree boundaries of wetlands (simulated).

eliminated. The farmlands, which are usually plowed, then have a characteristic spatial frequency. This permits them, as well as river areas where waves exist, to be discerned automatically. Thus, the areas marked F (farm) and R (river) would be deleted from the final printout by optical data processing (Fig. 12.17).

An example of optical data processing is shown in Fig. 12.18. A rectangular aperture was used with an He–Ne laser operating at a wavelength of 0.6328 μm. The amplitude diffraction patterns (a two dimensional Fourier transform of the original transparency) are of a forested area adjoining the wetlands at Little Lyons Creek. Pattern (a) has two intersecting lines resulting from the square aperture, but also a wedge-shaped decreased density area resulting from shadowing by the trees. This coherent shadow pattern produces the wedge-shaped diffraction pattern. There is a weak radial symmetry, seen better in diffraction pattern (b), which is the result of the diffraction of the images of the tree crowns.

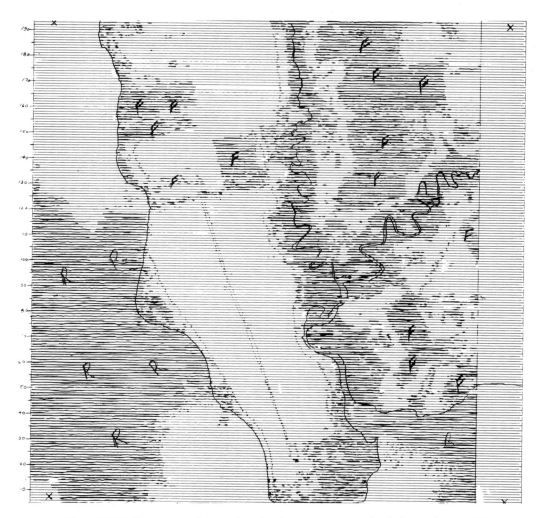

Figure 12.17 Computer printout of wetland areas; erroneously designated areas (R = river, F = farmland) are indicated, and would be deleted by optical image processing (simulated).

When Fourier transforms of an image are available, a "matched filter" technique may be used for recognition of ground features. In simple language, a remote sensing scanner transforms a spatial image (a ground feature) into a frequency spectrum as the scan proceeds. This frequency spectrum represents the same information that is in the spatial distribution of ground features; hence by constructing a matched electric filter that produces a peak signal (relative to noise) when the appropriate group of frequencies arrive from a scanner sensor, one may achieve automated real time recognition. The matched filter has been treated extensively in the technical literature (Van Vleck and Middleton, 1946; Turin, 1960; Schwartz, 1959; Goldman, 1953). A perfect matched filter is not always physically realizable and it then must be approximated (Craig, 1960).

A further benefit of a calibration program is tne possibility for prediction of true ground cover color (based on three spectral bands) using the imagery and the

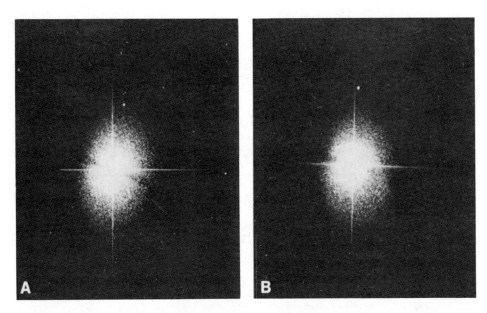

Figure 12.18 Two dimensional Fourier transforms of forested area adjoining wetland at Lyons Creek; (a) tree shadow effects, (b) tree crown symmetry effects.

calibration curves. This technique may be applied to color transparencies or color prints.

As an example, calibration curves are presented in Figs. 12.19 and 12.20. These curves are obtained by reflectance measurements on sets of true color and infrared color prints made from the 3000-ft-altitude imagery. These curves are valid only for the set of prints produced from the imagery taken on 25 September 1970, for one existing sun angle, the particular atmospheric scattering present, and film processing techniques. A set of curves such as these would result from a specific aerial survey when calibration panels are used.

Figure 12.19 True color print calibration. Print No. 5587; phase angle of 49.3°; altitude 3000 ft. Reflectances relative to $MgCO_3$ surface.

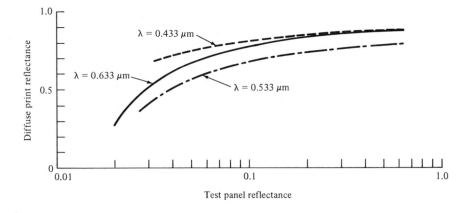

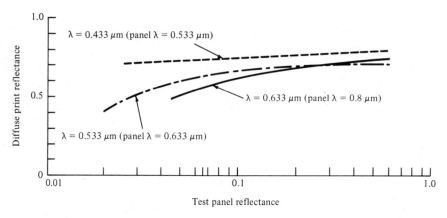

Figure 12.20 Infrared color print calibration. Print No. 2484; phase angle of 44.3°; altitude 3000 ft. Reflectances relative to MgCO$_3$ surface.

The curves of Figs. 12.19 and 12.20 are applied by measuring the diffuse print reflectance (either true color or false color infrared) for a particular ground object of interest at wavelengths of 0.433, 0.533, and 0.633 μm. The values measured correspond to an equivalent narrow band reflectance of an equivalent test panel on the ground, being the same wavelengths for true color imagery (Fig. 12.19) or shifted in wavelength for false color imagery (Fig. 12.20). The equivalent narrow band reflectances are read from the appropriate curves for the wavelength of interest and film used.

As a result of the analysis of imagery produced at 6000, 9000, and 12000 ft in comparison to the 3000-ft imagery, it is found that the system color degradation measured on the ground test panels increased with altitude, and was minimal to 6000 ft on true color prints and to 9000 ft on false color infrared transparencies. Color consistency across the frame was generally within 0.1 O.D. for the true color negatives and within 0.4 O.D. for the false color infrared transparencies, and differed for each emulsion. With altitude increase, the false color infrared transparencies become greener, apparently as the result of atmospheric scattered light.

Transforming Data into Information

While the aforementioned equipment assemblages acquire phenomenal amounts of data, the information content may be less. What is desired, ultimately, is information from the data. How do we transform data into information?

First, we must determine what information we want, and in what form. The information we want is invariably related to other factors. In terms of bioresources, there is a relation to ecology, perhaps industrial or recreational activity, meteorology, or pollution.

Secondly, the data acquisition system must be accurately calibrated, and precisely correlated, within the system limitations, to the ground truth. Then we may proceed with computer analyses of the data, optical image processing, or optical displays to obtain the desired information.

Computer analyses have been explored in great depth by the Purdue University Laboratory for Agricultural Remote Sensing in terms of agriculture (Purdue, 1968), and the USGS, Miami, for plant communities (Hartwell, 1969). Optical image processing has been described in the literature (see, for instance, NASA, 1970). Optical display techniques using multiband cameras have been described by Yost and Wenderoth (1968).

Rather than consider all possible systems, let us continue with the analysis of the data contained in the Timber Island type 8443 Ektachrome photograph. The original transparency has been subjected to a careful calibration and control program described in Chapter 8. An evaluation of the transparency densities was made on a Joyce Loebl recording microdensitometer at three wavelengths 0.633, 0.533, and 0.433 μm. A trace at a wavelength of $\lambda = 0.633$ μm of the ground transect V_0 (see Fig. 12.3) is shown in Fig. 12.7; the ground resolution is approximately 1 ft. Densities are plotted increasing to the right, and the ground truth labeled to the right of the curve. It can be seen that there are general density levels correlated with different species, but mainly because of effects of mixed varieties of ground cover, and physical state of the flora, variations exist (cf. marsh mallow). The ground truthing was made four months after the original aerial photography; since the fall season had been quite mild, there was still sufficient ground cover to permit ground truth to the determined reasonably well. For the diversified and mixed flora on Timber Island, a single aerial photograph, even though analyzed in three colors, is inadequate to characterize the ground cover. It is necessary to use more photographs at different viewing and illumination angles to acquire sufficient data to permit separating out the light scattering properties of the ground cover.

In general, ground truth should be acquired as near the time of aerial photography of bio- and hydroresources as is humanly possible, preferably concurrently.

Although the areas of color contrast shown on Fig. 12.3 are reasonably sharp, it is seen from the microdensitometer trace of Fig. 12.7 that the color contrast of general areas alone is indeed a poor means for classifying ground cover. However, the detailed use of aerial color (infrared and visible) photography (with a calibration and control program) appears to yield information that may be used for the optical remote sensing of even diversified and mixed varieties of ground cover.

Optical remote sensing of bioresources is indeed emerging as a reality. However, the present accuracy and reproducibility as a function of season, diversity of ground cover, physical state, time of day, instrumentation, and analysis technique is poor. The route to improved sensing involves the application of technology that already exists, together with appropriate information acquisition techniques.

Optical techniques on one hand have been highly developed, but the conversion of the technology to a precise information gathering means from a highly variable biological environment is difficult. Progress is being made and ultimately we envision optical remote sensing of shallow water areas in terms of the inherent bioresources, with spot ground truth calibrations frequently being made.

13

Marine Biology and Water Quality

The time has come for the user to take a critical look at exactly what *quantitative* water quality information can be obtained from Earth Resources Technology Satellites imagery, multicolor, true color, and false color aircraft photography; these are all considered in comparison to ground truth and in situ measurements of water quality. We emphasize *quantitative* in terms of an absolute reference, for it implies a calibration technique being used with the imagery to establish reference color levels, eliminate nonlinearities, and permit atmospheric effects to be subtracted.

To establish the boundaries of LANDSAT (ERTS) and aircraft data within which useful water quality information can be obtained, NASA had sponsored Experiment No. 589. This experiment was performed in the St. Thomas Harbor at Charlotte Amalie. The scope of NASA Experiment No. 589 included optical, biological, and chemical ground truthing of the harbor water, multicolor I^2S (International Imaging Systems) four band aerial photography, and the interpretation of ERTS 1 (LANDSAT 1) multispectral scanner (MSS) bands 4 (0.5–0.6 μm) and 5 (0.6–0.7 μm). A forerunner of this experiment was an effort including optical and physical ground truthing of the harbor water and aerial photography using both true color and false color Ektachrome film (Egan, 1972). As a necessary adjunct to both of these experiments, an optical ground calibration technique was developed and applied to the aerial and LANDSAT 1 imagery. Also included in the LANDSAT Experiment No. 589 was an extensive data handling computer program, which was applied to the in situ data and the LANDSAT 1 and I^2S imagery. Even though the multispectral I^2S imagery was not obtained concurrently with the true color and false color Ektachrome imagery, correlation may be made utilizing optical in situ measurements and calibrations. The techniques described are applicable to all LANDSAT imagery, although here applied to LANDSAT 1.

In planning the aerial photography for Experiment No. 589, multispectral imagery was selected because black and white negative film is used (with filters to select blue, green, red, and near infrared bands); the many problems associated with colorimetry using true color and false color Ektachrome films negated their use for precision photometric measurements (see Egan, 1969, for instance). Thus, the highly precise photometric calibration procedures are described in detail for the I^2S imagery because the inherent inaccuracies in Ektachrome film (due to interlayer effects, noncompletely automated development, film storage effects, etc., Egan, 1969) vitiate accurate absolute calibration. Photometric errors as large as 500% (a factor of 5 error) can occur with small targets imaged on Ektachrome film because of interlayer effects (see Table 3, Egan, 1972 for instance); but if large calibration areas are used, the absolute calibration errors are at best $\pm 20\%$ (see Table 3, Egan, 1972

Figure 13.1 Optical study area in the St. Thomas Harbor. The optical stations lie along the biological transect between Stations 102 and 120.

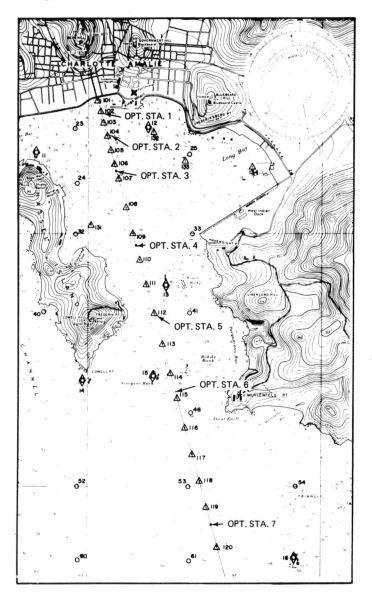

for instance). However, the Ektachrome films are useful for showing visual contrasts simply, quickly, and inexpensively without an elaborate viewer—required with I^2S imagery. The comparative absolute accuracies of I^2S and Ektachrome imagery will be summarized in the Conclusions.

Study Area

The optical study area lies along the ground truthing marine biological transect between stations 102–120 (Fig. 13.1). The optical stations, 1–7, were fewer than those used for the marine biology, mainly because of the length of time required for each in situ set of optical observations (location and measurement $\approx \frac{1}{2}$ h). There was also limited time for the entire optical transect because it was made in the afternoon following the 1016 hrs LANDSAT passage over St. Thomas Harbor on October 17, 1972, ERTS 1 SCENE #1086-14162.

A supplementary optical calibration area was situated at Brewers Bay Beach, located about 3 miles west of the optical study area.

The St. Thomas Harbor optical transect begins in a region of high pollution from sewage effluent (near the Coast Guard Dock) and terminates in a region of low pollution past Muhlenfels point (see Fig. 13.1). Raw sewage was discharged into the harbor at a peak rate of about 3 million gallons per day (Egan, 1969). Since there is no aquifer on the Island of St. Thomas, drinking water is obtained from catch basins and a desalination plant. Sanitary sewage (toilet waste) is flushed using bay water pumped inland through a separate water system. The raw sewage was discharged into the bay along the sea wall at the northern rim (Fig. 13.1).

There were additional sources of sediment in the harbor water. One source was the daily docking of cruise ships, generally at the West Indian Dock (east of the initial portion of the harbor optical transect, Fig. 13.1). Another cause of sediment was a dredge operating along the harbor entrance channel, acquiring coral sand for building construction.

Also, the ships docked at the West Indian pier discharged raw sewage, as well as did small craft docked nearby at the Yacht Haven Marina.

Optical Calibration Program

Why do we need an optical calibration program? Why can't we simply use contrasts in color to determine pollution?

These two rhetorical questions serve to introduce a treatment which is unique in the present thinking on interpretation of LANDSAT and aircraft imagery. There are techniques currently in use that employ contrast enhancement in imagery using false colors so that we may "see" pollution in brilliant color hues, but there is no quantitative assessment. It has been pointed out that sun illumination geometry, atmospheric filtration, imaging system response, and data reduction techniques all affect contrast (see Brezenski, 1972; Griggs, 1973, for instance). Court cases may be based on visual evidences of pollution, but the final decision always rests with the exact degree, with well defined specification; an image of water with discoloration does not quantitatively yield information on degree of pollution or even the nature. There must be optical calibration, and this calibration must include spectral (color) effects.

Brezenski (1972) pointed out the need for color calibration and control in true–false color photography; this is more difficult with color film than with panchromatic black and white film because of interlayer effects in color film. Therefore the preferred approach for aircraft imagery is the use of a four lens camera with selective color filtration (such as the I^2S), with automated film processing (as with the Kodak Versamat).

Satellite imagery (LANDSAT) suffers from the absolute calibration problem, with complicating factors resulting from atmospheric scattering and absorption combined with effects of reduced spatial resolution. Absolute calibration becomes necessary for LANDSAT imagery to quantify color levels absolutely and to determine the effects of atmospheric scattering and absorption.

The implementation of the optical calibration program involves four aspects: (1) in situ harbor water transect measurements, (2) Brewers Bay Beach and color panel calibration, (3) calibration of the I^2S photographic imagery, and (4) calibration of the LANDSAT data. These four aspects will now be described.

In Situ Harbor Water Measurements

The optical property of the harbor water that will be sensed will be its brightness. The brightness is the result of many factors that include scattering of particulates in the water, sun elevation, and waves which may reflect as sun glint along with the sky reflection. Where the water surface waves are small, and the sun glint does not significantly affect the water surface brightness, the brightness of the water will depend upon the light that is backscattered by macroscopic, microscopic, and submicroscopic particulates in the water. If the water is relatively uncolored by dissolved constituents, the transmission of light in the water is a good indicator of the concentration of scattering particles. The more scattering particles there are in the water, the lower will be the transmission. However, this is only an approximation. Further, the transmission will generally depend upon the wavelength of the light used in the transmission measurement.

The water optical transmission measurements for this program were made using a portable battery operated Minneapolis–Honeywell photometer (with interchangeable narrow band interference filters) mounted on a submarine viewing tube. The submarine tube had a water-tight Plexiglass window that permitted photometric

Table 13.1 Optical Data on St. Thomas Harbor Transect on 17 October 1972

Optical station	Transect station	$\alpha.433/cm^a$	$\alpha.533/cm^a$	$\alpha.633/cm^a$
1	102	0.019	0.014	0.019
2	104	0.019	0.014	0.019
3	106–107	0.019	0.014	0.019
4	109–110	0.0033	0.0044	0.0066
5	112	0.0018	0.0030	0.0052
6	115	0.0014	0.0019	0.0047
7	119–120	0.0008	0.0019	0.0041

[a] Half-bandwidth of filters: 20 nm.

measurements under water without interference from surface waves. The optical filters used in the measurements were peaked at 0.433, 0.533, and 0.633 μm; they were Optics Technology narrow band interference filters with a bandpass of 0.02 μm. These measurements for the seven optical stations are presented in Table 13.1; the location of the optical stations relative to the biological transect station is indicated in the second column. Where two biological stations are indicated, the optical station is located about midway between them. The quantities listed in Table 13.1 are in inverse centimeters, as defined by the relation

$$I = I_0^{-\alpha d},$$

where I_0 is the intensity of incident (sun)light, I the intensity of transmitted (sun)light, α the absorption coefficient, and d the path distance in centimeters. It is to be noted that at optical stations 1–3 the concentration of particulates was so great that a short path length had to be used, and the measurements have a lower accuracy ($\pm 30\%$) than the other stations which have a photometric accuracy of $\pm 3\%$ to 5%.

The absorption coefficients are listed in Table 13.1 for the ground truth transect described subsequently.

Brewers Bay Beach and Color Panel Calibration

The spectral reflectance of the coral beach sand used as a calibration standard for the LANDSAT and the aircraft imagery at a 6000-ft altitude is presented in Fig. 13.2 under illumination conditions closely analogous to those existing during the LANDSAT overpass. The sand in the central area is brighter because it is finer,

Figure 13.2 Laboratory spectral reflectance (relative to $MgCO_3$) of Brewers Bay Beach sand at an incident angle of 40° and a phase angle of 3°.

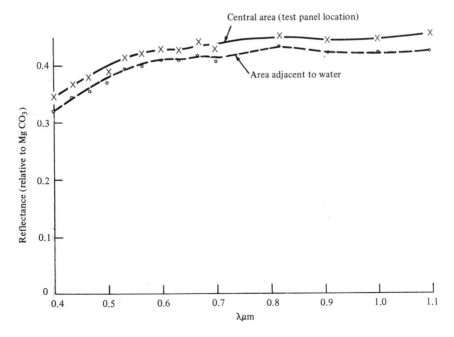

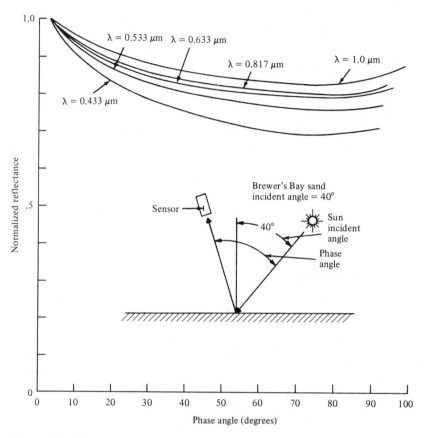

Figure 13.3 Laboratory spectrophotometric properties of Brewers Bay Beach sand from test panel location at an incident angle of 40°.

more firmly packed and free of larger shells that exist in the sand near the water. However, the reflectance from the sand also depends on the viewing angle and wavelength of the incident light; this is shown in Fig. 13.3 (the phase angle on the abscissa is the sum of the incident angle of 40° and a variable viewing angle). The reason that the spectral reflectance (Fig. 13.2) must be known is so that one can use any apparent spectral variation observed from the LANDSAT or from high altitude photography to determine the amount of atmospheric filtration and scattering. The angular dependence (Fig. 13.3) is used to determine the vignetting of the photographic lens and filter system on a selected photographic image where the Brewers Bay Beach area extends completely across the frame.

The spectral reflectance of the test panels on Brewers Bay Beach used for low altitude (2000-ft) aircraft imagery is presented in Fig. 13.4. The white (W) and gray panels (G1, G2, and G3) have reasonably uniform reflectance between 0.4 and 1.0 μm, but drop rapidly below 0.4 μm. The red (R) is almost saturated, but the blue (B) and green (G) are weaker in saturation; the important consideration is that the paints be "flat" (i.e., have uniform reflectance at all viewing angles in order not to introduce additional corrections). The actual "flatness" of the reflectance is shown in Fig. 13.5, which reveals small changes in reflectance between phase angles of 20° and

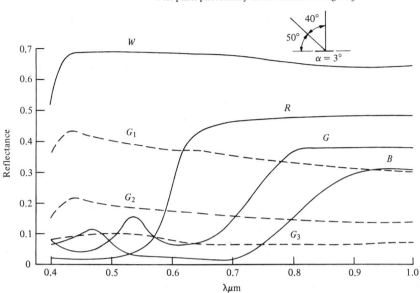

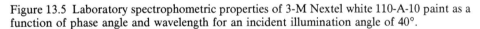

Figure 13.4 Laboratory spectral reflectance (relative to $MgCO_3$) of test panels at an incident angle of 40° and a phase angle of 3° (W = White; G_1, G_2, G_3 = gray shades; R = Red; G = Green; B = Blue).

110°. At phase angles below 20°, the pronounced increase in reflectance is characteristic of nearly all materials, and is generally much stronger.

The reason for using seven test panels is that they serve as an additional check on the film response in low altitude aircraft imagery, and as a cross calibration check on the sand photometry. As a bare minimum, a single panel (such as white) could be used, but this checks the response at only one reflectance level.

Figure 13.5 Laboratory spectrophometric properties of 3-M Nextel white 110-A-10 paint as a function of phase angle and wavelength for an incident illumination angle of 40°.

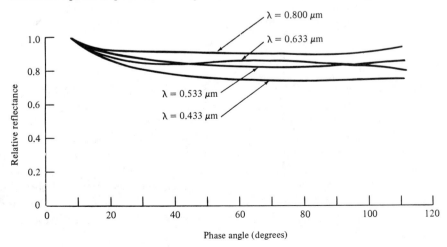

Calibration of the I²S Photographic Imagery

Photographic calibration of the imagery obtained from the I²S cameras involves quantification of the light absorption properties from the ground to the film emulsion. The elements involved are the radiance from the ground calibration areas (test panels and Brewers Bay Beach sand), the atmospheric absorption, the camera lens and filter absorption, the distribution of brightness in the film plane, and the spectral sensitivity of the film. These elements now will be discussed in that sequence.

Radiance Calibration

The ground truth radiance measurements were made with the same portable battery operated 3° acceptance angle Minneapolis–Honeywell photometer (with interchangeable narrow band interference filters) as was used for the in situ harbor water measurements. Measurements were made at Brewers Bay Beach adjacent to the test panel location. The photometer was calibrated with an Eppley thermopile in the laboratory. Measurements were made at center wavelengths of 0.433, 0.533, and 0.633 μm with the aforementioned filters having bandpasses of 0.02 μm. The radiance values on the white test panel, when corrected for the wider acceptance bands of the I²S cameras, are 0.69 mW/cm²/sr in the blue band, 1.25 mW/cm²/sr in the green band, and 2.00 mW/cm²/sr for the red band. The light level was subject to some variation because of clouds occasionally obscuring the sun. The observation time, of 0944 hrs, 17 October 1972, is earlier than the satellite over pass time of 1016 hrs, and later than the aircraft 2000-ft altitude overflight at 0920. These time variations will introduce a small correction to the observations.

Atmospheric Absorption

The atmospheric absorption for a 2000-ft altitude flight line would be expected to contain some atmospheric haze contribution, but those at a 6000-ft altitude would contain more. The actual quantitative evaluation of the atmospheric absorption and scattering will be discussed in the subsection on Data Analysis, and tabulated in Table 13.3.

Camera Filters and Lenses; Brightness in the Film Plane

The I²S camera consists of four Schneider Xenotar $f/2.8$, 100-mm focal length lenses. The four lenses produce four images 9.25 cm square on 9-in. wide aerial film. The field of view of the lens is 55.2° (half-angle of 27.6°). The relative spectral transmission of the lenses is shown in Fig. 13.6a. The transmission varies with wavelength, and this must be taken into account in calculating the attenuation by the lens. The filter attenuation curves are shown in Fig. 13.7. The Wratten filter transmission curves are shown for the red, green, and blue sensing cameras; since the infrared band will not be used in the present photometric analysis (because of the low information content due to water opacity, and the lack of a photometer for

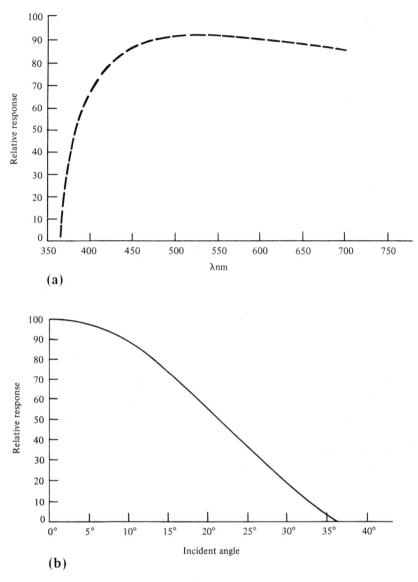

Figure 13.6 (a) Relative spectral transmission of Schneider Xenotar $f/2.8/100$-mm lenses at $f/2.8$ aperture. (b) Relative brightness in the image plane of Schneider Xenotar $f/2.8/100$-mm lenses at $f/2.8$ aperture.

ground truth calibrations in the infrared), the filter for this band was not included in the measurements. It is to be noted that the red, green, and blue Wratten filters all have passbands in the infrared; hence an infrared blocking filter is necessary in addition because type 2424 film, which was used in this work, has a response in the infrared portion of the spectrum (Fig. 13.8). These infrared filters must pass the red, green, or blue bands and yet block the infrared; the best type of filter for this application is an interference filter. However, interference filters have different characteristics for different angles of incidence.

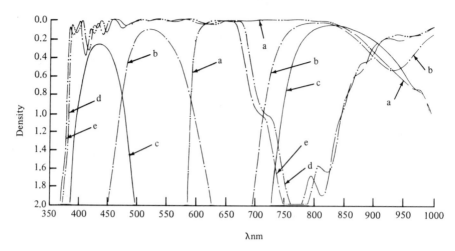

Figure 13.7 Laboratory spectral transmission curves for red, green, and blue camera band-pass filters and infrared interference filter; obtained on Cary 14 spectrophotometer. Curve a − · ·− red filter, Wratten No. 25; curve b −·− green filter, Wratten No. 57A; curve c ——— blue filter, Wratten No. 47; curve d ——·—— interference filter, normal incidence; curve e ——·— interference filter, 19° incidence angle.

In Fig. 13.7 are presented transmission curves for the green band interference filter. This curve is representative of the other interference filters. It is seen that the transmission at normal incidence is shifted toward shorter wavelengths at 19° incidence angle. This is important because the extreme rays incident through the camera lens enter the filter (which is placed in front of the lens) at an angle of 27.6° to normal, increasing this effect. This can then affect the bandpass of the combination of Wratten filter and interference filter so as to narrow or expand the bandpass, and thus the response of the photographic system will be angularly dependent.

Figure 13.8 Spectral response of type 2424 film. Source 2850 K; corrected to 5500 K.

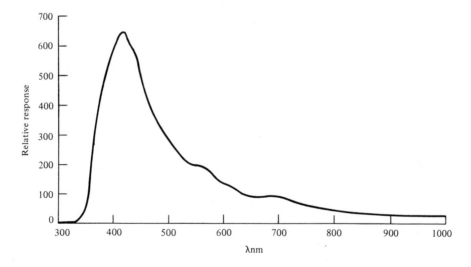

However, inherent in the imaging process of a lens is a $\cos^3$ dependence on incident angle. This is in addition to the angular dependence on the filter transmission properties just mentioned. This $\cos^3$ dependence is quite pronounced and produces a vignetting of the image. Antivignetting filters are sometimes employed to minimize this effect, but they slow down the effective speed of the imaging system. It should also be noted that the vignetting effect shown in Fig. 13.6b is aperture dependent; smaller f stops result in less vignetting. Ultimately the brightness distribution in the film plane must be checked experimentally against imagery having a known brightness distribution. This can be accomplished in the laboratory or in the field; one approach is to photograph a beach area (such as Brewers Bay Beach, St. Thomas) so as to cover an entire frame. Then the brightness variation in the photographic image can be used to calibrate the vignetting effect of the lens. Also, there may be differences between lenses, and the use of a curve such as Fig. 13.8 on one lens would not take these differences into account.

Film Sensitivity

The spectral response of type 2424 film is shown in Fig. 13.8. However, this does not reveal the density as a function of exposure. This requires a particular calibration program. Such a calibration program was carried out through the cooperation of Kennedy Space Center/NASA and Manned Space Center/NASA for the present work.

Essentially, the original undeveloped, exposed negative type 2424 film was sent to the Phototechnology Division, Houston. There the film was exposed to a 2850 K lamp using Corning C5900 and C2043 filters. The exposures are tabulated in Table 13.2. Through the use of the exposure information given in Table 13.2, the camera system response may be computed; these data are presented in the last column of the table. The exposures listed in the table have produced densities in type 2424 negative film from a calibrated step tablet; this calibrated step tablet was exposed on type 2424 negative film through the three filters. By plotting the measured densities on the negative against the original step tablet calibrated density, an H–D curve is obtained for three colors for the 2424 negative film.

Table 13.2 Exposures[a] and System Response on Type 2424 Film in I^{2}S Camera Used in St. Thomas Imagery

Filter	Exposure (sec)	Energy (erg)	Camera System Response (mW/cm^2/sr)
47B (blue) +IR	0.1002	7.9496−10	40
58 (green) +IR	0.0399	8.0743−10	67
25 (red) +IR	0.0399	7.9155−10	46

[a] By Phototechnology Division, Manned Space Center, NASA.

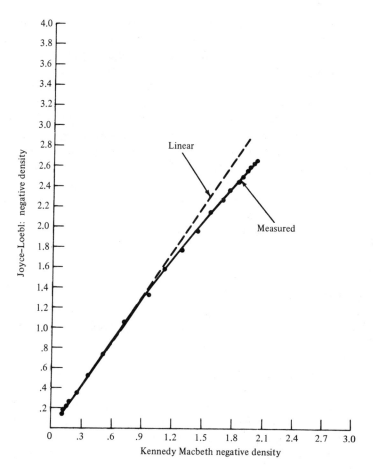

Figure 13.9 Macbeth to Joyce–Loebl densitometer conversion.

However, as mentioned in Egan (1969), the responses of different densitometers depend on their optical system and the amount of optical scattering in the film being measured. The measurements made in this program were made on a Joyce–Loebl recording microdensitometer because it permitted measurements to be made on microscopic areas such as the images of the test panels. The comparative response of the Macbeth and Joyce–Loebl densitometers is presented in Fig. 13.9. It is seen that the Macbeth reads lower densities than the Joyce–Loebl, and that the relation is nonlinear. The differences are the result of differences in the optical systems of the two units; if a nonscattering absorber were compared in the two densitometers, the readings would be the same. Because of scattering, some of the light is not sensed by the Joyce–Loebl optical system and this leads to the higher density readings on it. This does not lead to any problems as long as one densitometer is used consistently for a set of readings.

The H–D curves for red, green, and blue on type 2424 negative film are presented in Figs. 13.10, 13.11, and 13.12. The exposure is the reading on the Macbeth densitometer, and the density is read on the Joyce–Loebl microdensitometer with an

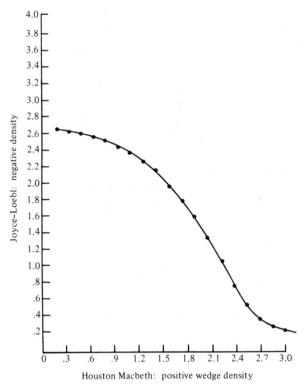

Figure 13.10 Red response of type 2424 film; Macbeth vs Joyce–Loebl (with No. 25 and IR filters).

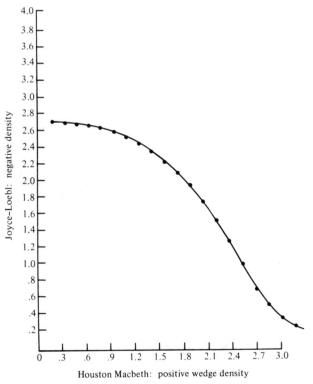

Figure 13.11 Green response of type 2424 film; Macbeth vs Joyce–Loebl (with No. 57A and IR filters).

effective aperture of 150 μm. This aperture was large enough to average over the photographic emulsion grains. It is to be noted that the images of the original step wedge were not constant in density variation, but increased slightly in density (due to scattering) toward the less dense end of the wedge.

The type 2424 negative was then used by NASA, Kennedy to make a duplicate positive print. By printing one of the step wedge images from the type 2424 film (No. 25 filter one) on the duplicate positive, the H–D curve for the duplicate positive is obtained (Fig. 13.13). There are two curves shown in Fig. 13.13, one for the head and tail of a duplicate positive roll prepared in June 1973, and those of another duplicate positive prepared in July 1973 of Frame 0004, containing the test panels. The second duplicate positive was necessitated by inferior resolution in the duplicate positive Frame 0004 of the red and green I^2S bands. Because of processing variations, the positive densities may vary by as much as 0.2 O.D. (Joyce–Loebl) for the same exposure. This is one of the limitations of photographic photometry.

The method of use of the sensitometry curves (Figs. 13.10–13.13) is as follows: first a density is measured with the Joyce–Loebl microdensitometer on the duplicate positive of an area of interest. This density of the positive on the ordinate of Fig. 13.13 yields the type 2424 negative density, as read on the Joyce–Loebl for the black and white image of the red, green, or blue bands. Then using either Fig. 13.10, 13.11, or 13.12 depending upon whether the red, green or blue original image was the one of interest that was measured, we determine what was the required Macbeth

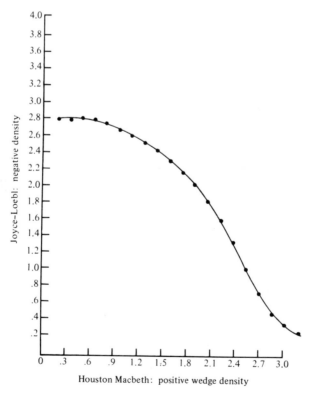

Figure 13.12 Blue response of type 2424 film; Macbeth vs Joyce–Loebl (with No. 47B and IR filters).

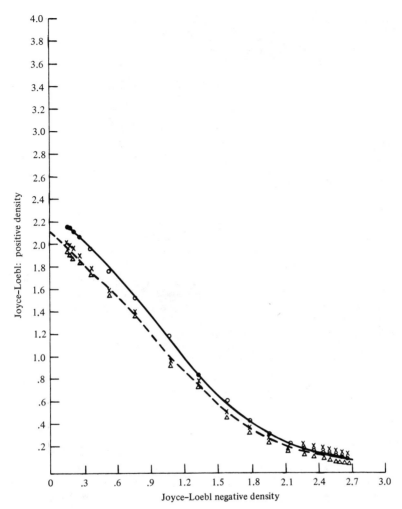

Figure 13.13 Response of duplicate positive print on June 1973 print, and July 1973 print (X = head, June 1973; △ = tail, June 1973; ⊙ = July 1973 print).

exposure. Then using this Macbeth exposure (density), and the values listed in Table 13.2 (last column), the radiance may be calculated. This procedure will be followed in the section on Data Analysis.

Calibration of the LANDSAT 1 Data

The IN ORBIT calibration of the LANDSAT 1 MSS scanner was inoperable because the calibration pulse had dropped considerably and appeared to have shifted. Therefore ground truth measurements must be depended on for calibration. However, the relative radiance levels from the LANDSAT 1 were still valid.

Since ground truth was obtained only in the green and red bands, the discussion will be limited to the corresponding LANDSAT bands 4 and 5.

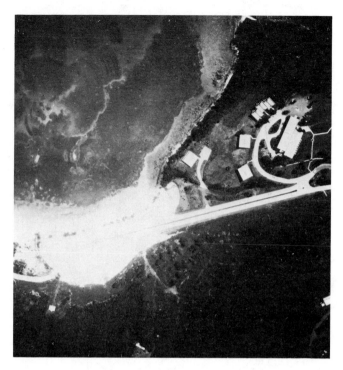

Figure 13.14 I²S aircraft photograph of Brewers Bay Beach calibration test site. Spectral band 0.410 to 0.470 μm.

The calibration area of Brewers Bay Beach is shown in the I²S Blue Band Photograph (Fig. 13.14). The corresponding computer printout from the bulk tape is shown in Figs. 13.15a and 13.15b for bands 4 and 5 respectively. The digital value used for computation is the highest value for the Coral Beach sand, which occurred along track line 1584 and across track line 846. This value is least likely to have been appreciably affected by the adjacent lower reflectance water or tree area in either the green (band 4) or red (band 5). This highest value is circled on Figs. 13.15a and 13.15b. There is the possibility of some error being introduced by this procedure.

In order to convert the digital levels on Figs. 13.15a and 13.15b to the true radiance of the Brewers Bay Beach sand, the atmospheric scattering and attenuation must be taken into account. These atmospheric corrections are listed in Table 13.3 for LANDSAT 1 bands 4 and 5. The clear atmospheric attenuation (from Allen, 1963) is listed in the second column, and the atmospheric scattered radiance in the third column. The atmospheric scattered radiance was derived from sky photometric measurements at Brewers Bay Beach, of clear blue sky in a direction away from the sun.

By using the corrections in Table 13.3, and the digital levels in Figs. 13.14 and 13.15, the radiance levels at the surface, based on LANDSAT 1 bulk digital data, may be computed. The results of these computations are presented in the last column of Table 13.4, with the intermediate computational steps shown. The comparisons of these levels with the ground measurements and photographic imagery will be made in the Data Analysis section following.

The LANDSAT 1 imagery was found unsuitable for microdensitometry purposes for various reasons. The band 5 imagery was unusable because of horizontal striations caused by nonuniform gain in the LANDSAT 1 sensor channels. There were also displacements of some lines horizontally. Band 4 posed problems as to the location of the shore lines, as fiducial points, to initiate microdensitometric scans. The I^2S viewer was used to superimpose the imagery, and some photographs were made; however, for quantitative work, the processed bulk tapes proved to be the most suitable.

Data Analysis

This section is concerned with an analysis of the calibration procedures, and the determination of the validity and the consistency of Brewers Bay Beach and test panel ground measurements, I^2S imagery densitometry of these objects, and the LANDSAT computer-processed bulk CCT data of Brewers Bay Beach. Also considered will be the validity of the atmospheric corrections.

Aircraft I^2S Camera Vignetting Correction

The first point to be checked is whether the I^2S vignetting is actually described by Fig. 13.6b in view of the fact that the optical bandwidth of the interference filters used with the I^2S cameras is a function of incident angle. This dependence of optical bandwidth on incident angle of rays into the camera lens will affect vignetting depending upon the properties of the red, green, and blue filters and the corresponding interference filter.

Frame 0178, taken at a 2000-ft altitude, images Brewers Bay Beach across the entire frame, and thus is suitable for checking vignetting, providing the viewing angle dependence of the sand reflectance (Fig. 13.3) and the calibrated sensitometry of Figs. 13.10–13.13 are utilized. The results of such an analysis are presented in Figs. 13.16, 13.17, and 13.18 for the red, green, and blue imagery, respectively. The blue correction is the least, the green is the greatest, and the red intermediate. There is still some residual viewing angle dependence of the Brewers Bay Beach sand in the curves, causing a displacement of the maxima to the northwest direction. This is the result of a variation between the experimental conditions for Fig. 13.3 and those actually existing at the time of the Frame 0178 imagery (i.e., sun angle different, and the data for Fig. 13.3 was taken in the plane of the sun and the normal to the sand surface). The blue variation (Fig. 13.18) is least because the shift in the interference filter bandpass compensates for the vignetting effect of the lens shown in Fig. 13.6b. The red vignetting (Fig. 13.16) is about what is expected with no effect from the interference filter, whereas the green vignetting (Fig. 13.17) is augmented. These vignetting curves must be used to correct for vignetting when the imagery does not lie in the center of the image field.

Microdensitometry of I^2S Imagery of Calibration Test Site (Panels and Beach)

The microdensitometry for the test panels on Frame 0004 ran into some problems. In the initial duplicate positive imagery furnished in June 1973 from a Kodak

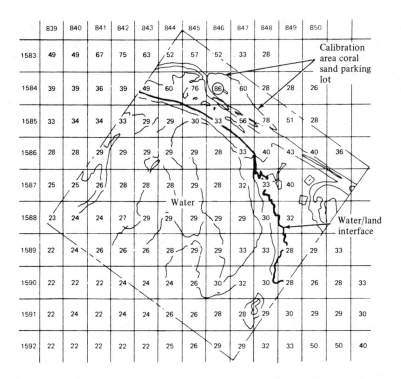

Band #4 scene 1086–14162

Scale $= \dfrac{1}{6,100}$

Cross track $\cong$ 58 m/character

Along track $\cong$ 78 m/character

(a)

Figure 13.15 (a) LANDSAT MSS bulk band 4 printout, scaled to overlay I^2S photograph of Brewers Bay Beach calibration test site. (b) LANDSAT MSS bulk band 5 printout, scaled to overlay I^2S photograph of Brewers Bay Beach calibration test site.

Versamat the blue (1) and infrared (4) bands produced sharp imagery of the test panels. But the test panel imagery of the green (2) and red (3) bands was blurred. This obviated any densitometry measurements. Inspection of the original negative of Frame 0004 revealed sharp imagery in all four bands. An improved remake of that frame was accomplished in July 1973 on the Kodak Versamat #11 CM (Developer VER 641). The red and green bands of the remake lacked resolution for the test panels; this can be seen by referring to Figs. 13.19 and 13.20, which are micro-densitometric traces of the test panels on Brewers Bay Beach in the blue and red bands. The effective microdensitometer aperture is 91 μm. (This aperture size was determined experimentally to be sufficiently large to average out over enough grains in the print to yield useful densitometry values.) Figure 13.19 clearly reveals the test panel densitometry in the sharp blue print; even the center junction in the 6-ft panel array is revealed. However, in the red band (Fig. 13.20) it was difficult to resolve the panels, much less the density values.

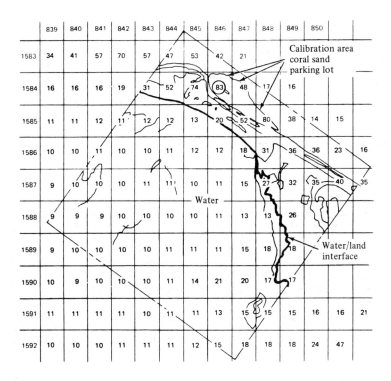

	839	840	841	842	843	844	845	846	847	848	849	850
1583	34	41	57	70	57	47	53	42	21			Calibration area coral sand parking lot
1584	16	16	16	19	31	52	74	83	48	17	16	
1585	11	11	12	11	12	12	13	20	52	80	38	14	15
1586	10	10	11	10	10	11	12	12	18	31	36	36	23	16
1587	9	10	10	10	11	11	10	11	15	27	32	35	40	35
1588	9	9	9	10	10	10	10	11	13	13	26		
1589	9	10	10	10	10	11	11	11	15	18	18	Water/land interface
1590	10	9	10	10	10	11	14	21	20	17	17	
1591	11	11	11	11	10	11	11	13	15	15	15	16	16	21
1592	10	10	10	11	11	11	12	15	18	18	18	24	47

Water

Band #5, scene 1086–14162

Scale $= \dfrac{1}{6,100}$

Cross track = 58 m/character

Along track = 78 m/character

(b)

The more dense direction in Figs. 13.19 and 13.20 is toward the right, and each major division is equivalent to 0.082 O.D.

Using the density levels indicated in Frame 0004 taken at a 2000-ft altitude and the procedure outlined in Table 13.5, we obtain the radiances for the test panels and Brewers Beach sand (Table 13.6). Also listed in Table 13.6 are the photographically determined radiances at a 6000 ft altitude of the Brewers Bay Beach sand at the panel location on the beach parking lot, and also along the beach area, both at 0945 hrs (Frame 0036) and at 1430 hrs (Frame 178). It is seen that both the red and green radiances decrease with altitude change from 2000 to 6000-ft, but the blue only increases slightly, indicating the dominant effect of atmospheric scattering in the blue. The sand near the water is slightly darker than the parking lot at 0945 hrs, but the illumination level increases at 1430 hrs for a 6000-ft altitude.

Table 13.3 Atmospheric Corrections

Band	Clear atmospheric attenuation[a] (%)	Atmospheric scattered radiance (mW/cm²/sr)
4	17	0.13
5	10	0.04

[a]*Source*: Allen (1963)

Table 13.4 Brewers Bay Beach Radiances from LANDSAT 1 Data for Bands 4 and 5

	1	2	3	4	5	6	7
	Count 127 base	Count 63 base (1/2 col. 1)	GSFC Calibr. Table (volt. sig.)	% radiance full scale (col. 3 × 25) (%)	Equivalent radiance at LANDSAT 1 (col. 4 × R) (mW/cm²/sr)	Atmospheric scattering correction (col. 5 − S) (mW/cm²/sr)	Radiance at surface (atmos. atten.) (correction) (col. 6 × A) (mW/cm²/sr)
					(R = 2.48)	(S = 0.13)	(A = 1.17)
Band 4	86	43	1.916	48.0	1.19	1.06	1.24
						(S = 0.04)	(A = 1.10)
Band 5	83	41-1/2	1.877	47.0	0.94	0.90	0.99

Figure 13.16 Red vignetting correction.

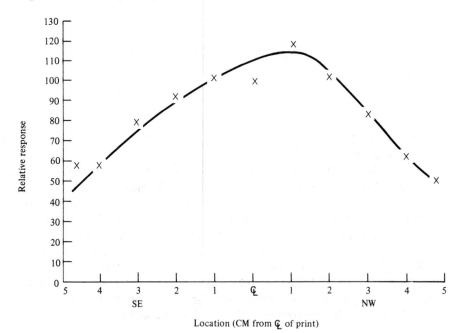

Location (CM from ₵ of print)

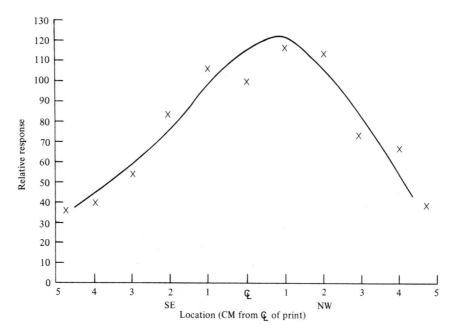

Figure 13.17 Green vignetting correction.

Figure 13.18 Blue vignetting correction.

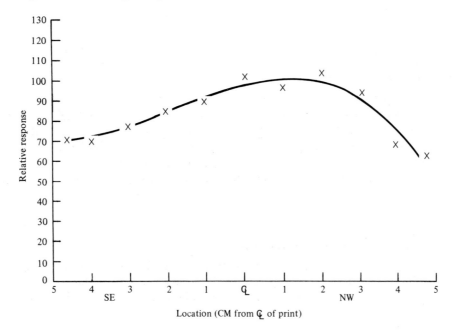

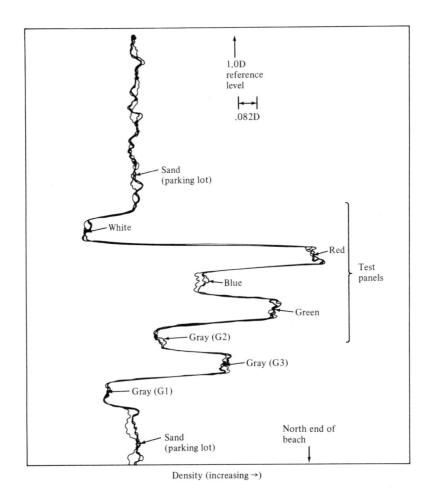

Figure 13.19 Test panel microdensitometry trace in the blue band.

Comparison of Radiance Measurements, Ground, Aircraft I^2S, and LANDSAT 1

A comparison of the radiance measurements by the various techniques of the Brewers Bay Beach sand in the parking lot is presented in Table 13.7. Agreement is excellent except for the LANDSAT 1 green band. The most probable source of error is that the green atmospheric scattering correction for the LANDSAT 1 should be greater than that used in Table 13.3. This is evident from the fact at that a 2000-ft altitude the photographic green photometric level is greater than at ground level (Table 13.7).

Observed Color Temperature Effects

Another item of atmospheric information may be obtained from the white test panel calibrations: the color temperature of the sun. During aerial photography, the apparent color temperature of the sun will change as a result of atmospheric

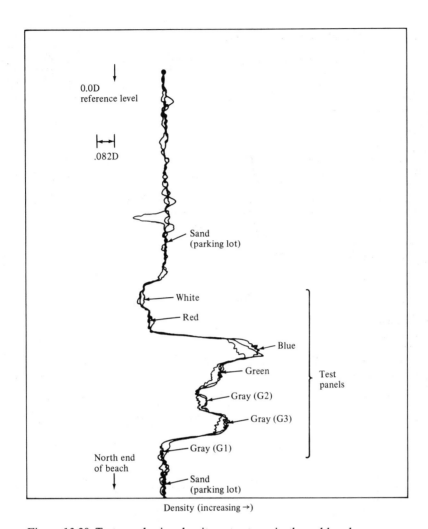

Figure 13.20 Test panel microdensitometry trace in the red band.

filtration. This color temperature is defined in terms of blackbody radiation curves. The usual assumed color temperature of the sun is 5500 K, but this color temperature is decreased at dawn and sunset by the longer sunlight path through the atmosphere; the color temperature may also be decreased by atmospheric aerosols and haze.

Table 13.5 Calculation Procedure for Test Panel and Brewers Bay Beach Data Reduction[a]

J–L Positive print density	J–L negative density	Macbeth positive density	Filter and lens correction	Corrected Macbeth density	Macbeth transmission (%)	Radiance conversion factor at 0 density (mW/cm²/sr)	Radiance (mW/cm²/sr)
0.279	0.202	1.80	−0.204	1.596	2.54	66.5	1.69

[a]Green ($\lambda = 0.52~\mu$m), frame 0004, positive print, white panel.

Table 13.6 Test Panel and Brewers Bay Beach Sand Photometry

Target	Blue ($\lambda = 0.4$ μm) J–L density	Blue Radiance (mW/cm^2/sr)	Green ($\lambda = 0.52$ μm) J–L density	Green Radiance (mW/cm^2/sr)	Red ($\lambda = 0.65$ μm) J–L density	Red Radiance (mW/cm^2/sr)
Test panels, 2000-ft alt. (Frame 0004)						
White	0.203	0.585	0.279	1.69	0.201	1.96
Red	1.237	0.091	1.203	0.345	0.244	1.22
Blue	0.739	0.151	1.135	0.379	0.742	0.418
Green	1.928	0.371	1.070	0.400	0.570	0.555
Gray (G1)	0.287	0.379	0.463	0.99	0.303	1.08
Gray (G2)	0.537	0.209	0.800	0.55	0.508	0.62
Gray (G3)	0.835	0.138	1.050	0.415	0.595	0.53
Brewers Bay Beach sand (parking lot) 2000-ft alt. (Frame 0004)	0.422	0.262	0.553	0.82	0.316	1.10
Brewers Bay Beach sand (parking lot) 6000-ft alt. (Frame 0036)	0.412	0.267	0.696	0.61[a]	0.376	0.85[a]
Brewers Bay Beach sand (near water) 6000-ft alt. (Frame 0036)	0.463	0.239	0.562	0.82	0.322	1.00
Brewers Bay Beach sand (near water) 6000-ft alt. (Frame 178)	0.209	0.454	0.400	1.05	0.264	1.20

[a] Light cloud haze suspect.

Table 13.8 indicates the observed color temperature effects; briefly, various targets (i.e., the test panel and Brewers Bay Beach sands) are compared to the blackbody power density ratios for various color temperatures. The effective color temperature of the sun may then be evaluated. The white panel laboratory reflectances are listed in order to indicate that the white panel is a useful constant reference as a function of wavelength for the spectral region under consideration.

Table 13.7 Comparison of Brewers Bay Beach Sand Radiance Data

Band	Ground measurement (mW/cm^2/sr)	Photographic determination (mW/cm^2/sr)	LANDSAT 1 determination (mW/cm^2/sr)
4	0.72	0.82	1.24
5	1.16	1.10	0.99

Table 13.8 Color Temperature Effects on I^2S Imagery

Item	Blue band (mW/cm²/sr)	Green band (mW/cm²/sr)	Red band (mW/cm²/sr)
White panel at 2000-ft alt. at 0920 hrs	0.585	1.69	1.96
White panel in laboratory (equal energy illumination)	0.68	0.68	0.66
Brewers Bay Beach parking lot sand at 2000-ft alt. at 0920 hrs	0.262	0.805	1.095
Brewers Bay Beach parking lot sand at 6000-ft alt. at 0945 hrs	0.267	0.61	0.845
Brewers Bay Beach front sand at 6000-ft alt. at 0945 hrs	0.239	0.82	1.00
Brewers Bay Beach front sand at 6000-ft alt. at 1430 hrs	0.454	1.05	1.20
Ground radiance of white panel at 0944 hrs	0.69	1.25	2.00
Blackbody at 5500 K	0.91	0.99	0.93
at 7000 K	0.98	0.87	0.70
at 5000 K	0.80	0.97	0.985

The apparent color temperature of the sand at a 2000-ft altitude is about 5000 K, whereas at a 6000-ft altitude, the apparent color temperature of the sand is about 7000 K. There is a measurable reddening of the incident sunlight at 0920 hrs as a result of atmospheric scattering.

Uniformity of Photographic Printing Process

One last point to be covered in the data analysis is the uniformity of the photographic printing process. On the duplicate positive received in June 1973 the cross frame Macbeth density variability was 0.03 O.D., and the head to tail 0.05 O.D. This was an improvement from 0.04 O.D. cross frame variability in the Nov. 72 print, but a deterioration of the head to tail variability above 0.04 O.D. Some of this variability is in the original negative, being 0.01 O.D. cross frame and 0.03 O.D. head to tail. All the previous measurements were made with a Macbeth TD102 densitometer, which has a published variability of ±0.02 O.D.; by careful checking of the zero of the instrument, this has been reduced.

The observations generally concur, with the Macbeth measurements indicating a cross frame variation of 0.05 O.D. read on the Joyce–Loebl and a head to tail variation of 0.06 O.D.

These process limitations form a limit on the accuracy of the photographic photometry since a 0.05 O.D. error produces an error of 19% in the density measurement. Since these are the maximum variations, and the effect of a variation in density is dependent on the overall photographic system response, the average error would be expected to be of the order of 5% in the present photometric procedures, for brighter images.

Results and Discussion

Having set forth the groundwork for the calibrations of the aircraft and LANDSAT 1 imagery, we may now apply them to a specific problem, the St. Thomas Harbor optical transect shown in Fig. 13.1. The radiance of the water in blue, green, and red bands has been determined photographically, and compared to previously determined in situ optical absorption, and LANDSAT 1 data corrected for atmospheric absorption.

Figure 13.21 Microdensitometry of St. Thomas Harbor Optical Stations 1–4 in green band.

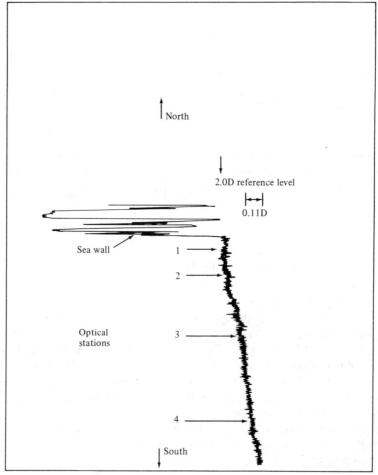

Density (increasing →)

Table 13.9 Calculation Procedure for Harbor Transect Data Reduction[a]

J–L positive print density	J–L negative density	Macbeth positive density	Macbeth transmission (%)	Print location correction	Shutter speed correction	Filter and lens correction	Radiance conversion factor at 0 density (mW/cm²/sr)	Radiance (mW/cm²/sr)
1.1741	0.37	2.93	0.117	2.63 (=1/0.38)	1.29	1.64	66.5	1.65

[a] Green ($\lambda = 0.52$ μm); Frame 0097, positive print, Station 1.

Comparison of Harbor Transect Radiance Based on I^2S, LANDSAT 1 and in situ Optical Data

The photographic densitometry was performed on the I^2S aircraft red, green, and blue bands along the harbor transect. A sample densitometric green band trace is presented in Fig. 13.21, with Stations 1–4 indicated. Using the data reduction procedure outlined in Table 13.9, we obtain the radiances indicated in Table 13.10 along the seven stations of the harbor transect in the red, green, and blue photographic bands.

The LANDSAT 1 bands 4 and 5 data reduction is presented in Table 13.11, along with the calculations to arrive at the radiances at the seven optical stations.

A comparison of the photographic and LANDSAT 1 radiances along the harbor transect is made in Table 13.12. The LANDSAT 1 band 4 radiances follow the photographic radiances in trend and magnitude, but the photographic radiances are higher than the LANDSAT 1 band 5 radiances because the type 2424 and duplicate positive film latitudes were exceeded. This is not a fault of the initial negative exposure or positive duplicate printing. It is the result of the fact that the red reflectance of the water is too low to produce an accurately readable density change; the density level of the red reflectance of the larger numbered optical stations is of the order of 0.01 O.D., which is less than the noise (the processing variations may amount to as much as 0.06 O.D.). Because of the processing variations, the density

Table 13.10 St. Thomas Harbor Transect Photometry

Photometric station	Location (Fig. 1)	Frame No.	Blue ($\lambda = 0.43$ μm) J–L density	Blue Radiance (mW/cm²/sr)	Green ($\lambda = 0.52$ μm) J–L density	Green Radiance (mW/cm²/sr)	Red ($\lambda = 0.65$ μm) J–L density	Red Radiance (mW/cm²/sr)
1	102	0097	1.427	0.281	1.741	0.165	1.889	0.160
2	104	0097	1.452	0.274	1.757	0.188	1.905	0.136
3	106 107	0097	1.573	0.227	1.842	0.108	1.942	0.088
4	109 110	0097	1.741	0.173	1.917	0.098	1.955	0.094
5	112	0098	1.509	0.245	1.796	0.101	1.895	0.123
6	115	0100	1.295	0.278	1.746	0.083	1.900	0.092
7	119 120	0101	1.405	0.248	1.845	0.061	1.895	0.092

Table 13.11 Bulk CCT Data Reduction for Bands 4 and 5 from LANDSAT 1 for St. Thomas Harbor Transect

		1	2	3	4	5	6	7
Optical station	Bio. sta.	Count 127 base	Count 63 base (1/2 col. 1)	GSFC Calibr. table (volt. sig.)	% radiance full scale (col. 3 × 25) (%)	Equivalent radiance at LANDSAT 1 (col. 4 × R) (mW/cm²/sr)	Atmospheric scattering correction (col. 5-S) (mW/cm²/sr)	Radiance at surface (atmos. atten.) correction (col. 6 × A) (mW/cm²/sr)
	Band 4					(R = 2.48)	(S = 0.13)	(A = 1.17)
1	102	32.4	16.2	0.487	12.17	0.302	0.172	0.201
2	104	29.8	14.7	0.439	10.98	0.272	0.14	0.164
3	106–107	26.8	13.4	0.385	9.64	0.238	0.10	0.117
4	109–110	26.2	13.1	0.375	9.38	0.232	0.10	0.117
5	112	26.25	13.12	0.375	9.37	0.232	0.102	0.119
6	115	24.67	12.33	0.348	8.70	0.216	0.086	0.101
7	119–120	20.3	10.1	0.274	6.85	0.170	0.040	0.047
	Band 5					(R = 2.00)	(S = 0.04)	(A = 1.10)
1	102	12.8	6.4	0.161	4.03	0.081	0.041	0.045
2	104	12.4	6.2	0.155	3.88	0.076	0.036	0.039
3	106–107	10.3	5.1	0.123	3.08	0.061	0.021	0.023
4	109–110	10.3	5.1	0.123	3.08	0.061	0.021	0.023
5	112	9.9	4.9	0.118	2.95	0.059	0.019	0.021
6	115	9.7	4.8	0.115	2.87	0.057	0.017	0.019
7	119–120	8.9	4.4	0.104	2.60	0.052	0.012	0.013

observations for the St. Thomas Harbor transect were referred to the maximum density rather than the clear transmission.

A graphical comparison of the bulk LANDSAT 1 bands 4 and 5 data with photographic and in situ water observations is made in Fig. 13.22, and the tabular data are presented in Table 13.12.

Table 13.12 Comparison of St. Thomas Harbor Transect Photographic and LANDSAT 1 Radiance

	Green band radiance		Red band radiance	
Optical station	photographic	LANDSAT 1 (band 4)	photographic	LANDSAT 1 (band 5)
1	0.165	0.201[a]	0.160	0.045[a]
2	0.188	0.164	0.136[b]	0.039
3	0.108	0.117	0.088[b]	0.023
4	0.098	0.117	0.094[b]	0.023
5	0.101	0.119	0.123[b]	0.021
6	0.083	0.101	0.092[b]	0.019
7	0.061	0.047	0.092[b]	0.013

[a]Accuracy limited by spatial resolution of LANDSAT 1 MSS.
[b]Less accurate values because film latitude was exceeded; see text for discussion.

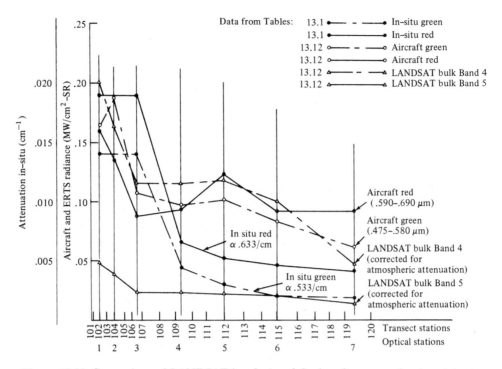

Figure 13.22 Comparison of LANDSAT bands 4 and 5, aircraft green and red, and in situ harbor transect optical data.

Figure 13.23 Printout of bulk band 4 showing cloud shadow effects on LANDSAT imagery.

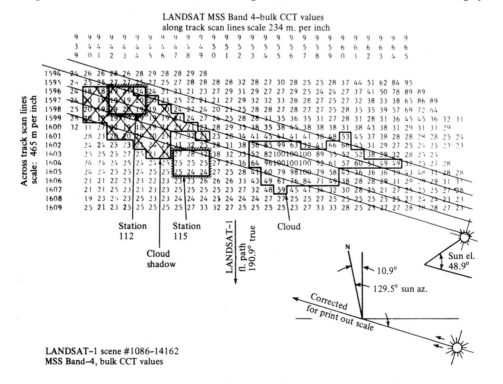

Effect of Clouds and Cloud Shadows

The LANDSAT 1 data in Table 13.12 and Fig. 13.22 have been corrected for the effect of a cloud shadow falling along the transect of Stations 112 and 115. This can be seen by referring to Fig. 13.23, which is a digital printout of bulk band 4; the sun elevation and azimuth are located to scale in the lower right hand corner of the figure, and the projected cloud shadow is seen to envelop Biological Station 112 (Optical Station 5). For the digital value of Optical Station 5, the pixels closely adjacent outside of the cloud shadow were used.

Effect of LANDSAT MSS Response Time Characteristics

Another effect on the digital pixels is shown by Fig. 13.24. The MSS detectors have a finite response time to a step function. The step function is introduced when the scanner passes from a bright area to a dark area or vice versa. In Fig. 13.24, the

Figure 13.24 Printout of bulk band 4 showing effect of MSS response time characteristics when scan passes from a bright area (sea wall) to a darker area (water).

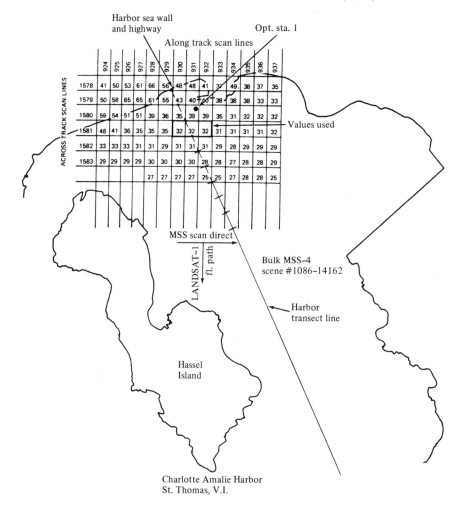

bright area is the cement sea wall at the St. Thomas Harbor edge; when the scanner passes from the sea wall to the much darker water, there is a finite time required to drop to the lower photometric level (response time). Thus the sea wall reads 61 (Line 1579 Band 4) and the water 40, and three pixels are involved in this transition; this effect is not the result of overlap in resolution elements of the scanner. In addition the town dock and Coast Guard dock appeared to contribute to the high values of line 1580; hence, we used the values of lines 1581-930 through 1581-932 inclusive as being most representative of the water at Optical Station 1.

Computer Correlation of LANDSAT 1, I^2S (Aircraft), and In Situ Optical, Biological, and Chemical Water Data

The most feasible technique for comparison is the use of the computer-generated matrix of Table 13.13. These matrices indicate the degree of correlation from -100

Table 13.13 Overall Variable Correlations (Optical and Biological)

	In situ red	Aircraft green	Aircraft red	Precision MSS 4	Bulk MSS 4	Bulk MSS 5	Turbidity	Chlorophyll a	Total chlorophyll	Pigment diversity	Total carotenoids	Bulk MSS 4 UC[a]	Bulk MSS 5 UC[a]	Bulk 4 MUC[b]	Bulk 5 MUC[b]	Chlorophyll C	Bulk MSS 6
In situ green	99.9	83	85	86	78	64	79	62	40	48	43	57	65	76	62	31	47
In situ red		82	83	86	78	64	79	60	38	49	41	56	65	75	62	29	51
Aircraft green			97	69	98	83	94	47	23	68	28	82	84	96	79	14	46
Aircraft red				78	90	71	88	61	35	54	41	70	72	90	66	27	28
Precision MSS 4					63	44	64	47	15	25	19	20	45	57	38	5	21
Bulk MSS 4						83	91	42	17	80	22	86	83	97	79	8	60
Bulk MSS 5							96	3	−11	49	−6	81	100	82	99.4	−16	53
Turbidity								26	5	53	10	80	96	90	90	−2	47
Chlorophyll A									93	38	94	34	4	49	1	88	11
Total chlorophyll										23	99.8	26	−11	29	−11	99.4	5
Pigment diversity											25	78	50	78	48	17	81
Total carotenoids												30	−6	34	−7	99	46
Bulk MSS 4 UC[a]													81	86	82	22	60
Bulk MSS 5 UC[a]														83	99.3	−15	53
Bulk MSS 4 MUC[b]															78	23	60
Bulk MSS 5 MUC[b]																−15	55
Chlorophyll C																	0

[a] MSS quantum value (0-127) uncorrected for atmospheric attenuation.
[b] Midpoint pixel MSS quantum value (0-127) uncorrected for atmospheric attenuation.

through 0 to +100. A value of +100 indicates a direct correlation and −100 an inverse correlation. A value of 0 indicates no correlation.

Table 13.13 is a representation of the correlation features, over the entire transect, of the optical in situ data, the aircraft photographic data, the LANDSAT 1 data, and the biological transect data of chlorophylls *a* and *c*, total chlorophyll, turbidity, pigment diversity, and total carotenoids.

It can be seen in Table 13.13 that the in situ green and red as well as the aircraft red and green photography correlate well with the bulk LANDSAT 1 bands 4 and 5 and turbidity. The pigment diversity shows some evidence of weak correlation with the in situ, aircraft photographic, and LANDSAT 1 data: chlorophylls *a* and *c*, total chlorophyll, and total carotenoids reveal a yet weaker correlation.

In order to investigate these variations more fully, we devised the three additional station matrices. The combinations are, sequentially: Optical Stations 1, 2, and 3; 2, 3, and 4; 4, 5, and 6; and 5, 6, and 7. All correlations will not be discussed because there is evidently good correlation between the in situ aircraft photographic and LANDSAT 1 red and green band data. However, the establishment of a correlation with the water biology and chemistry is less clear.

The turbidity correlates well with the in situ, aircraft photographic, LANDSAT 1 data; this is to be expected because the turbidity is the same as the weighted average of the in situ optical attenuation in red, green, and blue.

The more obscure relationship deals with the chlorophylls *a* and *c*, the total

Figure 13.25 Graphical correlation of turbidity, chlorophyll, and carotenoids along the harbor transect.

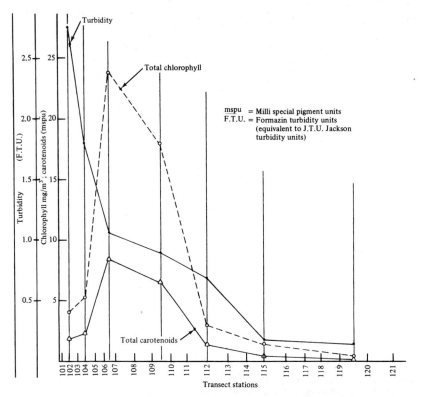

carotenoids, and total chlorophyll. The correlation of all of these parameters is generally negative (Fig. 13.25) near the sewage effluent area, and positive far from the sewage effluent area. This means that there is low biological level in terms of plant life near the sewage effluent area, where the light scattering properties of the harbor water is highest, as generally expected. However, the reverse trend is apparently the result of pigmentation in physical size of the biota in the St. Thomas Harbor entrance increasing the optical reflectance.

The strong correlation of pigment diversity with turbidity is an indication that there are strong variations in the factors making up the pigment diversity at the location where the turbidity is highest. The five factors considered in pigment diversity are chlorophylls a, b, and c, and the astacin and nonastacin carotenoids. This would be inferred to mean that the level of biological activity is highest in the turbid areas even though the level of pigments is not high.

As a further check on the optical properties of the St. Thomas Harbor transect, water spectral transmission measurements were made on samples taken near Optical Stations 1 and 6. These results are presented in Figs. 13.26a and 13.26b. It is observed that the water near the Coast Guard Dock (Optical Station 1) is highly absorbing throughout the visible spectrum, with the strongest absorption in the blue region due to Gelbstoff. Further out at Biological Station 46 (near Optical Station 6) the level of Gelbstoff drops, as expected. It is possible that humic acids and dissolved nitrates and phosphates also could be associated with the reduction in

Figure 13.26 Laboratory optical transmission measurements of St. Thomas Harbor transect water (a) off the Coast Guard dock (Optical Station 1), and (b) at Optical Station 6.

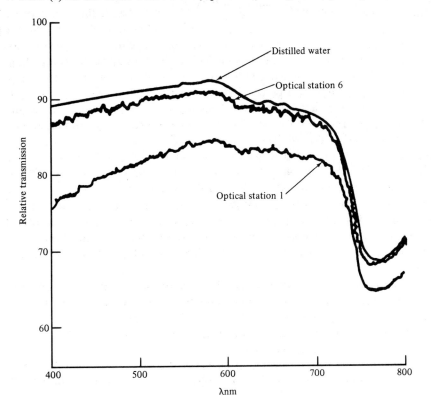

Table 13.14 Biological and Chemical Properties in St. Thomas Harbor

	Sewage effluent region	Outer harbor region
Turbidity	3.2 J.T.U.	0.1 J.T.U.
Total chlorophyll	4.0 mg/M^3	0.1 mg/M^3
Total carotenoids	2 mspu[a]	0.2 mspu
Phosphates (dissolved)	0.012 ppm	0.0046 ppm
Nitrates (dissolved)	0.03 ppm	Undetected
Humic acids (particulate)	1.5 ppm	0.3 ppm
Total coliform bacteria	8 per 100 ml	0 per 100 ml

[a] mspu = millispecial pigment units.

water quality near the Coast Guard Dock. As an example, in the St. Thomas Harbor, where turbidity is directly correlated with photometric brightness, the biological and chemical properties in Table 13.14 were evident.

Conclusions

There is a correlation between the optical in situ data, the aircraft multispectral imagery, the green and red emulsions of true color aircraft imagery, and the LANDSAT 1 MSS bands 4 and 5 data.

The water characteristic of turbidity also correlates well with these data. Chlorophyll and carotenoid pigments are inversely correlated with the falloff in turbidity near the sewage effluent area, and directly correlated with the falloff in turbidity toward the seaward direction. The photometric resolution of the LANDSAT 1 is ±1 digital level (i.e., 1.5%); that of the International Imaging Systems (I^2S) imagery is about 1%; and that of Roliflex cameras with high speed and false color Ektachrome is about 1%, in each emulsion, although color resolution to the eye is many times better than a fraction of a percent. The geometric location accuracy of the LANDSAT 1 bulk CCT data is ±1 picture element (i.e., ±80 M along the track and ± 57 M across the track for the St. Thomas Harbor area). The photographic spatial resolution is about 30 cm. The present optical ground truth in situ measurements have an absolute photometric accuracy of about ±10%, and a relative accuracy of about 5%. The absolute photometric calibration accuracy of the LANDSAT 1 and I^2S imagery is also limited to about 10% by the accuracy of the ground truth measurements. The absolute accuracy of true color Ektachrome is at best ±20% using large area calibrations; the same limitations apply to false color Ektachrome.

Computer processing assists in revealing correlations (i.e., calculating correlation matrices and factor analyses), and producing first look graphical representations of data.

Bulk LANDSAT computer compatible tape printouts rather than precision data, can be readily utilized in coastal zone investigations of water quality even for those objectives requiring high positional accuracy, by constructing a grid scaled to match the reference charts being used. MSS band 7 (0.8–1.1 μm) is useful in establishing the land–water interface location.

Registration of the corresponding pixels of the four MSS spectral bands to each other was within one picture element (pixel) as shown by comparing LANDSAT 1 bulk computer-compatible tape printouts with the aircraft I^2S imagery of the test site.

Temporal coverage by the LANDSAT is every 18 days for a given area, accepting the cloud cover and weather conditions occurring at the time of the overpassage; at the time of observation (because of recorder problems), coverage was limited to areas within real time contact from the satellite, except for high priority areas. Aircraft coverage is limited only by aircraft, equipment, and personnel availability for a flight.

The in situ measurements indicate that the blue ($\lambda = 0.433$ μm) band is best indicative of water quality, followed by the green ($\lambda = 0.533$ μm), and lastly red ($\lambda = 0.633$ μm).

There are very commendable display techniques that visually present contrasts, and relative changes in turbidity or scattered light, from water. However, these are only a first step in quantitatively assaying water quality. Turbidity, or equivalently, the backscattered radiance in the blue, green, or red bands, is the optical parameter that must be related to the biology. The level of remote detection of turbidity is about 0.2 Jackson turbidity units (J.T.U.). The radiance equivalent to this turbidity level is about 0.05 mW/cm^2/sr in the green, LANDSAT 1 band 4, and about 0.01 mW/cm^2/sr in the red, LANDSAT 1 band 5. Photographically, the multicolor photography green band will resolve this photometric level, but a conventionally exposed multicolor photography red photographic band will not because it will be underexposed in the dark water area.

The relation of turbidity (correlated to multispectral radiance) to chlorophyll, carotenoids, phosphates, nitrates, humic acids, sediments, and Gelbstoff and coliform bacteria must be established by ground truthing. The ground truthing establishes the ecological relationships between the turbidity and the various biological, chemical, and physical parameters that have a diurnal, tidal, weather, seasonal, wind, and temperature dependence.

The most promising long-term method for establishing these empirical relationships is by in situ recording or telemetering instrumentation, backed up ideally by a continuing biological, chemical, and physical study and monitoring program.

The physical scale of the area and the degree of repetitive coverage required will determine whether photographic and/or LANDSAT imagery is required.

Another bit of information that arises from the use of ground calibration is the attenuation of the atmosphere caused by aerosols and smoke. This information is of a general nature, and yields a reddening or shift in color temperature of the sun that must be supplemented by "ground truthing" in the atmosphere. Subsequent to the measurement program described, sewerage treatment plants were installed at Charlotte Amalie and the water quality was extremely improved.

14

Agriculture: Inventory, Identification

Introduction

Agricultural crop production statistics are important in national and worldwide status assessments of the major portion of food supplies. The USDA Statistical Reporting Service, operating through 43 state field offices, gathers national agricultural data on a voluntary basis: (1) each month from regular reporters; (2) March, June, and late fall with directed questionnaires to individual reporters; (3) randomly; and (4) from special purpose mailings. These data are not acquired on a daily or weekly basis and are thus limited in time coverage as well as area coverage. The international agricultural census is in far poorer condition.

The Purdue University Laboratory for Agricultural Remote Sensing has an extensive program involving aircraft, truck, field, and laboratory sensing of agricultural crops. Some very good crop classification techniques have been developed at Purdue, but they are restricted somewhat because associated computer programs and subsequent recognition must be based on corresponding aircraft, field, truck, or laboratory data. Thus, an aircraft flight over an area where good ground truth is available is used for training purposes in a corresponding computer recognition program. This program is then used for recognition of agricultural crops from other flight data. Similarly, data acquired in the laboratory on leaves of known species are used for training purposes in a computer program designed to recognize leaves corresponding to one of the known species.

Such laboratory training programs have not been applied successfully to actual crop determination measurements by aircraft or satellite. Laboratory optical effects on the spectral reflectance of cotton leaves have been used to indicate leaf age, moisture content, nitrogen fertilization extent, and soil salinity (Thomas et al., 1966), but this has not been done from an aircraft or satellite. Other subtle optical effects have been studied in the laboratory, such as the effect of pubescence on the reflectance of light from leaves as a means for leaf characterization (Gausman and Cardenas, 1968).

The question arises, "Why, in agricultural crop recognition, concern ourselves with the transition from laboratory data to aircraft (or satellite) data?" We must be concerned because the present accurate classification results that have been obtained both in the laboratory and from aircraft are coupled to a strong input from ground truth; this ground truth is used in the statistical analysis program that forms the training for subsequent recognition. For unknown or inaccessible areas, however, where ground truthing is unavailable, interpretation must be based on previous aircraft data or laboratory data. Generally, laboratory agricultural experiments are cheaper, more controllable, and easier to do than similar aircraft remote sensing experiments. It then behooves us to look to the laboratory, and put effort into this data transition problem.

The author became concerned with the transition problem following a series of flights over a Long Island agricultural area on April 9, 1968. The aircraft used was a Piper PA22-160 Tri-Pacer equipped with a low spatial resolution aircraft spectro-polarimeter/photometer, previously used in forestry remote sensing (Egan, 1968). The farm areas were selected with the help of the Suffolk County (New York) Agricultural Agent, in order to assure they were large and representative of readily accessible area crops. The objective was the simulation of sensing of agricultural crops by low resolution optical systems that would actually exist in a high altitude aircraft or satellite. The spectropolarimeter/photometer was intentionally arranged to have a ground resolution of 20 ft when flown at a 1000-ft altitude above the terrain.

As a result of 12 flights over the farm areas, which consisted of rye, wheat, potato, strawberry, corn, other vegetables, and peach orchard plantings, as well as bare soil and roads, it was evident that the data output from the sensor was highly diverse and formidable in quantity. Only one wavelength (0.566 μm) was used for measurement, and the photometrica and polarimetric data were recorded on a Sanborn Model 320 dual channel strip chart recorder. The data were more variable than those acquired over forest areas because of generally incomplete ground cover by the crops.

In order to interpret the data and to approach the problem of transition from laboratory training data to aircraft data analysis, laboratory measurements of agricultural specimens are necessary. The measurements are an extension of the approach using the bidirectional polarization characterization of various leafy specimens and minerals in the laboratory as described for forestry samples to determine the Stokes parameters I and Q.

The wavelengths were chosen to yield measurements at both ends of the spectral region sensed by the photomultipliers (0.350 and 1.0 μm) and at selected representative wavelengths between these limits. Two wavelengths were chosen to be near the chlorophyll-a bands at 0.435 and 0.665 μm (Tyler and Smith, 1967) (i.e., at 0.433 and 0.633 μm). The 0.633 μm wavelength was selected because it was near the 0.6328 μm He-Ne laser wavelength used in laboratory studies of the depolarization of laser radiation (Egan, Grusauskas, and Hallock, 1968). The green wavelengths of 0.533 and 0.566 μm were chosen for being within the high visual reflection region of agricultural crops, and the 0.8 μm wavelength was chosen for being in the transition region between the relatively low reflectance visible and high reflectance near infrared at 1.0 μm.

The source was positioned at a fixed elevation angle of 60° (30° to the surface normal, n), as representative of the summer noontime elevation of the sun at a latitude of 41° (i.e., the latitude of New York City).

As an additional check on the spectral reflectance of the samples, total reflectance measurements were made in the wavelength range between 0.25 and 2.3 μm on a Perkin–Elmer model 12 monochromator with a Gier–Dunkle absolute integrating sphere reflectometer (Model A1S-6B). The samples were mounted in calibrated quartz-covered containers and were compared to an MgO standard in the same container.

Generally, many reported total reflectance measurements are made on a Beckman DK-2A spectroreflectometer (e.g., Allen and Richardson, 1968). While not the rule, large calibration inaccuracies have been noted to exist to produce readings on the order of 10% low for high reflectance samples and 60% low for low reflectance samples for DK-2A units when compared to the system we used on standard samples distributed to various laboratories for calibration measurements (Geller and Johanson, 1967). Overall, the Gier–Dunkle system produced results that fall at the mean of all other systems. The systems compared were located at 20 government and industrial laboratories.

The crop samples that were selected for Stokes parameter determinations were fully developed alfalfa, Long Island potatoes, sweet corn, rye, and wheat obtained from the field in late summer and measured in the laboratory as quickly as possible. The sweet corn was subdivided into leaves and tassels, and the rye and wheat were subdivided into stalks and heads. Samples of a silty sand soil from a Long Island potato farm and a silty sand with 3% clay from a New York State alfalfa field were measured (analyses given in Fig. 14.1); the potato farm soil was measured both dried and with 21% tap water by weight.

Figure 14.1 Results of farm soil samples classification analyses.

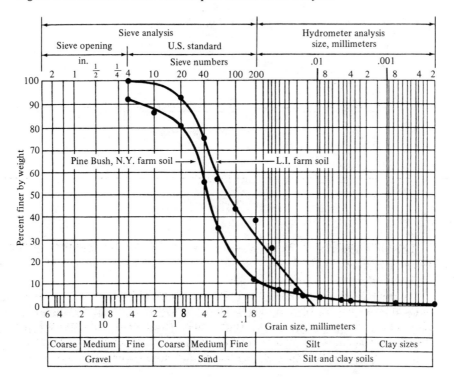

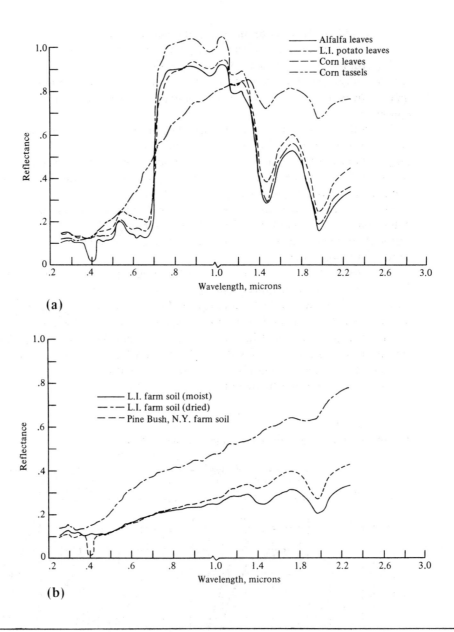

(a)

(b)

The crop selection is by no means inclusive, but is representative of species on large farm areas in the United States. The measurements did not attempt variety differentiation, but this may be possible.

Another point should be made: previous studies of rocks and particulate surfaces have shown that surface structure affects spectral bidirectional polarization and photometry (Egan and Nowatzki, 1967, 1968). Thus, measurements on leaves, stalks, tassels, and heads were made with them oriented parallel and perpendicular to the plane of vision (POV), that is, the leaves and stalks were oriented on the sample

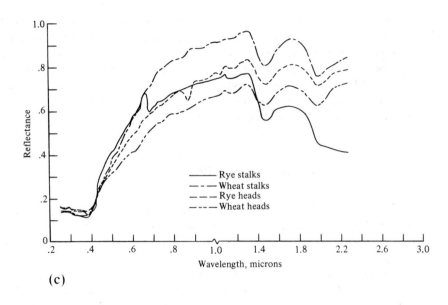

(c)

Figure 14.2 (a) Total reflectance curves (relative to MgO) for alfalfa, potato, and corn crops. Taken on Perkin–Elmer Model 12 monochromator with Gier–Dunkle absolute integrating sphere. (b) Total reflectance curves (relative to MgO) for farm soils. Taken on Perkin–Elmer Model 12 monochromator with Gier–Dunkle absolute integrating sphere. (c) Total reflectance curves (relative to MgO) for rye and wheat crops. Taken on Perkin–Elmer Model 12 monochromator with Gier–Dunkle absolute integrating sphere.

platen with their long axes parallel or perpendicular to the POV and the average surface located on the axis of rotation, as noted above. The rye and wheat heads, potato and alfalfa leaves, and the corn tassels were left on their stalks and they were oriented parallel or perpendicular to the POV.

Continuous bidirectional first and second Stokes parameter measurements were made at the wavelengths noted in the instrumentation discussions; they were done at phase angles between 3° and 96° for an incident light angle of 30°.

The first Stokes parameter I was measured in terms of a normalized "relative brightness" referred to unit brightness at 3° phase angle. The 3° phase angle was a physical apparatus limitation because this was the smallest phase angle possible when the source and sensor were adjacent. To supplement this relative brightness measurement, an "albedo" or brightness relative to $MgCO_3$ was made for the samples at a 3° phase angle. This would be π times the bidirectional distribution function for a perfect diffuser. However, none of the samples investigated were perfect diffusers at any wavelength.

The second Stokes parameter Q is given in terms of the relative polarization. The percent polarization (a frequently used representation of polarization) is then given by

$$\text{percent polarization} = \frac{Q}{I} \times 100. \tag{14-1}$$

The percent polarization does not define a particular material as uniquely as the second Stokes parameter. At a 3° phase angle (as a result of the normalization of the photometry at this angle), the second Stokes parameter is numerically equal to the percent polarization expressed as a decimal.

The total reflectance curves between 0.25 and 2.3 μm are presented in Figs. 14.2a, 14.2b, and 14.2c and are tabulated as "Diffuse Albedo" in Column 9 of Table 14.1. These curves are presented as a guide so that the overall spectral characteristics of the presently studied agricultural crops may be examined within the calibration accuracies of the spectrophotometric measurement instrumentation. The current multispectral aircraft agricultural remote sensing programs use spectrophotometry alone and do not consider polarization or explicit bidirectional photometric effects; these effects are significant because the wavelength dependence of the geometrical effects of reflectance can affect the spectrophotometry. These phenomena have been recognized in the laboratory for a tobacco and household vine leaf in a preliminary study (LARS, 1968, pp. 74–78).

The first group of spectra (Fig. 14.2a) compare alfalfa leaves, Long Island potato leaves, and sweet corn leaves and tassels. The central portion of the upper surface of the leaves are measured over a 2×12-mm illuminated sensing area, free from large veins. The detached tassels are oriented perpendicular to the length of the illuminated area in the integrating sphere sample holder. The alfalfa leaves have the lowest reflectance in the visual range, with an unusual dip at 0.4 μm (this same dip occurs in the spectrum of the soil where the alfalfa was planted, cf., Fig. 14.2b). The potato, corn, and alfalfa leaves have a maximum total reflectance in the green visual range ~ 0.54 μm, yielding the green color. It is noted that the infrared total reflectance between ~ 0.7 and 1.2 μm is high, with water absorption bands appearing at 0.97, 1.21, 1.45, and 1.95 μm, and chlorophyll absorption at 0.68 μm, similar to that observed in cotton leaves by Allen and Richardson (1968). The corn tassels, of flaxen hue, have a decidedly different total spectral reflectance variation, which would be superimposed upon the corn leaf total reflectance, more or less depending upon the actual viewing geometry and crop maturity in field observations.

The silty sand soil total reflectance (Fig. 14.2b) is a strong function of moisture content, the dry silty sand having the higher total reflectance. The wet Long Island farm soil (silty sand), as well as the moist (10.5% water) Pine Bush, New York farm soil (silty sand with 3% clay), has the water bands apparent at 1.45 and 1.95 μm; a dip at 0.4 μm, also evident for the alfalfa leaves (Fig. 14.2a), is an absorption band possibly due to spraying the alfalfa field or to the chemistry of the soil. Except for this dip, the general shapes of the curves are similar, being affected mainly by moisture, with a secondary effect presumably from the clay constituent. A comparison—not shown—of another farm soil sample from the same Long Island farm area does not reveal any significant spectral difference from that of the sample shown; slight spectral total reflectance differences do occur as expected, however, because of soil color differences. It is to be noted that laboratory measurements of soils may be at variance with field measurements because of compaction (porosity effects) (Myers and Allen, 1968).

The spectral total reflectance of rye and wheat stalks and heads is shown in Fig. 14.2c. The stalks and detached heads are arranged in a parallel orientation as in Fig. 14.3f–14.3i, perpendicular to the 2×12-mm illuminated area in the integrating sphere sample holder. The wheat stalks reflect most and the heads the least in the infrared region, with a field observation being a photometrically weighted average of the two, based on crop maturity and viewing geometry. The rye shows similar general total reflectance trends, with decidedly different detailed spectral characteristics. A comparison of the present reflectance data to that of LARS shows difference for soil, alfalfa, and wheat, due perhaps to geometrical or calibration factors.

Laboratory DK-2 spectrometer reflectance data of the type presented in Figs. 14.2a and 14.2b have been very successful in the laboratory identification of farm crops when used in association with an extensive maximum likelihood statistical program; in contrast, from aircraft, again with an extensive data analysis program, very successful aircraft classifications of crops have been made (LARS, 1968). But there is a tremendous gap between laboratory training as a basis of aircraft analyses: laboratory training data are used for laboratory analyses and aircraft training data are used for aircraft analyses (LARS, 1968). The University of Michigan aircraft, with a 12 wavelength scanner, uses a temporal (corresponding to positions) sampling technique, and it is possible on subsequent runs that identical areas on the terrain would not be sampled, and data would not correlate as well. Also, the geometry of the measurement, the lighting conditions, as well as the instrument performances affect correlation when using a given aircraft data run acquired under one set of conditions as training for another set of analysis conditions.

In comparison, bidirectional spectral polarization and photometry could be used in the construction of the n-dimensional vector matrices described in LARS (1968, p. 78–79). An electronic scanning technique that could be used for accurate determination of the Stokes parameters between 0.35 and 1.0 μm would use an image dissector. The image dissector is simply a scanning photomultiplier, and it may be used with built-in calibration to achieve photometric accuracies of the order of those attainable with photomultipliers (approximately 0.1%). In comparison, vidicons have shading errors dependent on beam current, scan speed, and focusing that may be reduced at best to about 5% with appropriate built-in calibration. A 1-in.-diameter image dissector with a 1-mil scanning aperture would require roughly 50 times more light as contrasted to a comparable high resolution vidicon with a scanning speed having a $1/30$-sec frame rate for the same signal-to-noise ratio. Low light levels, however, are no problem with terrestrial remote sensing, and different operating conditions will change the sensitivities. Also, observations may be made in microseconds with electronic scanners.

It should be noted that factors which influence field measurements, such as amount of soil between crops, crop planting geometry, physical arrangement of the crop leaves, and atmospheric haze, must be considered in the ultimate interpretation of field data.

Table 14.1 Synopsis of Spectrophotopolarimetric Characteristics of Farm Crops and Soils[a]

Material (wavelength region) (1)	Minimum percent (2)	Inversion location (3)	Maximum percent (4)	Maximum location (5)	% polarization at 96° (6)	Geometric albedo (7)	Microstructure factor 3°–6° (8)	Diffuse albedo (9)	Specularity reference to 3° (10)	Total roughness (11)	Reference to 3° macrostructure factor (96°) (12)
Alfalfa leaves											
0.350 μm ⊥	+0.42		+21.2	~96°	+21.2	0.059	0.101	0.103		15.6	0.44
=	−1.92	17.2°	+21.0	~96°	+21.0	0.054	0.101			17.4	0.45
0.433 μm ⊥	+0.58		+0.9	89.5°	+19.6	0.059	0.105	0.117		16.3	0.45
=	−0.89	15.4°	+22.8	~96°	+22.8	0.054	0.105			17.5	0.45
0.533 μm ⊥	+0.23	~	+6.0	87°	+5.9	0.127	0.074	0.197		17.9	0.53
=	−0.31	15.6°	+6.2	~96°	+6.2	0.121	0.074			19.5	0.57
0.566 μm ⊥	+0.42		+9.7	87°	+8.4	0.124	0.089	0.170		18.1	0.53
=	−0.36	14.7°	+9.9	~96°	+9.9	0.118	0.089			18.8	0.54
0.633 μm ⊥	+0.80			>96°	+13.3	0.077	0.100	0.137		16.5	0.49
=	−0.59	15.2°		>96°	+13.3	0.072	0.100			18.9	0.55
0.8 μm ⊥	+0.14			>96°	+1.8	0.522	0.038	0.897		23.2	0.69
=	−0.21	22.3°		>96°	+1.9	0.476	0.038			23.7	0.65
1.0 μm ⊥	+0.05			~96°	+1.6	0.510	0.055	0.900		22.1	0.69
=	−0.17	15.6°		~96°	+1.8	0.475	0.055			23.1	0.66
Potato leaves											
0.350 μm ⊥	0.00			>96°	+54.8	0.039	0.013	0.131		25.9	1.04
=	−2.03	20°	+43.8	~96°	+43.8	0.038	0.081			24.6	0.92
0.433 μm ⊥	+0.65			>96°	+59.2	0.039	0.034	0.135		23.7	0.97
=	−1.28	19.8°	+51.5	90°	+49.6	0.037	0.034			22.7	0.81
0.533 μm ⊥	+0.12			>96°	+33.5	0.089	0.053	0.205		20.8	0.77
=	−0.26	8.1°	+23.9	88°	+22.0	0.083	0.053			21.4	0.61
0.566 μm ⊥	+0.24			>96°	+35.1	0.085	0.047	0.183		20.3	0.72
=	−0.28	8.8°		>96°	+20.0	0.080	0.047			20.9	0.67
0.633 μm ⊥	+0.36			>96°	+82.2	0.056	0.050	0.167		23.4	0.84
=	−0.16	8°		>96°	+64.8	0.050	0.050			22.0	0.69
0.8 μm ⊥	+0.12			>96°	+4.29	0.580	0.021	1.010		25.2	0.72
=	−0.12	9.6°		>96°	+3.31	0.552	0.021			25.1	0.71
1.0 μm ⊥	+0.11			>96°	+3.69	0.527	0.017	1.000		25.1	0.73
=	−0.11	16.7°		>96°	+2.91	0.470	0.017			22.7	0.70

Corn leaves												
0.350 μm	⊥	+2.25			> 96°	+76.9	0.055	0.034	0.124		31.1	1.92
	∥	−3.12	19.6°		> 96°	+50.3	0.039	0.011		2.56	52.9	2.84
0.433 μm	⊥	+3.46			> 96°	+75.2	0.069	0.027	0.155		32.4	1.65
	∥	−2.03	14.2°		> 96°	+37.8	0.047	0.006		2.46	53.2	2.74
0.533 μm	⊥	+1.19			> 96°	+35.7	0.122	0.022	0.239		32.8	1.94
	∥	−1.44	18.7°		> 96°	+23.5	0.101	0.007		0.78	38.4	1.94
0.566 μm	⊥	+0.58		+18.8	> 96°	+14.2	0.108	0.017	0.242		44.8	2.34
	∥	−0.52	19.1°		~ 96°	+18.8	0.090	0.017		0.70	40.9	0.82
0.633 μm	⊥	+1.67	19.3°	+30.9	~ 96°	+49.2	0.076	0.023	0.212		58.5	2.92
	∥	−1.30	17.1°		~ 96°	+30.9	0.063	0.023		0.95	62.1	1.94
0.8 μm	⊥	−0.26	28.2°		> 96°	+20.6	0.375	0.002	0.891		31.8	1.20
	∥	−0.68	13.8°		> 96°	+ 9.3	0.396	0.002			31.9	1.33
1.0 μm	⊥	−0.47			> 96°	+17.6	0.394	0.005	0.908		30.6	1.29
	∥	−0.82	16.4°		> 96°	+10.9	0.402	0.004			31.2	1.14
Corn tassels												
0.350 μm	⊥	+2.05			> 96°	+12.0	0.149	0.082	0.119		18.2	0.71
	∥	−3.52	34.3°		> 96°	+22.5	0.081	0.024			37.4	1.67
0.433 μm	⊥	+3.17			> 96°	+15.4	0.137	0.087	0.167		17.8	0.68
	∥	−2.37	32.0°		> 96°	+21.9	0.095	0.007			38.2	1.67
0.533 μm	⊥	+1.27			> 96°	+ 4.4	0.254	0.051	0.243		20.9	0.76
	∥	−1.26	34.5°		> 96°	+ 7.0	0.177	0.008			37.3	1.52
0.566 μm	⊥	+1.77			> 96°	+ 5.7	0.268	0.053	0.278		20.8	0.78
	∥	−1.49	35.0		> 96°	+10.1	0.181	0.004			36.6	1.54
0.633 μm	⊥	+2.31			> 96°	+ 9.2	0.247	0.055	0.370		20.1	0.76
	∥	−2.34	35.5°		> 96°	+12.0	0.162	0.008			38.5	1.63
0.8 μm	⊥	+0.63			> 96°	+ 2.35	0.557	0.038	0.662		24.1	0.82
	∥	−1.39	30.7°		> 96°	+ 8.17	0.451	0.004			30.8	1.08
1.0 μm	⊥	+1.97			> 96°	+ 5.0	0.585	0.043	0.805		23.7	0.81
	∥	−1.25	30.5°		> 96°	+ 7.0	0.472	0.002			30.9	1.08
Wet farm soil 20.6% H$_2$O												
0.350 μm		−0.48	10.6°		> 96°	+56.8	0.033	0.063	0.115		19.5	0.65
0.433 μm		−0.48	9.7°		> 96°	+56.8	0.035	0.094	0.110		19.0	0.63
0.533 μm		−0.48	9.7°		> 96°	+33.1	0.061	0.057	0.133		20.1	0.65
0.566 μm		−0.29	8.3°		> 96°	+27.3	0.076	0.058	0.155		21.0	0.67
0.633 μm		−0.36	9.0°		> 96°	+26.3	0.097	0.057	0.175		21.6	0.66
0.8 μm		−0.52	15.0°		> 96°	+12.3	0.128	0.047	0.218		21.1	0.65
1.0 μm		0.00			> 96°	+13.4	0.147	0.036	0.245		22.8	0.71

[a]Angle of incidence = 30°.

(*continued*)

Table 14.1 Synopsis of Spectrophotopolarimetric Characteristics of Farm Crops and Soils[a] (*continued*)

Material (wavelength region) (1)		Minimum percent (2)	Inversion location (3)	Maximum percent (4)	Maximum location (5)	% polarization at 96° (6)	Geometric albedo (7)	Microstructure factor 3°–6° (8)	Diffuse albedo (9)	Specularity reference to 3° (10)	Total roughness (11)	Reference to 3° macrostructure factor (96°) (12)
Dry farm soil												
0.350 μm		−0.33	14.0°	+17.0	96°	+17.0	0.112	0.068	0.133		16.1	0.39
0.433 μm		−1.16	15.9°		>96°	+19.8	0.114	0.115	0.164		13.7	0.36
0.533 μm		−0.73	15.8°		>96°	+8.4	0.192	0.095	0.247		17.2	0.42
0.566 μm		−0.99	14.4°	+9.6	~96°	+9.6	0.216	0.083	0.282		16.6	0.44
0.633 μm		−1.12	15.0°		>96°	+9.1	0.258	0.075	0.333		16.9	0.47
0.8 μm		−0.99	17.1°		>96°	+7.2	0.320	0.045	0.410		19.6	0.48
1.0 μm		−1.20	17.3°	+4.3	~80°	+4.3	0.353	0.063	0.475		19.4	0.50
Rye stalks												
0.350 μm	⊥	+0.43			>96°	+20.4	0.180	0.031	0.119		16.5	0.47
	∥	−2.18	33.5°	+19.2	71°	+8.5	0.065	−0.003		3.24	49.9	1.6
0.433 μm	⊥	+1.76		+19.5	>96°	+30.9	0.271	0.062	0.280		15.2	0.29
	∥	−1.09	19.2°		66°	+11.5°	0.074	−0.007		4.43	55.7	1.40
0.533 μm	⊥	+2.00		+8.7	~60°	+8.7	0.237	0.044	0.464		19.5	0.41
	∥	−0.58	19.3°	+8.8	67°	+2.5	0.163	−0.001		2.76	44.4	1.22
0.566 μm	⊥	+1.83			~96°	+7.9	0.383	0.030	0.517		21.9	0.47
	∥	−0.73	19.6°	+12.8	66°	+2.6	0.196	−0.003		2.49	42.3	1.15
0.633 μm	⊥	+1.69			>96°	+11.5	0.538	0.028	0.618		18.7	0.39
	∥	−0.35	15.9°	+9.3	62°	+2.8	0.253	−0.007		2.18	39.4	1.06
0.8 μm	⊥	+1.80			>96°	+8.1	0.632	0.023	0.687		20.1	0.48
	∥	−0.65	13.5°	+13.4	61°	0.0	0.347	−0.004		2.05	38.4	1.10
1.0 μm	⊥	+0.59			>96°	+1.4	0.717	0.059	0.747		19.6	0.49
	∥	−0.18	{ 13.0° / 89.5°	+6.3	63°	−0.55	0.393	−0.002		2.08	40.3	0.98

Wheat stalks

0.350 μm	⊥	+2.1			>96°	+10.8	0.361	0.026	0.122	10.0	24.9	0.81
	∥	−1.8	28.5°	+16.8	62°	+ 8.3	0.063	−0.017			85.3	2.2
0.433 μm	⊥	+0.9			>96°	+47.1	0.227	0.050	0.227	9.6	19.5	0.56
	∥	−0.25	27.3°	+12.5	61°	+ 6.2	0.041	−0.007			88.0	2.2
0.533 μm	⊥	+3.3			>96°	+34.1	0.445	0.045	0.430		21.0	0.48
	∥	−0.56	{24.5° 93.0°	+13.8	60°	− 0.9	0.147	−0.015		5.7	58.5	0.64
0.566 μm	⊥	+3.3			>96°	+16.9	0.515	0.033	0.500		21.4	0.62
	∥	−1.1	{21.4° 86.5°	+18.1	61°	− 1.6	0.167	−0.005		4.8	53.1	1.44
0.633 μm	⊥	+2.9			>96°	+13.0	0.682	0.016	0.607		25.9	0.77
	∥	−2.3	{16.0° 77.8°	+10.2	60°	− 2.3	0.260	−0.010		3.91	51.1	1.54
0.8 μm	⊥	+2.3		+ 3.7	40°	+ 2.1	0.863	0.015	0.832		23.5	0.67
	∥	+0.3	86.5°	+ 5.2	59°	− 1.4	0.356	−0.003		2.87	47.1	1.45
1.0 μm	⊥	+2.2	80.5°		>96°	+ 4.9	0.950	0.028	0.916		23.8	0.65
	∥	+0.14	77.5°	+10.3	62°	− 3.0	0.387	−0.013		3.62	50.3	1.56

Rye heads

0.350 μm	⊥	+0.52			>96°	+19.5	0.093	0.095	0.146		17.9	0.52
	∥	−0.74	22.0°		>96°	+23.8	0.072	0.081			28.4	1.23
0.433 μm	⊥	+0.52			>96°	+12.6	0.111	0.069	0.240		17.9	0.52
	∥	−1.49	22.3°		>96°	+32.4	0.077	0.034			20.1	1.33
0.533 μm	⊥	+0.18		+ 5.5	~96°	+ 5.5	0.185	0.058	0.400		18.5	0.55
	∥	−1.15	20.3°		>96°	+15.6	0.142	0.019			29.4	1.09
0.566 μm	⊥	+0.34			>96°	+ 7.1	0.209	0.060	0.450		18.1	0.56
	∥	−1.00	20.7°		>96°	+15.8	0.159	0.026			28.6	1.03
0.633 μm	⊥	+0.23		+ 4.2	~96°	+14.4	0.253	0.057	0.524		18.7	0.54
	∥	−0.85	21.9°	+14.4	~96°	+ 2.5	0.192	0.017			28.1	0.91
0.8 μm	⊥	0.0		+ 2.5	~96°	+12.9	0.353	0.048	0.676		21.4	0.62
	∥	−0.79	20.3°		>96°	+ 3.6	0.277	0.023			29.1	1.05
1.0 μm	⊥	+0.16		+ 3.6	~96°	+ 9.8	0.371	0.053	0.774	1.08	21.2	0.66
	∥	−0.69	27.4°		>96°		0.294	0.013			30.0	1.01

aAngle of incidence = 30°

(continued)

Table 14.1 Synopsis of Spectrophotopolarimetric Characteristics of Farm Crops and Soils^a (*continued*)

Material (wavelength region) (1)	Minimum percent (2)	Inversion location (3)	Maximum percent (4)	Maximum location (5)	% polarization at 96° (6)	Geometric albedo (7)	Microstructure factor 3°–6° (8)	Diffuse albedo (9)	Specularity reference to 3° (10)	Total roughness (11)	Reference to 3° macrostructure factor (96°) (12)
Wheat heads											
0.350 μm ⊥	+0.73			>96°	+20.5	0.107	0.052	0.140		17.9	0.47
∥	−0.25	14.7°		>96°	+17.8	0.083	−0.003			30.5	1.34
0.433 μm ⊥	+0.07			>96°	+7.1	0.130	0.072	0.225		14.7	0.45
∥	−0.43	17.4°		>96°	+12.3	0.091	0.015			26.8	~1.1
0.533 μm ⊥	+0.06			>96°	+9.1	0.205	0.070	0.348		17.4	0.43
∥	−0.10	16.0°		>96°	+10.8	0.166	0.043			24.1	0.79
0.566 μm ⊥	−0.60	14.5°		>96°	+9.8	0.238	0.064	0.385	0.86	17.9	0.48
∥	+0.09		+15.0	>96°	+15.0	0.188	0.045			25.3	0.90
0.633 μm ⊥	0.00	18.4°		>96°	+5.7	0.308	0.051	0.444		18.0	0.55
∥	−0.69	20.0°		>96°	+11.7	0.243	0.013			36.6	0.94
0.8 μm ⊥	−0.13	25.4°		>96°	+4.6	0.396	0.039	0.589		20.6	0.62
∥	−0.60			>96°	+8.9	0.321	0.017			24.9	0.80
1.0 μm ⊥	−0.06	14.4°		>96°	+3.1	0.422	0.051	0.665		20.6	0.63
∥	−0.41	25.4°		>96°	+8.7	0.339	0.021			28.2	0.90

^aAngle of incidence = 30°.

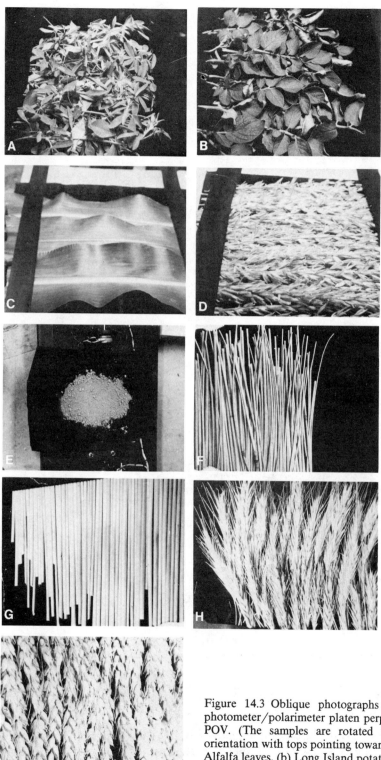

Figure 14.3 Oblique photographs of samples on photometer/polarimeter platen perpendicular to the POV. (The samples are rotated 90° for parallel orientation with tops pointing toward the source.) (a) Alfalfa leaves, (b) Long Island potato leaves, (c) corn leaves, (d) corn tassels, (e) farm soils, (f) rye stalks, (g) wheat stalks, (h) rye heads, and (i) wheat heads.

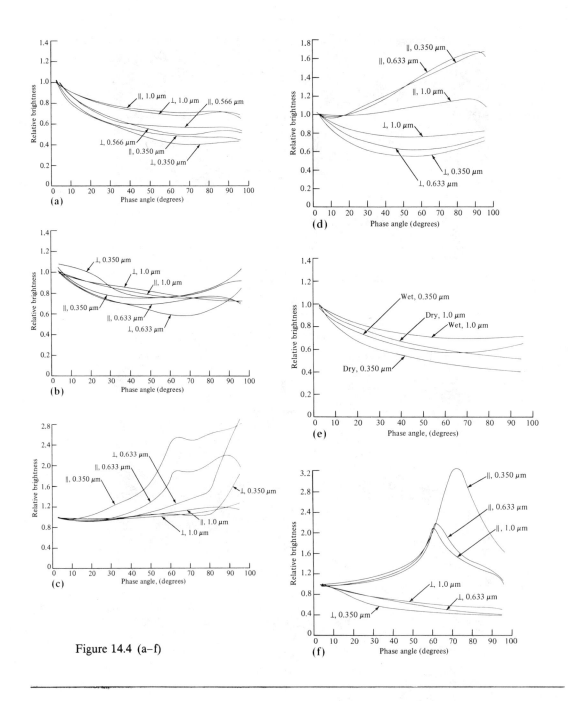

Figure 14.4 (a–f)

Spectral Photometry (First Stokes Parameter)

The bidirectional spectral photometry of the samples (Fig. 14.3) is graphically presented in Figs. 14.4a–14.4i and is summarized in Columns 7–12 of Table 14.1 for the conditions of the study. The graphs of Fig. 14.4 plot the relative brightness, normalized to unity at the minimum phase (α) of 3°, as a function of phase angle up

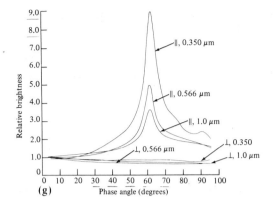
(g)

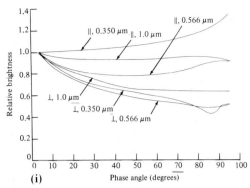
(i)

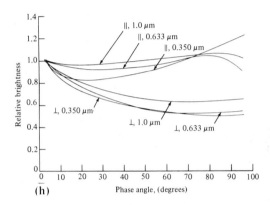
(h)

Figure 14.4 (a) Spectrophotometric curves for 30° incidence angle (first Stokes parameter) for alfalfa leaves. (b) Spectrophotometric curves for 30° incidence angle (first Stokes parameter) for Long Island potato leaves. (c) Spectrophotometric curves for 30° incidence angle (first Stokes parameter) for corn leaves. (d) Spectrophotometric curves for 30° incidence angle (first Stokes parameter) for corn tassels. (e) Spectrophotometric curves for 30° incidence angle (first Stokes parameter) for farm soils. (f) Spectrophotometric curves for 30° incidence angle (first Stokes parameter) for rye stalks. (g) Spectrophotometric curves for 30° incidence angle (first Stokes parameter) for wheat stalks. (h) Spectrophotometric curves for 30° incidence angle (first Stokes parameter) for rye heads. (i) Spectrophotometric curves for 30° incidence angle (first Stokes parameter) for wheat heads.

to the maximum 96°. The photometric curves are presented for the photomultiplier limiting wavelengths of 0.350 and 1.0 μm, and an intermediate wavelength of either 0.566 or 0.633 μm for the eight crop samples; the farm soil curves are presented only for the limiting wavelengths to simplify the figure.

Since leafy crop samples may be of arbitrary orientation in nature, the photometry in aircraft remote sensing will generally vary depending on these orientations. The procedure that would be used to cross the difficult gap between laboratory photometric data and aircraft field observations involves photometric weighting of crop and background first Stokes parameters at a particular wavelength, for a scene with i elements as in Chapter 5.

Thus, to explicitly determine the effect of orientation in the laboratory, previously noted as affecting photometry (Egan and Nowatzki, 1967, 1968), the leaves, stalks, or grain stems were oriented on the platen perpendicular or parallel to the POV. Oblique photographs of the samples presented in Fig. 14.3 are for the samples oriented perpendicular to the POV, except for the soil (Fig. 14.3e). The soil is symmetric with respect to the average normal to the surface within the resolution of the photometer. The samples are placed on a low reflectance background black flock cloth, although in most cases the background was not visible to the sensor field. The sensed areas are indicated by the fiducial marks.

The curves of Fig. 14.4 are discussed in the following paragraphs by crop type. The photometric characteristics in Table 14.1 have significance as indicated: geometric albedo of Column 7 is the reflectance of the sample compared to $MgCO_3$ at the wavelength and leaf or stalk orientation indicated in Column 1. The $MgCO_3$ is used in place of MgO because the large block necessary for the large photometric viewing area is more easily prepared and handled in the laboratory. This geometric albedo measurement is made at a phase angle of 3° and relates the photometric curves of Fig. 14.4, which have been normalized at 3°, to permit all curves to be shown to the same relative scale.

The microstructure factor of Column 8 is an index of the shadowing by the surface microstructural elements. The microstructure is that of a leaf, for instance, which cannot be seen with the unaided eye, but has an optical effect. It can be likened to a microscopic honeycomb, viewed end on. At near normal (perpendicular) viewing, the bottoms of the pores can be seen, but in off normal viewing, on a microscopic scale, the pore walls shadow the bottoms of the pores. Thus, the difference in the relative brightness between 3° and 6° phase angle is an index of this type of microstructural effect.

The diffuse albedo of Column 9 is simply a tabulation of the total reflectance graphically presented in Fig. 14.2, for comparison to the geometric albedo of Column 7.

The specularity of Column 10 is an indication of the specular (mirror) properties of a sample. If the sample were perfectly specular, like a mirror, a peak reflection would occur when the angle of incidence equaled the angle of reflection; in this case, a phase angle of 60° (angle of incidence = 30° = angle of reflection). However, the shiny specimens are not like mirrors, but are quasispecular, and produce a peaking at a phase angle of 60°. This peaking is expressed in terms of the relative increased brightness at 60° compared to the normalized brightness at 3°. When a sample is nonspecular, the column is left blank.

The total roughness of Column 11 is a relative index derived from the area under the photometric curves of Fig. 14.4 and is related to shadowing by macrostructural elements of the sample. Thus, by analogy, in a sun-illuminated forest where trees may be separated by approximately the crown diameter, as the viewing angle is increased from 0°, the shadows of the trees are seen in longer projection. This reduces the scene brightness, and lowers the photometric curve that would be obtained. Similarly, in this study, for leaves on the laboratory photometer platen, the broad-leafed specimens produce more visible shadowing at the larger phase angles than narrower-leafed specimens. The more shadowing that occurs at the larger phase angles, the lower the index (see Fig. 14.2); the macrostructure factor of Column 12 is another indication of this large scale shadowing in the sample, which decreases the relative brightness at 96° relative to that at 3°, and is so presented (an increase would indicate some degree of specularity). To put it another way, macrostructure is that surface roughness that produces shadowing at 96° phase angle, since the sensor would see no shadows at a 0° phase angle. The 3° phase angle is the closest to 0° obtained and is thus used as a reference for minimal shadowing.

The alfalfa leaves on the stalks (Fig. 14.3a) produce a higher relative reflectance in the infrared (1.0 μm) than at shorter wavelengths (Fig. 14.4a), with the leaves on the stems oriented parallel to the POV, at most phase angles. The higher relative 1.0-μm reflectance is probably the result of decreased microstructure effects of the surface at the longer wavelength (see Column 8 of Table 14.1) and dominates any macrostructure shadowing effect (Column 12). The generally increased relative reflectance for stalks with leaves parallel to the POV is probably the result of a macrostructure effect arising from the "V" angular shape of the leaves.

The potato leaves on the stalks (Fig. 14.3b) produce a more complicated photometry, with the parallel and perpendicular curves at the 0.633-μm wavelength tending to be lowest at the midrange of phase angles (Fig. 14.4b). This could be the result of contributions of macrostructure shadowing, roughness, and low albedo. The lack of a clear photometric differentiation as a function of wavelength would indicate that potato leaves do not show preferential structural effects. Slight differences in the leaf positions as a result of wilting during the measurement could vary the photometry.

The corn leaves (Fig. 14.3c) are quite different photometrically; they display a definite specularity (Column 10) for orientations parallel to the POV at the shorter wavelengths, with almost complete absence of microstructure and specularity at 1.0 μm (Fig. 14.4c). The flatness of the leaves could possibly produce multiple scattering because of the higher reflectance at 1.0 μm and wash out the specularity. The corn leaves oriented perpendicular to the POV have a greater roughness (lower roughness index, Column 11) and would not be expected to produce a specular effect. The corn leaves on the corn plant would produce a photometry lying between the perpendicular and the parallel extremes.

The corn tassels on stalks (Fig. 14.3d) display a distinct wavelength-dependent specularity for orientation parallel to the POV (Fig. 14.4d) as about half of them would ordinarily appear on mature corn plants when remotely viewed. This specularity would combine with that of the corn leaves, but because of color differences between the tassels, the combination would be highly wavelength-dependent. The roughness factor (Column 11) indicates the tassels to be smoother in the POV. The photometric curves for the tassels perpendicular to the POV would be approximately indicative of the other half of the tassels, and would be significant in data interpretation if the POV did not include the average normal to the surface (i.e., angular viewing).

The farm soil (Fig. 14.3e) shows a definite correlation for all wavelengths between moisture and photometry (Fig. 14.4e), geometric and diffuse albedo, and the macrostructure factor (also Columns 7, 9, and 12). Also, the microstructure (Column 8) and total roughness (Column 11) tend to decrease with moisture as expected. There is complete absence of any tendency toward specularity. The farm soil photometry enters into the overall photometry of a planted farm field remotely sensed, depending on the degree of crop ground cover.

The rye and wheat stalk samples presented a problem for laboratory photometric analysis; they normally occur almost vertically in crop fields, and form a relatively physically porous structure. It is expected that multiple scattering of light between the stalks contributes to the photometry. To examine the basic optical properties of the stalks, the two simple orientations of the stems parallel and perpendicular to the POV were used, with the average normal to the array lying in the POV.

The parallel rye and wheat stalks (Figs. 14.3f and 14.3g) are totally different photometrically from the previous crops in that they exhibit a pronounced specularity, decreasing with wavelength increase (Figs. 14.4f and 14.4g). The off 60° peak for the rye stalks at 0.350 μm is the result of the sample platen being tilted 5° off normal in the POV. The perpendicular orientation is of significance in interpretating data for tilted viewing or when the rye or wheat stalks have been damaged by wind or rain, producing off-vertical or near-horizontal orientation. The relative differences between the specular peaks (Column 10) serve to differentiate distinctly between the rye and wheat stalks in this orientation.

However, as noted, the stalks are nearly vertical as a crop, whereas the samples on the platen were horizontal. The high specularity here observed would lead to a low scattered radiation, and this together with shadowing effects would produce a very low brightness for rye or wheat stalks at phase angles higher than about 30°. For smaller phase angles, the vertical (parallel) component of scattered light would be a expected to be the highest, having the aforementioned wavelength dependence, with wheat being the brighter at shorter wavelengths. But at higher phase angles, the photometric contribution would be expected to be dominated by that of the mature heads because of their larger physical size and nearly horizontal position, and the contribution of the stalks would be small, so that we would not be too concerned with the photometric effect of the stalks for a healthy mature crop in good condition.

The orientation parallel to the POV of the rye and wheat heads (Figs. 14.3h and 14.3i, rotated to parallel orientation) are roughly similar, but differ in detail in the photometry as a function of wavelength (Figs. 14.4h and 14.4i); the parallel orientation rye relative brightness through the midrange of phase angles decreases with decreased wavelength whereas the parallel wheat is greatest for the shortest wavelength, with anomalies at other wavelengths (viz., 1.0 and 0.566 μm). These differences serve to distinguish the photometry of mature rye and wheat heads, which are vertical and would be combined photometrically in field observations with that of the stalks depending on the measurement geometry.

However, for the rye or wheat heads studied, which were obtained immediately prior to the harvest by combining, the heads were bent over to nearly horizontal positions varying between parallel and perpendicular to the POV. In remote sensing of such a field, the photometry would be a combination of the perpendicular orientation with the parallel, and the relative brightness as a function of phase angle would be decreased, but the 3° albedo increased (as a result of horizontal kernels). These characteristics then could serve as an index of a fully mature wheat or rye crop. There is also considerable variation in the parallel orientation microstructure factor (Column 8) between the rye and wheat that could be utilized in crop differentiation.

Spectral Polarimetry (Second and Third Stokes Parameters)

The bidirectional spectral polarimetry is graphically presented in Figs. 14.5a–14.5i and is summarized in Columns 2–6 of Table 14.1. The graphs of Fig. 14.5 plot the second Stokes (relative polarization) parameter as a function of phase angle referred to the POV including the average normal to the sample surface, thus making the third Stokes parameter U equal to zero. The range of phase angles, the wavelengths, and specimen orientations correspond to those used for the spectral photometry. Polarimetry as well as photometry depends on orientation for nonsymmetrical surfaces (Egan and Nowatzki, 1967, 1968).

As for photometry, an analogous procedure applies to the weighting of the equivalent polarimetric contributions of the i elements of a scene; this is described in Chapter 5.

In Table 14.1, the percent polarization [Eq. (14-1)] is tabulated; it is the quotient of the first and second Stokes parameters as obtained from the graphs. The polarimetric characteristics in Table 14.1 have significance as indicated: The minimum percent polarization (Column 2) is the value at 3° phase angle, with the convention that positive indicates that the dominant polarization intensity is perpendicular to the POV; the inversion angle (Column 3) is that angle where the percent polarization (also the second Stokes parameter) passes through zero; the maximum percent polarization (Column 4) and phase angle location (Column 5) denote the magnitude and location of the maximum percent polarization when a distinct peak exists; the percent polarization at 96°, the highest phase angle attained with the described equipment, is tabulated in Column 6, and this would be the maximum if no intermediate peak existed.

The alfalfa leaves (Fig. 14.3a) produce a maximum relative polarization that is decidedly wavelength dependent (Fig. 14.5a), and a minimum polarization that is positive for an orientation perpendicular to POV, and negative for an orientation parallel to POV, similar to an orientation effect noted on pumice by Egan and Nowatzki (1967). There is a correlation between albedo and relative polarization, the higher the geometric albedo (Column 7) and the lower the maximum polarization (Column 6); this is true within the experimental error for both the perpendicular and parallel components. Neither the inversion angles (Column 3) nor the minima (Column 2) show any clear trends with wavelength. The overall relative polarization from an alfalfa field is a photometrically weighted combination of perpendicular and parallel orientations.

The potato leaves (Fig. 14.3b) are different, producing an anomalously high relative polarization and percent polarization (Column 6) at 96° at 0.633 μm, compared to that at 0.350 μm (Fig. 14.5b). Here, the albedo–maximum-percent-polarization relationship does not hold in detail, probably the result of leaf structure. The structure phenomena (mentioned for the alfalfa leaves) is seen in Column 2, the minimum being positive for perpendicular orientation and negative for parallel orientation. The infrared polarization is low generally because of the high albedo. Here, too, the overall polarization from a potato field is a photometrically weighted combination of perpendicular and parallel orientations.

The corn leaves (Fig. 14.3c) have a considerable variation in relative polarization with wavelength (Fig. 14.5c), different from the previous specimens; it is noted that even though the second Stokes parameter at 96° at 0.350 μm for the parallel orientation is about the same as that at 0.350 μm for the perpendicular orientation, the percent polarization is less for 0.350 μm parallel than at 0.350 μm perpendicular because the photometry is higher for 0.350 μm parallel (Fig. 14.4c). The magnitude of relative polarization is twice as high at 0.350 μm for the corn leaves compared to the potato leaves, and 15 times as high as alfalfa at the higher phase angles. The structure factor which produces a positive polarization minimum for perpendicular orientation does not produce a negative one for perpendicular orientation at 0.8 and 1.0 μm, although the minimum is more positive for the perpendicular, showing a similar trend. A possible correlation may lie in the wavelength and orientation-dependent difference in the microstructure factor; the leaves are smoother at the

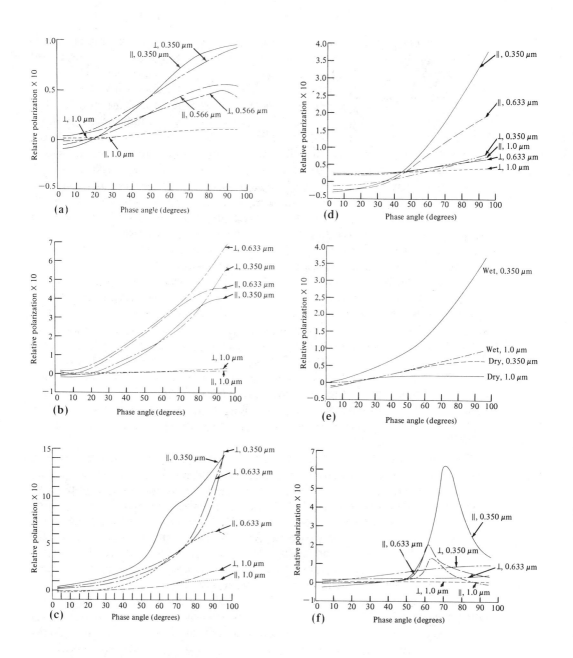

longer wavelengths for both orientations (resulting in roughly the same negative minima), but anisotropically rougher at the shorter wavelengths, being rougher for perpendicular than parallel. This greater roughness may orient high polarization facets in such a way as to produce a high overall percent polarization at high phase angles (see Egan and Nowatzki, 1968).

The corn tassels (Fig. 14.3d), appearing at the maturity of the crop, would be viewed in parallel and perpendicular orientations; a definite wavelength dependence

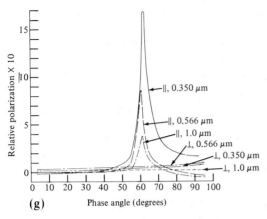

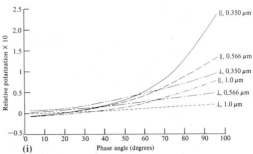

(i)

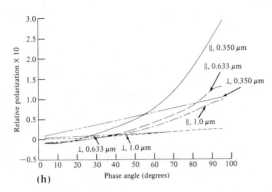

Figure 14.5 (a) Spectropolarimetric curves for 30° incidence angle (second Stokes parameter) for alfalfa leaves. (b) Spectropolarimetric curves for 30° incidence angle (second Stokes parameter) for Long Island potato leaves. (c) Spectropolarimetric curves for 30° incidence angle (second Stokes parameter) for corn leaves. (d) Spectropolarimetric curves for 30° incidence angle (second Stokes parameter) for corn tassels. (e) Spectropolarimetric curves for 30° incidence angle (second Stokes parameter) for farm soils. (f) Spectropolarimetric curves for 30° incidence angle (second Stokes parameter) for rye stalks. (g) Spectropolarimetric curves for 30° incidence angle (second Stokes parameter) for wheat stalks. (h) Spectropolarimetric curves for 30° incidence angle (second Stokes parameter) for rye heads. (i) Spectropolarimetric curves for 30° incidence angle (second Stokes parameter) for wheat heads.

of the maximum of the second Stokes parameter appears for the parallel orientation, inversely dependent on wavelength (Fig. 14.5d). The maximum of the photometry (Fig. 14.4d) did not indicate such a clearly defined trend. The maximum percent polarization (Column 6) is therefore a mixture of the two effects, and not the best indicator. The parallel orientation second Stokes parameter has a trend similar to alfalfa, but is about 5 times as great. The perpendicular orientation relative polarization is photometrically weighted with the parallel in remote sensing and would be useful in interpretation of angular viewing results. The relative optical contribution of the tassels (as compared to the corn leaves) in remote sensing would critically depend upon the geometry of viewing and the relative photometric weighting, being least for viewing at 0° to normal for the emergent rays for a field containing the crop.

The farm soil (Fig. 14.3e) shows the same analogous type of correlation between soil moisture as the photometry, but of much larger magnitude (Fig. 14.5e). The wet soil at 0.350 μm shows a second Stokes parameter which is 6 times that of the dry at the highest phase angle, and which is much greater than the difference evident in the first Stokes parameter between wet soil and dry soil in Fig. 14.4e. The wet soil would tend to have a highly polarizing water film on it to produce the higher polarization. Thus, the second Stokes parameter could possibly be used as a very sensitive indicator of soil moisture. The maximum percent polarization (Column 6) also correlates with soil moisture at all wavelengths, but is not quite as sensitive an indicator, as previously mentioned. The curves, of course, are not orientation dependent for an isotropic soil. However, if furrows are present, as in an actual field, the macrostructure will differ and both the photometry and polarimetry would not be isotropic. Also, such a surface structure, if not optically resolvable, could be discerned below the resolution limit of the viewing system. The polarimetry of a planted farm area will be affected by that of the soil in accordance with crop ground cover.

The comments regarding the photometry of the rye and wheat stalks in the Photometry section apply again; simple orientations of the stalks were used (parallel and perpendicular to the POV) to determine the fundamental polarimetric properties of the stalks.

The second Stokes parameter of the rye and wheat stalks (Figs. 14.5f and 14.5g) are roughly similar, but they are unique compared to any other sample in that they have two inversion angles for parallel orientation (Figs. 14.3f and 14.3g, rotated to parallel orientation) at certain of the wavelengths investigated. The two inversions differ between the rye and wheat stalks. This unusual polarization effect is evidently the result of a contribution of a refractive component from the stalks, which is generally of a negative sense. The structure, as well as the color, of the stalks affect this refractive component. It thus may be possible in remote sensing to use these effects to distinguish between varieties, as well as species, where such differentiation involves structural and spectral variations. Where wind or rain damage might bend over crops in a field, the perpendicular orientation would apply. The 10° shift of the 0.350 μm parallel peak on the rye stalks was previously mentioned as the result of a 5° tilt of the platen. The maximum relative polarization occurs for the parallel orientation of the wheat stalks (Fig. 14.5g) at 0.350 μm, whereas the rye under similar conditions is about half. The wheat stalks have more frequent inversions at higher angles than the rye at more of the wavelengths investigated (Column 3). With all of these unique differences, there would be little difficulty using the second Stokes parameter to distinguish these two grain crops from each other and from other crops if the parallel orientation existed naturally. (It appears that the percent polarization tends to cancel variations appearing in the second Stokes parameter.)

However, as with the photometry, the stalks are generally nearly vertical in nature. The photometric shadowing and specularity would influence the relative polarimetric effects. For phase angles less than about 30°, it would be expected that the major portion of the reflected radiation would have a low polarization, possibly entirely negative. For phase angles greater than 30°, there could be a scattered light component plus a refracted light component with a low polarization, possibly negative. The wheat and rye would be expected to exhibit relative polarization differences as a result of structural and color differences. As mentioned for the photometry, the polarimetric effect will be dominated by that of the mature kernels, and the stalks will contribute but a small amount polarimetrically under these conditions.

The rye and wheat heads second Stokes parameters (Figs. 14.5h and 14.5i), which appear at maturity, would be optically superimposed upon the characteristics of the stalks as remotely viewed, in accordance with the photometric weighting. The parallel orientation would dominate, with the relative polarization (while the heads are vertically oriented) being highest for the rye, but not much higher than the wheat at 0.350 μm. The lowest polarization exists for the 1.0-μm wavelength of the parallel components, the albedo being highest at this wavelength (Columns 7 and 9). The minima tend to be lower (more negative) for the rye heads than the wheat, and could serve as an additional identification index. However, as noted under the Photometry section, generally the heads are almost horizontal immediately prior to harvest by combining, and the orientations will vary between parallel and perpendicular to the POV. Under these conditions, the relative polarization will be decreased from the parallel, being a combination of the parallel and perpendicular, and thus lie somewhere between. This decreased relative polarization, too, can serve as an index of fully mature rye and wheat.

Comparison of Laboratory Data to Aircraft Data

As indicated in the Introduction, the Stokes parameters were recorded on a series of flights over Long Island farms on April 9, 1968. Some of the farm fields were showing initial signs of green, and other fields presented bare soil as evidenced by two aerial photographs taken during the measurements. Concurrent with the Stokes parameter measurements, a 16-mm triggered frame camera photographed the areas remotely sensed. Each frame on the 16-mm film corresponded to a marked position on the Stokes parameter recording. Color Ektachrome film was used in the camera. Because of lack of ground truth, the only agricultural areas that may be compared are the bare soil areas. The aircraft polarimeter/photometer operating at a wavelength of 0.566 μm with a 20-ft ground resolution recorded first and second Stokes parameters (at a phase angle of 78°, for instance) of 0.6 and 0.05, respectively. These aircraft data have a Stokes parameter accuracy of about $\pm 5\%$ because of a simplified electrical recording system necessitated by the limited space and power in the light aircraft. The laboratory first and second Stokes parameter data directly from laboratory graphs for a phase angle of 78° and source elevation angle of 60° are 0.46 dry (0.62 wet) and 0.040 dry (0.112 wet) for the same Long Island farm soil. One would deduce that the soil is somewhat moist, and more precise aircraft data would help. This is a simple example. However, to apply polarization to a highly diversified agricultural area, the data sampling rate must be increased. In the system described, the 13-Hz polarization sensing rate is too low for a light aircraft flying at 100 mph (161 km/hr); a sampling rate of 1000 Hz would be more appropriate, with the photometric response increased to 1 msec to yield a spatial resolution of 0.2 ft (6.1 cm). Further, the angular resolution of the photometer should be improved to 0.01° for this same aircraft flying at a 1000-ft altitude (305 m) to match the sampling rate to the spatial resolution. For a satellite, the polarimetric sampling could be accomplished concurrently with the photometric to yield a compatible system.

Subsequent to the aircraft polarimetric measurements of soil moisture, Curran (1978) used a photographic method to record polarized visible light, also to give an indication of soil moisture. Until that time, little was known of the effect of decreased soil moisture on decreased soil reflectance and the effect on polarization. Preliminary results on one soil type (Turbery Moor series peat) showed a strong correlation of the polarization of visible light with soil moisture; also, the maximum polarization occurred at phase angles between 90° to 100° in laboratory studies.

In further work on monitoring soil moisture, Curran (1979) used polarized panchromatic and false color infrared films. Laboratory spectral reflectance measurements of sandy loam and peat (Figs. 14.6a and 14.6b) suggest that the maximum differentiation of soil moisture states occurs at the longer wavelengths; the most

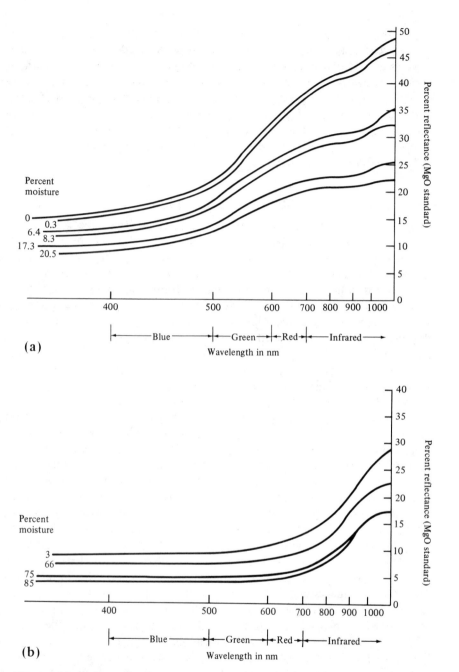

Figure 14.6 (a) Spectral reflectance curve for a sandy loam in six moisture states. (b) Spectral reflectance curve for peat in four moisture states.

sensitive region for sandy loam was at low moisture levels, and at high moisture levels in the peat. Polarized radiation successfully monitored a fairly wide range of soil moisture conditions (Fig. 14.7).

Another variant on the use of polarization to monitor agricultural crops was a model of plant canopy polarization response (Vanderbilt, 1980). The impetus was

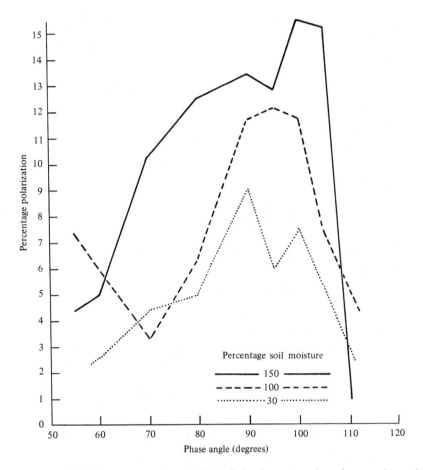

Figure 14.7 Laboratory study of light polarization at varying phase angles and soil moisture states.

given by the proposal to construct a multispectral resource sampler (MRS) using polarization to complement the thematic mapper. The study emphasizes the use of linear polarization from reflections off shiny leaves of crops such as wheat, corn, and sorghum. Although based on a model, it is concluded that crop development stage, leaf water content, leaf area index, hail damage, and certain plant diseases may be detected and monitored. Further, polarization may be used as an additional feature to discriminate between crops.

The model of Vanderbilt (1980), on which the aforementioned conclusions are based, extends the work of Egan (1970) on agricultural remote sensing by adding a probability density function $f_a(\theta, \phi)$ for leaves that defines the probability that a specific leaf area is oriented in a specific direction. The measurements of Egan (1970) yield the first and second Stokes parameters as a function of phase angle for the specific crops considered, and the technique for combining these to yield the overall photometry and polarization; however, weighting of the contributions of photometry and polarization is necessary, and would be given by the probability density function described by Vanderbilt (1980). To compute f_a, the incident light source

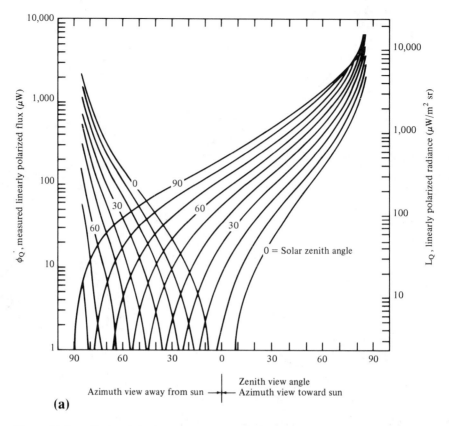

(a)

Figure 14.8 (a) Preheaded wheat canopy polarization response. Prior to heading the response is zero at the anti-solar point, the "hot spot," and increases with increasing zenith view angle. (b) Headed wheat canopy polarization response. After heading the response remains zero at the anti-solar point, is maximum at intermediate zenith view angles, and approaches zero for near-horizontal view directions, where heads and stems obstruct view of polarizing flag leaves.

beam is made polarized, and the canopy scattered and reflected radiation is measured with the analyzer oriented first parallel to the polarization direction of the incident beam, and then perpendicular. The specular flux is the difference between the two readings; in a particular direction,

$$f_a = |\mathbf{E}|(\text{LAI})\,k S_s P_s P_v\, \Delta\omega A_v / 2\phi_s \cos\theta_v$$

where $S_s = [(n-1)/(n+1)]^2$, LAI is the leaf area index, θ_v is the zenith angle of sensor, $\Delta\omega$ is the solid acceptance angle of sensor, P_s, P_v are the source and sensor canopy illumination probability, k is the proportionality factor, ϕ_s is the source azimuth angle, and A_v is the radiance scattered by canopy.

Although Vanderbilt (1980) gives explicit expressions for the specular and polarized components, the diffusely scattered is quite generally given as $A_v\,\Delta\omega$.

An important application is the use of *magnitude* of the polarized component (second Stokes parameter) to indicate the heading of wheat (percent polarization is a relatively insensitive indicator). This is shown in Figs. 14.8a and 14.8b; in the first

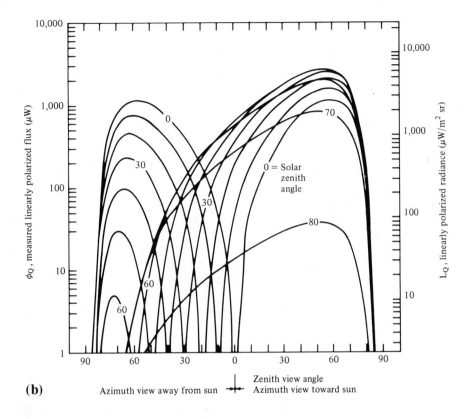

(b)

figure, prior to heading, the polarization response is zero at the anti-solar point (the "hot spot") and increases with increasing zenith sensor angle. After heading, the polarization at the anti-solar point remains zero, but the maximum occurs at intermediate zenith view angles, going to zero at the horizontal as a result of shadowing of the heads by the stems.

Vanderbilt (1980) expresses concern about the efficacy and feasibility of a satellite sensor measuring linear polarization through the earth's atmosphere; he notes that path radiance may (and will) alter the polarized radiation from a scene. Thus, the disentangling of the variables of polarization and photometry necessitate the use of a vector radiation transport model such as the Dave program described elsewhere in this text.

It might be noted that surface winds have the capability of reorienting the leaves in a plant canopy, thus having an effect on the interpretation of polarization.

Another point is that the leaf cuticle wax is birefringent, and therefore capable of producing circularly polarized light; Egan (1970) indicated that the amount of elliptically polarized light should be negligible because of the inherent random properties of vegetation.

Other plants besides wheat, corn, and sorghum have specularly reflecting leaves; these include coffee, sudan grass, banana, orange, sugar cane, and many forest species. The epicuticular wax causing the high polarization existed in half of the plant species tested (Schieferstern and Loomis, 1956).

Conclusions

The following conclusions are evident from the use of polarization and photometry for farm crop identifications: Spectral photometry and polarimetry yield information that can be used additionally to characterize farm crops with significantly large differentiation. The general representation of scattered light by Stokes parameters is the complete characterization of the photometric and polarimetric properties. With the use of the Stokes parameters, it appears that increased accuracy combined with increased area coverage is possible in a narrower wavelength band because additional information in the scattered light can be utilized. The sensing may aid crop identification with low resolution sensing systems.

The second Stokes parameter referred to a normalized first Stokes parameter, rather than the percent polarization, is more unique, informative, and easily obtained; the quotient of the two parameters, requiring a computer, generally diminishes the variation obtained. The second Stokes parameter may be better than the first in some cases as a more sensitive indicator, such as for soil moisture or between species.

Ultraviolet (0.350 μm) characteristics of the farm crops may in some instances be a better index than those obtained at longer wavelengths; however, atmospheric effects which affect all remote sensing, are greatest at the shorter wavelengths and would have to be taken into account. Finally, the reflectance spectra of farm crops are generally highly dependent upon viewing angle.

15

Forestry: Inventory, Identification

Remote sensing is becoming increasingly important in forest surveys. Forests cover nearly a third of the world's land area, and forest management relates to wood, forage, water, wildlife, and recreation. Since wood is the principal product of the forest, timber management by means of maintenance and improvement of existing stands is a significant requirement. The main approach to monitoring many of the world's forest conditions is by air photointerpretation of imagery. The photographic characteristics of shape, size, pattern, shadow, tone, and texture are used by photointerpreters for tree species identification. Stereo photography in scales ranging from 1:12,000 to 1:80,000 has been used for delineation of tree height; where images permitted evaluation of crown closure and diameter, an excellent correlation with cubic foot volume has been obtained for pine plots. Infrared imagery in the 4.5–5.5-μm band permitted tree species identification correct to 44% following a data reevaluation.

Another application of color and color infrared imagery is in the recognition and evaluation of tree vigor decline, such as that caused by attacks by bark beetles or Fomes annosus.

However, the problem of photointerpretation is more difficult in forestry for two reasons. A given farmland area is generally occupied by a single crop, and in combination with the soil (also visible), may form a uniform appearance on an aerial photograph (depending upon scale). In comparison, a forest stand similarly may be inhomogenous, often being occupied by an almost random mixture of many tree species; the result is that photoidentification of component species and timber classification becomes difficult. Further, most farm crops have been planted with a controlled spacing to receive a maximum sunlight, and thus exhibit a uniform appearance in small scale aerial photography. However, in forests many trees and plants are in the understory, being obscured in varying degrees by the overtopping tree crowns, and species identification is difficult if not impossible.

Thus, to date, optical imaging techniques (in particular, photography) has been used for the most part in forest foliage recognition. A fundamental departure was the successful use of nonimaging spectropolarimetry to delineate a red pine stand. Bidirectional spectropolarimetry and spectrophotometry appear to offer a new approach to forest inventory in low resolution nonimaging sensing systems.

In the following discussion, the results of laboratory studies of the Stokes parameters of various tree leaves will be presented (i.e., fresh and dried red pine needles, red pine bark, and leaves of black oak, sugar maple, and peach trees). Then the results of an aerial photometric/polarimetric survey of a red pine stand will be presented, and analyzed in terms of the laboratory data. Following will be an evaluation of the technique of combined spectrophotometry and polarimetry as a method to extend the value of remote sensing to nonhomogeneous stands.

Experimental Technique

A specially constructed large sample size spectropolarimetric/photometric goniometer was used to measure the bidirectional properties of selected leaf samples. Observations were made at wavelengths of 0.350, 0.433, 0.533, 0.566, 0.8 and 1.0 μm selected by narrow band interference filters with bandwidths (FWHM) of the order of 0.02 μm. The previously described airborne polarimeter/photometer was employed as the sensor.

The signal-to-noise ratio of the system was improved by using a Princeton Applied Research lock-in amplifier Model HR-8 to detect the ac portion of the photomultiplier signal.

The source of illumination (Fig. 15.1) was positioned at a fixed elevation angle of 60° (30° to the average surface normal n), selected as representative of the summer noontime sun at a latitude of 41°. The incident ray was collimated to about 1°, was of uniform brightness to 1% or less, and had a residual polarization of 0.1% or less. The incident ray illuminated the sample, and the scattered light was detected by the polarimeter/photometer sensor, which swung through an arc permitting a variation of the phase angle α between 3° and 96°.

The sample was located so that the average surface was at the system's center of rotation, and the average normal to the surface was in the plane of the phase angle.

Figure 15.1 Measurement geometry. All measurements made with the average normal to the surface *n* in the POV defined by the incident and emergent rays.

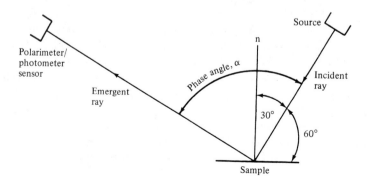

As indicated, the equipment was a large scale unit with a sensing area 60 mm in diameter and 3-m armlengths for the source and sensor.

The scattered light from the sample may be represented in the most general case by the bidirectional Stokes parameters (see Chapter 4). Here these parameters are

$$I = \langle A_x^2 + A_y^2 \rangle = A^2,$$

$$Q = \langle A_x^2 - A_y^2 \rangle,$$

$$U = \langle 2 A_x A_y \cos \gamma \rangle,$$

$$V = \langle 2 A_x A_y \sin \gamma \rangle,$$

where A_x, A_y are the amplitudes of the electromagnetic waves in mutually perpendicular directions, A^2 is the intensity, γ is the time phase angle between A_x and A_y, and $\langle \rangle$ indicates time averaging.

The Stokes parameters of tree leaves vary with the wavelength of the electromagnetic radiation and are generally dependent on the geometry of the source and sensor positions.

Briefly, the quantity I represents the intensity of the light reflected from the surface being viewed, and it is the result of the bidirectional photometric (reflectance) measurement. The quantity Q is the difference in the intensities of light measured in the mutually perpendicular directions used to specify A_x and A_y. The mutually perpendicular directions used in the experimental data are those parallel and perpendicular to the plane defined by the plane of the incident radiation and observational direction. The plane so defined is the plane of vision (Fig. 15.1). In choosing the reference plane as the POV, it has been found that, because of symmetry, $U = 0$. (The quantity U indicated the excess of light in the $+45°$ direction, over the $-45°$ direction, relative to the POV.) V is a quantity that indicates the amount of circularly polarized light; it is generally negligibly small except for reflection from good conductors such as metals, and as such can be neglected for foliage.

Measurements made with the polarimeter/photometer of the relative brightness I of the scattered light and of the polarization difference Q will be described. The relative brightness of the samples was referred to $MgCO_3$ at a $3°$ phase angle to yield a geometric albedo.

As an additional check on the spectral reflectance of the samples, measurements were made at 2% wavelength increments in the wavelength range between 0.25 and 2.3 μm on a Perkin–Elmer Model 12 monochrometer with a Gier–Dunkle absolute integrating sphere reflectometer (Model AlS-6B). Single layered samples were mounted on a diffuse white backing in calibrated quartz-covered containers and were compared to MgO standards in the same containers.

The samples were locally (Long Island) obtained fresh needles of red pine (*Pinus resinosa*) and fresh leaves of black oak (*Quercus velutina*), sugar maple (*Acer saccharum*), and peach (*Prunus persica*, var. Elberta), representative of conifer and deciduous species. Since the bark of the red pine may be visible through the crown, it was measured as well. Further, for the red pine needles, the optical effect of drying at about 40% relative humidity for 1 month was evaluated.

The number of samples needed of each species depended on their physical size, a sufficient number being required to cover approximately double the polarimeter/photometer sensing area.

Table 15.1 Synopsis of Spectrophotopolarimetric Characteristics of Foliage[a]

Material (wavelength region) (1)	Minimum percent (2)	Inversion location (3)	Maximum percent (4)	Maximum location (5)	% Polarization at 96° (6)	Geometrical albedo (7)	Microstructure factor 3° to 6° (8)	Diffuse albedo (9)	Specularity referenced to 3° (10)	Total Roughness (11)	Referenced to 3° macrostructure factor (96°) (12)
Fresh red pine needles											
0.350 μm ⊥	+0.1			>25°	+30.8	0.057	0.090	0.106		21.9	0.83
0.350 μm ∥	−1.6	28.5°	+10.5	67°	+6.5	0.030	0.033		2.2	40.7	1.29
0.433 μm ⊥	+0.05			>96°	+30.7	0.057	0.075	0.108		21.3	0.89
0.433 μm ∥	−0.8	20.0°	+15.3	67°	+13.2	0.031	0.035		2.04	38.5	1.25
0.533 μm ⊥	0.0			>96°	+44.6	0.099	0.110	0.147		20.2	0.75
0.533 μm ∥	−0.81	15.2°	+19.7	63°	+14.3	0.071	0.033		1.35	31.4	1.03
0.566 μm ⊥	0.0			>96°	+44.7	0.090	0.108	0.135		21.0	0.76
0.566 μm ∥	−0.67	14.2°	+19.9	63°	+14.8	0.062	0.027		1.40	32.2	1.07
0.633 μm ⊥	+2.59			96°	+53.8	0.067	0.118	0.124		19.9	0.77
0.633 μm ∥	−1.41	20.4°	+25.4	64°	+18.5	0.042	0.041		1.68	35.8	1.32
0.8 μm ⊥	−0.49	21.5°		>96°	+6.4	0.463	0.054	0.681		23.4	0.84
0.8 μm ∥	+0.33		+4.1	65°	+1.0	0.420	0.013		1.0	27.4	0.80
1.0 μm ⊥	−0.48	21.8°		>96°	+8.2	0.443	0.052	0.692		23.0	0.84
1.0 μm ∥	+0.12		+4.4	64°	+2.0	0.405	0.009		1.01	28.9	0.79
Dry red pine needles											
0.350 μm ⊥	+0.11			>96°	+66.3	0.072	0.096	0.118		20.1	0.67
0.350 μm ∥	−0.84	22.5°	+19.8	70°	+19.1	0.072	0.073		1.59	31.7	1.15
0.433 μm ⊥	+0.37			>96°	+22.1	0.086	0.090	0.167		18.5	0.62
0.433 μm ∥	−0.72	21.2°	+11.2	67°	+10.2	0.062	0.062		1.34	28.6	0.97
0.533 μm ⊥	−0.30	8.5°		>96°	+37.9	0.118	0.110	0.240		15.6	0.42
0.533 μm ∥	−0.61	15.8°	+7.5	69.5°	+9.2	0.097	0.040		1.13	27.2	0.87
0.566 μm ⊥	−0.26	7.5°		>96°	+44.6	0.132	0.160	0.295		12.6	0.47
0.566 μm ∥	−0.83	13.8°	+17.0	64°	+10.8	0.118	0.073		0.95	23.5	0.85

0.633 μm	⊥	−0.31			> 96°	+ 30.1	0.160	0.080	0.234		17.2	0.53
	∥	−0.57	7.0°	+ 15.6	63°	+ 9.9	0.136	0.023		1.01	27.4	0.88
0.8 μm	⊥	+ 0.36	18.9°		> 96°	+ 5.3	0.462	0.053	0.630		20.6	0.65
	∥	+ 0.20		+ 4.4	73.5°	+ 2.3	0.431	0.002		1.0	28.7	0.78
1.0 μm	⊥	−0.99			> 96°	+ 12.8	0.567	0.062	0.754		21.1	0.66
	∥	+ 0.27	19.4°	+ 4.6	65°	+ 0.8	0.502	0.007		1.0	28.1	0.80
Red pine bark												
0.350 μm	⊥	−0.76	10.8°	+ 36.8	~ 95°	+ 36.8	0.063	0.056	0.157		34.5	1.53
0.433 μm	∥	−0.91	10.5°	+ 40.0	92°	+ 38.9	0.062	0.060	0.162		33.5	1.51
0.533 μm	⊥	−0.70	10.0°		> 96°	+ 36.6	0.077	0.044	0.186		34.1	1.51
0.566 μm	∥	−0.87	10.3°		> 96°	+ 34.1	0.084	0.052	0.197		33.7	1.46
0.633 μm	⊥	−0.79	10.7°		> 96°	+ 31.0	0.104	0.042	0.256		32.7	1.36
0.8 μm	∥	−0.37	10.7°	+ 15.3	~ 96°	+ 15.3	0.227	0.034	0.545		28.9	0.99
1.0 μm	⊥	−0.85	15.5°	+ 12.4	~ 96°	+ 12.4	0.343	0.040	0.781		28.1	0.89
Black oak leaves												
0.350 μm	⊥	0.0			> 96°	+ 73.0	0.012	0.000	0.235		104.0	13.6
	∥	−1.9	16.0°		> 96°	+ 72.3	0.013	0.000			67.3	4.25
0.433 μm	⊥	+ 0.43			> 96°	+ 62.2	0.011	0.053	0.233		131.0	15.9
	∥	−1.12	13.0°		> 96°	+ 56.0	0.012	0.064			77.0	5.04
0.533 μm	⊥	−0.09	17.1°	+ 67.3	96°	+ 67.3	0.042	0.017	0.270		49.8	4.3
	∥	−1.05	16.5°		> 96°	+ 63.2	0.044	0.017		1.5	37.4	1.69
0.566 μm	⊥	−0.02	10.2°		> 96°	+ 68.9	0.041	0.019	0.263		48.8	4.7
	∥	−0.56	15.0°		> 96°	+ 60.4	0.043	0.019		1.55	38.3	1.72
0.633 μm	⊥	0.00			> 96°	+ 93.3	0.017	0.023	0.253		110.0	8.14
	∥	−0.47	11.0°		> 96°	+ 55.8	0.018	0.023		2.4	50.1	3.19
0.8 μm	⊥	−1.40	31.8°		> 96°	+ 21.1	0.178	0.010	0.953		31.3	1.28
	∥	−0.93	26.5°		> 96°	+ 9.1	0.173	0.010		1.0	29.8	0.72
1.0 μm	⊥	−0.91	29.0°		> 96°	+ 23.1	0.443	0.012	1.002		30.8	1.28
	∥	−0.96	29.8°		> 96°	+ 7.9	0.427	0.012		1.0	28.4	0.76

a Angle of incidence = 30°.

(continued)

Table 15.1 Synopsis of Spectrophotopolarimetric Characteristics of Foliage[a] (*continued*)

Material (wavelength region) (1)	Minimum percent (2)	Inversion location (3)	Maximum percent (4)	Maximum location (5)	% Polarization at 96° (6)	Geometrical albedo (7)	Microstructure factor 3° to 6° (8)	Diffuse albedo (9)	Specularity referenced to 3° (10)	Total Roughness (11)	Referenced to 3° macrostructure factor (96°) (12)
Sugar maple leaves											
0.350 μm ⊥	+3.47		+14.3	88.5°	+12.5	0.046	0.000	0.115		30.9	1.3
0.350 μm ∥	-2.00	24.3°	+31.8	90.5°	+26.5	0.039	0.000			37.7	1.9
0.433 μm ⊥	+1.48			>96°	+80.5	0.038	-0.008	0.117		36.0	1.5
0.433 μm ∥	+0.54			>96°	+83.5	0.022	-0.003			49.5	2.9
0.533 μm ⊥	+0.38		+33.5	~96°	+33.5	0.069	-0.005	0.163	1.1	32.4	1.05
0.533 μm ∥	-2.81	9.9°		>96°	+50.8	0.052	0.000			36.7	1.84
0.566 μm ⊥	+0.43			>96°	+41.1	0.078	-0.003	0.140	1.1	31.4	1.0
0.566 μm ∥	+1.70			>96°	+44.8	0.056	0.000			39.2	2.0
0.633 μm ⊥	+0.74			>96°	+41.2	0.039	-0.001	0.132	1.2	32.5	1.21
0.633 μm ∥	0.00	0°		>96°	+44.8	0.028	0.015			48.5	2.86
0.8 μm ⊥	0.00			>96°	+4.6	0.472	0.000	0.874		28.7	0.76
0.8 μm ∥	-0.01	14.5°		>96°	+6.4	0.422	0.000			30.2	1.09
1.0 μm ⊥	-0.42	18.9°		>96°	+10.5	0.380	0.000	0.886		29.7	0.79
1.0 μm ∥	-0.32	22.0°		>96°	+9.7	0.432	0.000			30.3	1.11
Fresh peach leaves											
0.350 μm ⊥	+1.38			>96°	+39.5	0.026	0.057	0.124		22.6	0.60
0.350 μm ∥	-3.15	31°		>96°	+41.3	0.023	0.004		1.48	37.9	1.22
0.433 μm ⊥	+6.42		+77.3	92.5°	+75.7	0.023	0.065	0.121		15.7	0.48
0.433 μm ∥	-4.25	25.3°	+59.3	~96°	+59.3	0.021	0.027		1.59	37.4	1.43
0.533 μm ⊥	+1.88			>96°	+13.0	0.073	0.042	0.174		18.3	0.47
0.533 μm ∥	-1.89	26.8°	+33.3	~96°	+33.3	0.067	0.019		0.97	26.7	0.61
0.566 μm ⊥	+1.38			>96°	+8.0	0.071	0.052	0.164		18.1	0.49
0.566 μm ∥	-1.03	29.5	+19.1	~96°	+19.1	0.066	0.020		1.03	27.9	0.76
0.633 μm ⊥	+3.84			>96°	+19.9	0.017	0.082	0.134		18.0	0.62
0.633 μm ∥	-3.19	24.7°	+37.7	90°	+34.5	0.020	0.020			30.6	1.32
0.8 μm ⊥	+0.61			>96°	+1.8	0.345	0.031	0.927		22.1	0.76
0.8 μm ∥	-0.39	34.3°		>96°	+6.7	0.417	0.002		0.95	26.7	0.51
1.0 μm ⊥	+0.18			>96°	+1.06	0.423	0.043	0.919		21.9	0.75
1.0 μm ∥	-0.38	28.7°	+6.1	~96°	+6.1	0.403	0.005			26.9	0.50

[a] Angle of incidence = 30°

Total Reflectance

The curves of total reflectance between 0.25 and 2.3 μm are presented in Figs. 15.2a and 15.2b and tabulated as diffuse albedo in Column 9 of Table 15.1. These curves permit examination of the overall spectral characteristics of the leaf specimens under these particular sample mounting and measurement system conditions.

The fresh red pine needles (Fig. 15.2a) have a relatively low reflectance in the visible range, with a slight increase in the green at about 0.55 μm, and a chlorophyll absorption band at 0.68 μm. There is a strong increase in the near infrared region, with dips at 1.45 and 1.95 μm, the result of water absorption bands. When the red

Figure 15.2 (a) Spectral reflectance curves for fresh and dried red pine needles and bark. (b) Spectral reflectance curves for black oak, sugar maple, and peach leaves.

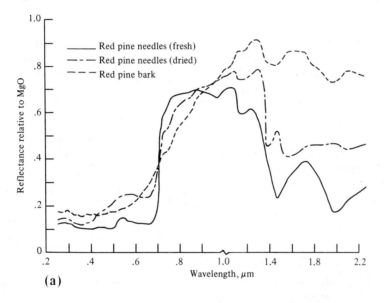

(a)

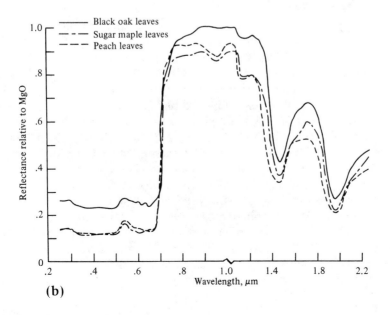

(b)

pine needle are dried, they tend to increase in visible and particularly in red reflectance and have a higher reflectance in the 1–2-μm region (Fig. 15.2a); the water absorption bands at 1.45 and 1.95 μm are still apparent but are less deep.

The red pine bark (Fig. 15.2a) has a trend similar to that of the dried needles, but the bark reflectance is higher than the needle reflectances between 1 and 2 μm.

The reflectances of sugar maple and peach leaves are almost the same (Fig. 15.2b) and present a severe problem in spectral differentiation based on spectrometry alone; the black oak leaves (Fig. 15.2b) have about twice as high reflectance as the sugar maple and peach in the visible and a higher reflectance at about 1 μm. Thus the deciduous leaves have a higher reflectance than the conifer studied. The water absorption bands for the former are clearly evident in Fig. 15.2b.

Spectral Photometry (First Stokes Parameter)

The bidirectional spectral photometry is shown in Figs. 15.3a–15.3f and summarized in Columns 7–12 of Table 15.1. The graphs of Fig. 15.3 show the relative brightness, normalized to unity at the minimum phase angle α of 3°, as a function of phase angle up to the maximum of 96°. Photometric curves are presented for the photomultiplier limiting wavelengths of 0.350 and 1.0 μm and an intermediate wavelength chosen to characterize species differentiation.

A previous study of red pine needles indicated that the photometry is affected by orientation (Egan, 1968). An orientation effect may also be expected for deciduous leaves. Hence measurements are made for orientations of the long axes of the needles or leaves (direction of the stem) parallel ($\parallel$) or perpendicular ($\perp$) to the POV defined by the plane of the phase angle α. The observed photometry of the leaves on a tree is a combination of the $\parallel$ and $\perp$ orientations.

The Fig. 15.3 curves are discussed below according to leaf type. The photometric characteristics shown in Table 15.1 have the following significance: the geometric albedo of Column 7 is the reflectance of the sample compared to $MgCO_3$ at the wavelength and orientation indicated in Column 1; the microstructure factor of Column 8 is an index of the shadowing by surface microstructural elements, which appears as a differential effect at near-normal incidence, being the difference between the relative brightness at 3° and 6°; the diffuse albedo of Column 9 was discussed above under "Total Reflectance"; the specularity of Column 10 indicates the specular properties of the sample, which would be a peaking of the photometric curves at a phase angle of about 60° (where the angle of incidence equals the angle of reflection), expressed in terms of the normalized relative brightness at 3°; the total roughness of Column 11 is a relative index, inversely proportional to the total roughness, derived from the area under the photometric curves (the rougher the sample, the more shadowing at large phase angles and, generally, the lower the index); the macrostructure factor of Column 12 indicates large scale shadowing in the sample, which decreases the relative brightness at 96° relative to 3°, and is so presented (an increase indicates some degree of specularity).

The $\parallel$ and $\perp$ orientations of fresh red pine needles (Fig. 15.3a) are distinctly different. The needles are analogous to cylinders and, when oriented $\parallel$, produce a mirror-type reflection at a 60° phase angle (see also Column 10), with the greatest effect at a wavelength where the needles have a relatively low reflectance (0.350 μm). Penetration of light into the needles at 1.0 μm could wash out the $\parallel$ specularity. The $\perp$ orientation is more diffuse, as is evident from the higher microstructure factor (Column 8), than is the $\parallel$. The diffuse albedo (Column 9) is higher than the

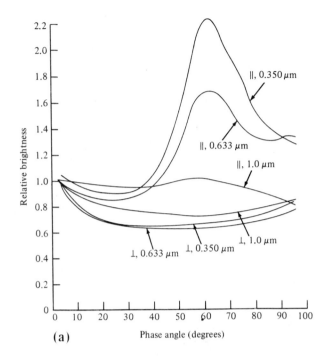

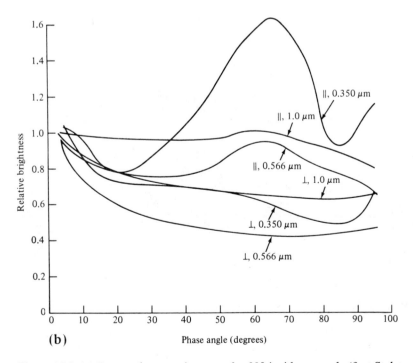

Figure 15.3 (a) Spectrophotometric curves for 30° incidence angle (first Stokes parameter) on fresh red pine needles. (b) Spectrophotometric curves for 30° incidence angle (first Stokes parameter) on dried red pine needles.

(*continued*)

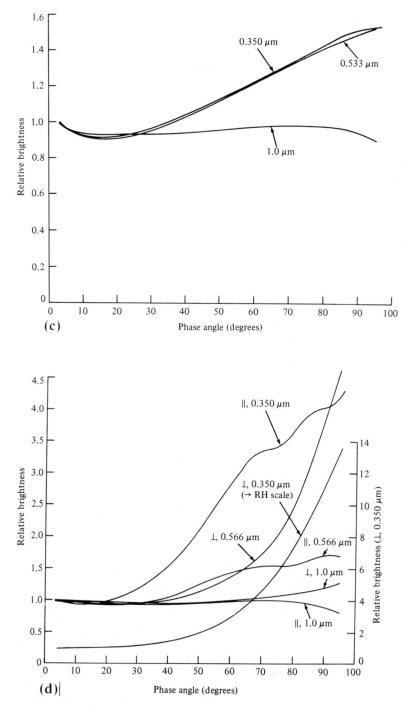

Figure 15.3 (c) Spectrophotometric curves for 30° incidence angle (first Stokes parameter) on red pine bark. (d) Spectrophotometric curves for 30° incidence angle (first Stokes parameter) on black oak leaves.

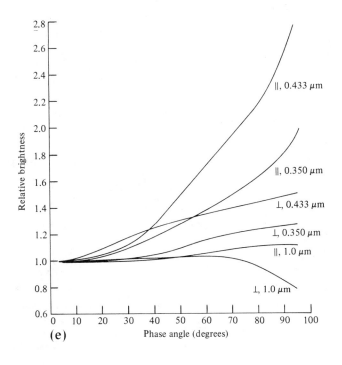

(e)

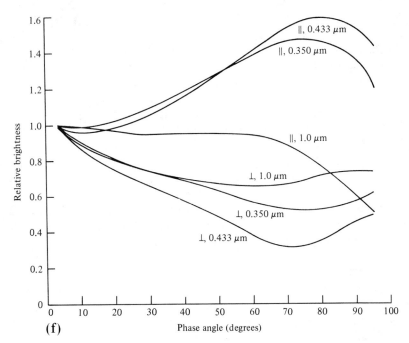

(f)

Figure 15.3 (e) Spectrophotometric curves for 30° incidence angle (first Stokes parameter) on maple leaves. (f) Spectrophotometric curves for 30° incidence angle (first Stokes parameter) on peach leaves.

geometric (Column 7) since all the scattered light is collected by the integrating sphere used for the former. The $\perp$ orientation has more shadowing at large phase angles, as is evident from the lower macrostructure factor (Column 12). The $\perp$ orientation is rougher than the $\parallel$ (lower values in Column 11). The $\perp$ geometric albedo is greater than the $\parallel$ (Column 7) because of more diffuse scattering by the $\perp$ orientation.

For a comparison to aircraft remote sensing observations, the same sensor used in the laboratory was mounted in a light aircraft and photometric and polarimetric measurements were made of a red pine stand.

In order that the initial measurements of ground foliage not be burdened by effects of inhomogeneous "ground truth" samples, a nearby pure stand of red pine was sought. Through the cooperation of various forestry groups, what was selected was a pure stand of red pine (Pinus resinosa) of approximately 20 acres located 7 miles northeast of New Haven, Connecticut in the New Haven watershed area. The area is shown in Fig. 15.4 as plots RP-1 through RP-7, and is owned by the New Haven Water Company. The seven areas were planted in 1928; plots 1–4 were planted on 8-ft centers, and 5 and 6 on 6-ft centers, and plot 7 on 8-ft centers. Plots 1–4 and plot 7 were lightly thinned (less than 10%) in 1961 and 1962, to remove trees having low quality form. The trees at the time of observation were 50–55 ft in height with a diameter breast high (DBH) averaging 8.5 in. The elevation of the area is 260 ft above sea level. During the life of the stand, there had been some root rot (Fomes annosus), which resulted in some spotted mortality. The forest area surrounding the red pine plantation is typical mixed hardwood composed of oak, beech, maple, yellow poplar, and black birch. Since the observations were made in December 1967 and January 1968, the mixed hardwoods were defoliated.

The observations of the red pine areas (Fig. 15.4) were made by flying over the various plots contained within the specified area, towards the sun and away from the sun. The sensor was oriented in the vertical plane by an attached level, resulting in the sensor optical axis being located in the POV. Subsequently, passes were flown back and forth across the red pine stand with the sensor oriented (tilted) in the POV to various viewing angles from 0° to about 70°. Data for I and Q were automatically recorded, as well as the photographic exposures and sensor angular information. The plane of polarization information was manually recorded from the oscilloscope by the operator. An example of the raw data and photographic display can be seen in Fig. 15.5 for the flight path shown in Fig. 15.4.

The photometric record (I) is at the right-hand side and the second Stokes parameter to the left with the computed percent polarization. The photographic record, at the extreme right (camera pips at the center), shows corresponding frames that were recorded by the polarimeter/photometer. A sensor angular pip occurs only when the sensor angle is changed, and does not appear for a fixed sensor tilt.

The aerial photometric behavior of the red pine stand observed at wavelengths of 0.533 and 1.0 μm is shown in Fig. 15.6. At that time (the winter sun elevation was 30°, with a specular peak expected at about 120°), the relative brightness was higher at 1.0 μm, and the overall photometric curve did not show a specular tendency. The 0.533-μm data tended to a specular peak, however, as expected from a contribution of $\parallel$ oriented needles. The low winter sun angle might be expected to produce more shadowing and a resultant lowering of the relative brightness at large phase angles.

In the laboratory studies, when the red pine needles are dried (Fig. 15.3b), the specularity (Column 10) for the $\parallel$ orientation decreases at wavelengths of 0.633 μm and shorter below that for the fresh needles, probably a result of decreased

Figure 15.4 Section of aerial photograph of red pine stand and adjacent area. (7 miles N.E. of New Haven, Connecticut). From USDA Agricultural Stabilization and Conservation Map CGN-5DD-185.

contribution by the water in the needles. The 0.350-μm photometric curve for $\parallel$ orientation has a dip at 20° and a specular maximum, a pattern that is followed in a diminished form at 0.566 μm. Similar trends appeared with the fresh pine needles but are less pronounced. There is nevertheless a differentiation between the $\parallel$ and $\perp$ curves for the dried pine needles, although it is less than for the fresh.

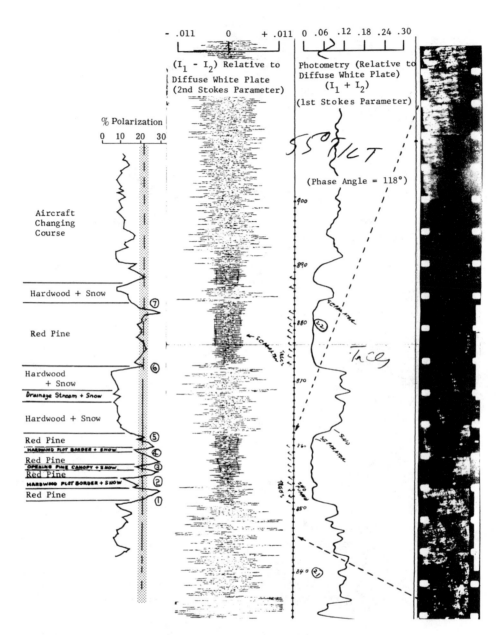

Figure 15.5 Example of raw data display, polarization reduction, and photographic correlation.

The dried red pine has higher geometric and diffuse albedoes (Columns 7 and 9) than the fresh, with the $\perp$ orientations, as above, having the higher geometric albedo, except at 0.350 μm. The total roughness and external microstructure factors (Columns 11 and 8) are, as might be expected, comparable for the fresh and the dried needles, but the macrostructure factor (Column 12) tends to be less. This indicates decreased shadowing, which might be the result of increased multiple scattering due to the increased albedo. Thus the dry red pine needles are significantly different photometrically from the fresh.

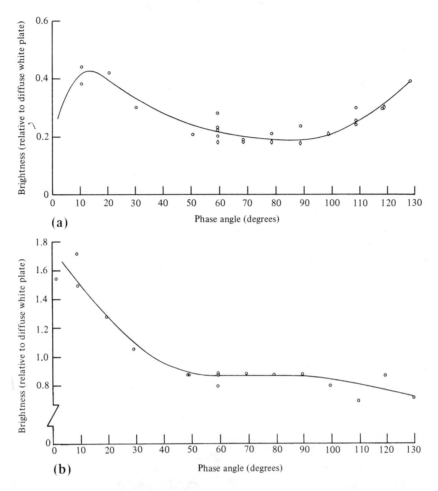

Figure 15.6 (a) Photometry of New Haven red pine area: green light, $\lambda = 0.533$ μm. (b) Photometry of New Haven red pine area: infrared light, $\lambda = 1.0$ μm.

The red pine bark (Fig. 15.3c) has a photometry decidedly different from that of the needles. Unusually high reflectance at the high phase angles (Column 12), indicating a lack of shadowing, suggests bark facets appropriately oriented to cause this effect. The effect is washed out at 1.0 μm, probably the result of the higher reflectance (Columns 7 and 9) and increased multiple scattering. The total roughness factor (Column 11) is large, indicating limited shadowing, and so is the microstructure factor (Column 8).

The overall photometry from a healthy red pine stand is dominated by the scattered radiation return from fresh needles, possibly, depending upon crown cover, with some contribution from the bark and ground cover. Another point: Since the photometry of the red pine needles varies as a function of wavelength, the spectrophotometry of the needles, when remotely viewed with a low resolution, nonimaging sensor, generally depends on the phase angle.

Photometric curves for the black oak leaves are shown in Fig. 15.3d; at the shorter wavelengths the shapes differ drastically from those for the fresh and dried red pine needles (Figs. 15.3a and 15.3b). The difference is a sheen (i.e., a strong specular type of reflection combined with diffuse scattering) that occurs at high phase angles at the shorter wavelengths and produces an apparent macrostructure factor (Column 12), indicating the complete opposite of shadowing. The sheen is probably the result of the extremely small microstructure factor (Column 8), particularly at the 0.350-μm wavelength. The microstructure factor is also relatively small at 1.0 μm, but the higher geometric and diffuse albedos (Columns 7 and 9) at this wavelength produce an increased scatter that washes out the sheen effect. Because of the sheen effect the apparent total roughness factor (Column 11) is extraordinarily high (indicating a smooth surface) for the shortest wavelengths. There is a trace of specularity at intermediate wavelengths (Column 10). The oak leaves were overlapped when they were placed on the sample platen to completely fill the field of view of the photometer. This overlap caused a tilt of approximately 20° away from the source in the ⊥ orientation, which resulted in an incident angle for the light of about 50°. A specular peak for the shorter illuminating wavelengths could then be expected to occur for the ⊥ orientation at about 100°. A trend to this peak, as a phase angle of 96° is approached, is exhibited in Fig. 15.3d.

The sheen effect is strongest for the ⊥ orientation because the veins in the leaves are dominantly ⊥ (in the direction of the stem, producing a smoother surface in that direction), and a shadowing effect apparently occurs for the ∥ orientation, reducing the sheen effect in the ∥ orientation.

Thus the photometric characteristics of the hardwood black oak leaves are radically different from those of the red pine needles.

The photometry of the sugar maple leaves (Fig. 15.3e) resembles that of the black oak leaves (Fig. 15.3d) in that the leaf sheen effect occurs at the shorter wavelengths. The amount of sheen for the sugar maple is considerably less, however, and it occurs for the ∥ orientation, not for the ⊥ as with oak. The microstructure factor for the sugar maple leaves (Column 8) is also characteristically small, as it is for the black oak leaves, in some instances even being negative. The reduced sheen effect for maple leaves appears to be the result of surface characteristics of the leaves in the ∥ orientation, and produces a large apparent macrostructure factor (Column 12) as well as a large roughness factor, indicating a smooth surface (Column 11). The large albedos (Columns 7 and 9) at the longer wavelengths (0.8 and 1.0 μm) again wash out the surface effects, presumably the result of external multiple scattering. There is a trace of specularity at the intermediate wavelengths (Column 10).

The photometry of the peach leaves (Fig. 15.3f) is a combination of that exhibited for the red pine (Fig. 15.3a) and for the hardwoods (Figs. 15.3d and 15.3e), showing characteristics of shadowing and reduced sheen effect at the shorter wavelengths. Thus the microstructure factor (Column 8) is greater than that of the hardwoods but less than that of the red pine. It is highest for the ⊥ orientation since the leaves are slightly folded parallel to the longitudinal vein. The specularity (Column 10) is a peaking of the curves rather than an indication of a specular maximum at a phase angle of 60°. The generally low macrostructure factor (Column 12) indicates the dominance of shadowing at large phase angles. The albedos (Columns 7 and 9) are highest at the longest wavelengths, which produces a relatively large multiple scattering. The 70° dip in the 0.433-μm ⊥ orientation photometric curve (Fig. 15.3f) is largely the result of shadowing.

Spectral Polarimetry (Second Stokes Parameter)

The bidirectional spectral polarimetry is illustrated in Figs. 15.7a–15.7f and summarized in Columns 2–6 of Table 15.1. The graphs of Fig. 15.7 plot the second Stokes parameter (relative polarization) as a function of phase angle. The range of

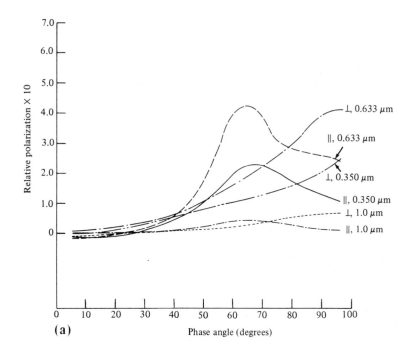

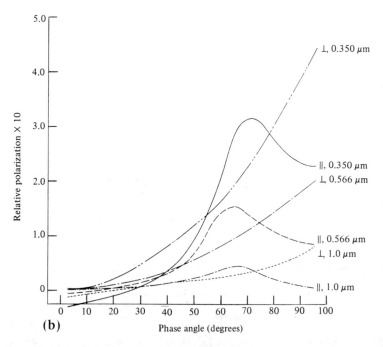

Figure 15.7 (a) Spectropolarimetric curves for 30° incidence angle (second Stokes parameter) on fresh red pine needles. (b) Spectropolarimetric curves for 30° incidence angle (second Stokes parameter) on dried red pine needles.

(*continued*)

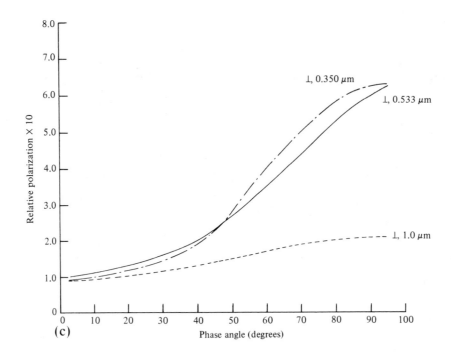

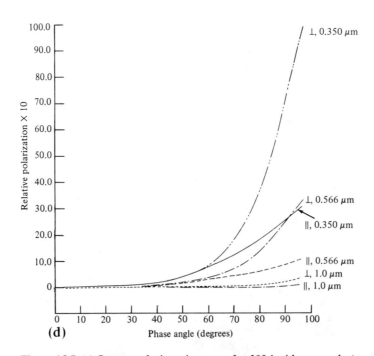

Figure 15.7 (c) Spectropolarimetric curves for 30° incidence angle (second Stokes parameter) on red pine bark. (d) Spectropolarimetric curves for 30° incidence angle (second Stokes parameter) on black oak leaves.

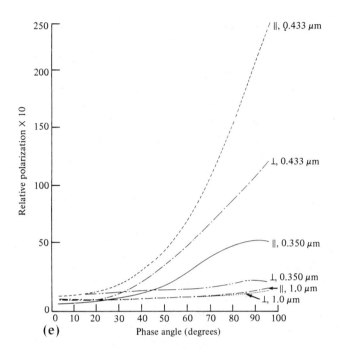

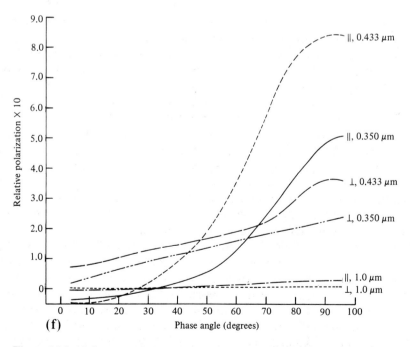

Figure 15.7 (e) Spectropolarimetric curves for 30° incidence angle (second Stokes parameter) on maple leaves. (f) Spectropolarimetric curves for 30° incidence angle (second Stokes parameter) on peach leaves.

phase angles, wavelengths, and specimen orientations correspond to those used for the spectral photometry. Polarimetry as well as photometry depend on orientation.

The percent polarization is tabulated in Table 15.1; this is the ratio, expressed as a percent, of the second to the first Stokes parameter. The polarimetric characteristics in Table 15.1 have the following significance: the minimum percent polarization (Column 2) is at a 3° phase angle, with the convention that positive indicates the dominant polarization intensity is perpendicular to the POV; the inversion angle (Column 3) is the angle at which the percent polarization (also the second Stokes parameter) passes through zero; the maximum percent polarization (Column 4) and phase angle location (Column 5) denote the magnitude and location of the maximum percent polarization at which a distinct peak exists; the percent polarization at 96°, the highest phase angle attained with current equipment, is tabulated in Column 6; this would be the maximum if no intermediate peak existed.

The fresh red pine needles (Fig. 15.7a) exhibit a maximum polarization for the ∥ orientation near the specular angle at all wavelengths as a result of the mirror effect of the needles (mentioned above in the section on photometry); this is also shown in Table 15.1, Column 5. Laboratory measurements at an effective wavelength of 0.54 μm produced a peak for the ∥ orientation at about 110°, the result of a 53.5° angle of incidence. The ⊥ orientation produced a polarization maximum near or above 96°. In general, the higher the albedos (Columns 7 and 9), the lower the maximum polarization since the multiple scattering of the more highly reflected longer wavelengths tends to reduce polarization.

It can also be seen from Table 15.1 that the ∥ orientation produces an inversion of the polarization from positive to negative for 0.633 μm and shorter wavelengths at the smallest phase angles, whereas the ⊥ orientation does not. This effect is reversed at 0.8 and 1.0 μm. It is possible that with the red pine needles a surface diffraction effect occurs at the longer wavelengths to cause this.

Aircraft polarization measurements of a red pine stand in Connecticut (Fig. 15.8) at a wavelength of 0.533 μm for an incident sun angle of 59° indicated a maximum polarization of about +25% at a 110° phase angle. Allowing for the incident angle of 30° in the present measurements, the mean polarization at 96° would be +29.5%, a value that compares well with the field observations. The infrared (1.0 μm) field observation indicated a maximum polarization at 105° of about +5% as against the +5.1% mean of the present observations, again a value that compares well with the field observations at the higher phase angles. For the 3° phase angle, the field observations indicated a value of about −1% for the 0.533- and 1.0-μm measurements; the present laboratory simulation yields mean values of −0.4% and −0.2%. The trend is correct, but the geometry differences, the weighting of ⊥ and ∥, and the effect of ground cover have not been taken into account.

The polarimetry of the dried red pine needles is presented in Fig. 15.7b. The main differences from the fresh red pine needles are that the 0.350-μm percent polarization has increased by a factor of about 2 above that for the fresh needles, and that at intermediate wavelengths it has decreased at the highest phase angles (see also Table 15.1, Column 6). The inversion characteristics for the ⊥ orientation at 0.533, 0.566, and 0.633 μm are different from the characteristics for fresh red pine needles (Columns 2 and 3). Reducing the water content of the needles by long term drying affects the chlorophyll and surface structure. Although polarization is a very complicated function of surface structure, it is quite a sensitive indicator of surface

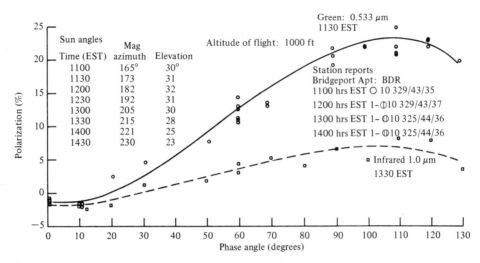

Figure 15.8 Polarimetry of New Haven red pine area: 1/31/68: green light, λ = 0.533 μm and infrared light, λ = 1.0 μm.

structure; this is probably the cause of the differences that clearly characterize the moisture state of the red pine needles.

The red pine bark (Fig. 15.7c) presents polarization characteristics quite different from those of the red pine needles, analogous to the differences in photometry. Inversions occur at all wavelengths, and the maximum polarization is near or above 96°. These characteristics are similar to those of absorbing particulate surfaces, and suggest that the optical behavior of the bark is similar.

The polarimetry of the black oak leaves (Fig. 15.7d) is quite different from that of the fresh red pine needles (Fig. 15.7a). This is also true of the photometry, with note being taken of the effect previously mentioned as resulting from leaf overlap on the platen. At the wavelengths shown the second Stokes parameter continues to increase with phase angle. An interesting effect can be observed in Table 15.1, Column 6: although the second Stokes parameter of the ⊥ orientation at 0.350 μm is higher than the ∥, the photometry enters into the percent polarization factor; and the percent polarization at 96° is nearly the same for the ⊥ and ∥ orientations. It may be seen, therefore, that the percent polarization tends to eliminate effects in the Stokes parameters, which is undesirable since the superior differentiation based on the Stokes parameters is thus decreased. Consequently, the Stokes parameters themselves are the best means of making optical comparisons.

It can be seen that the black oak leaves have a considerably higher polarization at 96° at all wavelengths than the fresh red pine needles, which is probably also the result of the photometric sheen effect mentioned above. The black oak leaves are generally characterized by inversions for the ⊥ and ∥ orientations for wavelengths of 0.533 μm and longer. This is perhaps the result of longer wavelength light penetration into and through the leaves, which possibly causes a refractive contribution that would generally be negative polarization.

The polarimetry of the sugar maple leaves (Fig. 15.7e) is generally similar to that of the black oak, except that at 0.350 μm the maple leaves have a lower second Stokes parameter than at 0.433 μm, which is an effect directly opposite to that of black oak leaves at 0.350 and 0.566 μm. The overall polarization maxima at 96° are generally less than those of the black oak leaves, and the inversion effects are mixed. These anomalies serve, however, to differentiate the two hardwood leaves on the basis of polarization alone.

The polarimetry of the peach leaves (Fig. 15.7f) is (as is the photometry) a distinctive cross between the hardwoods and conifer. The $\parallel$ orientation second Stokes parameters shown are highest at the largest phase angles, as they are with the maple leaves, in contrast to the parameters of the black oak leaves. The inversion angle trend is consistent at all wavelengths: at a 3° phase angle, $\parallel$ orientations produce negative polarization and $\perp$ orientations produce positive polarization. There are intermediate phase angle polarization maxima for most $\parallel$ orientations of the leaves.

Conclusions

The following conclusions are evident: Spectral bidirectional photometry and polarimetry yield information that can be used to characterize forest foliage with significantly large differentiation. The general representation of scattered light by Stokes parameters yields the most complete characterization of photometric and polarimetric differences. Forest remote sensing with Stokes parameters offers the possibility of extending the resolution limit of a system based on a particular set of hardware components. With Stokes parameters, increased accuracy may be possible in identifying forest cover as well as increased area coverage since the full capabilities of electronic scanning or photographic sensors can be used. The second Stokes parameter, referred to a normalized first Stokes parameter, is unique, more informative, and more readily obtained than percent polarization. The quotient of the two, requiring a computer, generally diminishes the differentiation obtained. The spectrophotometry of selected conifer and deciduous tree leaves, as seen by a low resolution nonimaging sensor, is generally highly dependent on the illumination-sensor geometry. The shorter wavelength characteristics of forest foliage are generally a better index than those obtained at longer wavelengths. Atmospheric effects, which affect all field remote sensing, are greatest at the shorter wavelengths, however, and have to be taken into account. The forest foliage data presented can be used as a guide in interpreting aircraft and satellite remote sensing data using Stokes parameters.

As a result of multispectral polarimetric and photometric observations of a pure stand of red pine, it appears that polarization may be used to remotely define the area, below the resolution limit of the sensor. Photometric observations may be used to augment the polarimetric measurements to improve the uniqueness of recognition. The morphology of the leaves of the trees appears to influence the resulting polarimetric signature strongly. Combinations of trees, or undergrowth may be analyzed polarimetrically and photometrically, based on multispectral sensing, below the resolution of the sensor; improved discrimination is achieved when the observed structure falls within the resolution limit of the sensor system.

In fact, a red pine area (unknown to the author) was located from the sensing data, using polarization. This adds credence to the technique by making it possible to locate a red pine stand without knowing ground truth.

16

Planetary Astronomy

Introduction

Conflicts still exist among the deduced properties of the Martian surface resulting from terrestrial simulation and in situ measurements. These conflicts might yet be resolved by a completely interdisciplinary approach utilizing optical polarization and photometry, thermal and microwave properties, aerodynamic theory, and soil mechanics. Only through such an interdisciplinary approach may we hope to utilize adequately the information available, and present a truly representative interpretation of the entire Martian surface and atmosphere.

The importance of large scale laboratory simulation is stressed using the observed optical polarimetric and photometric properties of the Martian surface. If it is assumed that a relatively small perturbation of the surface reflectance in the midspectral range is produced by the tenuous Martian atmosphere, then laboratory Martian surface simulation is facilitated. The samples examined are those that would be expected to exist in the bright desert areas on Mars; the perturbations of the surface polarization and photometry caused by aerodynamic and soil mechanics forces are subsequently discussed in light of the present theories about the Martian surface. Reference is made to infrared and microwave properties of the Martian surface insofar as they affect the interpretation of surface data.

The limonite hypothesis for the Martian bright areas was presented by Dollfus (1955, 1967), Dollfus and Focas (1966b), and is essentially based on polarization. Others, such as Tull (1966), Hovis (1965), and Sagan et al. (1965), have emphasized the reflectivity spectrum as a basis for Martian surface simulation. An approach using polarization and reflectivity coupled with new near-opposition measurements has been adopted by O'Leary (1967). Microwave and infrared observations have been brought to bear on interpretation by Rzhiga (1967). The importance of aerodynamic considerations has been emphasized by Egan and Foreman (1967).

The great amount of information available requires an integrated approach to Martian surface simulation. A summary of polarimetric and photometric properties of simulated Martian soils obtained on a large scale polarimeter/photometer furnishes basic information that can be utilized in conjuction with atmosphere/aerosol assumptions to formulate an adequate Martian surface hypothesis.

Previous Models of the Martian Surface

The telescopic observations that form the basis of Martian surface simulation are presented in the literature (Dollfus, 1955; de Vaucouleurs, 1964; Kuiper, 1964; Tull, 1966; O'Leary, 1967). Simulation attempts have been based on laboratory polarimetric and photometric observations with sample areas from about 0.5 cm to 1 cm in diameter. Unfortunately, the lack of large scale Martian surface simulation and the lack of detailed specification of the uniformity of the photometric and polarimetric viewing fields have limited the usefulness of measurements obtained. Observations of samples larger than 1 cm and nonuniform smaller samples can be biased compared to large scale observations with a viewing area an order of magnitude larger. Studies of large, composite samples on a polarimeter/photometer, with a viewing area of the order of 9 cm in diameter, indicate the importance of large scale investigations (Egan, 1967b). The uniformity of the source and sensor photometric fields in brightness should be of the order of 1% or better for highly accurate photometric observations, and spatial coherence should be 1° or less. Similarly, the residual polarization in the source and sensor fields should be of the order of 0.1% or better for highly accurate polarimetric observations, and the acceptance angle should be of the order of 1° or less. The 9-cm sample diameter polarimeter/photometer previously described (see Chapter 5) fulfills these requirements.

Previous simulation attempts have been made with limonite and goethite (Sagan et al., 1965; Dollfus and Focas, 1966b; Egan, 1967a; O'Leary, 1967), siderite (O'Leary, 1967), desert varnish coated rocks (Binder and Cruikshank, 1966a, b), and mixtures (Tull, 1966). New Idria serpentine (containing an iron hydrate) has been suggested as a material that undergoes a green to yellowish-brown color change in a moist CO_2 atmosphere, which would explain Martian color changes (Nowatzki, 1967). Overall, previous particle sizes studied have been predominantly powders because of the physical limitations of the photometric/polarimetric apparatus used.

Limonite, siderite, desert varnish rocks, and serpentine are investigated as representative, assuming that particle size is an important parameter in determining the polarization and photometric properties (Egan, 1967a; Egan and Foreman, 1967). Particle size is a fundamental parameter in determining polarization and photometry of surfaces, because the particle size affects of the surface roughness and thus the amount of multiple scattering of the incident light, as well as refractive versus reflective effects (Egan, Grusauskas, and Hallock, 1968). Other factors that could affect polarization and photometry are the porosity and surface structure of the composite particulate surface, the external particle shape (influenced by preparation), the particle structure (crystalline/noncrystalline), and particle composition (wavelength variation of complex index of refraction). The limonite (nodule type deposit) was obtained from Venango County, Pennsylvania, and the particles ranged in size from <1 μm to cm-sized chunks. The siderite was obtained from Roxbury, Connecticut, and ranged in size from <1 μm to 88 μm. The desert varnish rocks were obtained by A. Binder and D. Cruikshank from the Glove Mine and Catalina Mountain areas. The serpentine was obtained from an area northwest of Coalinga, California.

Polarimetric Measurements

The polarimetric curve obtained from Martian observations has a general characteristic form (see, for instance, Dollfus, 1955). Basically, the curve starts at zero phase angle (α) with zero percent polarization, usually drops to a negative minimum, and then rises to some positive percent polarization value at the highest attainable viewing angle from the earth (about 45°), and in so doing goes through zero percent polarization again at a phase angle of the order of 25° (the inversion angle). Hence, we can characterize a polarization curve of Mars as having the following essential features: (1) a minimum value and a corresponding phase angle location; (2) an inversion angle; (3) a value at some angle near the maximum attainable from earth orbit (here taken as 40°). Further, a maximum percent polarization may occur at a fairly large phase angle (of the order of 120°) and may be used to characterize the laboratory simulation. The effect of viewing angle (θ) is an additional parameter.

The characteristic, large scale, polarimetric data for limonite, siderite, desert varnish, and serpentine (before and after color change) are listed in Table 16.1, Columns 2–11. In addition, Columns 14 and 15 list the percent polarization at near the maximum phase angle attainable for 0° and 60° viewing angles to indicate the slope of the respective polarization curves. The polarimetric data was obtained in B(0.48 μm), G(0.54 μm), and I(1.0 μm) for the selected specimens of limonite and siderite. (The B, G, and I refer to general spectral regions, followed by the central wavelength; usage is explained in Chapter 17.) The data on the desert varnish and serpentine were obtained in G(0.54 μm) initially, to permit comparison to the "best" limonite and/or siderite sample, and establish whether further measurements were warranted. The three-significant-place polarization data contained in Table 16.1 have probable absolute errors of measurement as follows: B and G polarization $\pm 0.07\%$ and I polarization $\pm 0.1\%$; the remaining polarization data in B and G have errors of $\pm 0.1\%$ and I $\pm 0.15\%$. Angular measurements are accurate to $\pm 0.5°$ and albedos to ± 0.005. Since inversion, maximum, and minimum angles are derived from the curves, their location accuracies depend upon the curve shape and the corresponding errors of measurement of the polarization data. The standard of comparison used for matching is the 1965 observations of Dollfus and Focas (1966a, b). This initially neglects effects of atmosphere and aerosol contributions to the polarization; these effects will be mentioned subsequently.

Following a study of the polarimetric properties of the samples listed in Table 16.1, it can be seen that the 1.19–2.38-mm limonite is the closest match to the Dollfus data, and thus is the inferred dominant particle size of limonite on Mars. It should be mentioned that these size particles were not washed, but were cleaved in a rock crusher unit to produce a minimum of dust coating; thus a natural process of rock fracture is simulated, since a water washed rock is at present assumed rather unlikely on Mars. However, in order to determine the effect of washing, a second limonite sample, obtained from Venango County and fractured to 1.19–2.38 mm in size, was subjected to four thorough washings in tap water. It can be seen (Table 16.1) that the polarization at 60° phase angle (0° viewing angle) is increased as well as the maximum negative polarization, and the inversion angle shifted to a higher phase angle. Since any residual powder on the surface of the limonite is removed by washing, which causes the simultaneous dissolving of soluble surface materials as well, it is reasonable to expect the polarization difference.

Table 16.1 Summary of Polarimetric and Photometric Characteristics of Simulated Martian Soils

Spectral region and size specimen	Minimum Location				Inversion angle location		Maximum location Sensor angle = 60°		Percent polarization for phase angle = 40°		Geometric albedo for phase angle = 0°		Percent polarization	
	Sensor angle = 0° %	Location angle(°)	Sensor angle = 60° %	Location angle(°)	Sensor angle = 0°	Sensor angle = 60°	%	Location angle(°)	Sensor angle = 0°	Sensor angle = 60°	Sensor angle = 0°	Sensor angle = 60°	Phase angle = 60° Sensor angle = 0°	Phase angle = 120° Sensor angle = 60°
(1)	(2)	(3)	(4)	(5)	(6)	(7)	(8)	(9)	(10)	(11)	(12)	(13)	(14)	(15)
Limonite														
(I-1.0 μm)														
<1 μm	−0.72	27.5	−1.09	19.5	46.6	41.4	>+ 3.2	>125	−0.48	−0.10	0.58	.53	+ 1.0	+ 3.1
<37 μm	−0.85	15.0	−0.99	14.3	35.2	37.8	>+ 4.0	>125	+0.36	+0.17	0.39	.35	+ 1.5	+ 3.9
37–88 μm	−0.75	15.5	−0.89	10.0	37.6	28.8	>+ 4.0	>125	+0.15	+0.75	0.38	.35	+ 1.7	+ 3.9
1.19–2.38 mm	−0.70	16.5	−1.02	13.7	29.1	27.5	+ 8.8	113	+1.38	+1.68	0.27	.26	+ 3.9	+ 8.7
cm-sized chunks	+0.03[b]	3.5	+0.12[b]	3.5	None	None	+11.6	~115	+1.89	+2.20	0.25	.21	+ 4.3	+11.5
(G-0.54 μm)														
<1 μm	−1.01	20.3	−0.86	16.3	36.9	35.0	+ 5.9	120	+0.33	+0.48	0.24	.22	+ 2.7	+ 5.9
<37 μm	−1.17	19.5	−1.03	20.0	35.3	34.2	+ 9.5	116	+0.62	+0.83	0.17	.15	+ 4.1	+ 9.4
37–88 μm	−1.19	18.5	−0.95	19.5	34.5	33.8	+ 9.8	125	+0.71	+0.84	0.17	.16	+ 4.1	+ 9.8
0.42–1.19 mm	−0.75	10.5	−0.62	9.0	24.4	22.3	+15.6	113.5	+2.51	+3.07	0.17[a]	.15[a]	+ 6.9	+15.4
1.19–2.38 mm	−1.00	9.0	−0.78	9.5	23.7	21.7	+16.9	115	+2.80	+3.09	0.17	.15	+ 7.2	+16.5
1.19–2.38 mm (washed 4 ×)	−1.22	13.0	−1.41	15.2	26.0	26.7	+31.5	128	+2.50	+3.22	0.14[a]	.13[a]	+10.6	+31.0
cm-sized chunks	−0.94	8.0	−0.75	8.5	19.9	21.0	+23.5	116	+4.88	+3.78	0.12	.10	+10.1	+23.3
(B-0.48 μm)														
<1 μm	−1.05	18.8	−1.15	16.8	32.1	32.1	+14.7	>125	+1.25	+1.43	0.12	.13	+ 5.7	+14.3
<37 μm	−1.21	17.8	−1.17	16.0	31.5	29.8	+17.4	113	+1.83	+2.23	0.09	.10	+ 8.0	+17.2
37–88 μm	−1.24	15.8	−1.04	21.8	30.9	31.4	+17.7	125	+1.90	+2.17	0.11	.10	+ 7.9	+17.6
1.19–2.38 mm	−0.85	11.0	−0.73	12.7	22.4	22.0	+21.1	110	+4.07	+3.61	0.13	.11	+ 9.6	+20.6
cm-sized chunks	−1.07	9.0	−0.64	6.5	18.0	14.6	+34.3	110.5	+4.97	+5.54	0.08	.07	+14.6	+33.4
Siderite														
(I-1.0 μm)														
<1 μm	+0.00[b]	5.0	+0.01[b]	5.0	None	None	>+ 2.3	>125	+0.00	+0.57	0.69	.55	+ 0.8	+ 2.1
<37 μm	+0.01[b]	5.0	+0.10[b]	5.0	None	None	+ 4.0	>125	+0.24	+1.09	0.44	.36	+ 1.4	+ 3.9
37–88 μm			+0.53[b]	5.0		None	>15.1	>125		+1.02	0.22[a]	.16[a]		+14.6
88–210 μm			0.68[b]	5.0		None	+16.4	125		+2.17	0.20[a]	.17[a]		+16.3

	1	2	3	4	5	6	7	8	9	10	11	12	13	14
(G-0.54 μm)														
<1 μm	−0.52	10.0	−0.19	10.0	31.7	25.8	+ 4.0	125	+0.36	+0.61	0.41	.37	+ 1.4	+ 3.8
<37 μm	−0.67	9.5	−0.35	16.3	24.7	27.1	+ 8.1	111	+1.69	+1.28	0.17	.18	+ 4.3	+ 8.0
(B-0.48 μm)														
<1 μm	−0.61	12.2	−0.20	9.8	30.0	23.0	+ 5.3	125	+0.63	+0.78	0.33	.31	+ 2.0	+ 5.2
<37 μm	−0.62	9.8	−0.19	8.3	23.2	18.3	+11.6	111	+2.32	+2.80	0.13	.14	+ 5.7	+11.5
Desert varnish														
(G-0.54 μm)														
Glove Mine	−0.95	5.5	−2.97	7.5	16.5	25.9	+50.4	126	+5.61	+4.30	0.08[a]	.05[a]	+12.5	+49.5
Catalina Mt.	−0.10	7.3	−1.88	7.7	11.4	22.3	+30.5	130	+3.90	+3.68	0.08[a]	.05[a]	+ 9.0	+29.4
Glove Mine (mm pieces)	−1.01	3.8	−0.88	8.5	17.6	18.8	+18.8	122	+3.93	+3.64	0.14[a]	.14[a]	+ 8.5	+18.7
Serpentine														
(G-0.54 μm)														
Initial samples (Greenish)														
0.088–1.19 mm	−0.57	6.3			20.2				+1.79		0.46[a]		+ 3.7	
1.19–2.38 mm	−0.85	4.3			19.4				+2.03		0.39[a]		+ 5.0	
Final sample (Yellow-brown)														
0.088–1.19 mm	−0.50	12.0			22.5				+1.64		0.46[a]		+ 3.5	
Polarization of Mars: Dollfus (1966 a,b)														
I-1.05 μm	−0.98	10.2			24.6				+1.48					
G-0.53 μ	−0.95	11.4			23.2				+2.71					
B-0.50 μm	−0.96	10.6			22.5				+3.21					
Albedo of Mars: O'Leary (1967); de Vaucouleurs (1964)														
1.04 μm											0.342 [0.312]			
0.555 μm											0.194 [0.160]			
0.450 μm											0.107 [0.084]			

[a] Normal albedo extrapolated from $\alpha = 5°$ value.

[b] No minimum point; % polarization at corresponding angle in Columns 3 or 5.

Note: The abbreviation B, G and I refer to the general blue, green, and infrared spectral regions, followed by the effective wavelength of the measurements listed.

In making comparisons of the effect of particle size (Table 16.1), it should be noted that the mode of particle preparation can appreciably influence the results. Particles $< 37\,\mu$m, $37–88\,\mu$m, and $88–210\,\mu$m were obtained by ball milling larger particles, which were in turn obtained by cleaving the ore samples in a rock crusher. The $< 37\,\mu$m particles were obtained by fluid energy milling the $< 37\,\mu$m particles (Egan, 1967b).

Another point should be noted with regard to the sample preparation. Dollfus (1956) observed that powders of yellow ocre and aluminum produced different polarizations from smooth surfaces than when piled up from agitation, whereas volcanic ash showed no change. This is probably the effect of porosity, since surface porosity affects the inversion angle; a higher porosity basalt surface has a higher inversion angle and a less negative minimum (Egan and Nowatzki, 1968). A preferential surface orientation could result from agitation if the particles have a characteristic geometry, and this, too, can affect the inversion point and minimum (Egan and Nowatzki, 1967). The volcanic ash showed no effect, probably because of little porosity change after agitation and negligible preferential particle orientation.

No data is presented on $37–88\,\mu$m or $88–210\,\mu$m siderite at a viewing angle of $0°$ in I (nor any data in G or B) because the lack of an inversion excluded it as a single constituent of the Martian surface. Serpentine was observed only at $0°$ viewing angle before and after the color change because the initial results did not match the Martian data very well. The reverse color change was not obtained, but it is hoped that under more controlled conditions the color will revert back to the original greenish hue.

Figure 16.1 Comparison of polarization of particular limonite (Venango County, Pennsylvania), with Mars visual observations of Dollfus (1966 a,b).

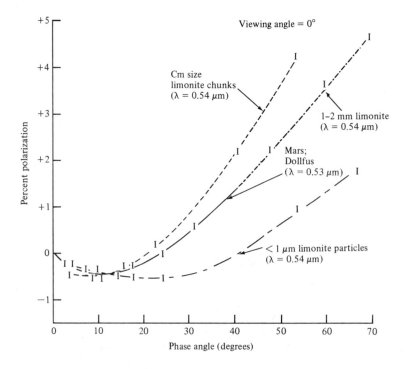

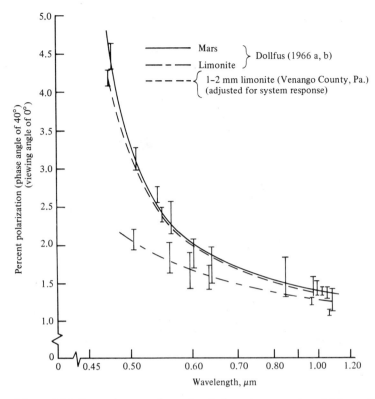

Figure 16.2 Spectral comparison of polarimetric properties of Mars with simulated limonite surfaces.

The comparison of the polarization of particulate Venango County limonite with the 1965 visual observations of Dollfus and Focas (1966a, b) are shown in Fig. 16.1. It can be seen that 1–2-mm limonite follows the Dollfus curve closely. If the spectral polarimetric properties of this size limonite with the Dollfus limonite powder and Martian observation are compared (Fig. 16.2), we observe that the 1–2-mm particulate limonite follows the Martian observations closer than the Dollfus limonite powder. The 1–2-mm limonite percent polarization at 0.54 μm has been corrected for the 0.072 μm bandpass of the polarimeter. This correction was required by the large change in polarization curve slope in this spectral range.

Photometric Measurements

Another group of factors in Martian surface simulation are the photometric properties; this is taken to include the broadband albedos, the reflectance as a function of phase angle, and the effect of viewing angle (limb darkening).

In Column 12 of Table 16.1, the geometric albedo of the various materials is shown; Column 13 lists the "albedo" of these materials at $\theta = 60°$, referred to a standard surface at $\theta = 0°$. The albedos are obtained at $\alpha = 0°$ with the aid of a beam splitter, and these albedos are compared with the albedos of $MgCO_3$ as the standard surface, which in turn is compared with the albedo of MgO. The albedos for the asterisked items in Table 16.1 were obtained at 5°, extrapolated to 0°, based

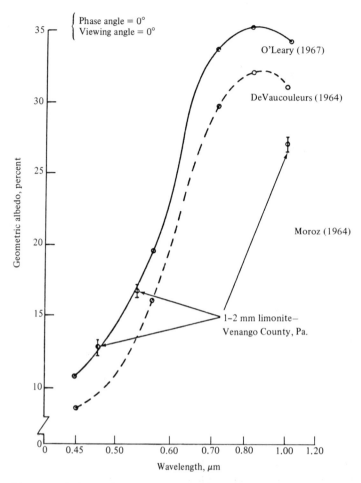

Figure 16.3 Spectral comparison of photometric properties of Mars with best simulated surface.

on an average opposition effect for the samples investigated. The photometer had a beam divergence of 1°. The Martian photometric data of O'Leary (1967), de Vaucouleurs (1964), and Moroz (1964) are used as an initial comparison in Table 16.1. The comparison is shown graphically in Fig. 16.3. The 1–2-mm limonite point falls below the O'Leary data by about 7% in I(1.0 μm). However, the absolute magnitude data for Mars are considerably lower than those of de Vaucouleurs and Moroz in R(0.70 μm), I(0.82 μm), and I'(1.04 μm), but O'Leary assumes a fit to the de Vaucouleurs data at $\alpha \gtrsim 10°$ at these wavelengths, thereby producing questionably high albedos. [These magnitude discrepancies do not occur at U(0.355 μm), B(0.450 μm), or V(0.555 μm).] The difference could be the result of the particular apparition, and a somewhat lower albedo [cf. 0.20 at 1.3 μm, from Moroz (1964)] might be more appropriate and more closely comparable to that of the 1–2-mm limonite.

The photometric properties of limonite as a function of phase angle at 0° viewing angle is shown in Fig. 16.4 for integrated visual (0.54 μm) light. It can be seen that the larger particles (1–2-mm and cm-sized chunks) follow the Mars curves of O'Leary and de Vaucouleurs. (The curves are arbitrarily normalized at 5°.) At the

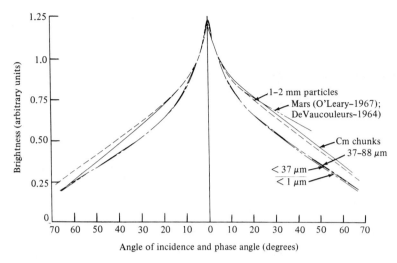

Figure 16.4 Photometric curves for Venango County limonite of various particle sizes (0° viewing angle).

highest phase angle these same sized particles tend to a decreased reflectivity. A 60° viewing angle, integrated visual light, photometric curve for limonite is shown in Fig. 16.5.

The limb darkening effect is inferred (Table 16.1) from the decrease in albedo as viewing angle is increased from 0° to 60°. It can be seen that the decrease is about 15% for 1.19–2.38-mm limonite in G, approximating that of the Martian surface within the limits of error of the tabular data (O'Leary, 1967).

Siderite was also investigated photometrically as a function of phase angle for 0° and 60° viewing angles. The results are shown in Figs. 16.6 and 16.7 for integrated visual light with an effective wavelength of 0.54 μm. The < 37 μm siderite is closer

Figure 16.5 Photometric curves for Venango County limonite of various particle sizes (60° viewing angle).

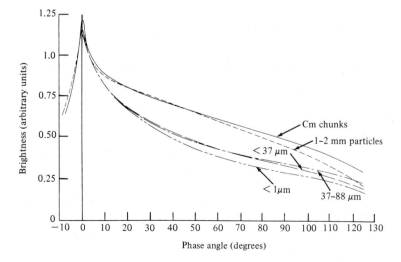

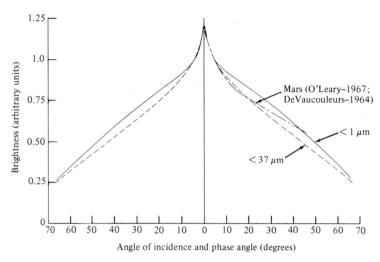

Figure 16.6 Photometric curves for siderite particles <1 and < 37 μm (0° viewing angle).

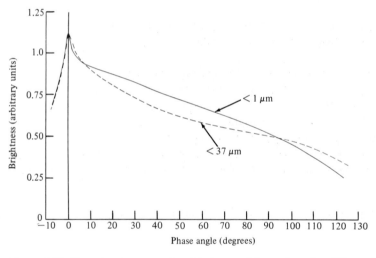

Figure 16.7 Photometric curves for siderite particles <1 and < 37 μm (60° viewing angle).

to the Mars reference curve between 10° and 30° phase angles than the <1 μm, and thus may contribute in a secondary way to the optical response.

Additional Measurements

In order to define the Venango County limonite adequately, a series of analyses were performed. These were spectroscopic analyses, differential thermal analyses (DTA), and thermogravimetric analysis (TGA). Also, integrating sphere measurements were made on the 1–2-mm and <1 μm limonite and < 37 μm siderite; these were similar to those made by Sagan et al. (1965). The porosities of these samples were measured, since porosity can possibly influence photometry and polarization (Egan and Nowatzki, 1968). Also, the radio frequency dielectric constants of these materials were measured at a wavelength of about 3 m.

The spectroscopic results indicated the following elements in the indicated quantities:

$$Fe \sim 90\%,$$
$$Si, Al, Mg < 10\%,$$
$$Cu, Ti, Mn, Ca, Na < 1\%.$$

The DTA runs made on a Deltatherm unit at a heating rate of 10°C/min indicated an exothermic reaction at about 470°C (probably a phase change). The net weight loss on three samples was 25.5%, 26.0%, and 23.2%; the loss is assumed to be H_2O, and the resulting residue has an appreciable positive magnetic susceptance, indicating largely Fe_2O_3.

The TGA was made using an Ainsworth BYR analytical balance, with the result shown in Fig. 16.8. The net loss in mass (also assumed to be H_2O) was 23.5%, and the resulting residue has an appreciable positive magnetic susceptance, indicating conversion to Fe_2O_3. Bog ore limonite, $Fe_2O_3 \cdot 3H_2O$, has 25.2% water of hydration, and this is comparable to the observed loss by the sample, thus indicating a closely similar composition. The average density of the limonite sample was 3.1, which is below the range of 3.6 to 4 for limonite (Hurlbut, 1966). This is taken to indicate the presence of interstitial water.

The integrating sphere measurement was made in a wavelength range from 0.25 to 2.3 μm with a Perkin–Elmer Model 12 monochrometer and Gier–Dunkle absolute integrating sphere reflectometer (Model A1S-6B). The samples were mounted in quartz covered containers and compared to an MgO standard in the same container. The results are presented in Fig. 16.9 for <1 μm and 1–2-mm limonite and < 37 μm siderite. The 0.87 μm iron oxide band (see O'Leary, 1967, for instance) is not present in the 1–2-mm limonite [or at least has shifted to longer wavelengths with the increased particle size (Hovis, 1965)] and a minimum exists near 1.0 μm, thus being consistent with Martian observations. However, the albedo at and above 1 μm is low, as noted previously (Fig. 16.3). It is possible that a small quantity of limonite (<1 μm) or siderite (< 37 μm) mixed in with the larger limonite particles

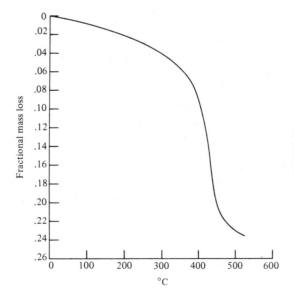

Figure 16.8 Thermogravimetric analysis of pulverized Venango County, Pa., limonite, performed in air, heating rate 13°C/min.

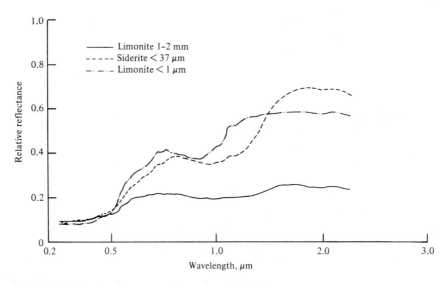

Figure 16.9 Total reflectance of limonite and siderite samples compared to MgO.

could contribute to the higher infrared albedo. Other materials in small quantities could also contribute to this. Since the integrating sphere measurements are neither geometrical nor Russell–Bond albedos, they can serve only as a guide to the Martian simulation (see, for instance Sagan et al., 1965). Differences between the results of Sagan et al. and the present sample are evident, insofar as the 1–2-mm limonite is a better spectral representation of the Martian surface.

The measured porosities and radio frequency (~ 80 mHz) dielectric constants are for respectively, 85% and 1.9 for limonite samples <1 μm, 50% and 3.0 for 1–2-mm samples, and 60% and 4.9 for samples < 37 μm. There can be no comparison of these porosities with previous data, as none exists, but the effect has been mentioned (O'Leary, 1967). However, Pollack and Sagan (1965) report the dielectric constant of limonite as 3.6 (at a frequency of 10 GHz), indicating a more consolidated limonite sample, or possibility a frequency dependence. The radio frequency dielectric constant of the present limonite samples are at both ends of the range indicated by Rzhiga (1967) for the bright regions (2.0–2.3). He indicates his low dielectric constant is probably caused by the porous structure of the soil. However, based on the present data, it can be proposed that the subsurface of the visually bright areas of Mars indicated by the radar measurements might be composed of a combination of <1 μm and 1–2-mm limonite to bring the dielectric constant into the 2.0–2.3 range. Subsurface ice (dielectric constant about 3.5 at 100 mHz) in a porous matrix could result in a dielectric constant in the 2.0–2.3 range. Siderite (< 37 μm), in a very porous state, would be possible.

Estimates for the thermal inertia parameter ($K\rho c$)$^{1/2}$ have been made (Rzhiga, 1967), but the only general conclusion is that the Martian surface is porous. The thermal inertia parameter can vary considerably depending on the surface composition and gas environment.

A thermal analysis of the Martian surface was made by Leovy (1966) utilizing an average diurnal variation of Martian light and dark areas, with the conclusion that the thermal conductivity of the Martian surface would be produced by particles less than 10 μm in size, independent of particle composition, depending only on atmospheric pressure. This was based on Smoluchowski's data presented in the International Critical Tables [National Research Council (1927)]. However, the original work of Smoluchowski (1910) describes a cylindrical glass apparatus with a thermal path in the powder of 2.4 ± 0.2 mm. The short path length in the apparatus can bias the sample volume results by the surface contact effects of the container. This is particularly true for the larger particles which were investigated. In comparison, smaller particles have a larger surface to volume ratio than larger particles, and would be especially susceptible to cumulative effects of surface contamination, which Smoluchowski sought to eliminate by "vacuum soaking." Thus they would be expected to show a relatively small but measurable (Smoluchowski, 1910, Tables I–XVI) change in average conductivity with material. However, the larger particles show considerable differentiation with material, because of the greater volume to surface effect. The particle sizes indicated by Smoluchowski are average values, and the distribution of sizes is not specified.

The larger particles have a higher conductivity for a given atmospheric pressure, but in the model already proposed (Egan and Foreman, 1967) containing smaller, less conductive limonite particles (<1 μm) under and between the larger optically dominant limonite particles (1–2 mm), the lower conductivity of the Martian atmosphere would lower the apparent thermal conductivity, approaching that noted for the smaller particles.

Interpretation of Polarimetric/Photometric Simulation Measurements

The inferences that may be drawn from the data presented must be tempered with the additional information presented by others. An open-mindedness is necessary, yet a depth and breadth of viewpoint is required.

Limonite may be the dominant constituent of the Martian bright areas. With the samples used in this report (Venango County, Pa.), 1–2-mm sizes (with whatever coating occurred in the rock crushing process under laboratory conditions) appear the best match optically, and therefore are the dominant particle sizes. It appears that a fractional amount of smaller particles with consistent polarimetric and photometric characteristics is required to increase the infrared albedo. Such particles could be <1 μm limonite and < 37 μm siderite. These particles are consistent with observed atmospheric dynamics phenomena (Egan and Foreman, 1967).

Dollfus (1967) indicated smaller sized limonite particles as an appropriate material to simulate the Martian surface based on his small scale polarimetry. Draper, Adamcek, and Gibson (1964), have obtained fair laboratory simulation with a mixture of materials; Binder and Cruikshank (1966a,b) suggested desert varnish; Rea and O'Leary (1965) have indicated the possibility of aerosols. O'Leary (1967) has widened the scope of possible materials based on small scale photometry, polarization, and spectrophotometry between U(0.355 μm) and I'(1.04 μm); he has further emphasized low resolution spectroscopy of Mars from outside the earth's atmosphere.

Notably lacking in all simulation is the appreciation of scale effects. This includes not only large scale laboratory polarimetry and photometry, but also surface structure effects such as porosity, lineation, and composite surface effects. Further, there is insufficient consideration of the implications of gasdynamic forces in a low pressure CO_2 environment in laboratory simulations. The chemical instability of goethite for a geologically long time scale has been evaluated (Fish, 1966), but limonite ($Fe_2O_3 \cdot 3H_2O$), its impurities, and range of possible thermodynamic properties have not been investigated.

Also, the combination of the optical determination with those from thermal and microwave measurements is required in the last analysis. For instance, a conclusion of Younkin (1966), based solely on reflectivity spectra in the range 0.34–1.1 μm, that there is no strong evidence for the existence of large amounts of limonite on the surface of Mars is not warranted. Reflectance spectra have been noted to be influenced by impurities and particle sizes, and reflectance is only one aspect of the entire problem. Moroz (1964), to the contrary, supports the limonite hypothesis with data in the wavelength range from 0.5 to 4 μm. Also, the limonite from Venango County appears to have a large amount of water of hydration, as would occur in the production of lateritic soils (Sagan, 1965).

Subsequently Pimentel et al. (1974) have used the data from the Mariner Mars 6 and 7 infrared spectrometer to show that a broad absorption band near 3 μm can be ascribed to either water of hydration of surface minerals or to ice. This is not in disagreement with the work described in this chapter. Further, Pollack et al. (1977, 1979) have analyzed the Viking Lander imaging data of the Martian sky, Phobos, and the sun to infer the properties of aerosols in the Martian atmosphere, the inference being that the aerosols arise from Martian surface material. The principal opaque material in the aerosols was proposed as $10\% \pm 5\%$ magnetite, yet not denying the existence of ferric compounds to cause the brownish color of the Martian surface. This work was extended in further work (Pollack et al., 1979).

However, Egan, Hargraves, and Pollack (1982) presented new data that revealed a ferric oxide absorption minimum at 0.75 μm in α- and γ-Fe_2O_3; also, additional data on magnetite showed a barely perceptible decrease in absorption at 0.75 μm, on which the original inference of magnetite in the Martian aerosols was based. The new magnetite data were determined in separate measurements of the refractive and absorptive components of the optical complex index of refraction, rather than in a dispersive analysis, as were the original data.

Heat treated nontronite, postulated to result from impact shock on the Martian surface, is also magnetic as required, and has an absorption minimum close to the 0.75 μm wavelength. The magnitude of the absorption of the nontronite is close to the revised values of Pollack; but the γ-Fe_2O_3 has absorption on the order of a magnitude higher, and must be mixed with a claylike material to bring it into the required range. Both nontronite and γ-Fe_2O_3 have the required increased absorption further into the infrared as required by the Viking lander observations (Pollack et al., 1977). These new data confirm the assertion that ferric oxide compounds exist in the Martian soil.

In essence, a *most probable* Martian soil determination is ultimately obtained, since uniqueness is always a question. There is no doubt that additional optical experiments in space will go a long way to improve the definition of parameters associated with the Martian surface. These experiments should include remote sensing through a wider range of phase and viewing angles in the UV, IR, and radio frequency spectral regions.

17

Stellar Astronomy

This chapter would more aptly be termed "Asteroid and Meteorite Astronomy," but the optical terminology and laboratory photometric/polarimetric spectral measurements and optical comparisons are immediately applicable to interplanetary and interstellar bodies with well defined surfaces of condensed material.

A short description of measurements and modeling of the interstellar medium is included to indicate the application of polarization.

Asteroid and Meteorite Astronomy

Recently, much astronomical effort has been devoted to the study of the physical properties of asteroid surfaces. This has resulted in an ever growing store of spectral, photometric, and polarimetric data that have not been sufficiently complemented by laboratory measurements. Because there are many indirect arguments (Anders, 1964) which indicate that most meteorites are fragments of asteroid bodies, and since McCord et al. (1970) have successfully identified the spectral reflectance curve of one asteroid, Vesta, with that of a basaltic achondrite, laboratory measurements on meteorite samples should prove useful in an interpretation of asteroid surfaces. The spectral, photometric, and polarimetric properties of the Bruderheim olivine–hypersthene chondrite, which is representative of the most common meteoritic material recovered on earth (Mason, 1962) and hence might also be the most common in interplanetary space, was the subject of a study.

A goniometer designed to measure the polarimetric and photometric properties of samples up to 60 mm in diameter was used for agricultural and forestry samples (see Chapters 14 and 15). The sensor arm normally traverses a phase angle range from 3.6° to a maximum; at 0°, 30°, and 60° incident illumination angles, the phase angle ranges are 3.6°–65.5°, 3.6°–95.5°, and 3.6°–125.5°, respectively.

The spectral filters in the sensor system are of the type suggested by Gehrels and Teska (1960), and the system wavelength response has been determined experimentally. The U(0.36 μm) analyses were made using the RCA 6199 (S-11) photomultiplier, an HNP'B polaroid, and a 3.0-mm thick Corning C-9863 filter. The infrared leak was found to be less than 2% of the observed signal levels for this filter–tube combination. The G(0.54 μm) analyses were made using the RCA 6199 (S-11) photomultiplier, an HN-22 polaroid, a 3.0-mm thick Corning C-3385 filter, and a 2.0-mm thick Schott–Jena BG-18 filter. The R(0.67 μm) analyses were made using the RCA 6199 (S-11) photomultiplier, an HN-22 polaroid, and a 3.0 mm thick Corning C-2030 filter. Here U, G, and R refer to the general ultraviolet, green, and red spectral regions, with the central wavelength following in parentheses.

The "normal" reflectance measurements were made at an incident angle of 0° and a scattering angle of 3.6°. Narrow band Optics Technology interference filters were used between 0.31 and 0.7 μm (at 0.02 μm spacings to 0.4 μm and at 0.03 μm spacings above 0.4 μm) with the RCA 6199 photomultiplier, and between 0.8 and 1.1 μm (at 0.1 μm spacings) with the RCA 7102 photomultiplier. The normal reflectances were calibrated against a freshly scraped magnesium carbonate surface. The relative reproducibility for each normal reflectance measurement is $\pm 2\%$.

The polarimetric and photometric observations were phase detected using a Princeton Applied Research lock-in amplifier Type 124 and recorded on a Houston Instrument Omnigraphic recorder. Polarimetric and photometric calibrations were made against $MgCO_3$ before and after each run.

The location of the polarization inversion angles were determined to an accuracy of $\pm 0.2°$.

Before proceeding further, some clarification of the astronomer's optical terminology is necessary. The term *magnitude* is used to designate brightness of stars, or solar reflected radiation from planetary surfaces. Magnitude is a logarithmic quantity, each magnitude being 2.512 (~ 2.5) times as bright as the one below it. Thus

$$\text{brightness} = C\,(2.5)^m, \tag{17-1}$$

where the "brightness" on the left hand side of the equation is in units determined by the constant C, m being the magnitude. For instance, a brightness difference of five magnitudes is a factor of 100; negative values of m are also allowed. The magnitude concept is a convention that has grown up with astronomy and is frequently used to quantify planetary surface brightness or the brightness of a point source (a star).

Another convention that exists in astronomy is the U, B, V photometric system; as mentioned previously, U refers to ultraviolet, B, to blue, and V to visual ($\sim$ yellow). The system was devised by Johnson and Morgan (1951, 1953) and accepted the use of broadband glass filters (either Corning or Schott–Jena) as the elements to define spectral bands. It is to be noted that at present interference filters are now readily available to define spectral bands rather precisely.

There are a number of variants on this system, depending on the astronomer and the filters used in his optical system. Another system is the R, G, U (Becker, 1938, 1946), which is defined photographically, as compared to the U, B, V system which is defined photoelectrically. The effective central wavelengths of these two systems and the half-widths are given in Table 17.1.

The Bruderheim meteorite sample was obtained from Dr. E. Fireman of the Smithsonian Astrophysical Observatory. It is part of sample B-74 which was recovered within a day of the fall; weathering was minimal. The sample was subsequently stored in tightly closed bottles. The sample is representative of olivine–hypersthene meteorites (Keil and Fredriksson, 1964), and, more specifically,

Table 17.1 Effective Central Wavelengths and Half-Widths of the Two Photometric Systems

Band	Effective wavelength (μm)	Bandwidth (μm) FWHM
U	0.3660	0.0700
B	0.4400	0.0970
V	0.5530	0.0850
R	0.6380[a]	0.0400
G	0.4680[a]	0.0490
U	0.3690[b]	0.0540

[a]Includes filtering effect of atmosphere.
[b]Approximate value.

it is representative of class L6 (Tandon and Wasson, 1968). The black fusion crust was chipped off, and the remaining meteorite was fractured in a rock crusher to produce a minimum of small particles. The fracturing process was continued until there were no pieces larger than 4.76 mm. The fragments were then sieved in 5-min. runs using a Combs Gyratory Sifting Machine with sieves having openings of 4.76 mm, 0.25 mm, 74 μm, 37 μm, and a residual catch pan for < 37 μm. The first aim was to obtain enough material in the 0.25–4.76-mm particle size range to cover the 60-mm diameter sensing area of the goniometer. Then the remaining < 0.25-mm particles were subjected to further pulverization in a Coors 90-mm-diameter porcelain mortar to obtain sufficient powder to coat the 0.25–4.76-mm particles.

The 0.25–4.76-mm sample, weighing 42 g, is shown in Fig. 17.1a. The particles are platelike with angular edges and tend to lie flat. Figure 17.1b shows the microscopic appearance of a typical fragment; chondrules and bits of nickel–iron are clearly visible.

Next, the "coated" sample (Fig. 17.1c) was made by dusting the 0.25–4.76-mm sample with 0.87 g of the < 37 μm-powder, and 0.12 g of the 37–74 μm-powder. The coating amounted to 2.4% by weight. A microscopic view of this sample is shown in Fig. 17.1d.

After these two coarse samples were measured, two other samples were prepared by further crushing in the mortar: 75–250 μm (Fig. 17.1e), and < 37 μm (Fig. 17.1f). The latter sample has the expected "fairy castle" texture.

The absolute normal reflectance of the four samples is shown in Fig. 17.2. As mentioned above, these "normal" reflectance measurements were actually made with an incident angle of 0° and an emission angle of 3.6°. No correction to convert these measurements to precisely zero phase was attempted.

As expected, of the two powders the < 37 μm sample is brighter than the 74–250 μm sample throughout our spectral range. The "coated" sample has a curve intermediate to those of the two powders, as it should, since it was produced by dusting the 0.25–4.76-mm sample with a coating of particles smaller than 74 μm. Incidentally, this illustrates the fact that for a sample of large particles that have been completely coated with smaller particles, the inherent photometric properties of the large particles are suppressed.

The spectral reflectance curve of the 0.25–4.76-mm sample seems anomalous. Its shape differs from that of the other curves, it crosses the "coated" sample curve at 0.6 μm, and it is higher than the 74–250-μm curve. Intuitively, one would not expect

Figure 17.1 Photographs of Bruderheim meteorite samples. (a) 0.25–4.76-mm particulates; (b) microscopic view of 0.25–4.76-mm particulates; (c) "coated" particulates; (d) microscopic view of "coated" particulates; (e) 74–250-μm grains; (f) < 37-μm powder.

the coated sample curve to fall below that for the 0.25–4.76-mm sample because the addition of fine particles should increase the importance of multiple scattering and therefore the brightness of the sample. This anomalous behavior may possibly be related to the presence of quasiplanar facets (Fig. 17.1a).

The 0.25–4.76-mm sample shows three distinct absorption features, two strong bands near 0.3 and 0.95 μm, both due to Fe^{2+} (Bancroft and Burns, 1967; White

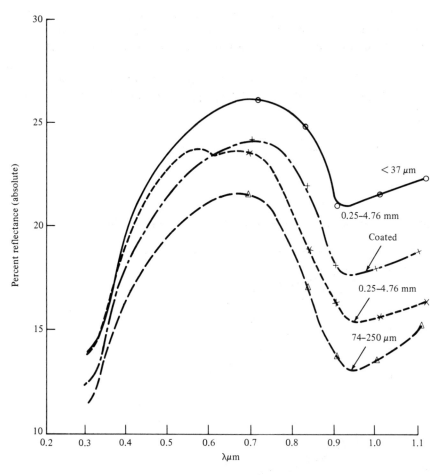

Figure 17.2 Absolute normal reflectance of the Bruderheim meteorite samples. (Data points are shown between 0.7 and 1.1 μm; points below 0.666 μm measured at intervals ≤ 0.03 μm are omitted for clarity.)

and Keester, 1967), and a weak feature near 0.6 μm that usually is attributed to Fe^{3+} (Hunt and Salisbury, 1970). In the case of Bruderheim, Egan and Hilgeman (1972) have shown this absorption feature to be due to FeS.

The 0.6-μm band is not evident in the spectra of the other samples. The ferrous 0.3-μm feature is generally similar in all the samples while that at 0.95 μm shows definite variations. The relative depth (minimum over maximum brightness in percent subtracted from 100) is 15%, 40%, 25%, and 30% for the < 37-μm, 74–250-μm, "coated", and 0.25–4.76-mm samples, respectively. It is reasonable that the band is deeper for the 74–250-μm sample than for the < 37-μm sample, and that the depth is intermediate for the "coated" sample. However, one would expect the 0.25–4.76-mm sample to have the deepest band, which is not the case. This is another indication that the reflectance curve for this sample is anomalous.

Photometric measurements were made as a function of phase angle in the U(0.36 μm), G(0.54 μm), and R(0.67 μm) bands, for 0°, 30°, and 60° incident illumination angle on the four particle size samples of Bruderheim. We have chosen to present these data in two ways: first, in terms of the usual phase curve, which gives the brightness (here normalized to that at a phase of 3.6°) as a function of the phase angle (*B* curve), and in terms of a function that minimizes the effects of geometry (*f* curve).

For a particulate surface in which multiple scattering is small, the intensity of the scattered light can be represented by an equation of the following type (Irvine, 1966; Veverka, 1970):

$$B(i, \epsilon, \alpha) \sim \tilde{\omega}_0 \left\{ \frac{\cos i}{\cos i + \cos \epsilon} \right\} f(i, \epsilon, \alpha), \qquad (17\text{-}2)$$

Figure 17.3 Photometric properties of 0.25–4.76-mm Bruderheim particulates.

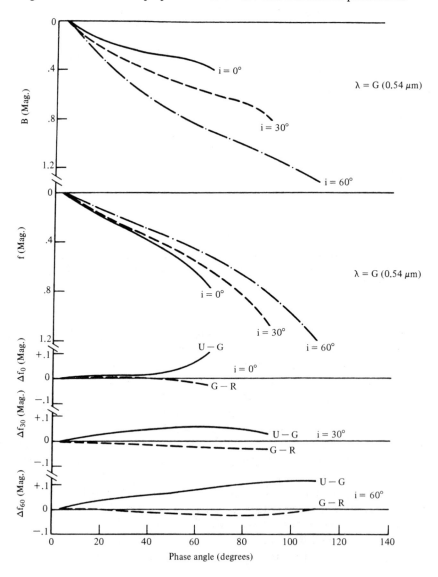

where $\tilde{\omega}_0$ is the single particle scattering albedo, and i, ϵ, α are the incidence, scattering, and phase angles, respectively. Here, f is a complicated function that depends on the phase function of a single particle and on the texture of the surface.

When Eq. (17-2) is strictly applicable, that is, when among other things multiple scattering is negligible, then f does not depend strongly on i or ϵ individually, but only on the phase angle α. In that case, Eq. (17-2) is approximated by

$$B(i, \epsilon, \alpha) \sim \tilde{\omega}_0 \left\{ \frac{\cos i}{\cos i + \cos \epsilon} \right\} f(\alpha). \tag{17-3}$$

Figure 17.4 Photometric properties of 0.25–4.76-mm Bruderheim particulates coated with < 74-μm Bruderheim powder.

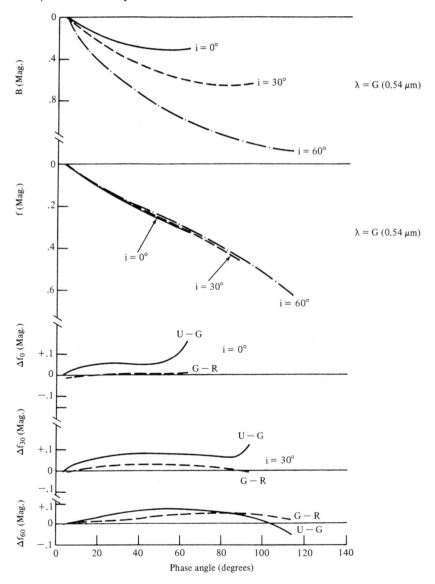

This is analogous to the lunar photometric function derived semiempirically by Hapke (1963; 1965); when $f(\alpha) =$ const, independent of phase angle, Eq. (17-3) reduces to the Lommel-Seeliger law.

The real photometric information about a surface is contained in the f function. In the usual phase curve (B plot), this information is obscured by the purely geometrical factor $\{\cos i/(\cos i + \cos \epsilon)\}$, which is, of course, the same for all surfaces. It is for this reason that the photometric results are presented in terms of both B plots, which are astronomically more convenient, and f plots, which are physically more informative.

The photometric characteristics of the four samples are shown in Figs. 17.3–17.6, and summarized in Tables 17.2–17.5. The figures show both the B curves and the f

Figure 17.5 Photometric properties of 74–250-μm Bruderheim grains.

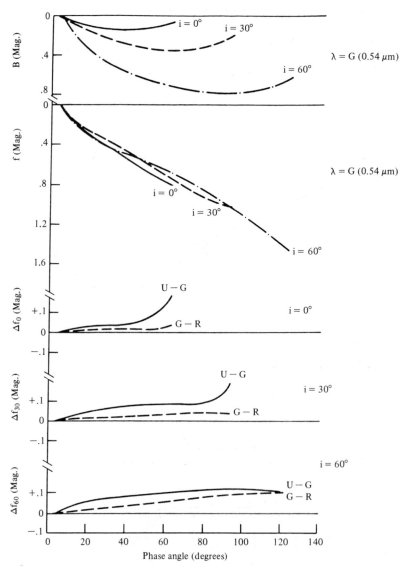

curves as well as the color dependence of the f curves at each of the three incidence angles studied $(0°, 30°, 60°)$. For instance, Δf_{30} give the $(U-G)$ and $(G-R)$ magnitude differences in the f function at an incident angle of $30°$ (note that $\Delta f \equiv \Delta B$). Here $(U-G)$ and $(G-R)$ represent the difference in magnitudes of the spectral regions within the parentheses.

For the 74–250-μm and the "coated" sample (Figs. 17.5 and 17.6), the f function is truly independent of i and is only a function of the phase angle α. This indicates that multiple scattering is not dominant in these two samples. For the < 37-μm sample, the f function shows a small dependence on the angle of incidence, especially at large phase angles (Fig. 17.6), probably due to the increased importance of multiple scattering. The f function for the 0.25–4.76-mm sample (Fig. 17.3) shows some dependence on i at all phase angles. This is not surprising because this sample

Figure 17.6 Photometric properties < 37-μm Bruderheim powder.

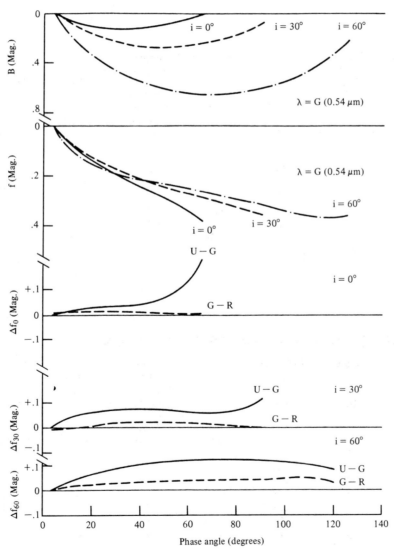

Table 17.2 Photometric Properties of the 0.25–4.76-mm Sample

Color	Angle of incidence (deg)	Reflectance at $\epsilon = 0°$ $i = 3.6°$	f at $\alpha = 60°$	Phase coefficient (B curve) (mag/deg)	Phase coefficient (f curve) (mag/deg)	Opposition effect Δm (B curve)
U	0	0.21	0.77	0.007	0.010	0.04
	30	0.20	0.63	0.012	0.011	0.05
	60	0.16	0.54	0.019	0.009	0.04
G	0	0.24	0.66	0.007	0.010	0.04
	30	0.23	0.58	0.011	0.009	0.05
	60	0.16	0.46	0.018	0.008	0.02
R	0	0.25	0.68	0.007	0.010	0.05
	30	0.23	0.62	0.011	0.009	0.06
	60	0.16	0.48	0.017	0.007	0.05

Table 17.3 Photometric Properties of the "Coated" Sample

Color	Angle of incidence (deg)	Reflectance at $\epsilon = 0°$ $i = 3.6°$	f at $\alpha = 60°$	Phase coefficient (B curve) (mag/deg)	Phase coefficient (f curve) (mag/deg)	Opposition effect Δm (B curve)
U	0	0.20	0.74	0.010	0.013	0.05
	30	0.18	0.69	0.015	0.013	0.07
	60	0.14	0.66	0.023	0.013	0.05
G	0	0.24	0.62	0.009	0.012	0.02
	30	0.23	0.61	0.013	0.011	0.04
	60	0.17	0.59	0.021	0.011	0.05
R	0	0.25	0.61	0.008	0.011	0.03
	30	0.24	0.59	0.012	0.010	0.05
	60	0.18	0.55	0.021	0.011	0.04

Table 17.4 Photometric Characteristics of the 74–250-μm Sample

Color	Angle of incidence (deg)	Reflectance at $\epsilon = 0°$ $i = 3.6°$	f at $\alpha = 60°$	Phase coefficient (B curve) (mag/deg)	Phase coefficient (f curve) (mag/deg)	Opposition effect Δm (B curve)
U	0	0.18	0.55	0.005	0.008	0.06
	30	0.17	0.45	0.010	0.009	0.06
	60	0.14	0.44	0.018	0.010	0.11
G	0	0.22	0.40	0.004	0.007	0.05
	30	0.20	0.37	0.008	0.006	0.05
	60	0.16	0.34	0.016	0.007	0.08
R	0	0.23	0.37	0.003	0.007	0.05
	30	0.21	0.34	0.008	0.006	0.05
	60	0.16	0.29	0.015	0.006	0.07

Table 17.5 Photometric Characteristics of the < 37-μm Sample

Color	Angle of incidence (deg)	Reflectance at $\epsilon = 0°$ $i = 3.6°$	f at $\alpha = 60°$	Phase coefficient (B curve) (mag/deg)	Phase coefficient (f curve) (mag/deg)	Opposition effect Δm (B curve)
U	0	0.22	0.49	0.004	0.008	0.04
	30	0.20	0.34	0.009	0.007	0.07
	60	0.17	0.38	0.017	0.008	0.10
G	0	0.26	0.35	0.003	0.006	0.04
	30	0.24	0.28	0.008	0.006	0.02
	60	0.19	0.26	0.015	0.005	0.08
R	0	0.27	0.35	0.003	0.006	0.04
	30	0.26	0.26	0.006	0.005	0.04
	60	0.19	0.23	0.014	0.004	0.08

consists of coarse, platelike chips, and Eq. (17-2) cannot be applied strictly (Veverka, 1970).

Most of the samples get redder with increasing phase. This is a common property of silicate powders discussed by Adams and Filice (1967), who also found that this trend often reverses at large phase angles (powders becoming bluer with increasing phase). There are indications of this happening in some of our samples at phase angles larger than 70°. The coarse sample, however, is anomalous in this respect. The $(U - G)$ color index shows a reddening with phase, but the $(G - R)$ index does not (Fig. 17.3).

In terms of astronomical applications we are most interested in the photometric properties up to phase angles of about 30°, because for asteroids, larger phase angles are not normally observable from the Earth. Therefore, using the above curves, we have evaluated the following parameters:

the *phase coefficient* (for either the B or f curves), defined to be the slope of the curve between 10° and 30°; and

the *opposition effect*, defined here to be the difference, in magnitude between the actual brightness at $\alpha = 3.6°$ and that inferred by extrapolating to this value from $\alpha = 10°$, using the phase coefficient.

These parameters are tabulated for the various samples in Tables 17.2–17.5. The magnitudes of the phase coefficients and their color dependence are typical of samples of such texture and reflectance. The values of the opposition effect given in the same Tables are estimates, as the measurements were not optimized to study this parameter. Other things being equal, the opposition effect for the powder and "coated" samples is largest at the shortest wavelengths where the albedo is the lowest. This is reasonable because multiple scattering hinders shadowing, and therefore, weakens opposition effects. Also, the < 37-μm and 74–250-μm samples have more pronounced opposition surges than the other two samples, in accordance with their more intricate microscopic texture (Fig. 17.1).

For each sample, polarization measurements were made in the U, G, and R as a function of phase angle for incidence angles 0°, 30°, and 60°. These results are shown in Figs. 17.7–17.10, and are summarized in Tables 17.6–17.9.

As first noted by Lyot (1929), one should expect the magnitude of the positive branch to be inversely correlated with sample albedo. This is because an increase in the importance of multiple scattering within a sample increases the albedo and decreases the polarization. This trend is present in the measurements. Both the amount of polarization at $\alpha = 60°$, and h, the slope of the positive branch near the crossover angle, tend to show an inverse correlation with surface reflectance (Tables 17.6–17.9).

The dependence of the positive branch on sample texture and color is indirect in that texture and color seem to be important only to the extent that they affect the reflectance of the sample.

Up to phase angles of about 60°, the dependence of the polarization curves on incidence angle is small, in all colors and for all samples. To first order, the

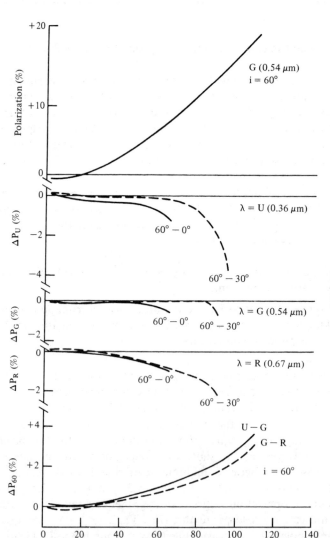

Figure 17.7.   Polarimetric properties of 0.25–4.76-mm Bruderheim particulates.

polarization curves are functions of α only, and do not depend strongly on i or ϵ separately. Only for the coarse sample does the dependence on i become strong, and then only at large phase angles (Fig. 17.7). This is probably due to the fact that the flat, almost flakelike particles of this sample (Fig. 17.1a) tend to scatter quasispecularly at large phase angles.

What are the effects of albedo, color, and scattering geometry on the negative branch? To first order, the negative branch depends only on the phase angle, and not on i or ϵ separately (Tables 17.6–17.9). The depth of the negative branch does not appear to vary systematically with either color or albedo. This is also true for the angle of minimum polarization, α_{min}. The crossover angle α_x appears to be inversely

Figure 17.8 Polarimetric properties of 0.25–4.76-mm Bruderheim particulates coated with < 74-μm Bruderheim powder.

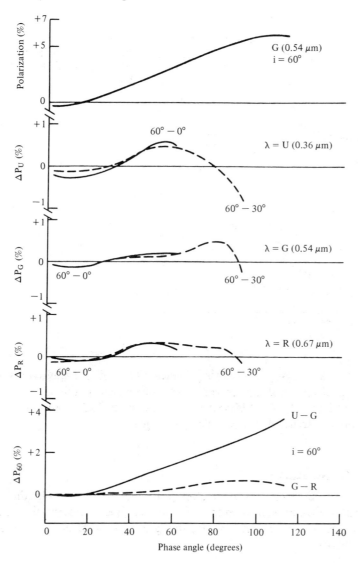

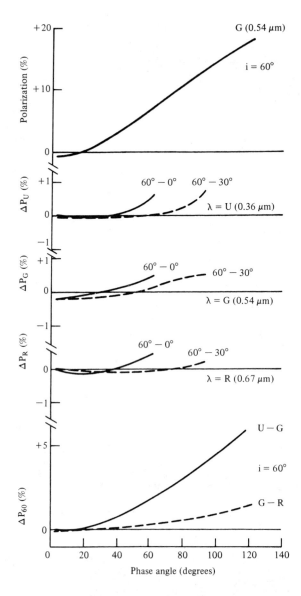

Figure 17.9. Polarimetric properties of 74–250-μm Bruderheim
grains.

correlated with particle size (comparing Tables 17.8 and 17.9) and thus directly
correlated with albedo (and consequently with wavelength). This is in contrast to
observations by Egan (1967) of Haleakala volcanic ashes and a furnace slag, which
showed an opposite trend in this particle size range.

Application to Asteroid Astronomy

An important astronomical task is to attempt to identify, if possible, the asteroid
parent body of the Bruderheim chondrite. Thus three types of information are at our
disposal: the spectral reflectance curves, the photometric phase curves, and the
polarization curves.

The spectral reflectance measurements are of primary importance for two main reasons: first, they are in many cases diagnostic of composition (Adams, 1968; McCord et al., 1970), and, secondly, there is no serious difficulty in going from proper laboratory measurements to disk integrated values. This latter advantage is shared by the polarization measurements because as Figs. 17.7–17.10 show, at small phase angles the degree of polarization would be roughly uniform across the disk. Polarization curves contain important information about the texture of the surface and its albedo, but are much less diagnostic of composition (Veverka, 1970) than spectral reflectance curves.

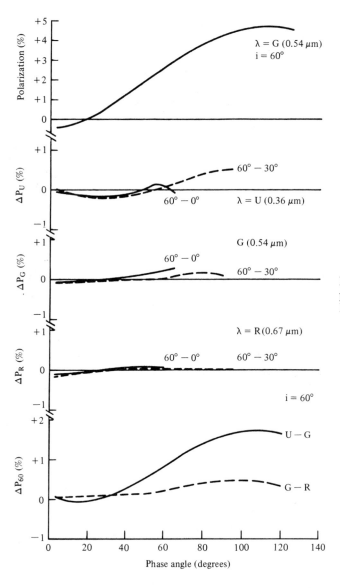

Figure 17.10. Polarimetric properties of < 37-μm Bruderheim powder.

Table 17.6 Polarization Characteristics of the 0.25–4.76-mm Sample

Color	Angle of incidence (deg)	Reflectance at $\epsilon = 0°$ $i = 3.6°$	Polarization at $\alpha = 60°$ (%)	h (%/deg)	Inversion angle (deg)	Minimum polarization (%)	Angle of minimum polarization (deg)
U	0	0.21	+6.6	0.12	15.3	−0.5	7.0
	30	0.20	+6.0	0.10	17.2	−0.5	5.5
	60	0.16	+5.8	0.09	16.5	−0.7	8.0
G	0	0.24	+5.2	0.09	15.8	−0.6	4.5
	30	0.23	+4.8	0.08	16.7	−0.4	3.5
	60	0.16	+4.7	0.09	16.2	−0.6	7.0
R	0	0.25	+4.8	0.08	16.0	−0.4	3.5
	30	0.23	+4.8	0.08	18.4	−0.5	3.5
	60	0.16	+4.0	0.07	17.4	−0.4	3.5

Table 17.7 Polarization Characteristics of the "Coated" Sample

Color	Angle of incidence (deg)	Reflectance at $\epsilon = 0°$ $i = 3.6°$	Polarization at $\alpha = 60°$ (%)	h (%/deg)	Inversion angle (deg)	Minimum polarization (%)	Angle of minimum polarization (deg)
U	0	0.20	+4.3	0.08	13.6	−0.2	3.5
	30	0.18	+4.4	0.09	15.1	−0.3	3.5
	60	0.14	+4.8	0.09	15.6	−0.4	3.5
G	0	0.24	+3.1	0.08	14.3	−0.3	3.5
	30	0.23	+3.1	0.06	14.7	−0.3	3.5
	60	0.17	+3.3	0.08	16.7	−0.4	5.5
R	0	0.25	+2.8	0.05	13.7	−0.3	3.5
	30	0.24	+2.8	0.05	15.8	−0.2	3.5
	60	0.18	+3.0	0.07	17.5	−0.3	3.5

Table 17.8 Polarization Characteristics of the 74–250-μm Sample

Color	Angle of incidence (deg)	Reflectance at $\epsilon = 0°$ $i = 3.6°$	Polarization at $\alpha = 60°$ (%)	h (%/deg)	Inversion angle (deg)	Minimum polarization (%)	Angle of minimum polarization (deg)
U	0	0.18	+6.2	0.12	15.7	−0.5	3.5
	30	0.17	+6.6	0.11	15.9	−0.4	3.5
	60	0.14	+6.7	0.12	16.0	−0.5	3.5
G	0	0.22	+4.4	0.09	15.1	−0.4	3.5
	30	0.20	+4.8	0.09	15.1	−0.4	3.5
	60	0.16	+4.9	0.10	16.8	−0.5	3.5
R	0	0.23	+4.2	0.09	15.4	−0.5	3.5
	30	0.21	+4.7	0.09	15.8	−0.5	3.5
	60	0.16	+4.6	0.09	17.7	−0.5	3.5

Table 17.9 Polarization Characteristics of the < 37-μm Sample

Color	Angle of incidence (deg)	Reflectance at $\epsilon = 0°$ $i = 3.6°$	Polarization at $\alpha = 60°$ (%)	h (%/deg)	Inversion angle (deg)	Minimum polarization (%)	Angle of minimum polarization (deg)
U	0	0.22	+3.4	0.07	16.5	−0.3	7.0
	30	0.20	+3.4	0.07	17.0	−0.4	4.5
	60	0.17	+3.5	0.07	19.3	−0.4	7.5
G	0	0.26	+2.4	0.05	17.0	−0.3	3.5
	30	0.24	+2.6	0.05	17.0	−0.4	7.0
	60	0.19	+2.6	0.05	18.3	−0.4	3.5
R	0	0.27	+2.3	0.05	17.0	−0.4	7.0
	30	0.26	+2.4	0.05	17.5	−0.4	7.3
	60	0.19	+2.4	0.05	20.3	−0.5	7.5

The photometric phase curves are the most difficult to use diagnostically. First, asteroid observations involve disk integrated quantities, whereas laboratory measurements simulate conditions at only a single point on the disk. Thus, an integration of the photometric function over the disk is required before any comparison with asteroid observations can be made. Even then there are complications as the phase coefficients may be significantly altered by large scale surface roughness (Veverka, 1971a) in a way that cannot be uniquely determined from the observations. In other words, phase coefficients calculated on the basis of laboratory measurements can only be compared with asteroids whose surfaces do not have significant large scale roughness. Secondly, there is no evidence to indicate that phase coefficients or their color dependence are diagnostic of composition. As a result, photometry becomes relevant only if the spectral reflectance curves of Bruderheim can first be successfully matched with those of some asteroid. Then it is possible to use the photometric data presented to construct phase coefficients for the disk integrated light, assuming various degrees of surface roughness (Veverka, 1971a), in order to check that these also agree with observations.

Thus, a reasonable plan of attack in trying to simulate asteroid observations with laboratory measurements is the following:

1. Verify agreement of the spectral reflectance curve.
2. Determine if polarization curves are compatible.
3. Check that the photometric properties of the material are consistent with those of the asteroid.

We shall now use this procedure to see if there are asteroids whose surface composition is comparable to Bruderheim. The most extensive study of asteroid spectral reflectance curves is that of Chapman (1972). Of the 32 asteroids analysed by Chapman, only Vesta has a spectral reflectance curve at all similar to Bruderheim (Fig. 17.11a). However, on the basis of more detailed study of the location and shape of the band near 1 μm, McCord et al. (1970) have concluded that this spectrum is better matched by a basaltic achondrite than by an ordinary hypersthene–olivine chondrite such as Bruderheim. In fact, even though the closest match with the data occurs for the < 37-μm Bruderheim powder, there is a distinct discrepancy in the near infrared, where the < 37-μm powder is too dark by about 25% at 1.1 μm. Also shown in Fig. 17.11b are the spectral reflectance curves for Juno and Hebe, both of

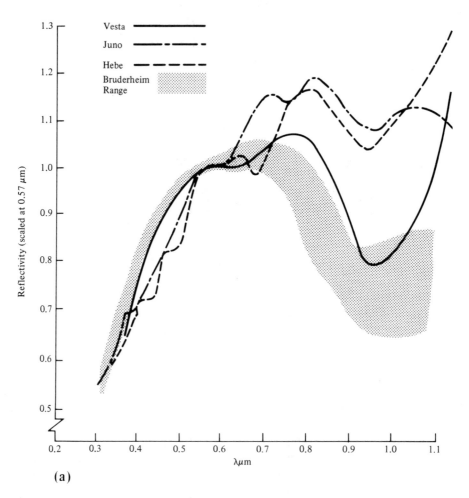

(a)

Figure 17.11 Comparison of the Bruderheim spectral reflectance curves with those of various asteroids; the curves have been normalized to unity at 0.56 μm. (a) Vesta, Juno, and Hebe (Chapman, 1972); (b) Icarus (Gehrels 1970).

which show absorption bands in the UV and near 1 μm, but which do not match the shape of the Bruderheim spectra. Many other asteroids such as Ceres and Pallas have spectra that are not at all similar to that of Bruderheim.

Polarization curves are available for about a dozen asteroids (Veverka, 1970, 1971b, 1971c). Because very few asteroids can be observed from earth at phase angles larger than about 30°, the characteristics of the negative branch (depth, angle of minimum polarization, and inversion angle) must serve as the primary comparison criteria. Generally speaking, asteroid negative branches are deeper and inversion angles larger than those for Bruderheim material. The one exception is Vesta, whose polarization curve is somewhat similar to that of Bruderheim.

It is noteworthy that the spectral reflectance curve of Icarus, an earth-crossing asteroid, bears some resemblance to Bruderheim (Fig. 17.11b). These data are taken from Gehrels et al. (1970) and are difficult to compare rigorously with our measurements as they were made using wide-band filters. To strengthen the argument, the polarization curve of Icarus is consistent with the curves for Bruderheim (Gehrels

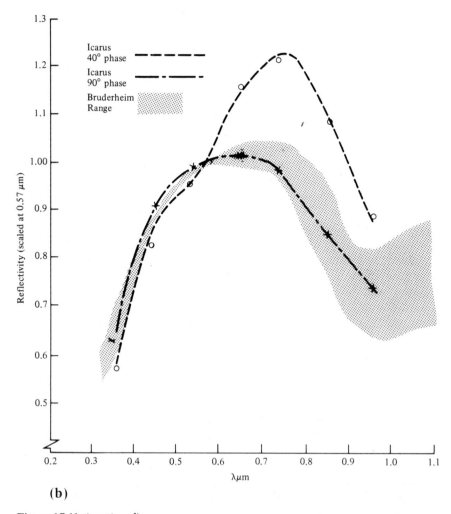

(b)

Figure 17.11 (*continued*)

et al. 1970). As Icarus was only observed over phase angles ranging from 40° to 100°, its negative branch cannot be compared with those of the samples; but at $\alpha = 60°$ the polarization of Icarus is about 3.5% (in the spectral range of the G filter), which compares favorably with the values for the "coated" sample given in Table 17.7.

Bruderheim is a sample of an ordinary olivine–hypersthene chondrite and as such is supposedly representative of the most common meteorite type which falls to earth. Therefore, unless the distribution of meteoroids by composition is very heterogeneous in space, Bruderheim should be representative of one of the most common meteorite types in the solar system. Of all the asteroids studied to date, however, only Icarus and Vesta resemble Bruderheim, and the similarity to Vesta is only qualitative.

Of the many asteroid spectral curves having an absorption band near 1 μm (for example Juno and Hebe), none match exactly, as this band is never deep enough (Fig. 17.11a).

There are at least two possibilities: first, most of the asteroids observed are composed of ordinary olivine–hypersthene chondritic material, but the details of the spectral reflectance curves (such as the depth of the 1-μm band) are selectively modified by some process to produce the observed variations; secondly, the surfaces of most asteroids in the belt do not consist of such material. The latter alternative has obvious important implications.

Consider the first alternative. A number of processes can modify the spectral reflectance curves so as to decrease the relative depth of the 1-μm band. These include: "solar wind darkening" (Hapke and van Horne, 1963; KenKnight et al., 1967; Wehner et al., 1963), decrease of mean particle size (Adams, 1968), and contamination with other materials, including dark glass derived from hypervelocity impacts (Adams and McCord, 1971, 1972). No matter which process is invoked, it must allow for the fact that Icarus and Vesta remain unaffected, and this we believe is impossible to achieve. Icarus and Vesta are on opposite ends of the asteroid size spectrum; consequently, any explanation based on size, such as the ability to retain a regolith, must be ruled out. Also, because Vesta, like Juno and Hebe, is a main belt asteroid, an explanation based on differences in orbital environment seems unlikely. Space weathering is hard to reconcile with the fact that the spectrum of Vesta has been successfully matched with that of a basaltic achondrite. Another process to be considered is that the Bruderheim meteorite has suffered substantial alteration since it fell to earth; however, as discussed previously, we feel that this is quite unlikely. Finally, it could be argued that the infrared portions of published asteroid spectral reflectance curves have a systematic error. However, not only does the data show internal consistency, but the instrumentation used by Chapman has been employed in many applications where this type of error would have been obvious. Thus, we must seriously consider the alternative hypothesis: none of the main belt asteroids studied to date consist of ordinary chondritic material.

It is certainly possible that the sample of meteorites intercepted by the earth need not be compositionally representative of the bulk of the asteroid belt. In fact there is strong evidence, both chemical (Anders, 1964) and orbital (Anders, 1971), that most meteorites are derived from a small number (probably less than a dozen) of parent bodies. An interesting possibility is that some of the Apollo asteroids, such as Icarus, may consist of ordinary chondritic material of this type and that this material, although rare as a *surface* constituent of large asteroids in the main belt, is very common in earth-crossing orbits.

The conclusions presented are based on one chondritic meteorite, more specifically of the L6 class. This does not mean that the results apply to a less metamorphosed sample (e.g., of the L3 class). These results are presented with this caution in mind. Clearly further detailed studies of chondritic meteorites are highly desirable.

The Interstellar Medium

It was established early that the interstellar medium was composed of solid particles. Schalen (1936) and Greenstein (1938) considered iron particles because at that time meteorites were believed to be of cosmic origin; the trend was then away from this belief, but again there is a very serious consideration that the original belief was correct. Lindblad (1935) suggested that the interstellar grains could accrete out of the interstellar medium, and could likely be ices of water, ammonia, or methane (van de Hulst, 1943). This hypothesis was supported by the correlation between interstellar gas concentration and extinction. Subsequently Hall (1949) and Hiltner (1949) discovered interstellar polarization. This caused difficulties in accounting for the observed polarization relative to extinction, because an anisotrophy was required for the interstellar grains. The anisotrophy could be produced by alignment of physically different shapes or optical properties. Others (Kamijo, 1963a, b) suggested silicates as a major constituent of grains and Egan and Hilgeman (1975) suggested that meteoritic material and the mineral bytownite were probable candidates for the interstellar medium.

A wide range of materials with very different optical properties may be used to explain the extinction and polarization because of the large range of free parameters defining the size distribution of the particles in the limited wavelength range from the near ultraviolet to the near infrared. For the best assessment of the chemical composition of the interstellar grains, one must consider the following (Greenberg and Hong, 1973):

the cosmic abundance of the elements,

the wavelength dependence of extinction and polarization,

the average total extinction,

the ratio of polarization to extinction,

the predominantly dielectric character of grains in the visible spectral region,

the infrared spectral characteristics of the grains.

Figure 17.12 Differential absorption versus wavelength for an interstellar medium consisting of various particle sizes of bytownite (Egan and Hilgeman, 1975b).

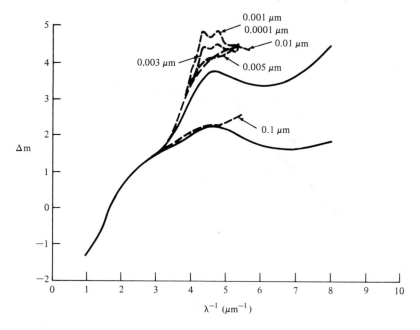

BN–KL spectrum

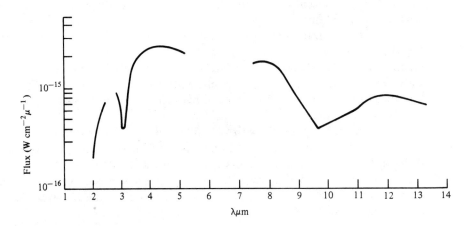

Figure 17.13 Infrared spectrum of the Becklin-Neugebauer object. (Adapted from Gillett and Forrest, 1973; originally published in *Astrophysical Journal*, Vol. 179, 1973.)

Figure 17.14 Comparison of calculated extinction by spinning core–mantle cylinders (solid curves) with observations (dashed curves):

$$R_{obs} = 3.20; \ a_c = 0.08 \ \mu m, \ n(a_m) = \exp\left\{5\left[\frac{a_m - 0.08}{0.12}\right]^3\right\}; \ R_{calc} = 3.19.$$

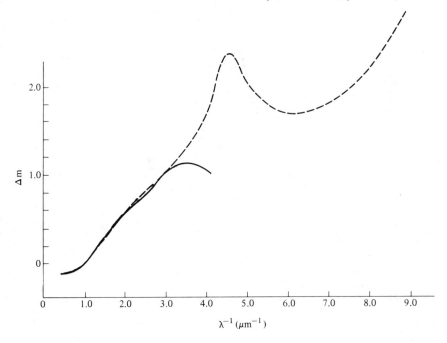

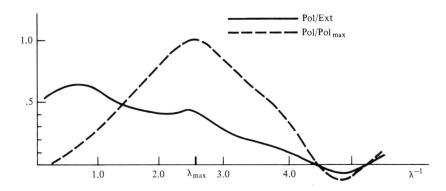

Figure 17.15 Calculated wavelength dependence of polarization and ratio of polarization to extinction by spinning core–mantle cylinders.

As an example, the range of differential extinction by the interstellar medium is delimited in Fig. 17.12 by the solid lines. On the dotted lines are indicated the particle sizes that produce each curve for the mineral bytownite; a reasonably good match is produced photometrically. Of particular interest is the 0.22-μm wavelength peak in the curve, although it is not specifically indicative of a silicate component in the interstellar medium; more important is a broad feature of emission and absorption centered at 9.7 μm. Concurrently, an absorption feature for ice should be present at 3.07 μm; Gillett and Forest (1973) found these to be present in the Becklin–Neugebauer object (Fig. 17.13).

Greenberg (1973a,b,c) has theoretically demonstrated that spinning core mantle cylinders have photometric and polarimetric properties that somewhat approximate the observations. Thus the calculated extinction is shown in Fig. 17.14 and the polarization/maximum polarization and corresponding polarization/extinction ratios are shown in Fig. 17.15 for one model. The visual observed ratio of polarization to extinction of ~ 0.4 (Greenberg and Hong, 1973) may be achieved.

Also, circular polarization has been observed in the intersteller medium (Kemp, 1973; Martin 1973a, Martin et al., 1973b). The inference is that metallic grains such as graphite are excluded because they would only produce linear polarization. However, Landaberry and Magalhaes (1977) explain interstellar circular polarization with a simple two cloud model using magnetite as the polarizing agent.

A recent study of the polarization of interstellar dust by UBVR photometry (Neto et al., 1982) concluded that no simple correlation can be obtained with UBVR photometry, but that finer wavelength resolution is required.

18

Atmospheric Constituents

Introduction

It is recognized that man can affect his environment in some ways obvious, in other ways subtle. One aspect is inadvertent modification of the earth's climate (Singer, 1970; Matthews, 1971; Landsberg, 1970; NCAR, 1980). In addition to localized thermal sources, two by-products of our industrialized society that are conjectured to produce long-term global climatic effects are suspended particulates (aerosols) and carbon dioxide. Climatic changes occur over large time scales, and controlled experiments in the atmosphere are difficult to perform. Thus, a practical way of assessing the effects of varying concentrations of these products is by computer modeling. The physical processes of atmospheric absorption and scattering of the incident solar radiation and the absorption and emission of the outgoing terrestrial and atmospheric thermal radiation define the thermal structure of the atmosphere and the resulting terrestrial surface temperatures.

The amount of atmospheric CO_2 has been noted to rise in recent decades. Because CO_2 is uniformly mixed and nonreactive in the atmosphere and relatively transparent to visual radiation (although strongly absorbing in the infrared), it is relatively easy to model. With the inclusion of the effect of clouds and the assumption of a reasonable tropospheric lapse rate, Manabe and Weatherald (1967) predicted a surface temperature increase of 2.4°C if the CO_2 atmospheric content is doubled. But Rasool and Schneider (1971) predicted a temperature increase of 0.7°C from doubling CO_2 using a slightly different model. The increase in CO_2 cannot account for decrease in annual mean terrestrial surface temperature observed since 1940 (Mitchell, 1961). There is much impressive evidence (Russel and Archibald, 1888; Meinel and Meinel, 1963) that indicates that there is an increase in atmospheric dust following a big volcanic eruption. If the eruption injects debris into the stratosphere, the atmospheric turbidity increase may last for several years. Also, fairly conclusive and well documented evidence (McCormick and Ludwig, 1967; Flowers, McCormick and Kurfis, 1969) indicates an increase of 70% in atmospheric turbidity from human activity over highly populated areas. Another interesting effect is that although the atmospheric turbidity over the North Atlantic has increased by 20% over a recent 60-year period, there is no detectable increase in the South Pacific atmospheric turbidity (Cobb and Wells, 1970), nor is there any detectable increase in turbidity over Mauna Loa (Ellis and Pueschel, 1971). They also found that an increase in atmospheric turbidity following the 1963 Mt. Agung eruption had dissipated in the 1963–1970 period and that the optical depth of the atmosphere over Mauna Loa was similar in 1958 and 1970. Thus, there is no conclusive evidence to show a general buildup of atmospheric aerosol content either from human activities or volcanic eruptions. The reason for the lack of buildup is the natural processes that purge aerosols from the atmosphere; in the troposphere, rainfall brings down all size aerosols, and the earth's gravitational force brings down the larger ones. In the stratosphere, photochemical reactions may break down the aerosols or the aerosols may descend into the troposphere by atmospheric mixing processes, or from gravitational forces, over a long time period.

Modeling of the effect of atmospheric aerosols is much more complex than modeling of atmospheric CO_2. The aerosols are not uniformly mixed, and vary spatially in concentration, composition, size distribution, shape, and optical complex index of refraction. Recent work has contributed significantly to the better definition of these parameters (Egan and Hilgeman, 1979; Egan 1982, 1983). Humidity can also affect the size and optical properties of aerosols (Hanel, 1968). The effect of atmospheric dust can produce atmospheric heating or cooling depending on the relative effects of scattering and absorption (Charlson and Pilat, 1969). Newell (1970) has presented evidence of warming of the stratosphere between the 60–150 mb levels, with negligible effect on temperatures near the earth's surface.

As an example of the effect of a well defined veil of aerosols, consider the region over Mauna Loa Observatory following the El Chichón volcanic eruption in 1982.

The Mauna Loa Observatory (MLO) on the island of Hawaii is almost ideally situated for establishing baseline measurements for aerosol monitoring. The site is far from landmasses, and local pollution is normally confined to the atmospheric layer below the trade wind regime temperature inversion. Further, there is a large complement of locally situated supporting instrumentation such as atmospheric LIDAR probes, nephelometers, and meterological sensors. Because of the normal clear atmospheric conditions, Mauna Loa Observatory is also an ideal site for the observation of the subsequent specific optical effects of the stratospheric volcanic dust from the El Chichón eruption that occurred on April 4, 1982. The dust has persisted in the stratosphere, but had moved irregularly to the north and south of the original range of latitudes (Strong et al., 1983).

The worldwide monthly location of the volcanic veil has been charted in considerable detail by Strong et al., May 1982 to March 1983 (Strong et al., 1983). They used the difference between the apparent sea surface temperature sensed by the NOAA-7 AVHRR and the actual to determine areas of IR absorption and thus locate the veil and the opacity. In May 1982, the veil was positioned mainly along 15°N latitude. Small regions appeared as far as 45°N latitude in June through October 1982. Also, portions appeared over the equator in June through September and December 1982. By November 1982, the veil was widely and nonuniformly distributed in the Northern Hemisphere with a high concentration in the region above MLO.

During the period spanning the MLO observations (November 5–7, 1982), the observed vertical visual optical depths of the atmosphere above MLO were high (0.3) as contrasted with the normal value of 0.01. The trade wind regime (normally an east wind) had broken down (strong westerly winds), and the sky was unusually bright.

There is a continuing interest in the effect of this particulate material in the atmosphere on short- and long-term climatic trends. Although the El Chichón veil is gradually dispersing, the question also persists as to whether, on a worldwide basis, aerosol particles in general are increasing or decreasing in the long term. The associated question of whether the aerosols contribute to a net heating or cooling of the earth depends on their optical complex index of refraction, size distribution, and number density.

A particularly sensitive technique for detecting the presence and approximate optical properties of aerosols is achieved by measuring the polarization and photometry of the light scattered and absorbed by them as a function of wavelength. The geometrical variation of the scattered radiation with illuminating and viewing geometry furnishes additional factors for correlation. The optical scattering and absorption effects of aerosols are significantly different from those of gaseous atmospheric constituents such as oxygen, nitrogen, and ozone, and at the high altitude of the MLO (3460 m), the effects normally are reduced.

There have been many observations under "clear" conditions of the polarization and photometry of the sky above MLO (Coulson, 1978; Herbert, 1979; DeLuisi, 1980; Bodhaine and Harris, 1981), particularly in the zenith direction during sunrise or sunset, to determine atmospheric turbidity. The Coulson (1978) measurements were made at wavelengths from 0.365 to 0.8 μm (beyond the Fesenkov, 1965) at one-thirteenth the optical depths of Fesenkov (1965). The maximum sky polarization was found by Coulson (1978) to be dependent on the surface albedo and was of the order of 80%; for a low surface albedo at a wavelength of 0.80 μm, the polarization approaches 100%.

Observational Programs

The verification of the applicability of the Dave model to the specific optical phenomena of the El Chichón veil requires measurements of the photometric and polarimetric properties of the sky at a variety of wavelengths. The necessary observations were begun at sunrise at MLO on November 6, 1982 and were continued until 1 p.m. local time. The MLO temperature and relative humidity at 7 a.m. local time were 41°F and 12%, respectively, becoming 55.5°F and 55% at noon. A previously described spectropolarimeter/photometer was used in the program (Egan, 1968). Observations were made at wavelengths of 0.36, 0.40, and 0.5 μm at solar elevation angles between 10° and 54°, sensor zenith angles of 90°, 85°, 80°, 70°, 45°, and 10° at four compass azimuthal directions of 0°, 90°, 180°, and 270°.

The interference filters for wavelength selection of 0.36, 0.40, and 0.50 μm had bandwidths of 0.023, 0.015, and 0.013 μm respectively (FWHM). Photometric and polarimetric calibrations were made from a solar illuminated diffuse white Nextel painted panel concurrently optically sensed with a Soligor spot photometer configured to permit spectral measurements at wavelengths of 0.400, 0.433, 0.500, 0.533, 0.600, 0.633, 0.7, and 1.0 μm. Further, the specular incident solar radiation was measured at these wavelengths, as well as at 1.4 and 1.5 μm with an Eppley Model NIP normal incidence photometer. The total diffuse-plus-specular incident radiation was measured with an Eppley Model 8-48 black and white pyranometer. In addition, a Volz photometer was used to determine optical depths at wavelengths of 0.342, 0.380, 0.500, 0.868, 0.946, and 1.67 μm and also to permit a determination of the total precipitable water.

The temperature, humidity, and wind speed and direction were also recorded, as well as the scattering coefficients from the four wavelength GMCC nephelometer. Data on the height of the volcanic aerosol layer were obtained from the LIDAR.

Rawinsonde data on the temperature, dew point depression, and wind speed and direction were obtained from the National Weather Service at Hilo, Hawaii, as well as GOES West visible, Hawaiian Island area and full earth disk 11 μm infrared imagery.

Results

The development and presentation of the results involves basically a threefold process: (1) describing the observational data on the aerosols as input to the vector atmospheric model; (2) exercising the model with appropriate representations of the underlying terrestrial surface; and (3) comparing the model output with both the observed optical depths of the atmosphere and the polarimetric and photometric properties of the atmosphere (sky) as a function of wavelength. Following a reasonable validation of the model with available experimental data, an inference may then be made as to the composition, quantity, and optical effects, if any, of the El Chichón aerosol layer. The results will now be described in the order indicated.

The size range of the modeled aerosols are of radii between 0.005 and 4 μm, and the assumed size distribution will be a modified gamma function, representative of natural ambient aerosols (Egan, 1983; Deirmendjian, 1969). A subsidiary distribution of small particles is included to account for asperities and edges on the naturally occurring aerosols (Egan, 1983). Initially, in the modeling, the optical properties of the El Chichón ash were assumed to be close to those of the Mt. St. Helens' ash that remained aloft long enough to be collected at a large distance from the eruption in 1980 (Spokane, Washington). The Mt. St. Helens' ash is colorless andesitic glass with the composition and optical properties differing as the distance from the volcano to the collection site is increased; the material collected at Spokane had a lower refractive and absorption component of index than that collected at Kelso, Washington, and near Mt. Ranier (Egan and Selby, 1980). These are the indices of refraction presented in Table 18.1. However, the material in the El Chichón aerosol layer consists mainly of sulfuric acid (Rosen and Hoffman, 1983; Harder et al., 1983) and

Table 18.1 Input Parameters for Vector Atmospheric Programs: Mt. St. Helens' Ash[a] and 75% Sulfuric Acid

Wavelength (μm)	Index of refraction	
	Mt. St. Helens' ash	75% sulfuric acid
0.36	$1.435 - i0.0002218$	$1.452 - i\,(<10^{-8})$
0.400	$1.419 - i0.000200$	$1.438 - i\,(<10^{-8})$
0.500	$1.422 - i0.000176$	$1.432 - i\,(<10^{-8})$

[a] Collected at Spokane, Washington.

the real portion of the indices of refraction of 75% sulfuric acid fortunately are quite close to andesitic glass (Table 18.1); thus, the scattering effect of the El Chichón aerosol will be almost the same as that for andesitic glass, even though it is composed of sulfuric acid. The effect of the small absorption of sulfuric acid compared to andesitic glass must be evaluated from the Dave atmospheric model.

Three wavelengths are selected to be checked in the vector model, and these correspond to the polarimetric/photometric wavelengths of observations: 0.36, 0.400, and 0.500 μm (Table 18.1). The atmospheric particle size distribution was taken as the Deirmendjian Haze H ($\alpha = 2$, $\gamma = 1$, $b = 20$, $r_c(\mu) = 0.100$, $\alpha = 1.18E + 07$) to represent coastal or marine aerosols, as characteristic of the region around MLO. Augmentation was based on matching the model to the observed volume scattering coefficients at MLO for the Haze H model. The actual aerosol vertical number density profile as used in Parts IV or V of the Dave program is determined by adjusting the Dave number density distribution above the height of MLO to match the observed aerosol number density at MLO. The number density of condensation nuclei during the observation period ranged from 158 to 224/cm^3; the number of the larger aerosols was then somewhat smaller (100/cm^3).

However, the vertical aerosol number density of necessity was increased between 16 and 30 km altitudes with a peak at 23 km to account for the El Chichón aerosol layer. The measured volume scattering coefficient at MLO as given by the GMCC four wavelength nephelometer for a wavelength of 0.455 μm ranged from 3.33×10^{-7} to 3.96×10^{-7}/m. The calculated value, at 0.500 μm using the Dave program and the andesitic glass optical properties, was 2.3×10^{-7}/m, thus agreeing with the observed at 0.455 μm within a factor of 2. The AFGL pressure, ozone, and H$_2$O vertical profiles for the tropical (Palmer and Williams, 1975) were used for the remaining variables in the radiative transfer model. The surfaces underlying the atmosphere were represented as diffuse clouds of 80% reflectivity, the condition present in the valley areas below MLO. The output of the vector programs will be presented on the observational data graphs to follow.

The comparison observational data for the model consists of three groups: (1) the vertical optical depth of the atmosphere (measured times secant of the zenith angle) between 0.342 and 1.67 μm (Table 18.2); (2) the direct and diffuse sky radiation properties and (3) the photometric and polarimetric observations of the sky at 0.36-, 0.400-, and 0.500-μm wavelengths in the cardinal directions (north, south, east and west) at viewing angles between horizontal and 80° elevation (Figs. 18.1–18.4). Figure 18.5 shows the final vertical aerosol profile that produced a reasonable match of the observational data.

In detail, Table 18.2 lists the observed vertical optical depths of the atmosphere above MLO at wavelengths of 0.342, 0.380, 0.500, 0.868, 0.946, and 1.67 μm. These data were acquired with a Volz photometer. The vertical optical depths are obtained by correcting the photometer reading for the solar zenith angle at the time of observation. There is a general increase in optical depth with increasing solar zenith angle; the amount of precipitable water also increases (as determined by the ratio of readings at $\lambda = 0.946\mu$m to that at $\lambda = 0.868\mu$m). The data of Strong et al. (1983) indicates the November localized position of a portion of the El Chichón aerosol over MLO, which would corroborate the greater optical depths nearer vertical viewing of the sun. For the observation at the solar zenith angle of 36°, at the time of the Hilo radiosonde, the optically calculated precipitable water above MLO (0.58

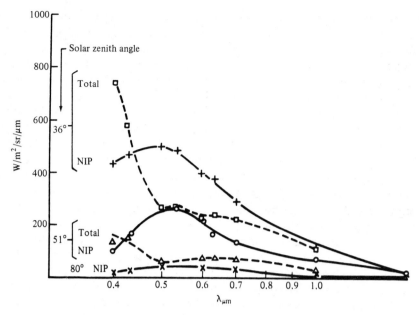

Figure 18.1 Horizontal direct and total diffuse radiation at Mauna Loa Observatory on 11/6/82 as a function of wavelength and solar zenith angle.

cm) was greater than that deduced from the radiosonde observations by 0.40 cm. Since sulfuric acid in varying concentrations absorbs in roughly the same spectral regions as water, although much more strongly, the large optical depth of 0.841 is attributed to the aerosol veil. The normal visual optical depth is < 0.01, in the absence of the El Chichón aerosol veil, and the additional optical depth also is

Table 18.2 Vertical Optical Depths of Atmosphere

λ (μm)	Solar zenith angle		
	10°	50°	36°
0.342	0.584	0.667	0.794
0.36 (model)	0.535	0.535	0.535
0.380	0.441	0.565	0.654
0.4 (model)	0.397	0.397	0.397
0.500	0.243	0.356	0.428
0.5 (model)	0.209	0.209	0.209
0.868	0.111	0.133	0.188
0.946	0.277	0.569	0.841
1.67	0.0556	0.142	0.181
Optically calculated precipitable water	0.29 cm	0.48 cm	0.58 cm
Rawinsonde water vapor	. . .	. . .	0.40 cm

Table 18.3 Characteristics of Incident Radiation

Sensor elevation angle (deg)	Observations			Model		
	Surface direct irradiance (W/m^2)	Total irradiance (W/m^2)	Ratio: $\left(\dfrac{direct}{total}\right)$	Surface direct irradiance (W/m^2)	Total irradiance (W/m^2)	Ratio: $\left(\dfrac{direct}{total}\right)$
10	93	201	0.460	79	184	0.429
35	595	750	0.793	571	726	0.787
55	779	847	0.920	904	1068	0.876

inferred to be caused by the nonuniformly distributed aerosol veil. There were no cirrus clouds during the morning measurements.

The inverse wavelength dependence of the optical depths would indicate that the particles are smaller than the radiation wavelength (perhaps of average size ~ 0.01 μm).

The observed properties of the incident radiation are shown in Fig. 18.1 and Table 18.3. In Fig. 18.1, the specular radiation on a horizontal surface (solid lines) increases with solar elevation. These readings were made using the Eppley pyrheliometer, corrected for solar zenith angle, and appropriate narrow band interference filters. The total radiation of a horizontal reference standard (a white Nextel painted panel) is also shown in Fig. 18.1 (dotted lines). These determinations were made with calibrated photometers (Soligor and Honeywell). Measurements beyond a 1-μm wavelength were not possible because of lack of sensor response. The rise of the curves at 0.4 μm is the result of Rayleigh sky scattering.

The numerical values for the total observed irradiance are given in Table 18.3. The ratio of specular to total is seen to increase for lower solar zenith angles because of decreased scattering path through the volcanic aerosol and attendant lower scattering.

The photometric and polarimetric observations are shown in Figs. 18.2–18.4; part (a) of each figure presents the photometry, and the second part (b) the polarization. Measurements were made at wavelengths of 0.36, 0.400, and 0.500 μm, both photometric and polarimetric measurements being made simultaneously for a given geometry and wavelength. The photometric results will be discussed first and then the polarimetry.

Photometry

Figure 18.2a presents the sky radiance at a wavelength of 0.36 μm; the observations are shown as lines through the data points in the four cardinal magnetic compass directions. The solar zenith angle was $40° \pm 3°$ during measurements; however, only one model with a solar zenith angle of 40° was used because of the small range of solar zenith angles. The model results are shown as indicated in the legend. In general, the model results are below the observations by a factor of 2. The southerly direction has the highest radiance because of the southerly direction of the winter sun. Although the models show the same brightness at zenith, the south and west observations are larger than those north and east at a sensor elevation angle of 80°;

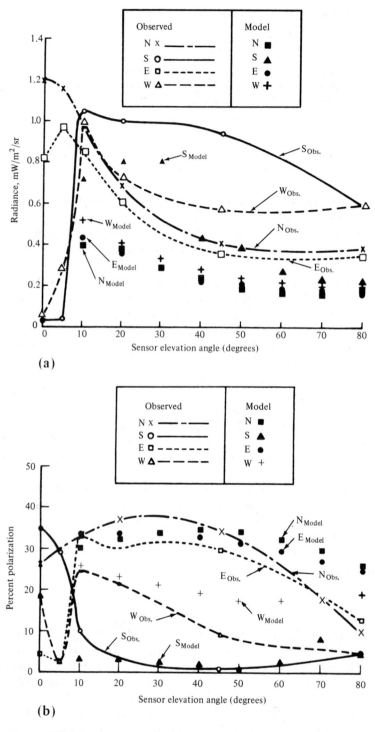

Figure 18.2 Optical properties of sky above Mauna Loa Observatory 11/6/82; $\lambda = 0.36 \ \mu$m and comparison to Dave vector model; solar zenith angle of $40° \pm 3°$ in four azimuthal directions. Lines denote observations. (a) Photometric; (b) polarimetric.

this appears to be the result of a nonuniform sky near zenith and caused by afternoon thin cirrus clouds and time lapse between the measurements. The 0.36-μm measurements were made shortly after noon, as clouds began to form.

Near the horizon (sensor elevation angle of 10°), the radiance increased from increased atmospheric scattering, being highest in the sun direction of south. The south and west directions reflectivity at 0° elevation angle is low because of low reflectivity volcano fields that were viewed by the sensor, whereas in the north and east the sensor viewed in part some high reflectance cumulus clouds in the adjoining valley.

In Fig. 18.3a, the photometry for a wavelength of 0.400 μm is shown; the solar zenith angle was 55° $\pm$ 3°, characterizing an early morning observation. Again, the observational data is represented by lines through the data points, and the model (for 55° solar zenith angle) as indicated in the legend. Here the easterly direction has the highest radiance because of the sunrise in the east. It is seen that near zenith (80° sensor elevation angle), all four observations lie close together, indicating a uniform sky for these observations; there is also close agreement with the model. Again, toward the horizon (sensor elevation angle of 10°), the increased scattering causes the sensed radiance to increase. The model shows close agreement in the north, south, and west directions.

The photometric observations at $\lambda = 0.500$ μm are presented in Fig. 18.4a, again as solid lines through the data points for a solar zenith angle of 45° $\pm$ 3°; the model (solar zenith angle of 45°) results are shown designated in accordance with the symbols of the legend. The near zenith observations (sensor elevation angle of 80°) show some evidence of a slight sky nonuniformity, with the southerly observed radiance being decreased because of distant cirrus clouds (evident in the satellite imagery). Since the sun was in the southerly direction, the observed radiance should have increased as the sensor elevation angle decreased. The sky radiance at $\lambda = 0.500$ μm is less than that at 0.36 μm in the north and east directions. The model at $\lambda = 0.500$ μm is quite good for all directions, with the exception of south at the lower sensor elevation angles.

Polarimetry

The polarimetric trends are more sensitively dependent on the model than radiance. As indicated by the results in the photometry section, it appears that we have a reasonable modeling approach to account for the aerosol size distribution that accounts for aerosol edges and asperities of the El Chichón veil. In Fig. 18.2b ($\lambda = 0.36$ μm), there is remarkable agreement between the observations and the model; the west model produces twice the observed polarization and there is a considerable difference between the near zenith model polarizations (80° sensor elevation angle). The model polarizations are high, which would indicate inadequate near zenith depolarization (scattering) in the model compared to the actual observational condition. This conclusion is borne out in the fact that, as noted in the corresponding photometry section, thin cirrus clouds were developing that served to increase atmospheric scattering near the zenith and thus reduce polarization as shown. The polarization increases near the horizon, which is to be expected from the phase function behavior in Mie scattering angles (between 120° and 180°), the

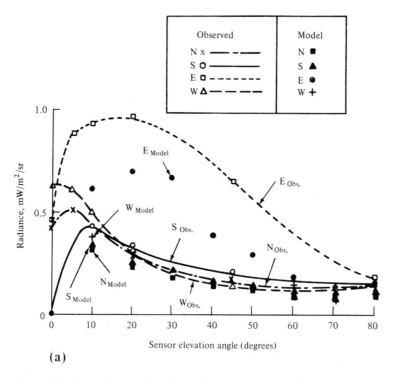

Figure 18.3 Optical properties of sky above Mauna Loa Observatory 11/6/82: $\lambda = 0.400 \ \mu$m and comparison to Dave vector model; solar zenith angle of $55° \pm 3°$ in four azimuthal directions. Lines denote observations. (a) Photometric; (b) polarimetric.

polarization generally is large for nonabsorbing or weakly absorbing spheres. For smaller scattering angles, and for usual distribution of particle sizes, the polarization decreases. The north, south, and west polarizations are high at the horizon because surface polarization increases with the phase angle (the angle between the sun and sensor measured in the plane containing the incident solar direction and the sensor viewing direction). The easterly direction has the smaller phase angle.

The polarization for the $\lambda = 0.400 \ \mu$m wavelength is presented in Fig. 18.3b. The absolute agreement between the model and the observations is not as good as for $\lambda = 0.36 \ \mu$m. The problem appears to lie in the inadequacy of the plane parallel geometry model to represent the actual physical conditions at high solar zenith angles, such as that existing at the time of observation ($55° \pm 3°$). The lack of inclusion of horizontal atmospheric inhomogenieties and the curvature of the earth appear to be the major deficiencies in the model. However, the observational trends are followed in the model for the most part.

The polarimetric properties of the sky at the $\lambda = 0.500 \ \mu$m wavelength are shown in Fig. 18.4b. Here the model depicts higher polarization near the zenith (sensor elevation angle of $80°$), indicative of possible increased atmospheric scattering at the time of observation from thin cirrus clouds. At lower sensor elevation angles, the agreement between the model and the observations becomes better, even near the horizon. The north and east polarizations are higher than the south and west because of the higher scattering phase angles of the former.

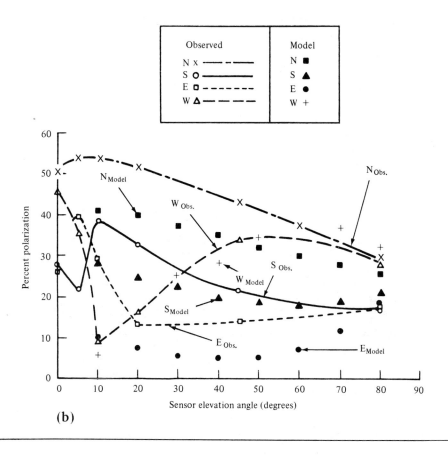

(b)

Discussion

Thus, it is seen that the Dave vector model reasonably represents the atmospheric photometric and polarimetric properties when accurate input parameters are employed. The inferred particle vertical number density distribution for the El Chichón aerosol that produced the matches to the observations is shown in Fig. 18.5, and the associated Rayleigh and Mie scattering and absorption in Fig. 18.6. As was indicated in Fig. 18.5, the number density at 23 km was increased to give the best fit to all photometric and polarimetric observations, with an ordinary Braslau and Dave distribution below 16 km and above 30 km; also, as previously indicated the overall distribution was adjusted to produce a number density of $100/cm^3$ at the altitude of MLO. From the number density and the size distribution, one may obtain a rough index of the amount of volcanic ash in the stratosphere; a rough calculation shows it to be on the order of the 4×10^6 tons previously calculated on a worldwide basis (Ashok et al., 1982).

As mentioned previously, the optical properties of the El Chichón ash were taken to be either sulfuric acid or andesitic glass, both which have nearly the same real index of refraction. Fortunately, a small change in the real index of refraction has negligible effect on the scattering (Egan, 1983), but would contribute to the absorption. The higher absorption portion of the index of refraction of the andesitic glass (which may compose 5%–10% of the dust veil) is not large enough to cause an

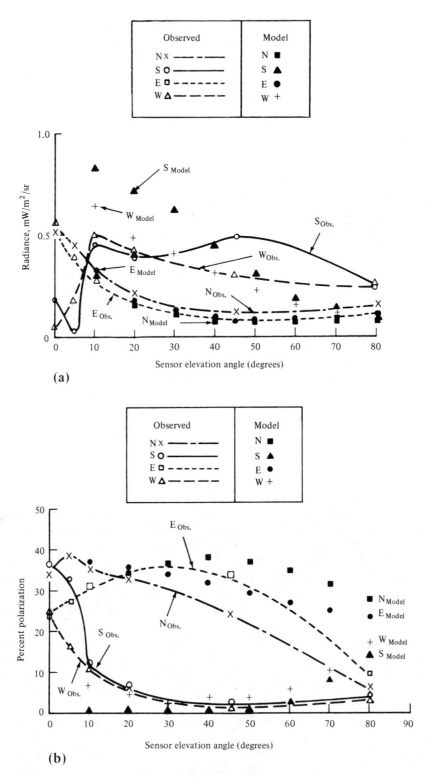

Figure 18.4 Optical properties of sky above Mauna Loa Observatory 11/6/82; $\lambda = 0.500\ \mu$m and comparison to Dave vector model; solar zenith angle of $45° \pm 3°$ in four azimuthal directions. Lines denote observations. (a) Photometric; (b) polarimetric.

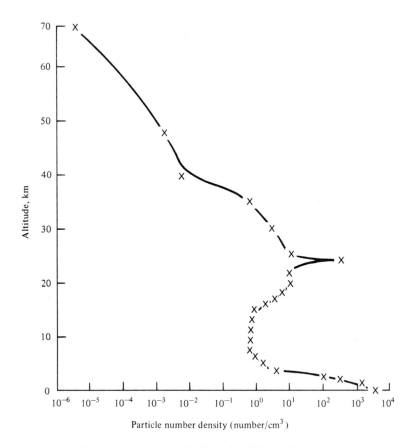

Figure 18.5 Vertical aerosol profile from the 32 layer Dave vector radiative transfer program.

appreciable effect on the aerosol scattering (Egan, 1983), but would contribute to the absorption. Thus, the calculated scattering properties of the aerosol cloud are valid. Near the surface of the earth, there will be a slight error in assuming adesitic glass as the local aerosol, but from the computer model, it is evident that the major atmospheric effect on photometry and polarization is the result of the El Chichón high altitude aerosol, and not the low concentration of near surface aerosols.

As seen in Fig. 18.6 at a 0.500-μm wavelength, the aerosol veil maximum at the 23-km altitude is a strong Mie scatterer, but only a weak absorber, and one could speculate that a relationship could exist between the El Chichón veil and El Niño, the reverse trade wind phenomena. There is no question that the large scattered veil from El Chichón will effect a redistribution of the incident solar radiation mainly by scattering. This scattering radiation will be directed primarily south of the veil as well as in the retrodirection toward the sun. Preliminary modeling of the radiation indicates that direct solar radiation reaches the surface in the equatorial region together with radiation scattered from the El Chichón veil, generally positioned north of the equator. For a uniformly diffusely reflecting veil located at 20°N latitude, the effect of the scattered radiation is negligible (0.02%–0.2%). For cloud locations just north of the equator (and depending upon location), the increase could be between 4% and 8%. The amount of this additional scattered energy that could

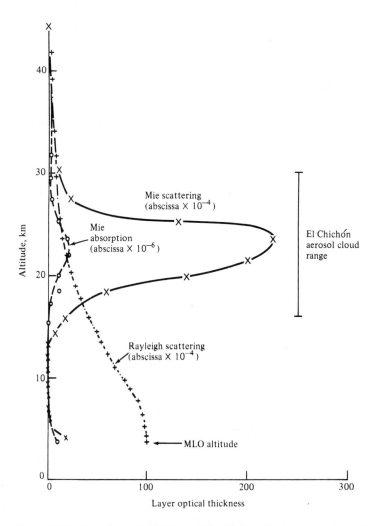

Figure 18.6 Plot of modeled Mie and Rayleigh scattering, and absorption (at a wavelength of 0.500 μm) versus altitude showing scattering and absorption by the El Chichón aerosol cloud.

reach and be absorbed in part by tropical cloud layers, haze, atmospheric ozone, or water vapor could be as much as 15%. The net result would be a generally increased incident radiation in equatorial regions and a decreased radiation beneath the veil.

An examination of the radiosonde temperature data shows that the El Chichón aerosol layer was heated sufficiently in its location above the tropopause at 38 mb level (22.4 km) to prevent the normal descent of the air mass within which it was located (see Fig. 18.7). The normal lapse rate, for November 5–7, 1981, is to be compared to the daytime November 5–7, 1982. The effect of the El Chichón aerosol veil (peaking just below the 30 mb level) is to cause a stratospheric temperature rise of 20°C, enough to disrupt the normal stratospheric circulation pattern that initiates easterly trade winds. As expected, at night (November 6, 1982), the temperature decreases somewhat because of the lack of solar heating of the aerosol layer. The higher temperatures caused the air mass to rise, concurrrently lifting the air above it and allowing the air mass below it to expand upward. An approximate indication of

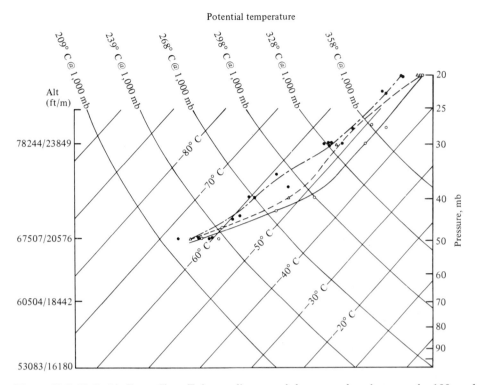

Figure 18.7 U. S. Air Force Skew T, log p diagram of the atmosphere between the 100- and 20-mb levels with Hilo, Hawaii radiosonde data for November 5–7, 1981, and 1982. November 5–7, 1981 (— · —); November 5–7, 1982 (——); November 6, 1982, at night (---).

the relative vertical air mass motion is indicated by the vertical transect times of the radisonde balloon: the average time to traverse from the 100 mb to 20 mb levels was 54 sec longer on 11/4–7/81 than 11/5–7/82; the average total time to rise from the ground to the 20 mb level was correspondingly 230 sec longer. This rising air accompanied a breakdown of the trade wind regime at that time and could be instrumental in producing and prolonging the El Niño phenomena. It is interesting to note that the average precipitable water vapor, derived from the radiosonde observations, was 3.76 cm on 11/4–7/81 and 3.08 cm on 11/5–7/82 (i.e., less precipitable water vapor under the El Chichón veil).

19

Oceanography

Introduction

The global distribution of sea state is of great interest because it determines local ocean and coastal processes and is indirectly an indicator of the wind field above the sea. The sea state also is affected by surface flotsam such as oil slicks and thus has an environmental interest.

Present techniques for remotely measuring sea state include passive microwave radiometry, radar scatterometry, and laser altimetry. A new system has been developed that indicates the wave height by means of the polarization of the sea's emitted and reflected infrared radiation.

It has been recognized for a long time that emitted radiation from surfaces is polarized. In 1895 R. A. Millikan published two papers explaining the polarization of light emitted by incandescent solid and liquid surfaces. Earlier, Arago (1824) had shown that light from incandescent solids and liquids was partially polarized. Solids need not be highly heated in order to emit polarized radiation. Solids at room temperature emit infrared radiation peaking near a 10-μm wavelength; their radiation also would be expected to be polarized and this fact was confirmed with experiments by Sandus (1965).

The emitted radiation from bodies of water, in the wavelength range 2–200 μm, appears to be polarized (Hall, 1964). In scanning from the nadir to the horizon Clark and Frank (1963) observed a decrease in the apparent radiometric temperatures of the sea. Hall (1964) used a theoretical analysis based on Fresnel's equations to show that this decrease is caused by the polarization of the radiation emitted from a calm sea.

Because the emission of polarized radiation has been established, there arises the question as to whether surface roughness (such as sea state) can affect the polarization. Roughness of solid surfaces was found to decrease polarization (Sandus, 1965). Thus, the polarization of the emitted infrared radiation would be expected to decrease with sea surface roughness (wind waves or foam).

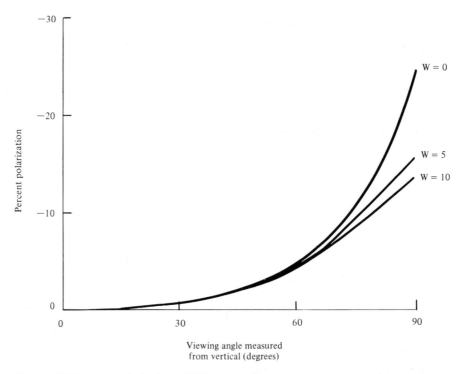

Figure 19.1 Percent polarization of light emitted by the sea; $m = 1.28 - i.294$; $\lambda = 12.6 \ \mu$m; W = wind speed in m sec^{-1}.

Figure 19.2 Percent polarization of light emitted by sea; $m = 1.28 - i.051$; $\lambda = 8.5 \ \mu$m; W = wind speed in m sec^{-1}.

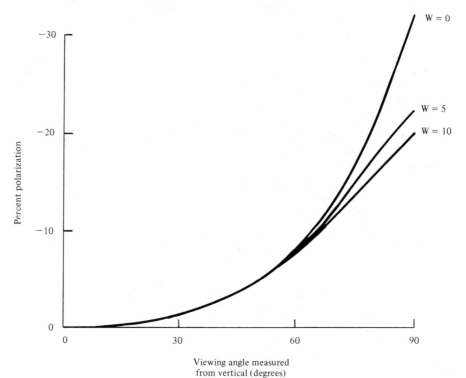

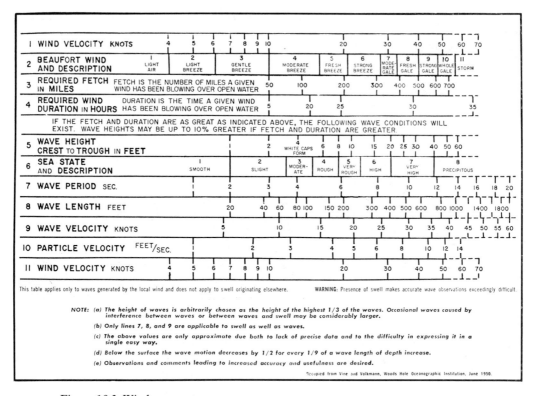

Figure 19.3 Wind waves at sea.

We first consider the polarized emitted radiation from the sea and then the effect of polarized reflected infrared sky radiation reflected from the sea.

A theoretical investigation has been conducted of the effect of crosswind speed (causing waves) on the polarization of infrared light emitted by the sea (Basener and McCoyd, 1967). The results of their analyses are presented in Figs. 19.1 and 19.2 for 12.6 and 8.5 μm, which are chosen above and below the maximum infrared emission wavelength of 10 μm. These wavelengths were selected to illustrate the effect of a variation in absorption coefficient. The polarization increased with nadir angle and absorption, but decreased with wind speed. Since wind speed is related to wave height for a given fetch (Fig. 19.3), wave height is related to the polarization of the sea's emitted radiation.

The basis of the polarization effect of the emitted radiation from the sea can be understood by referring to Fig. 19.4. These curves present typical results of a theoretical investigation by Hall (1964) for the parallel and perpendicular components of 12.6-μm emitted radiation as a function of emission angles. These curves are calculated using the optical complex index of refraction of water at 12.6 μm. The published values of the infrared optical complex indices of refraction are similar between authors (Irvine and Pollack, 1968; Kislovskii, 1959). The values used in the present calculations are close to those published. At normal viewing, the apparent emission temperature for both the parallel and perpendicular components of emitted radiation is about 270 K. The actual sea temperature (T_w) is assumed to be 280 K and the air above it (T_s) at 290 K. (The parallel and perpendicular components refer to the direction of vibration of the electric vector relative to the plane formed by the emitted ray direction and the nadir.) The percent polarization is given by the relation

$$\% \text{ polarization} = 100 \times \left(E_\perp - E_\parallel\right) / \left(E_\perp + E_\parallel\right),$$

$$E_\parallel = \frac{\text{intensity of the parallel component}}{\text{of } \textit{emitted} \text{ infrared radiation,}}$$

$$E_\perp = \frac{\text{intensity of the perpendicular component}}{\text{of } \textit{emitted} \text{ infrared radiation.}}$$

At a 12.6-μm wavelength, the percent polarization shown in Fig. 19.1 may be calculated from the curves of Fig. 19.4 for smooth water. (The negative sign for polarization signifies that the dominant electromagnetic wave vibration is parallel to a plane defined by the emitted ray and the surface normal.)

A water wave can be modeled as shown on the inset of Fig. 19.4 using small facets on the wave tilted at various angles to the viewing direction. These individual facets still emit polarized radiation relative to their local normal in accordance with the curves of Fig. 19.2. However, the projected area of these facets is such that it emphasizes the lower local nadir angles. In effect, the curves of Fig. 19.4 are displaced to the right (shown qualitatively as the dashed curves), decreasing the percent polarization. An exact representation depends on the model of the facet distribution. The results of using the Cox and Munk (1954) crosswind wave

Figure 19.4 Polarized components of infrared radiation emitted by the sea at 12.6 μm; $m = 1.28 - i\,0.294$.

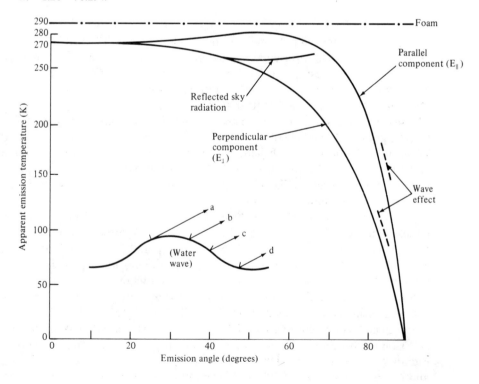

distribution are as shown in Fig. 19.1 and 19.2. For upwind or downwind, there is a small difference in the probability distributions of the water waves, and the cross-wind components are generally smaller than the upwind mean square slope components. Thus, if an upwind wave distribution were used to calculate Figs. 19.1 and 19.2, a greater depolarization would be expected for the same corresponding wind velocities.

When foam occurs, the brightness temperature of the perpendicular and parallel components is equal and independent of viewing angle (Fig. 19.4).

As a prelude to describing the effect of reflected infrared sky radiation from the sea, there will be a reflected parallel component of infrared radiation from the water, shown in Fig. 19.4. The amount of reflected radiation will depend on how high the sky temperature is. The reflected radiation will generally contribute to both the parallel and perpendicular emitted components, with the perpendicular component contribution dominating. This will result in positive polarization of the emitted radiation, from reflection.

We again may use the results of Hall (1964) to calculate the polarized components of reflected infrared radiation from the sea at 12.6 μm, for instance. The results are presented in Fig. 19.5 as a function of reflection angle (equal to the incidence angle) for the two components in terms of the components shown on Fig. 19.4. The intensities are normalized in order to show relative effects. The calculated percent polarization using Fig. 19.5 is presented in Fig. 19.6 for smooth water; the polarization is seen to be positive with a peak of 94.4% at the Brewster angle (53°).

Figure 19.5 Polarized components of infrared radiation reflected by the sea at 12.6 μm; $m = 1.28 - i0.294$.

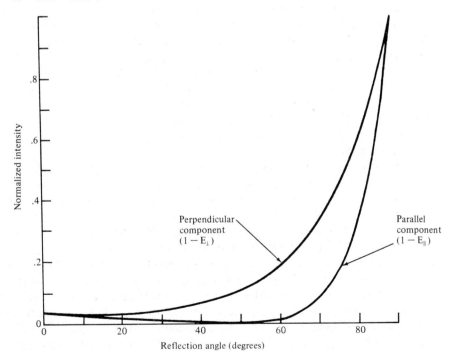

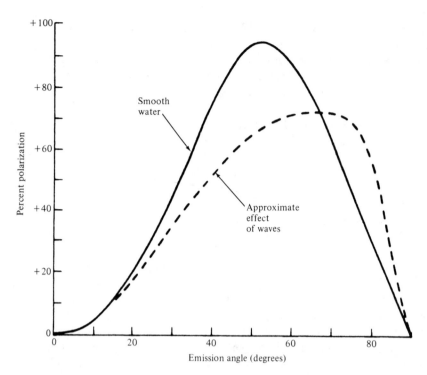

Figure 19.6 Percent polarization of infrared radiation reflected by the sea; $m = 1.28 - i0.294$; $\lambda = 12.6 \mu$m.

When waves perturb the surface of the water, the percent polarization will increase for emission angles above the Brewster angle (the amount of increase will depend upon the emission angle and the wave height). However, below and at the Brewster angle, the polarization will decrease. The approximate effect of surface roughness from waves is schematically shown as the dashed curve in Fig. 19.6.

A similar polarization curve occurs (Fig. 19.7 dashed curve) at the 8.5-μm wavelength where the water is much less absorbing (i.e., the optical complex index of refraction is $1.28 - i0.051$). Now, because of the low absorption, the peak polarization is almost 100% at 53°. Here the effect of surface roughness from waves is schematically shown as the dashed curve.

At the Brewster angle and at larger angles, the sea surface strongly reflects the sky radiation, which in general may be at a temperature different from the sea. When the sky temperature is greater than the sea (as when an aerosol layer exists), the magnitude of the reflected perpendicular component will increase (Fig. 19.4) and decrease the polarization at the Brewster angle. When the water is more absorbing, perhaps because of pollutants, the surface reflection is greater, producing stronger positive polarization. Thus, we see that the polarized infrared emitted radiation from the sea is dependent on the sea's emitted as well as reflected infrared radiation. The disentangling of the two variables requires a knowledge of the sky infrared background as a function of elevation angle and azimuth, as well as the measurement of the polarized components of the infrared radiation from the sea as a function of azimuth and elevation angle. A suitable radiative transfer model is necessary.

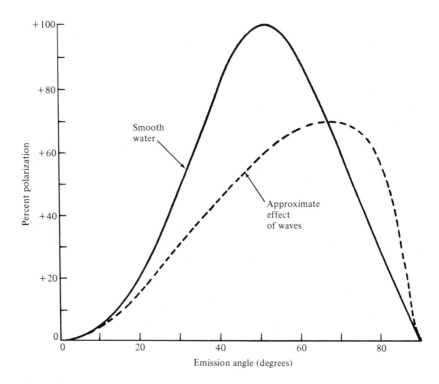

Figure 19.7 Percent polarization of infrared radiation reflected by the sea; $m = 1.28 - i0.051$, $\lambda = 8.5\ \mu$m.

Models

Radiative transfer models based on Monte Carlo calculations have been constructed for the earth's atmosphere and ocean at wavelengths of 0.46 and 0.70 μm (Plass et al., 1975, 1976); these models considered the radiance distribution over a wind ruffled sea with contributions from sun glitter, the sky, and ocean. Further modeling of the photometry and polarization of the visible sun glitter was done by Plass et al. (1976) and Guinn et al. (1979). Infrared photometric observations and modeling of the sun glitter at 2–3 μm and 8–13 μm were made by Gambling (1975), with mixed results.

An infrared analytical model was developed for the 7–12-μm region in the following manner:

1. Using Fresnel's equations for the reflection from a smooth surface, the bidirectional infrared reflectance properties of the sea as a function of wavelength were derived.
2. Using Fresnel's equations and Planck's radiation law, the emitted infrared radiation from a "smooth" sea was determined as a function of viewing angle and sea temperature.
3. Using LOWTRAN 4 (Selby et al., 1977), a program developed at the Air Force Geophysical Laboratories, the effect of atmospheric emission and transmission in the infrared was determined.

4. The sea surface roughness was introduced based on the measurements and analyses of Cox and Munk (1954, 1956), to include effects of wind speed and direction.

5. Refinements were introduced to consider such effects as swell, foam, spray, and whitecaps.

6. Finally, the effect on polarization of cloud and haze emission and obscuration were considered.

The analytical program is a formidable one, but is a necessary precursor to the understanding of the results of a full scale observational evaluation program with an infrared polarimeter system.

The computer modeling involved a number of subroutines linked together to calculate three infrared sea emission contributions: (1) the *atmospheric emission*, (2) the *sea emission*, and (3) the *sun glint*. The conceptual descriptions are presented in Fig. 19.8. In this work, viewing from space is considered an application for a remote sensing satellite, but the computer programs permit viewing conditions for low and high altitude aircraft as well as for sensors near the sea surface.

Since the infrared atmospheric emission (and transmission) enters into all contributions (Fig. 19.8) it will be discussed first. The LOWTRAN 4 program (Selby et al., 1977) was adapted to run on a HP3000 computer. LOWTRAN 4, 5, or 6 calculates the transmittance and/or radiance of the earth's atmosphere in the wavelength range from 350 to 40 000 cm^{-1} (0.25–28.57-μm wavelength) at 20 cm^{-1} spectral resolution on a linear wavenumber scale. Refraction and earth curvature effects are included. The atmosphere is layered in 1-km intervals between 0 and 25 km, 5-km intervals to 50 km, a 20-km interval to 70 km, and a 30-km interval to 100 km. A midlatitude summer atmospheric model was used in the calculations for this work although topical, midlatitude winter, subarctic summer, subarctic winter or the 1962 U.S. Standard may be used. A haze model with a visual range of 23 km was included in the program.

In Fig. 19.8a the methodology is shown for computing the infrared atmospheric emission assuming a set of atmospheric parameters and a geometry (1). There are two contributions, one large contribution (6) from the atmosphere intervening between the sea and the sensor, and the other from the sky light (2) that is reflected by the sea surface (5). In the concept presented, the emitted atmospheric radiation (2) is taken as unpolarized on the average, although each individual molecule of air emits polarized radiation, the amount depending upon its polarizability. Any polarized contribution would occur from the reflection of the unpolarized sky radiation (2) by the sea (5). Thus, the reflectivities of the sea surface (3) for the two components of reflected radiation (one component parallel to the plane of incidence of the radiation and the other component perpendicular to the plane of incidence) are calculated from the input of the optical complex indices of refraction of water (4) (from Irvine and Pollack, 1968). The intensity of sky emission to the sea surface is given by LOWTRAN 4 as well as the atmospheric emission in the sea sensor line of sight; also the attenuation of the reflected sky radiation is given by LOWTRAN 4. A sea surface temperature of 300 K was assumed for the calculations. The sea surface roughness could be included in (5) and would require a summation of contributions over appropriate directions to a distribution of reflection facets of the rough sea surface.

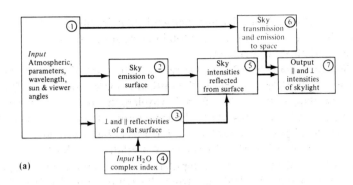

(a)

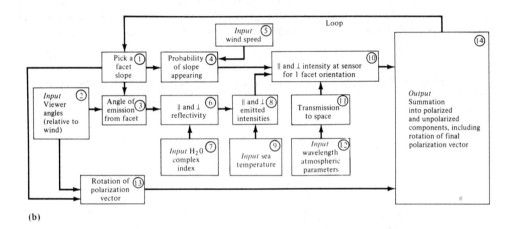

(b)

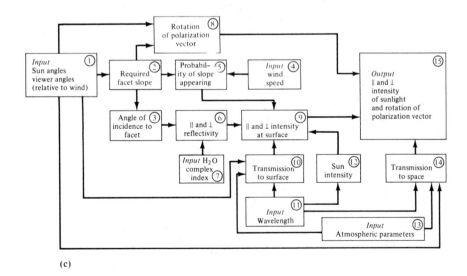

(c)

Figure 19.8 Conceptual flowcharts for the calculation of the intensity and polarization of long wavelength infrared radiation, (a) Atmospheric emission; (b) sea emission; (c) sun glint.

The output of parallel and perpendicular intensities of sky light (7) should be included as contributions to the total polarization arising from (14) and (15) in Figs. 19.8b and 19.8c, respectively.

Sea emission (Fig. 19.8b) requires as input a set of viewing angles (2) (relative to the wind direction) as well as the optical complex index of refraction of water (7), the sea temperature (9), infrared wavelength, and atmospheric parameters (12). In the general case, emission to the sensor arises from a sequence of facets (1) chosen to include all possible orientations. Each facet has a probability of occurring (4) and has an associated angle of emission (3) relative to the facet normal. The emitted intensity (8) of each facet is computed from the parallel and perpendicular reflectivity (6) by Kirchoff's law and from the temperature by Planck's law. There is a weighting (10) for a given facet orientation, which includes the atmospheric attenuation to the sensor (11) and the probability of occurrence (4). Since each facet is oriented differently, and the plane of polarization lies in, or perpendicular to, the

Figure 19.9 Intensities as seen from space for sun and sensor at 0° zenith angles.

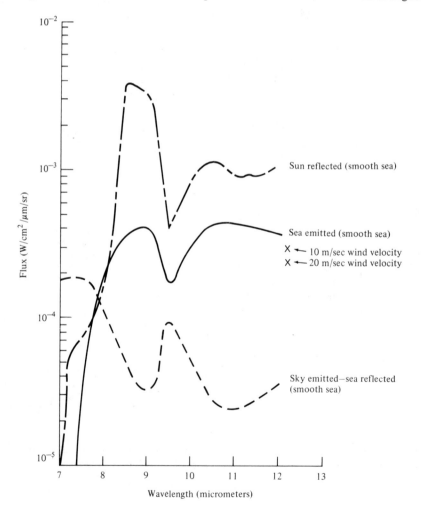

plane of emission, a calculation (13) is made to reference the individual polarization planes to the sensor before summing all the facets (14).

The output (14) of percent polarization, as well as photometry, should be combined with the atmospheric emission (and sun glint if required) shown in (7) and (15) of Figs. 19.8a and 19.8b.

Sun glint (Fig. 19.8c) may occur under certain orientations of the sensor relative to the sea surface that permit specular reflection from properly oriented facets of the rough or smooth sea surface. Inputs are also required of geometry (1), optical complex index of refraction of water (7), infrared wavelength (11), wind speed (4), and atmospheric parameters (13).

From the geometry input, the required facet slope (2) is selected. This slope with additional geometrical considerations then determines the rotation of the polarization vector from vertical (8), the probability of the slope appearing (5), and the angle

Figure 19.10 Intensities as seen from space for parallel component with sun and sensor at 60° zenith angles.

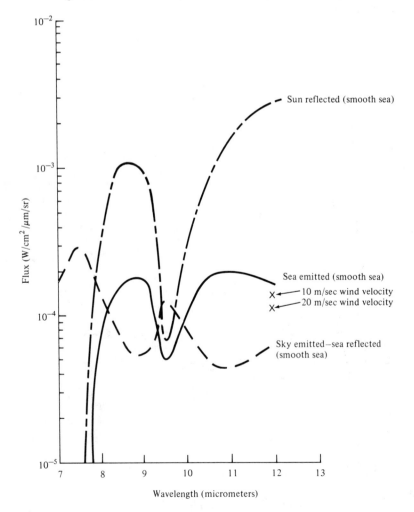

of incidence and reflection relative to the facet normal (3). The sun intensity (12), assumed to be from a blackbody at 5700 K, is reflected from the facet (9) as a result of parallel and perpendicular reflections (6) for a particular incidence angle (3). The reflected intensity depends on the probability (5) of the appearance of the necessary sloped facet. The transmission of the atmosphere (14) enters into the calculation both for the incident and reflected sunlight. The rotation of the polarization vector out of the plane of the vertical (8) is calculated as in Fig. 19.8b (13).

The output for sun glint (15), when it occurs, should be combined with the polarization and photometry calculations of (7) and (14) of Figs. 19.8a and 19.8b.

Initial results from the computer modeling programs of Fig. 19.8 are shown in Figs. 19.9–19.12 for the wavelength region 7–12 μm. Two sun–sensor zenith angle combinations (0° solar incident, 0° sensor zenith angles, and 60° solar incident, 60° sensor zenith angles) were used in the calculations of the sun-reflected, sea-emitted, and sky-emitted–sea-reflected intensities. Figure 19.9 presents the smooth sea calculated results for 0° zenith angles (the parallel component of the emitted radiation is

Figure 19.11 Intensities as seen from space for perpendicular component with sun and sensor at 60° zenith angles.

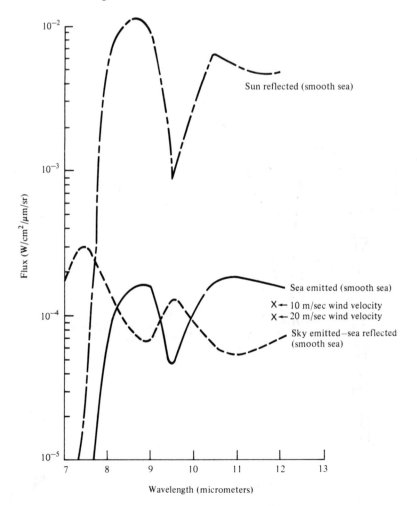

identical to the perpendicular component, by symmetry), and Figs. 19.10 and 19.11 present the corresponding smooth sea results for 60° zenith angles (the parallel and perpendicular emitted components, respectively). It can be seen that the sun-reflected radiation is greater than the sea-emitted radiation in the 8–12-μm region. This would suggest that sun glint could cause an effect when the sun was in a position to produce a specular reflection into the sensor; however, a wind ruffled sea does not produce any observed sun glint (Maxwell, 1978), indicating that the number of facets at the proper tilt angle for a specular sun reflection is quite small. This has been verified in calculations for near-normal incidence, but at higher zenith angles with a wind ruffled sea, the sun-reflected radiation could reach a value close to that emitted by the sea.

It is observed that the sea reflectivity enters into the calculation of sun-reflected, sea-emitted, and sky-emitted–sea-reflected intensities through the optical complex index of refraction of the sea water. For the computer modeling presented, the optical complex index of refraction of pure water (Irvine and Pollack, 1968) was used; the values were compared to those indicated for sea water (Wolfe, 1965), and agreement was quite close for the 7–12-μm region, thus validating the use of the pure water values.

Another point to be noted (Figs. 19.10 and 19.11) is that the parallel component of sea-emitted radiation is always greater than the perpendicular component. The difference is small and not easily seen in the figures, but gives rise to the polarization shown in Fig. 19.12. The percent polarization of the smooth sea is negative (Fig. 19.12), and the magnitude increases with decrease in wavelength within the 8–12-μm atmospheric window.

Note that for a wind ruffled sea, the polarization components are not limited to the vertical and horizontal directions, and a more general expression for polarization must be used.

Figure 19.12 Percent polarization for 60° zenith angles.

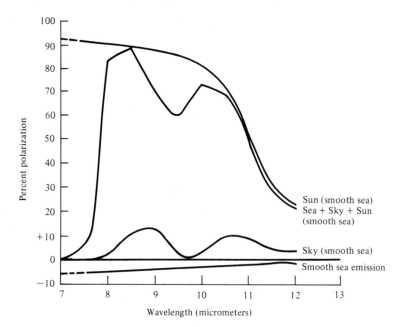

If the percent polarization is to be used as an indicator of the wave action of the sea, it is desirable to have the sea polarization as large as possible by appropriate selection of polarimeter operating wavelength. At a wavelength of 8.5 μm, the percent polarization is -4.5% (Fig. 19.2) and still well within the 8–13-μm atmospheric transmission window. However, for a smooth sea at 60° zenith angle, the sun reflection (if present) and the sky reflection have positive polarizations, and when all contributions are included (sea + sun + sky), the percent polarization is $+68.5\%$ (Fig. 19.7). However, in the presence of a rough sea, this percentage is decreased.

Large scale sea surface roughness (facets larger than a wavelength) would be expected to reduce the photometric brightness of the sea-emitted and sun-reflected intensity below that of the smooth sea because of the facet distribution. Because of the large optical depth (effective temperature) of the atmosphere near the horizon, the opposite effect could be expected for the sky-emitted–sea-reflected intensity. This is borne out by the modeling described in Fig. 19.8. Thus, in Figs. 19.9–19.12, the effect of wind velocities of 10 and 20 m/sec is a decrease in the sea-emitted radiation (shown by crosses at 12-μm wavelength) for downwind velocity. This results in a decrease in percent polarization as a function of wind velocity. The decrease depends on whether viewing is in the upwind, downwind, or crosswind direction.

Infrared Polarimeter System

The optical configuration for an infrared emission polarimeter is shown in Fig. 19.13, the actual implementation in Figs. 19.14 and 19.15, and the associated electronics block diagram in Fig. 19.16. The entrance optics (Fig. 19.13) consists of a 39-mm diameter, 63-mm focal length KRS-5 lens that passes infrared radiation between wavelengths of 1 and 40 μm, followed by an aperature to restrict the acceptance angle to 1°. Reflective optics are generally undesirable for polarimeters even though they achieve large light gathering with a minimum of emission by the elements; while having the right curvature, they may not have their center of symmetry exactly along their geometrical centers. This displacement of the true axis

Figure 19.13 Infrared emission polarimeter: present optical configuration.

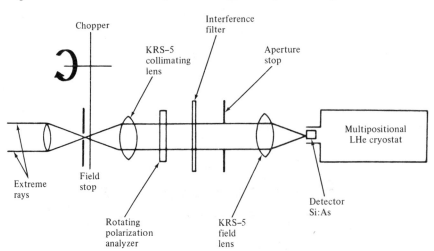

Figure 19.14 Close-up view of polarimeter with LHe cryostat at the right; chopper and polarization analyzer motors are located on top of the optical housing.

can cause extraneous polarization to be introduced into the optical system. The field stop (Fig. 19.13) may be chosen to limit the system field of view from 1° to 1' of arc; a 1° field of view is appropriate for most low altitude observations; a 1' field of view would yield a sea level resolution element of about 200 m for a satellite at an altitude of 1000 km. The radiation is chopped at a frequency of 450 Hz (nonmultiple of the power line frequency) to minimize interference from line frequency noise, with the reference being the reflected radiation from the housing by the gold plated rear chopper surface. The polarization analyzer is a wire grid rotating at a speed of 6.9 Hz (also a nonsubmultiple of the power line frequency). All lenses are KRS-5. A specially designed multipositional cryostat is used so that when the system is flown in an aircraft, or views the sky, the cryostat will maintain the detectors in a cooled condition. The interference filter in association with the Si : As detector determines the sensed wavelength and bandwidth. The aperture stop serves to limit the majority of the rays to near axial. With the use of large field stops, this is necessary for proper bandwidth limitation on the interference filters and for effective operation of the polarization analyzer.

The optical system is about 50 cm long. The major assemblies can be seen in Fig. 19.14 corresponding to those elements indicated in Fig. 19.13.

The electronic assembly (seen in Fig. 19.15) is a straightforward combination of readily available units (Fig. 19.16). The infrared signal from the detector consists of an intensity signal (450 Hz) amplitude modulated by the polarization component at 6.9 Hz. Phase sensitive detection of both signal components is accomplished by the Princeton Applied Research Model 124 and 129 phase sensitive detectors. The low pass filter in the Par 124 detector filters out the 450-Hz and 6.9-Hz signals, and the output to the three- channel recorder is proportional to infrared signal radiance. The

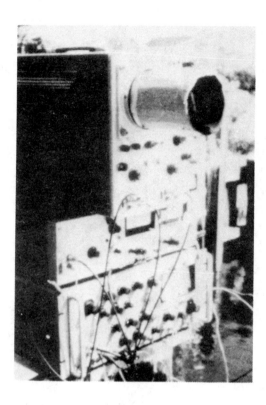

Figure 19.15 Electronics assembly: Par 124 at bottom, Par 129 center, and oscilloscope top.

PAR Model 129 Phase/Vector lock-in amplifier detects the polarization vector of the signal and resolves it into components about a reference direction (which may be recorded).

This approach has the advantage of determining the plane of polarization when it is not apparent from geometrical considerations. It is not anticipated that the sea would have significant circular polarization, which would require another sensor modification.

Figure 19.16 Infrared emission polarimeter: system electronics.

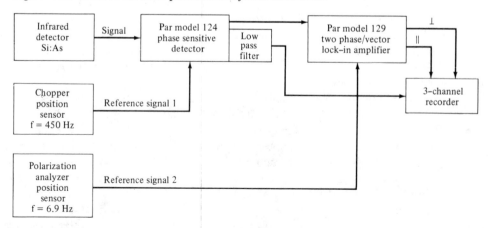

Measurements

Observations of the polarization of the radiation emitted from the sea at Great South Bay, Long Island (Egan and Hilgeman, 1967) indicated that the radiation was indeed polarized, and that the polarization varied with wave height (wind velocity). It was found that at a zenith angle of 83.5°, the relative polarization (relative to a 299 K chopper reference temperature) increased to +50%. The polarization referred to absolute zero, as shown in the graphs, is smaller, but the trends would be similar. To explain the observations, we refer to Fig. 19.12. It is seen that positive polarization can only arise from a combination of sky-emitted–sea-reflected and sun-reflected radiation. The polarimeter was pointed south during the summer afternoon observations. As a result, the sun was above and to the right of the polarimeter, so that if the sea were smooth, a specular sun reflection would not be seen. However, "glassy water" specular conditions did not exactly exist during the observing period, but instead the surface was mildly perturbed. In addition, a thick haze layer existed on the horizon. It appears that the higher temperature haze layer caused the sky-emitted–sea-reflected intensity to produce the initial positive polarization; subsequently, with increased wind, some facets of the waves become appropriately oriented to increase the sun-reflected contributions to cause the polarization to rise. Thus, although the analysis of the polarization emitted from the surface of a wind ruffled sea is not easy, a useful interpretation can be made.

The effect of swell on the model is to produce a subsidiary low frequency distribution of wave facets; the swell is the remaining wave motion when the higher frequency wave components are damped out with distance.

Spray and whitecaps produce attenuation and scattering of both incident and emitted infrared radiation, both from the sky and sun, and also the sea. The general effect would be to add a diffuse scattering component to decrease the percent polarization.

The effect of cloud and haze emission and obscuration can be approximated by a suitable selection of input parameters for the LOWTRAN 4 program.

An infrared emission polarimeter thus appears capable of resolving and measuring wave roughness. A comparison is now made to other approaches.

Passive microwave radiometry as a means for sensing the ocean surface is discussed by Hollinger (1971), with experimental results between 1.41 and 19.34 GHz. It is observed that the microwave brightness temperature of the ocean is significantly dependent on the surface wind field, which causes ocean surface roughness; this ocean surface roughness affects the horizontal polarization of the emitted radiation at larger incidence angles. The problem with passive microwave radiometry is the requirement for extremely large antenna arrays for high spatial resolution from space. In contrast, polarimetric radiometry in the infrared can achieve this resolution with small entrance apertures. The infrared polarimeter is related to microwave radiometry, with the advantage of high spatial resolution, but with the disadvantage of decreased ability to penetrate cloud cover.

The use of radar altimetry to measure ocean height is discussed by Yaplee et al. (1971). They describe a 1-nsec (X-band) radar (which is applicable in principle to a fast pulsed laser); the reflected pulse is unfortunately quite broad and has a positive time shift resulting from increased reflectivity from the wave troughs. The shift is negligible for wave heights below 2 ft, but can cause an inaccuracy equivalent to 10 cm for wave heights above 6 ft. A major problem with the laser and microwave altimetry systems is that they are limited to the area directly beneath the sensor. Other problems are high power requirements, large microwave antenna systems, and nonpenetration of cloud cover by the laser reflection.

The radar scatterometer approach described by Krishen (1971) indicates that at 13.3 GHz the backscatter is wind dependent, resulting from the change in small scale structure of the sea. However, the problems of antenna size for adequate resolution and problems in achieving a high signal-to-noise ratio from space make this approach questionable.

The above-described active systems have a considerable power, size, and data handling requirement that does not accrue to the infrared polarimeter. The polarimeter becomes more sensitive at large nadir angles, whereas the altimeters exhibit the reverse behavior.

The principle of emitted polarized infrared radiation being related to wave height has been demonstrated. However, there still remains an extended verification and modeling program to validate the observations under varying wave, weather, and wind conditions as a function of emission angle, observational direction, and sensed wavelength. Initial measurements should be made from a stable platform, such as an oceanographic tower, and when such a program is completed, the system may be adapted to aircraft or satellites.

Sea temperature, salinity, contaminants, and oil slicks could have an effect as well as the sky infrared background.

20

Depolarization

Introduction

For every remote sensing application of incoherent radiation, there is probably a laser counterpart. Therefore, it is assumed that laser systems will be used for remote sensing of targets in noisy backgrounds in which depolarization will be an important parameter, and that lasers will also be used for remote sensing of the backgrounds themselves for geophysical, meteorological, and other scientific purposes in which depolarization is important. Calibration standards and their properties for depolarization calibration then become important. Details of measurements of photometric standards will be mentioned together with the results.

Experimental studies (Egan and Hallock, 1966; Egan, 1967) have emphasized the fact that the polarization produced by the directional scattering of incident unpolarized radiation is a variable that depends on the scatterer. Although completely plane polarized light does not occur naturally, it can be of use in the analysis of directional surface scattering such as that occurring with the optical (or infrared) laser or microwave radar sources. This work permits the further elaboration of scattering theory through correlation with the nature of the scattering surface, and the analysis of the effects of coherence. The degree of coherence (both spatial and temporal) would be expected to influence the results, particularly diffraction effects.

The depolarization of linearly polarized electromagnetic waves scattered from rough surfaces has been studied theoretically by Beckmann (1963), Mitzner (1966), Fung (1966), and Kodis (1966) by assuming single scattering. An experimental study of backscattered radiation by Renau et al. (1967) indicated the necessity of including multiple scattering. Theoretical approaches to multiple scattering of electromagnetic waves have been suggested by Twersky (1967) and Valenzuela (1969).

Apparatus

The experimental arrangement used a Spectra-Physics Model 125 He–Ne laser of wavelength 0.6328 μm as the source; its polarization was determined by the orientation of the Brewster-angle windows at the ends of the gas discharge tube. The orientation was set within a fraction of a degree with the electric vector either parallel ($E_{\parallel}$) or perpendicular ($E_{\perp}$) to the plane common to the incident and viewing directions. The laser beam, with an output power of about 80 mW, was expanded by use of a telescope to produce a 4-in.-diameter beam incident upon a sample about 10 ft away. The scattered radiation was sensed by either of two polarimeters having a viewing angle of 0° or 60°. These were located at about 12.5 ft from the sample area, and they examined a circular spot 3.5 in. in diameter, delineated by a field stop of approximately 1.5° at the conjugate focus of a 12-in., $f/2.5$ objective lens.

The minimum achievable phase angle was 14°, for the 0° polarimeter, and 10° for the 60° polarimeter. The sample table was located to allow the average surface of the sample to be on the axis of rotation of the arm carrying the source.

Table 20.1 Samples Used in Depolarization Study

Sample	Physical characteristics
Haleakala volcanic ash	Three sizes: <1, < 37, and 37–88 μm; dull vesicular gray brown particles
Silica Beach sand (Rockaway Beach, NY)	Two samples: wet, dry; wet contained 5.2% sea water evenly distributed; sample was white and predominantly silica particles about 0.5 mm in size
Gravel (Oakdale, NY)	Two samples: wet, dry; wet contained 4.8% sea water evenly distributed; sample was light brown and sizes ranging from 1 to 10 mm
Limonite (Venango County, PA)	Two sizes: 37–88 μm, 1.18–2.38 mm; light brown particles
Basalt (Chimney Rock, NJ)	Two samples: 26% porosity and 40% porosity; gray particles < 37 μm in size
Silt (Albany Sand, NY)	Two samples: dry, wet; wet contained 2.6% tap water; fine, dark brown powder
White pine leaves (Pinus strobus)	Two samples: fresh, dry; fresh were less than 24 h old, and were short branches with needles attached; dry were 9–10 days old, and underwent a loss of 46.2% evaporable material (mostly water)
Rosebay rhododendron leaves (Rhododendron maximum)	Two samples: fresh, dry; fresh were less than 24 h old and were short branches with leaves attached; dry were 9–10 days old, and underwent a loss of 45.3%, predominantly moisture
Japanese holly	Two samples: fresh, dry; fresh were less than 24 h old and were short branches with leaves attached; dry were 9–10 days old, and underwent a loss of 42.4%, predominantly moisture

The sensor was a Type 6199, end window photomultiplier (S-11), with a rotatable HN-22 polarization analyzer located in a collimated beam before it. The effective surface radiance scattered directionally into the objective and transmitted by the analyzer was measured by the amplified dc output of the phototube.

The samples used are listed in Table 20.1. The soils were chosen to represent those that might exist over large areas on the earth, moon, or Mars, whereas the vegetation is representative of evergreen types only.

Measurements and Data

The laser illumination was repositioned at 10° intervals to produce phase angles (angle between incident and viewing directions) between 14° and 60° for the 0° viewing sensor, and phase angles between 10° and 124° for the 60° sensor. At each selected phase angle, the polaroid analyzer was rotated at ~ 4 rpm for several revolutions and the maximum intensity (I_{max}) and minimum intensity (I_{min}) were recorded. Variations in laser output were averaged by repetitive observations.

The magnitude of the residual polarization in the scattered radiation is expressed conventionally, i.e.,

$$\% \text{ polarization} = \left| \frac{I_{max} - I_{min}}{I_{max} + I_{min}} \right| \times 100, \tag{20-1}$$

and these values are plotted (Figs. 20.1–20.5) as a function of phase angle. The accuracy of the polarization measurements is of the order of a few tenths of a percent. The plane of polarization of the residual was found to lie in the plane common to the incident and viewing directions (containing the phase angle) or perpendicular to that plane, and the I_{max} and I_{min} occur in mutually perpendicular directions. $P_{\parallel}$ is the magnitude of the % polarization for $E_{\parallel}$, and $P_{\perp}$ is the magnitude of the % polarization for $E_{\perp}$. The magnitude of the % depolarization is defined as

$$\% \text{ depolarization} = 100 - (\% \text{ polarization}). \tag{20-2}$$

Results

Effect of Particle Size on Depolarization

The depolarization curves for a Haleakala volcanic ash at both 0° and 60° viewing angles are shown in Figs. 20.1a and 20.1b for particles ranging in sizes from <1 μm, < 38 μm, and 37–88 μm. For both viewing angles, the depolarization for both $E_{\perp}$ and $E_{\parallel}$ decreases with the particle size, the greater effect being between the 37–88-μm particles and the < 37-μm particles. For both 0° and 60° viewing angles, an angle occurs where the depolarization for $E_{\perp}$ and $E_{\parallel}$ is the same. Also, a trend toward the equality of the depolarization of $E_{\perp}$ and $E_{\parallel}$ can be clearly seen (Fig. 20.1b) for the 60° viewing angle as a 0° phase angle (backscattering angle) is approached; this is in concurrence with the conclusions of Renau et al. (1967). Thus, there appear to be two angles where the depolarization of $E_{\perp}$ equals that for $E_{\parallel}$ for these samples.

The greatest difference in depolarization $P_{\parallel} - P_{\perp}$ occurs for the largest particle size for both viewing angles. It is observed that at the 14° phase angle, $P_{\parallel} > P_{\perp}$ for

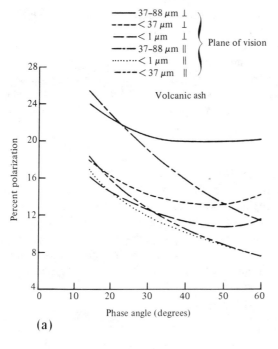

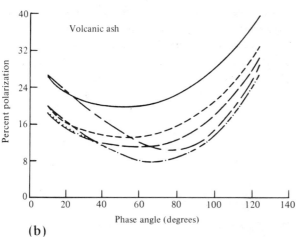

Figure 20.1. Haleakala volcanic ash: effect of particle size on depolarization (<1 μm, < 37 μm, 37–88 μm). (a) 0° viewing angle; (b) 60° viewing angle.

the 0° viewing angle; and at the 10° phase angle, $P_{\parallel} \geq P_{\perp}$ for the 60° viewing angle. The depolarization is greater at normal viewing than at 60° for the three particle sizes. The widest separation between the $P_{\parallel}$ and $P_{\perp}$ depolarization curves (Fig. 20.1) occurs at the higher phase angles for both viewing angles (above about 45°). The depolarization tends to be greater for normal viewing than at 60°.

If the depolarization caused by Haleakala volcanic ash (Figs. 20.1a and 20.1b) is compared with that of Venango County limonite (Figs. 20.2a and 20.2b), the particle size of both being 37–88 μm, we observe that the depolarizations of each are comparable at the 60° and 0° viewing angles for a phase angle of about 15°; however, the angle where the depolarization of $P_{\parallel}$ equals $P_{\perp}$ is at an appreciably

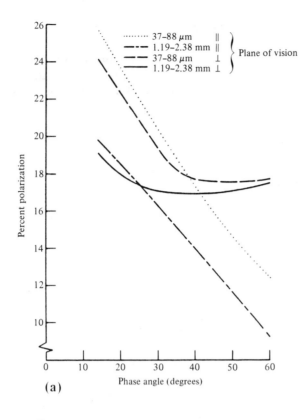

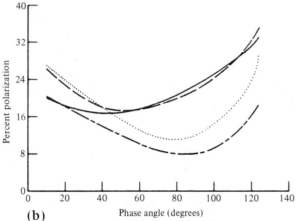

(a)

Figure 20.2. Limonite (Venango County, PA): effect of particle size on depolarization. (37–88 μm; 1.19–2.38 mm). (a) 0° viewing angle; (b) 60° viewing angle.

(b)

higher phase angle—about 40° for the limonite. At both viewing angles, $P_{\parallel} > P_{\perp}$ for the lowest phase angles. A clear trend for equality of depolarization at 0° phase angle can be seen for the 60° viewing angle curves.

Effect of Porosity on Depolarization

For the largest particles of limonite (1.19–2.38 mm), depolarization is greater than for smaller particles at the smaller phase angles (below about 60°), but the depolarizations of $P_{\parallel}$ approach each other at and above 60°. However, the phase angle for equal parallel and perpendicular depolarization occurs at about half that for the smaller particles.

The surface porosity of a basalt powder affects the amount of depolarization (Figs. 20.3a and 20.3b). A sample of < 37-μm-diameter particles of basalt from the Chimney Rock region of the Watchung Mountain Range of New Jersey were

Figure 20.3 Basalt (Chimney Rock, NJ): effect of surface porosity on depolarization. (a) 0° viewing angle; (b) 60° viewing angle.

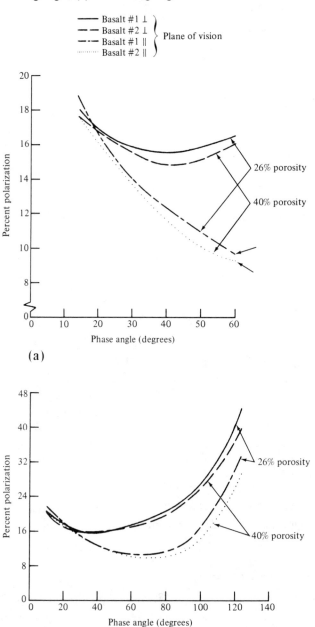

prepared to two surface porosities (26% and 40%) by sieving and compaction. It is evident from Fig. 20.3 that the coarser, higher porosity (40%) sample produces the greater depolarization for $E_{\parallel}$ and $E_{\perp}$ at both 0° and 60° viewing angles.

For 0° viewing, the depolarization of $E_{\parallel}$ is equal to $E_{\perp}$ at a lower phase angle (14°) for the higher porosity sample than for the lower porosity sample (18°). For 60° viewing, the equality for both porosities occurs at about the same phase angle (27°); in addition, there is evident a trend toward equal depolarization again at 0° phase angle.

At the 60° viewing angle, the difference between the $P_{\perp}$ and $P_{\parallel}$ curves is greatest for the more porous sample at a phase angle in the region of 100°. Also, the depolarization is greater for 0° viewing than 60° viewing at a phase angle of 14°.

Effect of Moisture on Depolarization

Samples of moist and dry silica sand, gravel, and silt (generally known as Albany Sand) were measured for depolarization, and the results are presented in Figs. 20.4 and 20.5. In general, the moist samples were poorer depolarizers than the dry for both $E_{\perp}$ and $E_{\parallel}$ at the higher phase angles.

Moist silica sand (Figs. 20.4a and 20.4b) at a viewing angle of 0° produces approximately 4% more residual polarization at low phase angles than dry sand. The

Figure 20.4 Effect of moisture on depolarization of various soils (beach sand, gravel, silt). (a) Beach sand, 0° viewing angle; (b) beach sand, 60° viewing angle; (c) gravel, 0° viewing angle; (d) gravel, 60° viewing angle; (e) silt, 0° viewing angle; (f) silt, 60° viewing angle.

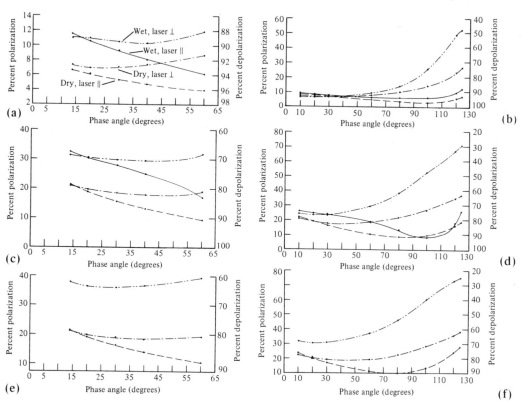

362

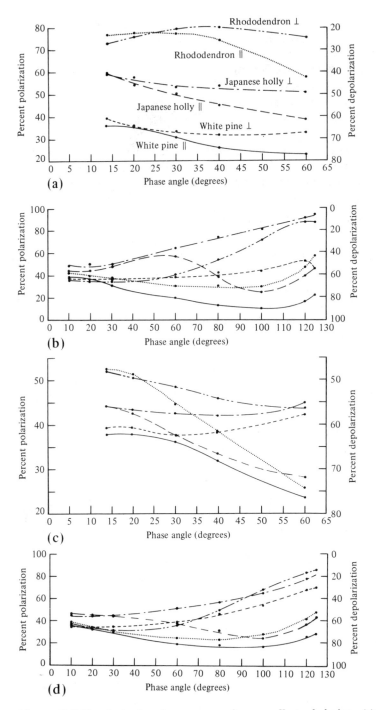

Figure 20.5 Depolarization by evergreen leaves; effect of drying. (a) Fresh pine, rhododendron, and holly, 0° viewing angle; (b) fresh pine, rhododendron, and holly, 60° viewing angle; (c) dried pine, rhododendron, and holly, 0° viewing angle; (d) dried pine, rhododendron, and holly, 60° viewing angle.

depolarization of $E_\parallel$ is equal to $E_\perp$ at a phase angle of 17° for the moist sample, but this equality did not occur within the range measured for the dry sample at 0° viewing, but a trend toward equality at 0° phase angle appears. The differences between $P_\perp$ and $P_\parallel$ are nearly equal at a phase angle of 60°, 0° viewing. At 60° viewing, the dry sand has a greater depolarization for $E_\parallel$ at all phase angles, but the wet depolarizes more for $E_\perp$ below a phase angle of 30° (above 30°, the wet depolarizes more). At the highest phase angle, 60° viewing, the difference between $P_\parallel$ and $P_\perp$ is greatest for the moist sample.

Gravel (Figs. 20.4c and 20.4d) has similar trends at a 0° viewing angle as sand, except that depolarization equality for the dry gravel is seen at 15°, compared to 18° for the moist gravel. The moist gravel has a larger difference between $P_\parallel$ and $P_\perp$ at the 60° phase angle than the dry at 0° viewing. At the 60° viewing angle, the depolarization equality occurs at 20° for dry and 30° for wet gravel, with a trend for equality at the 0° phase angles. The depolarization is approximately equal at the 14° phase angle for both viewing angles of the dry gravel, but greater depolarization was measured for 60° viewing than 0° degree viewing at 14° phase angle for the moist gravel. Also, the moist gravel displays a rather low depolarization for $E_\perp$ at 120° phase angle (possibly a specularity effect with the moisture on the gravel), and a large difference between $P_\perp$ and $P_\parallel$ results. Gravel does not depolarize as effectively as sand.

The silt (Fig. 20.4e) results are for $E_\perp$ and $E_\parallel$ illumination of a dry sample, but for $E_\perp$ only for the wet sample. It was not possible to obtain consistent results for $E_\parallel$ because surface changes during the measurements produced considerable optical variations that did not occur for $E_\perp$. It is evident that the moist silt (for $E_\perp$) depolarizes less than the dry. The depolarization is approximately the same at 0° and 60° viewing angles below a 60° phase angle. The wet silt (at 60° viewing) produces considerably lower depolarization at 120° phase angle; this is similar to the results for gravel.

Foliage: Effect of Species and Freshness

In Figs. 20.5a and 20.5b the depolarization observations for fresh white pine, Japanese holly, and rhododendron leaves are presented. At 0° viewing, there is a distinct difference between the three species with the rhododendron depolarizing least, white pine the most, and Japanese holly lying about midway between. At 0° viewing, there is equality in the depolarization of $E_\perp$ and $E_\parallel$ of the rhododendron (25°) and holly (15°), but only close approach for the pine at 20°. There is no doubt some geometrical effect is caused by the leaf arrangement, but this is essentially part of the depolarization characterization.

At 60° viewing, the poorest depolarizer is generally the holly, and the pine is the best, although the differentiation is not as clear as at 0° viewing. The interchange in the depolarization effectiveness of the holly and rhododendron leaves may be the result of a combination of leaf geometry and shadowing. It can be seen that the difference between $E_\perp$ and $E_\parallel$ illumination is about the same at a phase angle of 100°. There is no equality of depolarization for the holly, but there is for the pine (20°) and rhododendron (40°), and a clear trend toward equality at 0° phase angle only for the pine.

In Figs. 20.5c and 20.5d the data for the three species following a drying interval of 9–10 days are presented. At 0° viewing, the differentiation of species still exists, but not as clearly as for leaves in the fresh state (Figs. 20.5a and 20.5b). The general trend of the depolarization is increased with dryness, similar to the results obtained for dry soils. However, at larger phase angles (~ 60°) and 0° viewing angle, the dry rhododendron and dry holly exhibit relatively more depolarization for $E_{\parallel}$ than for $E_{\perp}$. Furthermore, pine presents a unique depolarization trend at 0° viewing angle and phase angles below 20° (where the depolarizations are about the same), that is, the curves for fresh and dry pine do not intersect, whereas the curves for rhododendron and holly do.

At 60° viewing angle, the depolarization trends between fresh and dry foliage are not as clearly delineated as at 0° viewing angle; however, Japanese holly depolarizes slightly more when dry than when fresh. Fresh and dry holly generally tend to depolarize less than the other samples for phase angles below 70°, and fresh and dry pine tend to depolarize most of all samples for $E_{\parallel}$ at phase angles above about 20°. Dryness in general tends to generally increase the depolarization somewhat. The dry holly has equality of depolarization for $E_{\perp}$ and $E_{\parallel}$ at a phase angle of about 30°, whereas this does not occur for the fresh leaves; the equality occurs at 15° for the pine and at 25° for the rhododendron. A clear trend toward equal depolarization at 0° phase angle is not evident.

Polarization and Depolarization Mechanisms

Depolarization has been found to be quite strong for the lower phase angles in the samples observed. However, at the highest phase angles and at 60° viewing (where the specular angle is 120°), there may be considerably reduced depolarization, particularly for moist samples. The strong depolarization would essentially be expected to be produced by multiple scattering (Renau et al., 1967), whereas retention of the polarization of the incident beam is generally characteristic of single scattering. The predominance of single scattering under certain conditions and the relationship to porosity have been previously discussed by Egan and Nowatzki (1968).

It should not be inferred that multiple scattering is the only depolarization mechanism, because for particles of different sizes, where light could be transmitted through the smaller particles and give greater depolarization, the effect of refraction is evident. The effect of moisture is to produce a thin specular type of film on the base material and modify the overall reflection and depolarization properties to that of a good dielectric, corresponding to the water in the visible region of the spectrum.

From the experimental data, the following conclusions may be drawn regarding the depolarization of the 6328 Å laser radiation by the surfaces investigated:

1. For both 0° and 60° viewing angles, the depolarization for $E_{\parallel}$ and $E_{\perp}$ increases inversely with particle size for Haleakala volcanic ash particles in the range below 88-μm sizes.
2. Limonite particles of sizes 1.19–2.38 mm produce a greater depolarization than particles in the range 37–88 μm at phase angles below about 60°.
3. The higher surface porosity (40%) basalt sample produces a greater depolarization than the lower surface porosity (26%) sample for both $E_{\perp}$ and $E_{\parallel}$ at 0° and 60° viewing angles.
4. For both $E_{\perp}$ and $E_{\parallel}$ at the higher phase angles, moist soil samples of sand, gravel, and silt are poorer depolarizers than the dry samples.

5. Moist rhododendron, holly, and pine leaves are clearly differentiated by depolarization at 0° viewing angle, but less discernable at 60° viewing angle.

6. Drying of the leaves generally increases the depolarization; this is similar to the effect observed for soils.

7. The depolarization of $E_\perp$ is small at the specular angle (60° viewing) for wet sand, gravel, and silt.

8. Shiny leaves, such as those of rhododendron, depolarize less than irregular leaves such as pine needles.

9. Additional surface characterization or signature can be obtained through measurement of depolarization characteristics.

10. Multiple scattering is an important depolarization mechanism, but refraction can also contribute strongly, as well as diffraction.

11. There appears to be an angle, in addition to the backscattering direction, where equality of depolarization occurs for coherent incident radiation that is parallel or perpendicular to the plane of incidence.

21

Radiative Transfer

The fundamentals of modeling the photometric and polarimetric atmospheric effects were presented in Chapter 9 and an application to stratospheric atmospheric constituents in Chapter 18. However, at lower altitudes (in the trophosphere), the modeling is difficult because of the generally existing vertical and horizontal nonhomogenieties (because of nonuniformly mixed water vapor and local air pollution sources) and wide range of aerosols. An application will now be presented of the polarimetric/photometric modeling of a variety of areas ranging from marine to mountain areas, as well as an industrial complex.

Additionally, a short section on high resolution photometric modeling will be described with the Dave radiation transfer model. The application is to the atmosphere of Mars.

General Approach

A considerable number of observations are necessary to furnish the required inputs to the computer program (see Chapter 9), and these observations should involve a time span sufficient to achieve an adequate experimental validation of the model. The instruments for the experimental program described are listed in Table 21.1.

The spectropolarimeter/photometer has been described by Egan (1968). The photometric accuracy is $\pm 1\%$ and the polarimetric accuracy is $\pm 0.1\%$ at wavelengths between 0.185 and 1.1 μm. The Volz photometer is a commercially available instrument calibrated to an accuracy of between $\pm 1\%$ to $\pm 3\%$ depending on scale. The pyrheliometer is a commercial Eppley instrument linear to $\pm 1/2\%$. Filters may be used on the pyrheliometer to measure solar radiance within narrow spectral bands. The Soligor and Honeywell photometers are of the photographic type and are calibrated to $\pm 5\%$ and may be used with narrow band interference filters. The MET (meteorological) conditions were determined with thermometers readable to $\pm 0.2°F$ and a commercial anemometer. The Royco Model 203 particle size analyzer classifies

Table 21.1 Instrumentation and Data Acquired

Instrument	Data acquired
Spectropolarimeter/photometer	Sky (and surface) radiance and plane polarization at wavelengths between 0.4 and 1.0 μm
Volz photometer	Optical depths of the atmosphere at wavelengths of 0.342, 0.380, 0.500, 0.868, 0.946 and 1.67 μm, as well as precipitable water and sun halo along the almucantar
Pyrheliometer, Eppley	Specular incident radiation, total; plus wavelengths of photometers, and 1.4 and 1.5 μm
Photometers (Soligor and Honeywell)	Brightness of white Nextel reference panel at wavelengths of 0.400, 0.433, 0.500, 0.533, 0.600, 0.633, 0.700, 1.0 μm
MET	Dry and wet bulb temperatures (relative humidity) barometric pressure, wind speed and direction
Royco, Model 203 particle size analyzer	Size distribution of atmospheric aerosols in range of 0.1–9-μm diameters, on ground and in aircraft
Aerosol collection	Collection on millipore/polycarbonate filter pads for scanning electron microscopy; energy dispersive analysis by x-rays, and Auger spectroscopy

particles of known composition to ±0.05-μm diameter. Aerosols were collected on millipore filters with a flow of 300 cm³/min, produced by a vacuum pump. In addition, during most observations, the standard airport Service "A" teletype data giving station reports, which gave temperature, dew point, barometric pressure, and visibility were obtained from airports located nearby. The National Weather Service also furnished Rawinsond data (Service "C") that gave temperature and dew point data at heights into the stratosphere from a regional balloon launch location.

Measurements were made on 20 days extending from February to October 1982 at five locations. The locations were atop an 18.5-m tower in an urban industrial complex at Bethpage, Long Island (TWR), at the Research Laboratory in Bethpage (LAB), at Republic Airport, Farmingdale, Long Island (APT), or airborne above the airport, at a representative marine area on the south shore of Long Island (LI Beach) and in a representative rural forested area in northern Alabama (ALABAMA).

Table 21.2 shows the dates of observations with the instrumentation described above. As seen in this Table, all possible data were not acquired on each day; however, on each occasion a sufficient data set existed for analysis, and further cross correlation was possible between different observations. It is to be noted that the volume of data possible from a complete set of one day's observations, much less over a period of many days, can be considerable. Thus a perspective must be maintained as to which data are most significant, and those of less significance given lesser weight.

Program Development

The development and presentation of the program results is basically a threefold process: (1) describing the observational data on the aerosols as input to the vector atmospheric model; (2) exercising the model with appropriate Stokes parameter

Table 21.2 Observations and Instrumentation (1982)

Instrument/ service	Dates of Observations																			
	2/25	2/26	3/5	3/18	4/1	4/2	5/21	5/25	5/27	6/30	7/1	7/2	7/7	7/12	7/16	9/27	9/28	9/29	9/30	10/1
Spectro polarimeter/ photometer																				
0.4 μm	✓					✓											✓			
0.5 μm	✓				✓	✓														
0.7 μm		✓				✓					✓	✓	✓		✓	✓	✓	✓	✓	
1.0 μm		✓		✓		✓														
Volz photometer						✓		✓	✓		✓	✓	✓			✓	✓	✓	✓	
MET	✓	✓	✓	✓	✓	✓			✓		✓	✓	✓			✓	✓	✓	✓	
Soligor photometer			✓	✓	✓	✓					✓	✓	✓		✓	✓	✓	✓	✓	
Honeywell photometer		✓		✓								✓			✓	✓				
Royco Model 203																				
Ground							✓		✓	✓	✓	✓	✓	✓		✓	✓	✓	✓	✓
Air							✓		✓	✓	✓	✓	✓	✓		✓	✓	✓	✓	✓
Location	TWR	TWR	LAB	TWR	TWR	TWR	APT	TWR	APT	TWR	TWR	TWR	TWR	LAB	BEACH	ALABAMA				
Pyrheliometer 5 filters	✓	✓	✓	✓	✓	✓					✓	✓	✓		✓	✓	✓	✓	✓	✓
Additional filters			✓	✓	✓	✓					✓	✓	✓			✓		✓	✓	
Aerosol collection							✓		✓									✓		
National Weather Service			✓	✓					✓	✓	✓	✓	✓			✓	✓	✓	✓	✓
Teletype reports	✓	✓	✓	✓	✓	✓	✓			✓	✓	✓	✓			✓	✓	✓	✓	✓

representations of the terrain; and (3) comparing the model output with both the observed optical depths of the atmosphere, and the polarimetric and photometric properties of the atmosphere (sky) as a function of wavelength. The correlation of atmospheric aerosol extinction with visibility and water content may also be made at the surface for the representative urban, rural, and marine environments. The results now will be described in the order indicated.

The size range of the modeled aerosols are of radii between 0.005 and 4 μm, and the size distribution will be assumed to be a modified gamma function as representative of natural ambient aerosols or road dust (Egan, 1982). A subsidiary distribution of small particles is included to account for asperities and edges on the naturally occurring aerosols (Egan, 1982). Where there is a large moisture content in the air, as in the marine areas, the particles are assumed to be spherical water droplets. The rural indices of refraction selected based on the elemental analyses are those for a Haven loam (Warner et al., 1975) farm soil. The indices of refraction for summer and winter urban Long Island are a combination of the refractive portion of the Haven loam and absorption portion determined on rainwater samples collected over Long Island during a corresponding summer and winter. The appropriate refractive index data are presented in Table 21.3. Three wavelengths are selected to be checked in the model: 0.4, 0.5, and 1.0 μm (Table 21.3); the parameters for the modified gamma distribution functions used in Part II of the Dave program are listed in Columns 3–7. The urban Long Island summer atmosphere is taken as the Deirmendjian (1969) Haze L. Because the Royco aerosol particle analyzer did not give a complete particle size spectrum, and because the Dave code does not permit an arbitrary distribution, and because the urban Long Island aerosols were solids, an augmented Deirmendjian number density was required. The one picked was the Haze L as representative of continental aerosols. Augmentation was based on matching the volume scattering coefficients at 0.5 μm for the Haze L and Haze H models assuming the same surface aerosol density. The Haze H model was aug-

Table 21.3 Input Parameters for Vector Atmospheric Programs

Wavelength,(μm)	Index of refraction	α	γ	b	$r_c(\mu)$	a	Surface aerosols (no/cm^3)
Urban Long Island (summer) — haze L							
0.4	$1.501 + i0.00344$						
0.5	$1.474 + i0.00302$	2	1/2	15	0.07	4.98E06	945
1.0	$1.464 + i0.00277$						
Urban Long Island (winter) — haze H							
0.4	$1.501 + i0.00235$						
0.5	$1.474 + i0.00215$	2	1	20	0.10	1.18E07	378
1.0	$1.464 + i0.00209$						
Rural farmland — haze L/haze H							
0.4	$1.501 + i0.00211$						
0.5	$1.474 + i0.00130$	2	1/2	15	0.07	4.98E06	3780
1.0	$1.464 + 0.000732$	2	1	20	0.100	1.18E07	3780
Marine area — haze M							
0.4	$1.343 + i0.00706$						
0.5	$1.336 + i0.00609$	1	1/2	8.9	0.05	5.33E04	3780
1.0	$1.326 + i0.00646$						

mented as for Arizona road dust as described by Egan (1982). The augmentation for the Haze L model was adjusted to match the volume scattering coefficients. This procedure is admittedly weak, but represents the best that can be done at present. The urban Long Island winter, because of its clarity, is represented as submicron dust particles [Deirmendjian (1969) Haze H plus the subsidiary augmentation described by Egan (1982) for the Arizona road dust]. Since autumn was a transition between summer and winter, the rural farmland model was exercised using both the augmented Haze L and Haze H models; however, a comparison of the optical depths of both models (Fig. 21.3), shows the Haze H model to be best representative of the actual observed conditions. Finally, the marine area was represented as a Deirmendjian Haze M model that reproduces coastal or marine aerosols. The surface aerosol number per cm^3 (last column, Table 21.3) as used in Parts IV or V of the Dave Program is determined by adjusting the Dave vertical number density distribution up to a height of 4 km to match the observed ground aerosol number density. The AFGL pressure, ozone, and H_2O vertical profiles for summer and winter (McClatchey et al., 1972) were used in the radiative transfer model. The surfaces underlying the atmosphere were represented as diffuse soil for Long Island (20% reflectivity), tree covered ground for the rural area (Stokes parameters from Egan, 1968), and diffuse sand for the beach area (40% reflectivity). The output of the vector programs will be presented on the observational data graphs to follow.

Observational Data Comparisons

The comparison observational data for the model consists of two groups: (1) the vertical optical depth of the atmosphere (measured times secant of the zenith angle) between 0.342 and 1.67 μm on cloudless days (Figs. 21.1–21.3), and (2) photometric and polarimetric observations of the sky at selected wavelengths between 0.4 and 1.0 μm in the cardinal directions (north, south, east, and west) at viewing angles

Table 21.4 Visibility/Vertical Optical Depth at 0.500 μm/Volume Extinction

Date	Local Time	Vertical optical depth (0.500 μm)	Airport visibility (km)	Vertical atmospheric water content (cm)	Surface relative humidity (%)
5/25/82	9:20 a.m.	0.886	6.5, hazy	1.7	68
5/25/82	11:00 a.m.	0.621	6.5, hazy	1.4	68
5/27/82	11:15 a.m.	1.52	4.9, hazy	3.6	68
5/27/82	12:15 p.m.	1.69	4.9, hazy	2.0	93
5/27/82	1:00 p.m.	0.886	8.1, hazy	1.5	93
7/02/82	9:25 a.m.	0.411	32.5	1.0	39
7/02/82	11:05 a.m.	0.737	32.5	1.3	44
7/07/82	1:05 p.m.	0.777	16.2	2.5	72
7/16/82	2:10 p.m.	1.73	9.7, hazy	4.2	62
9/27/82	11:34 a.m.	0.635	8.1, hazy	1.6	45
9/28/82	8:15 a.m.	0.629	9.7, hazy	1.6	52
9/28/82	2:42 p.m.	0.464	11.4	1.5	32
9/29/82	10:11 a.m.	0.588	11.4	1.7	45
9/29/82	1:16 p.m.	0.648	11.4	1.7	39
9/30/82	10:14 a.m.	0.878	13.0	1.6	44

between horizontal and 80° elevation (Figs. 21.4–21.7). Table 21.4 summarizes the interrelationships between the observables of vertical optical depth, nearby airport weather station reported visibility, vertical atmospheric water content, and local relative humidity.

In detail, Fig. 21.1 shows a comparison of the winter and summer optical depths of the atmosphere accurate to 0.03 in optical depth as a function of wavelength. Rayleigh scattering dominates the UV portion, with aerosol scattering becoming more important than Rayleigh at wavelengths longer than ~ 0.5 μm ($\tau \sim 1$ at $\lambda = 0.4$ μm). The very high extinction level ($\tau \sim 2.5$ at $\lambda = 0.4$ μm) of the maritime observations (7/17/82) is caused mainly by the H_2O based aerosols located near the earth's surface. The winter data (4/2/82) are based on the atmospheric transmission measured with a pyrheliometer. The summer and other data are based on the Volz photometer measurements. A clear normal summer day (7/7/82) has a generally low optical depth ($\tau \sim 0.7$ at $\lambda = 0.4$ μm); it is to be noted that the optical depth of the atmosphere may change by 0.3 (at $\lambda = 0.5$ μm) between morning and afternoon observations (Fig. 21.1, 7/2/82 a.m. and p.m.). The Deirmendjian Haze M model (shown as a star on the figure) matches reasonably in the UV, but not at the longer wavelengths for the very hazy maritime atmosphere; however, the Haze L shows a reasonable agreement at the longer wavelengths to the urban summer observations, but disagrees at the shorter wavelengths. The Haze H model appears to follow the trend of the winter measurement but is approximately 0.25 less in optical depth.

Figure 21.2 presents the spring urban optical depths of the atmosphere. The observed data trends are similar to those in Fig. 21.1, lying slightly above the clear summer day by 0.3 (at 0.5 μm) but well below the hazy maritime optical depths. The amount of vertical H_2O vapor was in the same range as in Fig. 21.1 (see Table 21.4).

Figure 21.3 presents the autumn rural optical depths of the atmosphere; they are generally lower than the spring urban values by 0.2 (at $\lambda = 0.5$ μm). The Haze L

Figure 21.1. Comparison of winter and summer vertical optical depths of atmosphere; dashed and solid lines show trends for maritime and urban atmospheres, respectively.

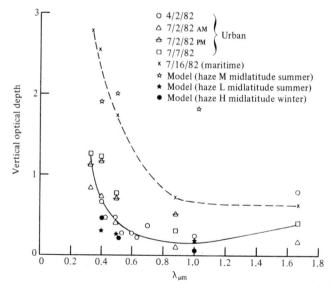

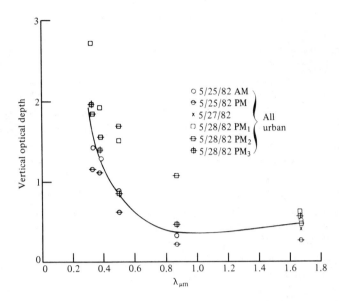

Figure 21.2 Spring optical depths of atmosphere; solid line shows trend for spring and urban atmospheres.

model (shown by open stars) matches the UV observations, but not the longer wavelength point. However, the Haze H model (shown by the solid stars) is a good match; the reason for the strong wavelength dependence is the relative shapes of the curves for particle size distribution and aerosol extinction coefficient. The extinction coefficient of the larger particles is increasing in the same region as the particle

Figure 21.3 Autumn optical depths of atmosphere; solid line shows trend for autumn rural atmospheres.

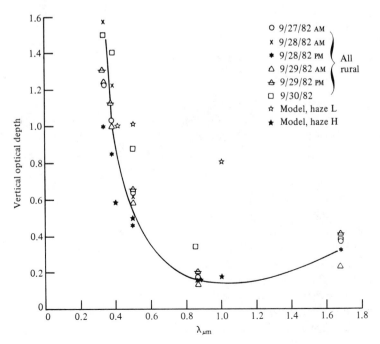

concentration for this model, whereas the comparative variation is opposite in the other models.

Table 21.4 summarizes the observed vertical optical depth data at a wavelength of 0.500 μm (with date and time), the reported visibility at an airport located within 4.9 km of the observation, the calculated equivalent vertical precipitable H_2O column (based on Volz photometer measurements at 0.946 μm), and local ground level relative humidity. A perusal of the table reveals no clear trends to correlate the reported airport visibility with vertical optical depth, atmospheric water content, or surface relative humidity. The three correlation coefficients are small, being -0.007, -0.004, and -0.023, respectively. Visibility is an important variable for correlation because it is conveniently available at many locations and is an essential input to a number of atmospheric codes such as LOWTRAN (Kneizys et al., 1980). One would not expect a strong visibility correlation with relative humidity or vertical water content because none of the humidities were above 95%, where condensation would occur on condensation nuclei. However, it might be expected that correlation would occur between vertical "visibility" (vertical optical depth) and horizontal visibility; this lack of correlation is a notable result. Relative humidities above 50% generally give rise to a hazy, low visibility condition, but the relative humidity does not correlate well with the vertical water amounts (correlation coefficient of $+0.0014$); nor does it correlate with the vertical optical depth (correlation coefficient of $+0.0006$). A slight correlation exists between the vertical optical depth and vertical atmospheric water content ($+0.144$). All the latter results are not too surprising because water is not uniformly mixed in the atmosphere. One can see that it is not the H_2O alone that produces absorption, but H_2O in combination with aerosols. One explanation of the lack of correlation is that the present accepted calculation of equivalent column of H_2O is incorrect. The present technique for calculating H_2O is based on ratioing the absorption in an H_2O band to a nearby region where there is negligible absorption; because scattering is additive, not multiplicative (see Egan, Fischbein, Smith and Hilgeman, 1980), this technique may be subject to errors.

The photometric and polarimetric observations are shown in Figs. 21.4–21.7; part (a) of each figure presents the photometry, and the second part (b) the polarization. Measurements were made at wavelengths between 0.400 and 1.0 μm with both photometric and polarimetric measurements being made simultaneously for a given geometry and wavelength. The photometric results will be discussed first, and then the polarimetry.

For the photometry, Fig. 21.4a characterizes the observed urban winter sky radiance in a northward direction for a clear sky. The measurements at 0.400 and 0.500 μm were made on 2/25/82 and those at 0.700 and 1.0 μm were made on 2/26/82. The MET conditions were similar: temperature and humidity were $25°$ F$\pm 5°$ and $33\pm 8\%$, respectively, and both days were clear. Although only one direction is presented, there are similarities in the east and west directions, with the southward location being strongly affected by the sun (in the south). The solar elevation during the observations was $45°\pm 5°$ and the azimuth $180°\pm 15°$. One model with a solar elevation of $45°$ was used for the modeling, because of the small range in solar elevation angles. In general, the model is low by a factor of 2 or more. The observed sky radiance is inversely proportional to wavelength; in the model, the 0.400- and 0.500-μm radiances are almost equal. The cause of the pronounced increase in sky radiance at $\lambda = 0.400$ μm is uncertain; the phenomena requires atmospheric aerosol particles comparable to or somewhat larger than the radiation wavelength. This larger particle is required because the increased scattering must be

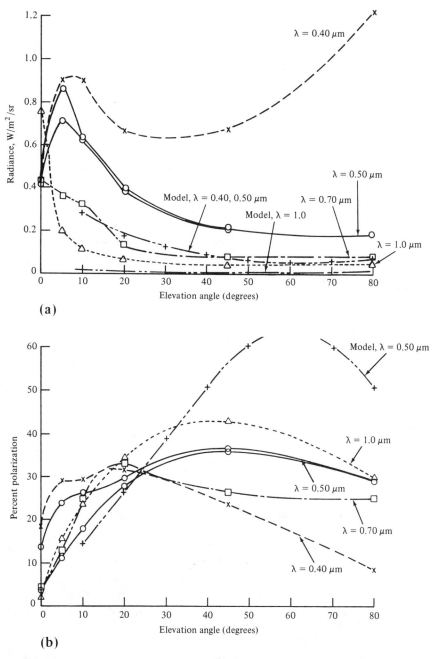

(a)

(b)

Figure 21.4 Optical properties of urban winter north sky 2/25–26/82 (Long Island); four wavelengths and comparison to Haze H, midlatitude winter model; solar elevation angle between 40° and 50°; azimuthal angles between 165° and 195°. (a) Photometric; (b) polarimetric.

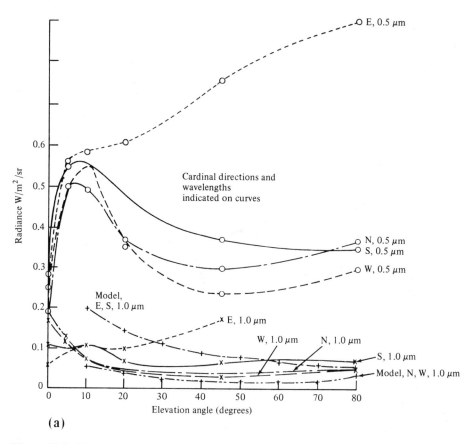

Figure 21.5. (a)

attributed to a phase function effect, which could be caused by high altitude ice crystals.

The property of the model that predicts radiance less than that observed is generally true for the other modeled conditions. In Fig. 21.5, the north radiance is low, although the east and west cross. In Fig. 21.6, the north model is lower than the observed. The trend in Fig. 21.7 is not apparent. The cause of these disagreements in the model appears to be that the volume scattering and absorption coefficients do not represent the actual experimental conditions.

The observed urban summer photometric properties in the north, south, and west direction generally increase toward the horizon between elevation angles of 30° and 10° (Fig. 21.5). This is true at both the 0.500- and 1.0-μm wavelengths. However, in the east direction at both wavelengths, a different trend is seen, explainable only in terms of horizontal inhomogeneities (clouds) in the southerly direction. The general trend of radiance increase toward the horizon occurs when the optical thickness is less than 1 (Fig. 21.1); the radiance initially increases toward the horizon and then drops as the optical thickness becomes greater and the radiation does not penetrate.

The general curve trends for the maritime summer sky (Fig. 21.6) indicates that the optical depth is thicker (the observed radiance slopes upward with increased

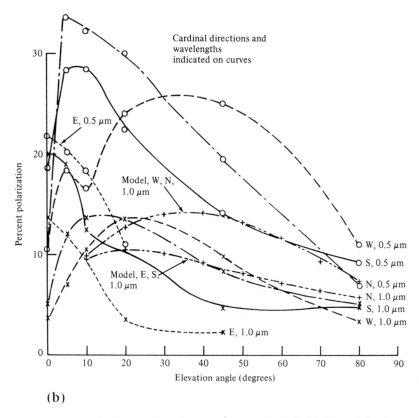

Figure 21.5 Optical properties of urban summer sky 7/2/82 (Long Island); two wavelengths and comparisons to Haze L, midlatitude summer model; solar elevation angle between 58° and 64°. (a) Photometric; (b) polarimetric.

elevation angle). However, the models do not show this trend. Therefore there are not enough aerosols in the model to produce the low radiance near the horizon. In this model, the volume scattering and absorption coefficients were not increased to account for edges and asperities; the aerosols were assumed to be spherical. It is possible that the scatterers in the Long Island area have a greater scattering effect.

The photometric properties at $\lambda = 0.500$ μm of the rural autumn sky on two days are presented in Fig. 21.7 in the south, east, and west directions. The model geometry corresponds to the morning conditions (1, 3, 4) with a sun elevation angle of 40°. The MET conditions were similar on both 9/29/82 and 9/30/82, temperatures being $80° \pm 3°$F and humidities $50 \pm 10\%$. The data definitely suggest validity, with a strong effect from the sun in the southerly direction (compare model $S_{1,3,4}$ with observed S_1, S_3, S_4; the effect of the noon sun is seen in the peak of S_2). The fact that E_2 is lower than W_2 and E_4 higher than W_4 suggests horizontal inhomogeneity. There is limited success with this model, within a factor of 2 in radiance in absolute intensities with the general photometric trends followed. The sequential photometric trends indicate an optically thin atmosphere overhead, becoming thicker toward the horizon, decreasing the radiance.

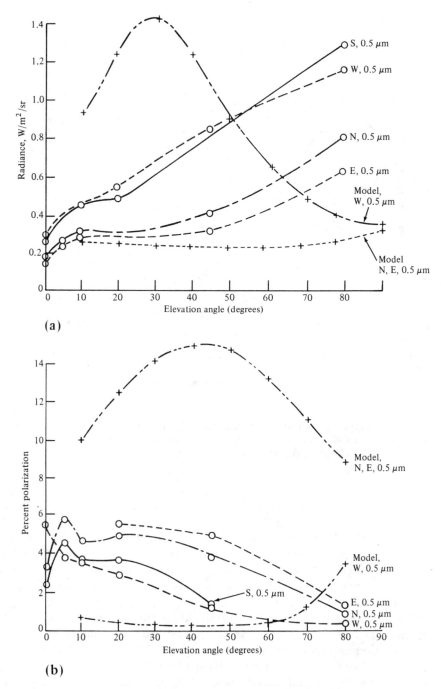

Figure 21.6 Optical properties of maritime summer sky 7/16/82 (south shore of Long Island); two wavelengths and comparisons to Haze M, midlatitude summer model; solar elevation angle between 64° and 68°. (a) Photometric; (b) polarimetric.

In the case of the polarimetry, the trends are more sensitively dependent on the model than radiance. As indicated in the photometry section, we do not as yet have good models for the aerosol size distribution that account for aerosol edges and asperities, particularly for the urban and maritime models. As a result, even though the observed polarization in Fig. 21.4b shows a definite wavelength dependence, the model yields no wavelength effect. Thus, something is not working properly in the model even though the sky radiance and optical depths may be correct. The cause could be incorrect input data.

The problems of polarimetry are borne out in Fig. 21.5b, with measurements of polarization from 5% to 35%, with the model in the middle and no clear trends.

However, in Fig. 21.6b, even though the photometry model is poor (Fig. 21.6a), there appears to be some agreement with the polarimetric trends. Thus the north and east 0.500-μm polarimetric models follow the observed data trends, although being high by a factor of 5.

In Fig. 21.7b, the general data behavior is roughly modeled following the trend of good photometric modeling of Fig. 21.7a. Thus the Dave code appears valid, although the polarization appears extremely sensitive to aerosol size distribution. The $S_{1,3,4}$ polarimetry model follows the trend of S_3, S_4, and even S_2, although differing in magnitude by a factor of up to 2 depending upon elevation angle. However, the observed east and west polarizations have large excursions; E_4 and W_2 have a large disagreement with the model although E_1, E_2, and W_3 match the model more closely.

Thus from the data, it is apparent that the aerosol models for the urban and maritime conditions are in need of improvement. It appears that there is not enough aerosol optical thickness. The optical model (Fig. 21.1) is lower than the trend, although nearly in agreement. The maritime optical depths are satisfactory at 0.400 and 0.500 μm but not at 1.0-μm wavelengths (Fig. 21.1); thus there is a problem with the optical depth at the larger wavelengths.

The rural optical depth trends (Fig. 21.3) are correctly matched, and thus the Haze H model produces a good photometric model (Fig. 21.7).

Discussion

From the comparison of the polarimetric and photometric data presented, it is evident that perfect agreement does not occur between observations and the models. Whereas the observations do show some slight variation as the result of the low level haze and high level cirrus, or ice clouds, a major problem occurs when an obviously unusual condition exists, such as strong low level maritime haze (Figs. 21.6a, 21.6b). The polarimetric and photometric modeling presented incorporating Stokes parameters, is not the first of its kind. However, the model described is one of the most advanced research polarimetric and photometric types. Because of the large number of input parameters possible (as described in the Observations and Instrumentation section), many parametric variations exist. Thus validation checks with observational data are possible at every step in the modeling, beginning with the optical complex indices of refraction and particle size range, through the particle size distribution (with the associated volume scattering and absorption coefficients), to the ultimate quantities of polarization and photometry. Circular polarization is also characterized by the model. We may thus check the adequacy of the model in detail as well as overall.

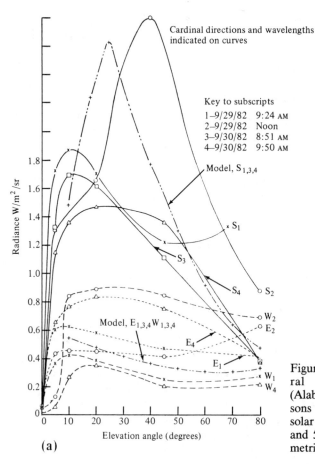

Figure 21.7 Optical properties of rural autumn sky 9/29–30/82 (Alabama), $\lambda = 0.50$ μm; comparisons to Haze H, midlatitude summer; solar elevation angles between 39° and 54°. (a) Photometric; (b) polarimetric.

The problem areas that appear as a result of the disagreement of the model with observational data may be determined by detailed examination of each conflict sequentially.

To begin with, the wavelength dependence of the optical depth observed with the Volz photometer at large optical depths generally does not agree with the model. Where good agreement occurs (within 0.1) at all optical depths (Fig. 21.3), it is the result of a unique combination of a specific particle size range (0.01–8-μm diameter) and a specific particle size distribution (Haze H). This unique condition is not generally borne out by nature. Thus, in general, either the model is deficient at large optical depths (the variation of optical complex index of refraction with wavelength, or the calculation technique is incorrect) or the Volz photometer readings somehow are not being correctly adjusted (using the aforementioned ratio technique for Rayleigh scattering). The complex indices of refraction are determined by a proven technique (Egan and Hilgeman, 1979); the real portion is correct to ±10% for most aerosols, but the absorption portion may be at variance with the actual by a factor of

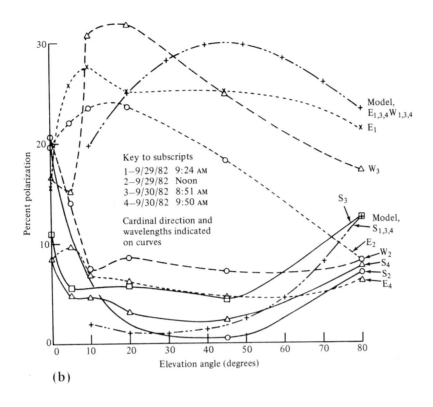

Key to subscripts
1–9/29/82 9:24 AM
2–9/29/82 Noon
3–9/30/82 8:51 AM
4–9/30/82 9:50 AM

Cardinal direction and
wavelengths indicated
on curves

Percent polarization

Elevation angle (degrees)

(b)

10. The absorption, which is generally small for aerosols, does not contribute to the primary source of atmospheric attenuation-scatter. The Mie scattering calculation used in the present model has been shown previously to be valid within a factor of 2 for naturally occurring aerosols with the addition of a supplementary particle distribution (Egan, 1982). It is quite possible that the model's mathematical limitation of not including low altitude horizontal inhomogeneities from low level haze and clouds, could introduce an error as great as 10 in the simple sec Z correction (Z is the solar zenith angle). As mentioned previously, because of the additive nature of scattering, simple ratio corrections do not properly account for scattering (Egan, Hilgeman and Smith, 1978).

The photometric and polarimetric data (Fig. 21.4–21.7) are affected by the horizontal homogeneity of the earth's atmosphere. If the earth's atmosphere were horizontally inhomogeneous, as it is in almost all parts of the world, the Dave horizontally stratified homogeneous atmospheric model is not representative of actual conditions near the horizon; this was seen in comparisons of the model with actual observations. Clouds were almost always present on the distant horizons during measurements, and the urban Long Island area is frequently overspread with haze.

The model does work reasonably well under lighter atmospheric aerosol loadings, and thus may be suitably applied to correcting aircraft and satellite photometric and polarimetric data for atmospheric absorption.

However, more work needs to be done on the validation of the intermediate properties, specifically the variation of the volume absorption and scattering coefficients with wavelength. These quantities depend on the optical complex index of refraction of the aerosols, their size range, and number density. This first theoretical link between the volume scattering and absorption coefficients and the fundamental optical and physical properties must be more firmly established. The initial efforts of Egan (1982) to modify the aerosol size distribution to account for asperities need to be developed on a firm theoretical basis in order for the technique to be applied to a variety of conditions.

Subsequently, more elaborate radiative transfer models must be developed to account for the horizontal inhomogeneities of the atmosphere, such as clouds, when low solar elevation angles occur because low level cumulus clouds appear to significantly modify radiation transfer.

Also, a more elaborate radiative transfer model is required for determining the total precipitable H_2O in the presence of scattering; the simple ratio technique of adjacent optical depths in an H_2O absorption band and a clear band is inaccurate. It is technically incorrect to take the ratio of two smoothed (low resolution) spectra, since this is not always identical to the smoothing of the ratio of two high resolution spectra (Egan, Fischbein, Smith and Hilgeman, 1980).

High Resolution Radiative Transfer Modeling

A multilayer radiative transfer, high spectral resolution infrared model of the lower atmosphere of Mars has been constructed to assess the effect of scattering on line profiles. The model takes into account aerosol scattering and absorption and includes a line by line treatment of scattering and absorption by CO_2 and H_2O. The aerosol complex indices of refraction used were those measured on montmorillonite and basalt chosen on the basis of Mars IR data from the NASA Lear Airborne Observatory. The particle sizes and distribution were estimated using Viking data. The molecular line treatment employs the AFGL line parameters and Voigt profiles. The modeling results that will be described indicate that the line profiles are only slightly affected by normal aerosol scattering and absorption, but the effect could be appreciable for heavy loading. The technique to be described permits a quantitative approach to assessing and correcting for the effect of aerosols on line shapes in planetary atmospheres.

High resolution spectra of planetary atmospheres are of considerable importance in remote sensing, both astronomically and geophysically. The aerosol effects on such spectra have been acknowledged but not described quantitatively, in part because of the complexity of appropriate models. Compared to the earth, the Martian atmosphere is relatively simple in structure and composition. A high spectral resolution Martian atmospheric optical model including aerosols could yield more accurate gas parameters, for instance, the determination of pressure–temperature profiles. There is also a possibility of determining some of the properties of the aerosols, although this is not as straightforward. The initial impetus for high resolution modeling was given by the availability of high resolution (0.08 cm^{-1}) Fourier transform spectroscopic (FTS) observations of the planets Mars (non-dust-storm conditions), Venus, Jupiter, and Saturn (Connes et al., 1969). More recently lower resolution (17 cm^{-1}) FTS observations of Mars (also non-dust-storm conditions) using the NASA Lear Jet Airborne Observatory have been reported, and were employed in characterizing the Martian surface material and thus the aerosol composition (Egan et al., 1978).

The Viking Lander data (Huck et al., 1977) is utilized and the Dave (1970) scattering program is used for obtaining a measure of the aerosol loading. Having this aerosol loading, the same program is then applied to the atmosphere in combination with a line by line calculation treating the molecules. The detailed results of the Martian atmosphere modeling calculation are presented in the wavelength region of 4975 cm^{-1} with a calculated wavelength resolution grid of 0.02 cm^{-1}. In addition, results smoothed to 0.08 cm^{-1} are presented with comparisons to the Connes et al. (1969) observations.

Modeling Procedure

In order to represent the Martian atmosphere by a mixture of CO_2, H_2O, and aerosols as major constituents, a suitable modeling program must be employed. This requires a radiative transfer model, which for simplicity, could be based on Mie and Rayleigh scattering. A sufficient number of layers must be used to make the model realistic. Then the details of the CO_2, H_2O, and aerosol distributions must be specified and inserted into the model. The high resolution requirements of the modeling necessitate the inclusion of suitable line profiles for CO_2 and H_2O. We may parametrically vary the aerosol loading and viewing geometry to determine the effect on line shape.

The basis of the Martian atmosphere modeling program is the Dave (1972) four part scalar program for computing Mie and Rayleigh scattering of optical radiation. This program was developed for application to the terrestrial atmosphere, but because of the very general nature of the code, it can be readily adapted to Martian atmosphere modeling.

Scattering and absorption by both aerosols and molecules can be included in the model extending to an altitude of 70 km. In order to establish an appropriate aerosol vertical distribution, comparisons were made of the Viking Lander multispectral imaging camera results (Huck et al., 1977) with the predictions of the model for various aerosol loadings.

The Viking Lander camera observations (Fig. 21.8) in the wavelength range $0.4–1.0$ μm (solid and dashed lines) indicate a tan color for the Martian surface and a similar color for the sky, the sky being brighter. The conclusions were based on six channel multispectral data of the surface and sky from Viking camera events 11A147/026, 11A149/026, 12A168/028, and 12A170/028 (Huck et al., 1977); the images were taken near local noon, and two sky and two ground patches were used in the analysis as indicated in Fig. 1 of the Huck et al. (1977) article.

The Dave (1972) radiative transfer code was used for the comparison with the Lander data by incorporating the measured average optical complex indices of refraction between 0.4 and 2.5 μm of a 50–50 combination of basalt (New Jersey) and montmorillonite (Mississippi, see Table 21.5). This composition agrees with one

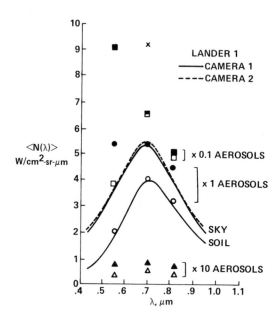

Figure 21.8 Comparison of spectral radiance of sky and soil patches with those estimated with a radiative transfer model described in the text (Sky and soil radiances from Huck, 1977). Solid symbols designate sky predictions, and open symbols designate soil predictions. The cross at the 0.700 μm wavelength represents the surface intensity assuming no atmosphere.

possible configuration suggested by Toon et al. (1977) as well as the Lear Jet observations (Egan et al., 1978). A trace of iron oxide required by the Toon et al. (1977) model will not significantly affect the optical complex indices of refraction, but would reduce the reflectivity toward the blue slightly more than for montmorillonite. Rayleigh molecular scattering is included as a function of pressure.

The aerosol size distribution, which used the optical complex index of refraction, was a modified version of Deirmendjian's Haze Model C of the form $n(r) \propto r^{-4}$, assumed characteristic of continental aerosols (Deirmendjian, 1969). This distribution matched the size distribution "one" described as fitting the IRIS data by Toon et al. (1977). The distribution also fitted the Viking Lander camera data (Pollack et al., 1979).

The distribution was cut off at 5 μm because the large amount of additional computing time required for the larger sizes produces a negligible result on the

Table 21.5 Optical Complex Indices of Refraction for 50–50 Mixture of Basalt and Montmorillonite

Wavelength (μm)	n_r	n_i
0.4	1.581	0.00269
0.5	1.571	0.00186
0.566	1.560	0.00128
0.6	1.565	0.00125
0.7	1.557	0.00125
0.817	1.553	0.00137
1.0	1.554	0.00205
1.5	1.537	0.00179
2.0	1.538	0.00226
2.5	1.527	0.00509

modeling (Selby, 1977). The model of Toon et al. (1977) requires uniform vertical and horizontal mixing; this is true under dust-storm conditions, but for quasistatic, non-dust-storm conditions, the vertical particle number density might be expected to be analogous to a terrestrial one. One such distribution is shown in Fig. 21.9, based on one described by Braslau and Dave (1972). The ordinate was initially arbitrarily chosen, but the actual value presented in Fig. 21.9 was determined by comparison of the Lander photometry to the model predictions using 40 layers. For an incident solar angle of 10° from the zenith, a viewing angle of 10° from the horizon (simulating the noon viewing of the Lander cameras), and 20% surface albedo, the results shown in Fig. 21.8 are obtained. The visual optical depths for the ×0.1, ×1.0, and ×10 aerosol loadings (relative to Fig. 21.9) are 0.05, 0.62, and 2.58; the optical depth for ×1.0 loading is close to that found for morning sun by Pollack et al. (1977) from the Viking Lander imaging data on sol 150. The cross at the 0.700-μm wavelength designates the intensity that would have been at the surface if there were no scattering or absorption in the Martian atmosphere. The surface intensity is most closely matched by the ×1.0 aerosol loading but the predicted sky brightness is high at 0.566 μm. It is noted that the surface albedo drops between 0.7 and 0.4 μm (Huck et al., 1977) and incorporation of this would have slightly lowered the calculated values of the sky and soil patch spectral radiance at 0.566 and 0.700 μm. This

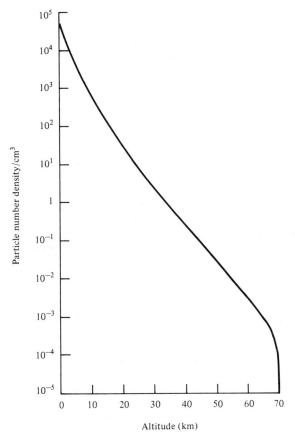

Figure 21.9. Assumed vertical distribution of particle density used for radiative transfer model.

surface albedo change could possibly be the result of magnetite (Pollack et al., 1979), not included in the analyses.

For the high spectral resolution portion of the modeling of the Martian atmosphere in this work, the number of layers in the radiation transport code has been reduced to a minimum of 16 so that the problem would be tractable for computer analysis. Also available was a high resolution modeling program created independently of the Dave (1972) model and made up of 10 layers of equal optical depths; these layers contained essentially all the CO_2 and H_2O in the Martian atmosphere. In order to dovetail the 10 layer line by line model with the 16 layer radiation transport model without rewriting either of the proven programs, an expedient, approximation approach was adopted. The lower 10 layers of the radiation transport model were made to include the 10 layer line by line model by incorporating the line shapes into the subroutine that was originally intended for ozone. The 10 equal mass layers of the line by line model closely approximated the selection of heights for the layers in the radiation transport model. In the course of the calculations, the upper six layers of aerosols had no effect (i.e., rounded off to zero by computer accuracy). The published temperature and pressure profile of the Martian atmosphere (Seiff and Kirk, 1976) was used in the 10 layer line by line model. The reduction in the number of layers to effectively 10 still provides a reasonable approximation of the Martian atmosphere for the purposes of this analysis. The line model program requires temperature and pressure, as well as CO_2 and H_2O concentrations, as a function of altitude (i.e., layer) to produce opacities. The Dave program accepts these optical depths for the 10 lowest layers, which extend from the Martian surface up to 29.7 km, above which the CO_2 and H_2O is assumed to be negligible. The change in the flux balance (i.e., the temperature–pressure profile) caused by including the aerosols is neglected. This procedure is approximately valid for the ×1.0 aerosol loading because the optical depth is less than 1. For the ×10 loading this is not true, and temperature changes in the aerosol region would be expected. The aerosol scattering and absorption and Rayleigh scattering are assumed constant over the calculated CO_2 and H_2O bands in the small wavelength range analyzed. However, in general, any variations must be taken into account.

Voigt line profiles were used in this model. The amount of CO_2 used was 6.84×10^{23} molecules/cm^2, corresponding to an average full disk absorption of 3.8 Martian air masses, and the amount of H_2O used was 40 precipitable μm during northern midsummer at midlatitude (Farmer et al., 1976). The vertical profile was adapted from their data, with the assumption of saturation at all levels. The Air Force Geophysical Laboratories line parameters compilation (McClatchey et al., 1973) was used in the modeling. The 10 layer line by line model was exercised at 0.02-cm^{-1} resolution in the wavelength region near 4970 cm^{-1}. The outputs at each wavelength were the vertical optical depths of each layer of the line by line model as well as the total vertical optical depth.

The output of the Dave program is the intensity as a function of viewing direction either upward or downward, and as a function of selected heights from the Martian surface upward, and as a function of the diffuse reflectance of the Martian surface. Aerosol loadings can be varied parametrically in order to show the effect on line shape. Values of ×0, ×0.1, ×1.0, and ×10, relative to Fig. 21.8, were chosen for this parametric analysis. Surface reflectances selected were 0%, 5%, 20%, 50%, and 100%. A solar incident angle of 58° was chosen for the model to correspond to 3.8 air masses for the total passage of a ray in and out of the Martian atmosphere (corresponding to the average atmospheric absorption for the entire Martian disk). Also selected for comparison were 1.9 air masses (looking toward the sun) and 2.9 air masses (radiation entering at 58° solar incident angle relative to the vertical and viewed vertically downward).

In the illustration to follow, only the region within a few wavenumbers of 4970 cm^{-1} will be shown to demonstrate typical results.

Results

The results of the atmospheric computer modeling are presented graphically in Figs. 21.10 and 21.11. In the following, the effects of varying aerosol loading, path lengths, and surface diffuse reflectance on line shapes is discussed.

Figure 21.10 presents the line shape effect associated with varying the aerosol loading from that of Fig. 21.9 to ×10, ×0.1, and ×0.0. A diffuse surface reflectance of 100% is assumed for this parametric comparison so that no flux would be lost at the surface. There is a negligible change in line shape for a change in aerosol loading from 0 to ×1. However, as the loading is increased to ×10, the depth of the absorption lines decreases, and the width narrows. The flux through the transmission windows between the lines is simultaneously decreased. The overall effect on effective transmission depends upon the counterbalancing effects of line narrowing and transmission window flux decrease. In essence, multiple scattering and aerosol absorption decrease the transmission between the lines, and multiple scattering contributes to filling in the line centers. The ×10 loading condition accentuates the

Figure 21.10 Effect of aerosol loading on unsmoothed (0.02-cm^{-1} resolution) 3.8 air mass Martian atmospheric transmission models with a surface reflectance of 100%: ——, 0 loading; -----, ×0.1 of normal loading (nearly coincident with 0 loading); – · – · –, normal loading; ------, ×10 normal loading.

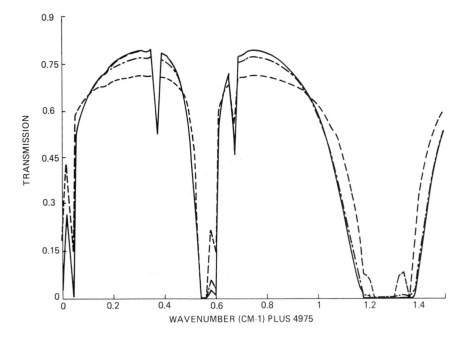

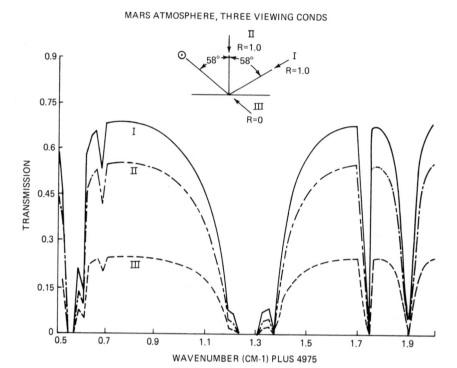

Figure 21.11 Effect of viewing direction (amount of Martian air mass) on Martian atmospheric transmission (includes effect of backscattered radiation) for solar nadir angle of 58°, ×10 loading. Condition I: 3.8 Martian air masses, viewing direction of 58° to vertical, diffuse surface reflection of 100%. Condition II: 2.9 Martian air masses, viewing direction vertically downward, diffuse surface reflection of 100%. Condition III: 1.9 Martian air masses, viewing direction directly at sun.

effects of aerosols, demonstrating their significance. In addition, it has significance under dust-storm conditions. Although the temperature profile changes in the atmosphere caused by the dust are not taken into account, an appreciable effect on the line shape is seen; this effect is not likely to be exactly counterbalanced by the atmospheric temperature profile change.

The effect on line shape of increasing the viewing path length in the Martian atmosphere might be expected to be analogous to the increased atmospheric aerosol loading; however, this is not so. For the ×10 aerosol loading considered, the greatest path length as seen outside at the top of the atmosphere (air mass of 3.8 and 100% surface reflectivity: Fig. 21.11, Curve I, Geometry I) produces the greatest transmission (in the regions where transmission windows occur). Transmission is defined as an apparent transmission (i.e., the intensity of the scattered radiation relative to the incident radiation, including the amount of radiation scattered back into the beam by the aerosols). This definition is in accordance with the actual experimental observations of a planetary atmosphere where the sources of the observed scattered planetary radiation are not separately observed. Vertical viewing downward from the top of the atmosphere at a 58° solar incident angle and with 100% surface reflectance results in an air mass of 2.9 (Curve II, Geometry II). If viewing is directed upward at the sun from the surface, the air mass is 1.9 (Curve III, Geometry III); however, the surface reflectance was made 0 in this case so that the surface backscattered radiation is not included. This was done to determine the effect of the atmosphere alone on the atmospheric transmission. As the diffuse surface reflectance is changed from 0% to 100% (not shown), the amplitude of the transmission portions of Curve III is approximately doubled. The curve for an air mass of 2.9 lies between the 3.8 and 1.9 air mass curves. The only effect of air mass decrease appears to be a decrease of the apparent transmission. This seemingly paradoxical behavior is understandable in terms of the definition of apparent transmission, which includes scattering into the beam.

The effect on line shapes caused by reducing the surface reflectance from 100% to 50%, 20%, and 5%, for 3.8 Martian air masses also depends upon the aerosol loading. For zero loading, the flux in the transmission windows is found to be linearly dependent on the surface reflectance because the molecular Rayleigh scattering is negligibly small. However, for ×10 normal aerosol loading, the flux in the transmission window is found to be nearly independent (within about 1% of Curve I, Fig. 21.11) of the surface reflectance. The transmission windows of the Martian atmosphere, as a result of aerosol loading between these extremes, will lie between the 0 and ×10 normal transmission limits. The filling in at the line centers, relative to the continuum, will be greater with decreasing surface reflectance.

Discussion

Line by line modeling of atmospheric spectra is essential when lower resolution spectra must be divided out to eliminate the earth's atmospheric absorption. This requirement is one of the main justifications for line by line modeling. For example, in terrestrial or airborne observations of Mars, a standard spectrum such as that of the moon is taken and used to divide the spectrum of Mars. Such a procedure is not valid with low resolution spectra of Mars. The nondistributive property of division of low resolution spectra is frequently misunderstood where absorption lines are common to an optical path through two atmospheres. That is to say, it is technically incorrect to take the ratio of two smoothed (low resolution) spectra since this is not always identical to the smoothing of the ratio of two high resolution spectra. One approach that has been used (see Egan et al., 1978) is the modeling of the results of dividing smoothed CO_2 spectra (telluric and telluric plus Martian), as well as smoothing the results of dividing high resolution spectra. A comparison of these two calculations allows making an estimate of the effect within the bands we considered, and hence the correction needed.

It is interesting to compare the computed spectrum with observational data. Figure 21.12 shows the transmission for 3.8 air masses and ×1.0 aerosol loading over a 20% reflectance Martian surface in the wavelength range between 4975.5 and 4977.0 cm^{-1}. This has been smoothed to 0.08 cm^{-1} (solid line) using a Hamming function (Blackman and Tukey, 1958) to permit comparison with the interferometric data of Connes et al. (1969) at the same resolution (dashed line). The existing Connes et al. (1969) observations are full disk of Mars and have been degraded by a variation in the Doppler shift occurring during the observations. The smoothing of the calculated data is seen to smear out the deeper absorption lines and weaken the narrow line at 4976.7 as expected. The match to the Martian spectrum is reasonable,

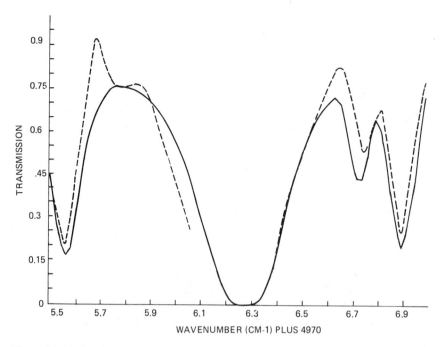

Figure 21.12 Results of line by line modeling of transmission of the Martian atmosphere. The model contains the normal quantity of aerosols, H_2O, and CO_2 and assumes a 20% reflecting surface. Observational data are shown for comparison between 4975.5 and 4977 cm^{-1}.——smoothed to 0.08 cm^{-1};----- Connes et al. (1969) observations.

with a significant distortion at ~ 4975.7. Improvements in the AFGL line parameters atlas (McClatchey et al., 1973) may enable resolution of these differences.

An immediate application of the high resolution modeling of the Martian atmosphere presented is in the validation of the calculations of CO_2 abundance. Young (1971) concluded that the surface pressure on Mars is 5.16 ± 0.64 mb based on line by line modeling of CO_2, and comparison with the Connes atlas (1969). However, referring back to Fig. 21.10 at a 0.02-cm^{-1} resolution, we see that the effect of the atmosphere of the $\times 1$ aerosol loading over a surface with 100% reflectance is to narrow the wider absorption bands slightly and lower the continuum by a few percent from the nonaerosol condition. This results in a decrease of the line half-widths by a few percent, and even more over the low reflectance Martian surface (5% to 20%; Egan et al., 1978) which would lead to an increase in abundance and surface pressure by several percent (Young, 1971). The effect of normal ($\times 1.0$) aerosol loading does not appear to be discernible within the limits of the error specified by Young (1971).

Although line by line modeling of the Martian atmosphere is feasible using Voigt profiles, the AFGL line atlas, and a suitable atmospheric radiative transfer program based on Mie scattering, the amount of computing time for a 1.5-cm^{-1} interval for the Mie scattering portion of the program was about 500 CPU min on an IBM 370/168 computer, plus 20 h of CPU time on an HP3000 computer for the line modeling. Computational efficiency was a prime consideration in the layer modeling program to reduce computing time, yet the amount of computing is quite extensive, particularly if an extended spectral range is required. It is best, therefore, to choose particular short interesting intervals for modeling to conserve computer time.

The indices of refraction used, with absorption portions close to those inferred theoretically, were those of terrestrial analogs; the trend in the updated inferred indices of refraction has been consistently a decreasing imaginary portion closely converging to a value of ~ 0.001 in the ultraviolet (Chylek and Grams, 1978) from a value of ~ 0.01 (Pang et al., 1976). The decreasing trend has been the result of using an improved theoretical representation of irregularly shaped aerosol particles.

Similarly, Pollack et al. (1979) have confirmed the mode radius RM of Toon et al. (1977) of 0.4 μm, updating a value of 0.1 μm (Pollack et al., 1977) as well as an order of magnitude decrease in the inferred absorption portion of the index of refraction to ~ 0.005; this updating and confirmation were based on the use of a recently developed algorithm that allows for the possible nonspherical shape of the particles. Two parameters ALFO (the maximum value of the parameter α which is the ratio of the particle circumference, $2\pi r$, to the wavelength λ) and FTB (the integrated front to back scattering) serve to characterize the aerosols. Pollack et al. (1979) conclude that the particles are platelike in shape, typical of claylike particles. In the analyses presented in this work the nonspherical nature of the particles has not been taken into account; an additional effect on the line shapes could occur from this cause.

As a result of the modeling described, it can be seen that atmospheric aerosols contained in the Martian atmosphere can affect line shapes at high resolution (0.02 cm^{-1}). It appears that for normal aerosol loading, the line shapes fortunately are only slightly changed. But for 10 times normal (dust storm), there could be an appreciable change in absorption linewidth for full disk viewing. With higher spatial resolution combined with the high spectral resolution (~ 0.02 cm^{-1}), dust vertical profile conditions on Mars as well as CO_2 vertical profile variations could be remotely monitored, as well as surface conditions in the transmission windows. Such vertical profile probing could not be obtained with visual imagery. If the atmospheric temperature changes caused by the presence of dust are not too great, the effect of a dust layer (in excess of $\times 10$) should be apparent by a comparison of the measured line profiles with those for $\times 0.1$ or $\times 1.0$; if there is a very narrow observed linewidth, one would infer a very heavy dust layer based on the modeling presented. Supplementary information from a Martian lander would be essential to reduce the ambiguities between the dust vertical profile and the CO_2 vertical profile. Additional high spectral resolution atmospheric data would be useful to compare and develop better modeling in order to analyze planetary atmospheric processes.

Line by Line

The previous material presented in this book dealt with photopolarimetry, whereby the state of radiation polarization was determined with broad spectral resolution, perhaps of the order of 10^3 cm^{-1} or larger. In this section, the concept of moderate to high resolution polarimetry is introduced, that is, within a spectral line or band of lines with a resolution of a few wavenumbers or less. This concept was introduced by Fymat in 1979. The theory was developed along with the appropriate instrumentation and applied to the observation of the first polarization spectra of Venus to a resolution of 0.5 cm^{-1}. The observations made use of a Fourier transform spectrometer (FTS) modified to sense polarization.

In order to understand the concept of spectral line polarization and the relationship between the line polarization and depth of formation in a scattering medium, a review of some basic concepts is necessary. The propagation of radiation in a medium involves both absorption and scattering. Scattering produces significant polarization, whereas single particle absorption (and emission) does not. If the particles are oddly shaped, a significant overall emitted polarization can occur. Because scattering and absorption vary with wavelength, so does the polarization; thus, the variation of absorption and polarization can yield much information about the composition and structure of the medium.

When scattering occurs, as in all planetary atmospheres, there are distinct limitations on absorption/emission spectroscopy. Figure 21.13 shows a typical spectral line profile in the absence of scattering. The lines are identified by the positions of the line centers (ν_0), and their abundances calculated from spectroscopic theory by the area between the line and the continuum. The abundance also can be calculated from the effective width (W) of the shaded rectangle of area equal to that of the spectral line. When scattering occurs, the continuum level is depressed (Fig. 21.13), because scattering is equivalent to additional absorption in the viewing direction. The line shape is generally unchanged, as well as the location of ν_0. However, the equivalent width W is less, and an estimation of the elemental abundance will be too large. Also, if fluorescence occurs within the line (Fig. 21.13), the equivalent width is further reduced. Thus spectropolarization measurements can yield scattering information and permit accurate remote sensing of abundance of atmospheric constituents.

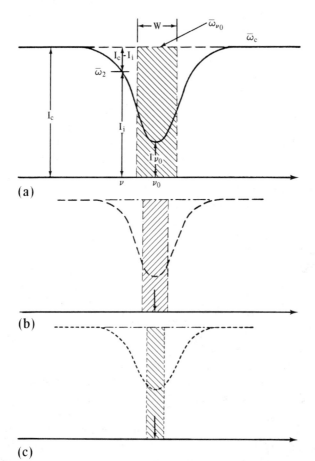

Figure 21.13 Schematic representation of an absorption line profile. (a) Case without scattering; (b) case with continuum scattering; (c) case with continuum plus line scattering. I = radiance; W = equivalent width; ω = single scattering albedo. Subscripts c, ν or l, ν_0: for continuum, line frequency, line center.

The line center (Fig. 21.13) is formed at the higher levels in a scattering medium. Frequencies away from the center are formed at lower levels in the atmosphere, with the continuum formed at the lowest level or deep within the medium. Also, weaker spectral bands are formed deep within the medium, while stronger bands are formed at higher levels. For a given frequency within a line, the larger the absorption, the weaker the scattering and thus the stronger the polarization; weak scattering in a highly absorbing medium only occurs a few times before being attenuated. Thus polarization is proportional to absorption, and polarization will vary as the line strength within a band. Thus, by appropriate selection of either line strengths or line frequencies, different levels of an atmosphere may be remotely probed. This depth probing technique has been applied to intensity (which is sensitive to the atmo-

Figure 21.14 Proposed two-beam optical interferometers for measuring the state of polarization of incident radiation.

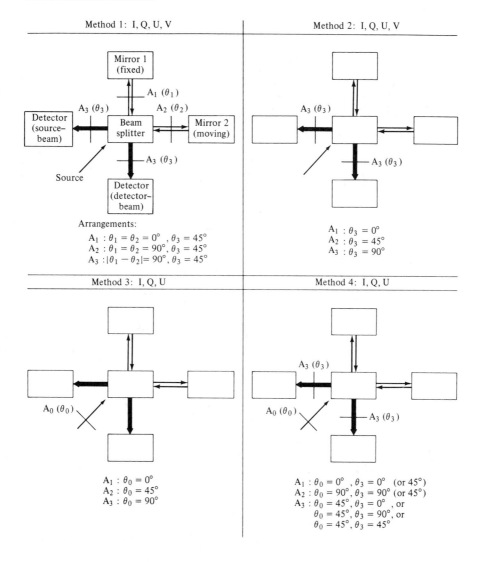

spheric gaseous component) and may be applied to spectropolarimetry (which is sensitive to the aerosol component).

The basic instrument used for spectropolarimetry is a two-beam Michelson type FTS. An interferogram is recorded, and the spectrum obtained from the Fourier transform of the interferogram. The interferogram $I(l)$ is given by

$$I(l) = \int_0^\infty B(\nu)\cos 2\pi\nu l\, d\nu,$$

where ν is the wavenumber, l is the path difference between the two arms of the interferometer, and $B(\nu)$ is the required intensity spectrum. For a perfectly symmetrical instrument, the spectrum is given by the Fourier transform:

$$B(\nu) = F_{\cos}[I(l)],$$

where

$$F_{\cos}(\cdots) = \int_0^\infty (\cdots)\cos 2\pi\nu l\, dl.$$

Figure 21.15 Schematic diagram of the Fourier interferometer polarimeter (FIP). Light from the telescope enters the instrument at the two Cassegrain focal-plane flat field mirrors at point F. The light from these two 13-arcsec portions of field is directed into the interferometer by means of two collimator mirrors marked M_1 and flat mirror M_2. The interferometer mirrors M_3 (stationary) and M_4 (movable 1 cm) together with the beamsplitter BS comprise the interferometer. The beams emergent from the interferometer are directed into the liquid-nitrogen-cooled PbS detectors D by flat mirrors M_5 and transfer mirrors M_6 and M_7. The polarization analyzer P is positioned ahead of one of the M_5 flats. M_8 is inserted for guiding purposes and for whenever the instrument is used for spectropolarization. (From F. F. Forbes and A. L. Fymat, 1974).

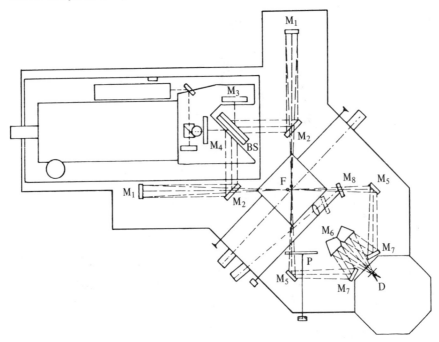

The Fourier transform technique has an accuracy limitation because in practice, the interferogram is recorded only over a restricted range of wavelengths (i.e., it is truncated). The band limited Fourier transform exhibits secondary maxima of moderate intensity that are smoothed out by a mathematical technique termed *apodization*.

The Fourier transform approach requires significant computation, but the technique has four distinct advantages:

1. The interferometer spectrometer is equivalent to a large number of detectors, the number being proportional to the wavelength range swept out and inversely proportional to the path length l. This multiplexing factor can be applied to a gain in signal-to-noise ratio, decrease in integration time, increase in spectral resolution, or increase in spectral range.

2. There is a larger throughput than a conventional spectrometer which is limited by the slit size.

3. The resolution obtainable is limited only by the path difference l: the larger the path difference, the higher the resolution.

4. Calibration is built in because a single line is used as a standard, and accurate line identification is then possible.

Figure 21.16 Spectra at 0.5-cm^{-1} resolution of Stokes parameters I_l, I_r and their combination with U and V, obtained with the Fourier interferometer polarimeter for Venus on July 12 and 13, 1974 at the National Mexican Observatory, Baja California, Mexico. (The three orientations of the polarization analyzer are indicated. A comparison solar spectrum also shown.) Note the polarization effects on the three features labeled A, B, and C. (From F. F. Forbes and A. L. Fymat, 1974).

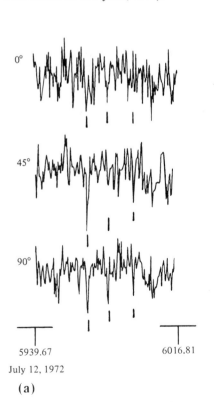

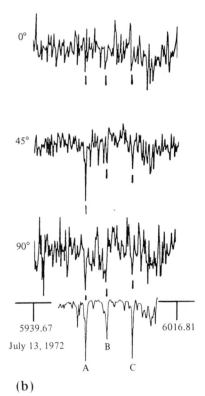

(a)   (b)

Fymat has proposed four alternate spectropolarimetric methods, referred to as Methods 1, 2, 3, and 4. These are summarized in Fig. 21.14. A basic Michelson interferometer is used with the addition of appropriate polarization analyzers. The differences lie in the number and locations of the polarizers used and the number of Stokes parameters provided. The source beam or sensor beam or both may involve polarization analyzers, and the same considerations apply to either beam.

Method 1: One linear polarizer in each interferometer arm, and a linear polarizer–analyzer, before the detector. There are different possible orientations for the arrangements of the transmission axes of the polarizers. Let $A_i(\theta_1, \theta_2, \theta_3)$ represent the ith arrangement and $i = 1, 2, 3$ and $\theta_1, \theta_2, \theta_3$ as the azimuths of the two arm polarizer and analyzer, then Method 1 can be represented as

$$A_1\,(0°,0°,45°),\ A_2\,(90°,90°,45°),\ A_3\,(0°,90°,45°);$$

Method 2: one analyzer of variable orientation in front of the detector:

$$A_3\,(0°),\ A_3\,(45°),\ A_3\,(90°);$$

Method 3: analyzer placed in path of incident beam:

$$A_0\,(0°),\ A_0\,(45°),\ A_0\,(90°);$$

Method 4: a hybrid of Methods 2 and 3:

$$A_1\,(0°,0°\ \text{or}\ 45°),\ A_2\,(90°,90°\ \text{or}\ 45°),$$
$$A_3\,(45°,0°;\ \text{or}\ 45°,\ 90°,\ \text{or}\ 45°,\ 45°)$$

The interferometer spectrometer schematic is shown in Fig. 21.15 in an astronomical application to measure the IR (0.8–2.7 μm) spectropolarimetric properties of Venus. At most, three interferograms are recorded, where the coefficients of the interferogram are linear combinations of the four Stokes parameters. Details of the calculations are given in Fymet (1972).

Results of observations of Venus using Method 2 are shown in Fig. 21.16; lines of the solar spectrum are shown at the bottom for comparison. Features A and B are Al lines and C is an Fe and Si line, shown for illustration.

Appendix
Optical Complex Indices of Refraction— Ultraviolet, Visual, and Near Infrared

Both atmospheric aerosol and surface modeling require as a basic input parameter the complex index of refraction as a function of wavelength. Very little information has been available in the technical and scientific literature on appropriate values. A listing of the complex index of refraction of 24 naturally occurring rocks and minerals has been presented (Egan and Hilgeman, 1979). Subsequently, additional data on samples have become available. These are presented in the following pages. Included are: data on sand samples from the Sahara Desert, Israel, and various U.S. beach areas; volcanic cinders from Vesuvius, Mt. St. Helens' eruption, and Haleakala; acid rain samples from the New York City area; Nextel diffuse paints; coral; maple leaf; carbon; photometric standards ($BaSO_4$, $MgCO_3$); and miscellaneous other rocks and minerals.

The samples were pulverized and compressed within a surface layer of a KBr pellet (Egan and Hilgeman, 1979); total diffuse transmission and reflection measurements were made on these pellets with an integrating sphere. The data obtained were analyzed to determine the absorption (imaginary) portion of the index of refraction separate from the scattering. When the appropriate refractive (real) index was available, the modified Kubelka–Munk (MKM) theory was used to determine the absorption portion as developed by Egan, Hilgeman, and Reichman (1973); when the real index was not available, the Kubelka–Munk (KM) theory was applied to yield the absorption. The effective thicknesses of the samples are indicated in the table legends.

Table A.1 Chemical Analyses of Israel, Sinai, and Sahara Sand Samples, and Mt. St. Helens' Ash

Sample No.	Israel samples			
	Israel 1	Israel 2	Israel 3	Israel 4[a]
SiO_2	31.45	73.06	86.54	50.63
Al_2O_3	8.31	3.11	3.18	19.12
Fe_2O_3	4.28	1.23	0.78	10.23
FeO	0.16	0.04	0.12	0.36
MgO	2.78	0.74	0.20	0.98
CaO	23.98	9.84	3.77	1.85
Na_2O	0.36	0.28	0.41	0.11
K_2O	1.07	0.79	0.74	1.53
H_2O^+ H_2O	7.86	2.36	0.91	11.64
CO_2	18.62	7.42	2.73	0.66
TiO_2	0.66	0.29	0.44	1.34
P_2O_5	0.47	0.33	0.02	0.21
MnO	0.09	0.01	0.03	0.11
S				0.82
F				
TOTAL	100.09	99.50	99.87	99.59

[a] Other organic carbon as C (= 0.82).

The refractive portion of the index of refraction was determined using the Brewster angle technique (Egan and Hilgeman, 1979). For powder samples, a compressed pellet served for the measurements, and where larger consolidated polished samples were available, they were used.

The wavelengths of observations were chosen using interference filters between 0.185 and 2.7 μm wavelengths. All samples were measured between 0.185 and 1.1 μm, and a number of them to 2.7 μm.

The Sahara Sand, Israel Sinai sand, and Mt. St. Helens (#1 and #2) samples were chemically analyzed and the results are presented in Table A.1. The Sahara sands were obtained from the areas of Niger as indicated below:

1. Azelik: flood plain;
2a. Maradi: surface dune sample;
2b. Maradi: 30 cm below the dune surface;
3. Amakon: dune sand from Tenere Desert, East from Aïr;
4. N'G'UIGMI (diatomite from the Lake Chad area); and
5. Dune sand from Niger valley close to Niamez.

Table A.1 (*continued*)

	Sahara samples						Sinai	Mt. St. Helens[b]
1	2a	2b	3	4	5			
54.75	98.70	95.90	95.30	51.20	95.95	88.80	63.40	
15.84	0.18	2.09	2.44	1.97	2.03	0.82	16.92	
6.74	0.12	0.32	0.33	0.50	0.57	0.16	2.46	
0.25	0.06	0.10	0.17	0.15	0.16	0.11	2.72	
2.83	0.027	0.045	0.087	1.85	0.036	0.343	2.14	
2.08	0.049	0.050	0.124	23.54	0.061	5.02	5.07	
0.96	0.015	0.032	0.288	0.266	0.024	0.116	4.65	
2.51	0.11	0.12	0.63	0.35	0.06	0.04	1.43	
6.03	0.24	0.46	0.28	0.76	0.56	0.22		
5.45	0.14	0.20	0.19	0.55	0.22	0.11		
1.38	0.07	0.45	0.14	18.52	0.18	4.26		
0.95	0.04	0.14	0.10	0.07	0.19	0.14	0.68	
0.17	0.03	0.02	0.02	0.04	0.01	0.03		
0.112	0.003	0.005	0.007	0.119	0.009	0.006		
							0.397	
							0.0389	
100.05	99.78	99.93	100.11	99.89	100.06	99.95	99.91	

[b] $Cu = 54$ ppm; $Zn = 61$ ppm.

The Israel sands were as follows:

1. calcareous serozems, location 4 km east from Beit-Sheàn near kibbutz Ma'oz Haim;
2. loessial serozems, location near Tel-Malkhata, 23 km east from Beer-Sheva;
3. sand dunes location near Yamit northern coast of Sinai Peninsula;
4. terra rossas, Brown Rendzinas and Pale Rendzinas;
   and Sinai sand from 10 km south of Elat, Sinai Peninsula.

The acid rain samples were collected in Woodhaven, Long Island (New York City), and the dates of acquisition, meteorology and chemical analyses are listed in Table A.2. Of the acid rain samples, the ones indicated were prepared as KBr pellet samples to permit determination of the absorption portion using the KM technique. There does not appear to be any clear correlation of the chemical composition with the meteorology; the low pH could be caused by sulfates, chlorides, or dissolved CO_2.

The Nextel paint samples were prepared from dried paint and as such include the

Table A.2 Collection Conditions and Chemical Properties of Rainwater Samples

| Rainwater Sample | Collection times | | Ambient temperature (°F) | Amount of precipitation (inches) | Wind | |
	Start $\left(\begin{array}{c} E.S.T. \\ date \end{array}\right)$	End $\left(\begin{array}{c} E.S.T. \\ date \end{array}\right)$			Direction	Velocity (mph)
A	0600 3/5/80	1800 3/5/80	38–50	0.10	S/SW/NW	15–20
B	0700 3/7/80	0930 3/8/80	44–55	0.04	S/SE/SW	10–20
C	1700 3/8/80	2300 3/9/80	37–52	0.04	S/SW	10–20
D	2200 3/10/80	0700 3/11/80	30–47	0.39	S/SW/NW	15–20
E	2230 3/20/80	1700 3/22/80	47–55	1.78	E/SE	15–20
F	0730 3/24/80	1800 3/25/80	37–43	0.41	N/NW	15–20
G	0700 3/28/80	1200 3/30/80	45–48	0.63	SE/E/NE	10–20
H	1400 3/30/80	0700 4/1/80	35–47	0.83	NE	15–20
I	0700 4/2/80	1020 4/5/80	50–64	0.58	SE/NW	15–20
J	2000 4/8/80	0700 4/10/80	45–64	1.36	SE/SW	15–25
K	1800 4/13/80	0700 4/15/80	44–56	0.30	E/SE	15–25
L	0500 5/7/80	1630 5/8/80	48–56	0.41	W/NW	10–15
M	0600 5/12/80	1630 5/12/80	56–64	0.23	S	10–15
N	1700 5/30/80	0600 5/31/80	63–76	0.05	S/SW	10–15

pigment, binder, extender, and dried solvents; the indices of refraction presented are thus appropriate for modeling the dried paint. The refractive portion was obtained using the Brewster angle technique on paint samples that were compressed between highly polished dies to produce a shiny (glossy) surface on the paint.

The Mt. St. Helens samples included ones obtained during the eruption, as well as samples collected further away at Kelso and Spokane. The volcanic material was generally andesitic glass and the heavier larger particulate material fell out first (during the eruption). The finer material remained aloft longer (Spokane sample).

The silicate sands of Long Island (from the ocean front at Jones Beach, and the Connequot River of Great South Bay) are white. Those from North Carolina (Cape Fear, Hatteras, Ocracoke) are brownish, containing shells of marine animals.

The Miami coral sample was prepared from a variety of white staghorn coral.

Table A.2 (*continued*)

pH[a]	NO$_2$[a]	NO$_3$[a]	Cl$^-$ [a]	SO$_4$[a]	Na[a]	TOC[a]	TIC[a]	CO$_2$[a]	Pollen	CO$_3$[a]
5.2	0.72	< 0.05	1.4	< 5.0	0.57	30	< 1	< 4	· · ·	· · ·
4.9	0.17	< 0.05	1.4	< 5.0	0.39	5	< 1	< 4	No	2.4
4.8	0.17	< 0.05	2.0	< 5.0	0.67	1	< 1	< 4	No	4.0
4.9	0.25	< 0.05	1.2	< 5.0	0.19	2	< 1	< 4	No	2.0
6.2	0.21	< 0.05	.79	< 5.0	0.12	3	< 1	< 4	Yes	4.0
4.7	0.15	< 0.05	.60	< 5.0	0.10	2	< 1	< 4	No	0.8
5.3	0.10	< 0.05	2.8	< 5.0	1.34	4	< 1	< 4	Yes	1.6
4.4	2.0	< 0.05	19.3	9	0.29	3	< 1	< 4	Yes	4.0
4.8	0.32	< 0.05	1.2	5.0	0.11	2	< 1	< 4	Yes	2.0
5.2	1.3	< 0.05	2.4	· · ·	1.06	13	< 1	< 4	· · ·	9.6

[a] Parts per million (ppm).

The maple leaf (Acer saccharum) sample was obtained and measured during the summer to determine a typical complex index of refraction for leaf chlorophyll modeling. Since many leaves have a waxy coating, the refractive and absorptive indices of refraction obtained would be representative of the waxy coating, not the chlorophyll. The maple leaf has a dull appearance (top surface), and was thus chosen as a reasonable representative to determine the index of refraction of chlorophyll.

Various carbon samples are included because carbon is a major atmospheric constituent in industrial areas. The lampblack (Fisher chemical C-198) was prepared by burning oil and collecting the soot; it was found to contain considerable sulfur. Two aerosol samples of propane soot itself are included as well as two pure graphite samples, one bulk "Graphite" and one prepared as a KBr pellet ("Graphite I").

Methylene blue is included to represent a strong absorber.

Ammonium sulfate $(NH_4)_2SO_4$ is included and represents a product of human activity, and two samples of it mixed with propane soot as an aerosol are also listed.

Anatase and rutile complex indices of refraction data obtained by both the KM and MKM theories are provided for comparison. The MKM is more elaborate and generally more accurate. The real portion of the indices have been corrected to the bulk for the MKM calculations using the "empirical theory" (Egan and Aspnes, 1982).

The various iron compounds are representative of weathering products such as those that exist on the earth's surface or might exist on Mars. The gamma Fe_2O_3 is magnetic, whereas alpha Fe_2O_3 is nonmagnetic. Two calculations are presented, the KM and MKM for the alpha and gamma phases.

Various samples of Arizona road dust are included to characterize particulate aerosols that exist over arid regions of the southwest United States.

Table A.3 Optical Constants for Aerosol 1. nk Determined from KM Theory.
Effective Sample Thickness = 0.0137 cm

LAMBDA	T	R	NK
0.185	0.293E-04	0.060	0.102E-02
0.190	0.136E-03	0.046	0.951E-03
0.200	0.150F-04	0.050	0.115E-02
0.210	0.418E-05	0.046	0.127E-02
0.215	0.152E-04	0.053	0.126E-02
0.220	0.353E-05	0.056	0.139E-02
0.225	0.481E-05	0.047	0.129E-02
0.233	0.277E-05	0.045	0.135E-02
0.240	0.345E-05	0.048	0.141E-02
0.260	0.800E-07	0.056	0.152E-02
0.280	0.127E-06	0.044	0.166E-02
0.300	0.162E-05	0.054	0.175E-02
0.325	0.971E-06	0.060	0.194E-02
0.360	0.352E-05	0.062	0.208E-02
0.370	0.746E-05	0.063	0.221E-02
0.400	0.653E-04	0.065	0.235E-02
0.433	0.122E-03	0.071	0.219E-02
0.466	0.295E-03	0.073	0.210E-02
0.500	0.604E-03	0.076	0.215E-02
0.533	0.121E-02	0.076	0.207E-02
0.566	0.226E-02	0.076	0.200E-02
0.600	0.258E-02	0.081	0.202E-02
0.633	0.437E-02	0.083	0.199E-02
0.666	0.482E-02	0.083	0.206E-02
0.700	0.475E-02	0.102	0.216E-02
0.817	0.132E-01	0.080	0.203E-02
0.907	0.208E-01	0.085	0.204E-02
1.000	0.273E-01	0.051	0.205E-02
1.105	0.328E-01	0.085	0.219E-02

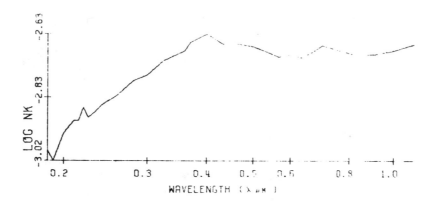

Table A.4 Optical Constants for Aerosol 2 (Aerosol Composition B).
nk Determined from KM Theory. Effective Sample Thickness = 0.00893 cm

LAMBDA	T	R	NK
0.185	0.159E-04	0.057	0.164E-02
0.190	0.391E-03	0.046	0.132E-02
0.200	0.431E-04	0.048	0.166E-02
0.210	0.329E-05	0.046	0.187E-02
0.215	0.223E-05	0.043	0.204E-02
0.220	0.244E-05	0.037	0.196E-02
0.225	0.144E-05	0.046	0.205E-02
0.233	0.203E-05	0.048	0.210E-02
0.240	0.406E-05	0.045	0.214E-02
0.260	0.169E-05	0.057	0.242E-02
0.280	0.459E-05	0.045	0.253E-02
0.300	0.355E-04	0.060	0.252E-02
0.325	0.111E-03	0.071	0.256E-02
0.360	0.666E-03	0.084	0.233E-02
0.370	0.124E-02	0.098	0.220E-02
0.400	0.479E-03	0.103	0.270E-02
0.433	0.108E-02	0.123	0.262E-02
0.466	0.173E-02	0.134	0.263E-02
0.500	0.133E-01	0.150	0.192E-02
0.533	0.331E-01	0.155	0.161E-02
0.566	0.250E-01	0.167	0.185E-02
0.600	0.293E-01	0.172	0.187E-02
0.633	0.328E-01	0.176	0.191E-02
0.666	0.974E-02	0.183	0.273E-02
0.700	0.517E-02	0.211	0.326E-02
0.817	0.152E-01	0.150	0.302E-02
0.907	0.166E-01	0.153	0.327E-02
1.000	0.183E-01	0.200	0.353E-02
1.105	0.328E-01	0.155	0.334E-02

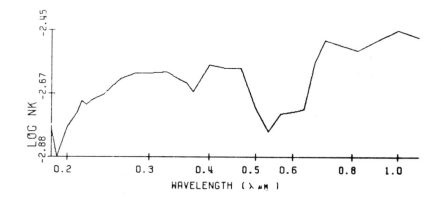

Table A.5 Optical Constants for Aerosol 3 (Aerosol Composition H).
          nk Determined from KM Theory. Effective Sample Thickness = 0.0494 cm

LAMBDA	T	R	NK
0.185	0.266E-04	0.074	0.285E-03
0.190	0.391E-03	0.060	0.240E-03
0.200	0.144E-04	0.061	0.315E-03
0.210	0.338E-06	0.058	0.353E-03
0.215	0.165E-04	0.044	0.342E-03
0.220	0.238E-04	0.044	0.353E-03
0.225	0.245E-06	0.058	0.362E-03
0.233	0.285E-05	0.057	0.385E-03
0.240	0.150E-05	0.057	0.398E-03
0.260	0.166E-06	0.062	0.422E-03
0.280	0.970E-07	0.050	0.465E-03
0.300	0.798E-06	0.055	0.493E-03
0.325	0.130E-07	0.063	0.532E-03
0.360	0.101E-05	0.064	0.612E-03
0.370	0.167E-05	0.063	0.643E-03
0.400	0.199E-04	0.066	0.630E-03
0.433	0.511E-05	0.055	0.708E-03
0.466	0.167E-04	0.074	0.728E-03
0.500	0.279E-04	0.087	0.834E-03
0.533	0.825E-04	0.101	0.852E-03
0.566	0.153E-03	0.101	0.789E-03
0.600	0.206E-03	0.116	0.810E-03
0.633	0.331E-03	0.121	0.806E-03
0.666	0.272E-03	0.123	0.875E-03
0.700	0.286E-02	0.116	0.659E-03
0.817	0.231E-02	0.118	0.796E-03
0.907	0.259E-02	0.125	0.867E-03
1.000	0.456E-02	0.133	0.865E-03
1.105	0.674E-02	0.142	0.886E-03

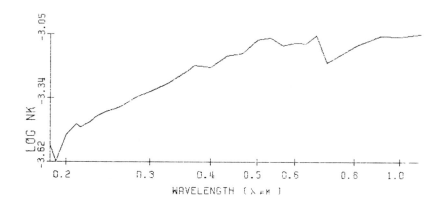

Table A.6 Optical Constants for Aerosol 4 (Aerosol Composition J).
nk Determined from KM Theory. Effective Sample Thickness = 0.00986 cm

LAMBDA	T	R	NK
0.185	0.250E-04	0.081	0.153E-02
0.190	0.205E-03	0.058	0.130E-02
0.200	0.317E-04	0.062	0.153E-02
0.210	0.110E-04	0.060	0.167E-02
0.215	0.135E-04	0.073	0.173E-02
0.220	0.275E-04	0.073	0.175E-02
0.225	0.653E-05	0.058	0.182E-02
0.233	0.144E-04	0.057	0.184E-02
0.240	0.137E-04	0.056	0.185E-02
0.260	0.545E-05	0.054	0.207E-02
0.280	0.627E-05	0.045	0.239E-02
0.300	0.464E-05	0.061	0.240E-02
0.325	0.165E-04	0.055	0.255E-02
0.360	0.494E-04	0.060	0.265E-02
0.370	0.821E-04	0.068	0.265E-02
0.400	0.727E-04	0.076	0.304E-02
0.433	0.283E-03	0.065	0.281E-02
0.466	0.611E-03	0.097	0.276E-02
0.500	0.164E-02	0.107	0.258E-02
0.533	0.247E-02	0.124	0.257E-02
0.566	0.358E-02	0.128	0.251E-02
0.600	0.535E-02	0.142	0.252E-02
0.633	0.666E-02	0.144	0.255E-02
0.666	0.485E-02	0.146	0.285E-02
0.700	0.115E-01	0.153	0.249E-02
0.817	0.158E-01	0.146	0.272E-02
0.907	0.208E-01	0.156	0.282E-02
1.000	0.224E-01	0.155	0.305E-02
1.105	0.290E-01	0.157	0.314E-02

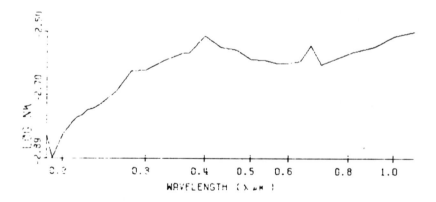

Table A.7 Optical Constants for Aerosol 5 (Aerosol Composition J).
nk Determined from KM Theory. Effective Sample Thickness = 0.0109 cm

LAMBDA	T	R	NK
0.185	0.896E-05	0.070	0.140E-02
0.190	0.855E-04	0.073	0.127E-02
0.200	0.426E-04	0.073	0.212E-02
0.210	0.114E-04	0.045	0.153E-02
0.215	0.155E-04	0.045	0.158E-02
0.220	0.230E-04	0.046	0.155E-02
0.225	0.383E-05	0.048	0.170E-02
0.233	0.324E-05	0.045	0.172E-02
0.240	0.676E-05	0.046	0.177E-02
0.260	0.329E-05	0.046	0.154E-02
0.280	0.165E-04	0.045	0.204E-02
0.300	0.513E-04	0.046	0.204E-02
0.325	0.130E-03	0.053	0.209E-02
0.360	0.582E-03	0.062	0.165E-02
0.370	0.805E-03	0.066	0.192E-02
0.400	0.264E-02	0.081	0.238E-02
0.433	0.852E-02	0.075	0.220E-02
0.466	0.206E-02	0.121	0.209E-02
0.500	0.221E-01	0.155	0.138E-02
0.533	0.346E-01	0.471	0.121E-02
0.566	0.499E-01	0.352	0.117E-02
0.600	0.575E-01	0.466	0.114E-02
0.633	0.713E-01	0.481	0.110E-02
0.666	0.813E-01	0.496	0.131E-02
0.700	0.955E-01	0.551	0.101E-02
0.817	0.128E+00	0.523	0.103E-02
0.907	0.140E+00	0.571	0.103E-02
1.000	0.139E+00	0.550	0.112E-02
1.105	0.124E+00	0.557	0.133E-02

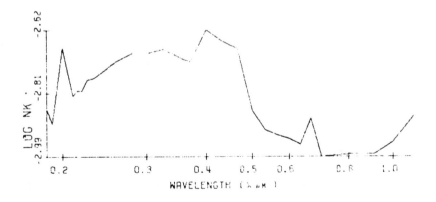

Table A.8 Optical Constants for Aerosol 6. nk Determined from KM Theory.
Effective Sample Thickness = 0.0123 cm

LAMBDA	T	R	NK
0.185	0.215E-04	0.064	0.677E-03
0.190	0.152E-03	0.063	0.686E-03
0.200	0.663E-04	0.066	0.703E-03
0.210	0.356E-05	0.061	0.800E-03
0.215	0.428E-05	0.060	0.806E-03
0.220	0.120E-04	0.057	0.855E-03
0.225	0.216E-05	0.062	0.893E-03
0.233	0.311E-05	0.063	0.858E-03
0.240	0.177E-05	0.062	0.917E-03
0.260	0.365E-06	0.057	0.972E-03
0.280	0.267E-06	0.066	0.106E-02
0.300	0.143E-05	0.057	0.112E-02
0.325	0.112E-05	0.072	0.126E-02
0.360	0.155E-05	0.080	0.137E-02
0.370	0.299E-05	0.083	0.146E-02
0.400	0.215E-05	0.092	0.150E-02
0.433	0.164E-04	0.092	0.162E-02
0.466	0.331E-04	0.134	0.167E-02
0.500	0.112E-02	0.155	0.126E-02
0.533	0.101E-02	0.241	0.136E-02
0.566	0.172E-02	0.247	0.133E-02
0.600	0.219E-02	0.267	0.135E-02
0.633	0.261E-02	0.255	0.136E-02
0.666	0.887E-02	0.307	0.115E-02
0.700	0.166E-01	0.336	0.104E-02
0.817	0.271E-01	0.320	0.107E-02
0.907	0.346E-01	0.370	0.109E-02
1.000	0.356E-01	0.366	0.119E-02
1.105	0.945E-02	0.372	0.186E-02

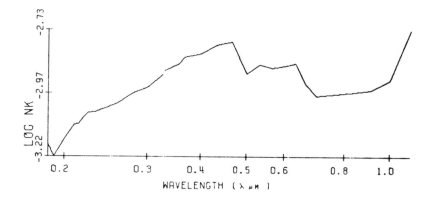

Table A.9 Optical Constants for Aerosol 7 (Aerosol Composition G).
nk Determined from KM Theory. Effective Sample Thickness = 0.0150 cm

LAMBDA	T	R	NK
0.185	0.515E-05	0.055	0.903E-03
0.190	0.241F-03	0.051	0.032E-03
0.200	0.410F-04	0.051	0.100E-02
0.210	0.557F-05	0.051	0.110E-02
0.215	0.125F-05	0.051	0.116E-02
0.220	0.117F-04	0.050	0.117E-02
0.225	0.175F-05	0.050	0.120F-02
0.233	0.157E-05	0.050	0.127F-02
0.240	0.255F-05	0.045	0.132E-02
0.260	0.205F-06	0.050	0.152F-02
0.280	0.430E-06	0.048	0.149E-02
0.300	0.631E-06	0.050	0.159E-02
0.325	0.240E-06	0.050	0.170E-02
0.360	0.312E-06	0.054	0.195E-02
0.370	0.156F-06	0.055	0.205F-02
0.400	0.304F-05	0.057	0.217E-02
0.433	0.770E-05	0.055	0.242E-02
0.466	0.663E-05	0.054	0.254F-02
0.500	0.745F-03	0.107	0.150F-02
0.533	0.111E-02	0.166	0.191E-02
0.566	0.276F-02	0.151	0.176F-02
0.600	0.422E-02	0.520	0.172F-02
0.633	0.643E-02	0.260	0.167E-02
0.666	0.161E-01	0.311	0.142F-02
0.700	0.272E-01	0.353	0.129F-02
0.817	0.473E-01	0.364	0.126E-02
0.907	0.553E-01	0.433	0.127F-02
1.000	0.356E-01	0.448	0.155E-02
1.105	0.356E-01	0.444	0.176E-02

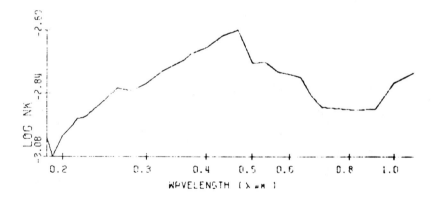

Table A.10 Optical Constants for Aerosol 9 (Aerosol Composition M).
nk Determined from KM Theory. Effective Sample Thickness = 0.00358 cm

LAMBDA	T	R	NK
0.185	0.221E-03	0.051	0.343E-02
0.190	0.147E-03	0.051	0.362E-02
0.200	0.347E-04	0.051	0.431E-02
0.210	0.181E-04	0.048	0.505E-02
0.215	0.126E-04	0.051	0.480E-02
0.220	0.755E-05	0.053	0.497E-02
0.225	0.578E-05	0.051	0.528E-02
0.233	0.291E-05	0.052	0.529E-02
0.240	0.108E-04	0.050	0.523E-02
0.260	0.341E-05	0.053	0.591E-02
0.280	0.741E-05	0.054	0.631E-02
0.300	0.157E-04	0.056	0.648E-02
0.325	0.375E-04	0.057	0.698E-02
0.360	0.122E-03	0.066	0.706E-02
0.370	0.169E-03	0.067	0.699E-02
0.400	0.374E-03	0.061	0.706E-02
0.433	0.264E-04	0.068	0.564E-02
0.466	0.140E-02	0.073	0.677E-02
0.500	0.410E-02	0.076	0.609E-02
0.533	0.619E-02	0.084	0.601E-02
0.566	0.940E-02	0.090	0.586E-02
0.600	0.109E-01	0.094	0.601E-02
0.633	0.141E-01	0.103	0.598E-02
0.666	0.100E-01	0.099	0.680E-02
0.700	0.233E-01	0.104	0.583E-02
0.817	0.335E-01	0.108	0.615E-02
0.907	0.467E-01	0.114	0.615E-02
1.000	0.538E-01	0.122	0.646E-02
1.105	0.595E-01	0.118	0.690E-02

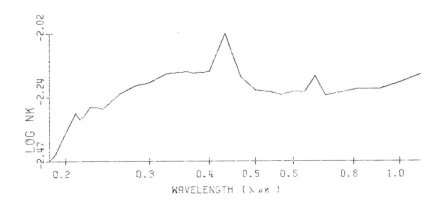

Table A.11 Optical Constants for Aerosol 10. nk Determined from KM Theory.
Effective Sample Thickness = 0.0116 cm

LAMBDA	T	R	NK
0.185	0.263F-03	0.056	0.753E-03
0.190	0.201F-03	0.054	0.109F-02
0.200	0.396F-04	0.054	0.132F-02
0.210	0.218E-04	0.051	0.139E-02
0.215	0.310F-04	0.054	0.141F-02
0.220	0.114E-04	0.054	0.149E-02
0.225	0.349E-05	0.053	0.156F-02
0.233	0.238F-06	0.054	0.164E-02
0.240	0.385E-06	0.051	0.165E-02
0.260	0.532E-06	0.052	0.186F-02
0.280	0.322E-06	0.053	0.196F-02
0.300	0.200E-08	0.049	0.209F-02
0.325	0.460E-07	0.051	0.229E-02
0.360	0.428F-06	0.053	0.258F-02
0.370	0.116F-05	0.054	0.261F-02
0.400	0.285E-05	0.055	0.273E-02
0.433	0.822E-03	0.062	0.210F-02
0.466	0.862E-04	0.065	0.258F-02
0.500	0.139E-02	0.073	0.225F-02
0.533	0.270F-02	0.082	0.216F-02
0.566	0.527E-02	0.101	0.203E-02
0.600	0.699E-02	0.101	0.204E-02
0.633	0.101E-01	0.123	0.199E-02
0.666	0.961E-02	0.115	0.212F-02
0.700	0.176E-01	0.133	0.193E-02
0.817	0.261E-01	0.154	0.203F-02
0.907	0.325F-01	0.164	0.211E-02
1.000	0.294E-01	0.172	0.240E-02
1.105	0.356F-01	0.165	0.251E-02

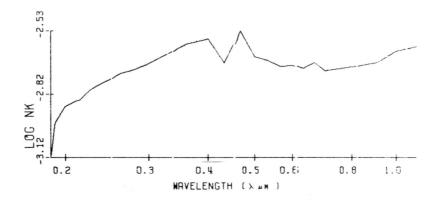

Table A.12 Optical Constants for Aerosol 11 (Aerosol Composition N).
nk Determined from KM Theory. Effective Samples Thickness = 0.00745 cm

LAMBDA	T	R	NK
0.185	0.129E-02	0.063	0.131E-02
0.190	0.456E-03	0.063	0.154E-02
0.200	0.355E-04	0.062	0.206E-02
0.210	0.111E-04	0.055	0.221E-02
0.215	0.840E-07	0.063	0.230E-02
0.220	0.657E-05	0.062	0.234E-02
0.225	0.659E-05	0.064	0.238E-02
0.233	0.775E-05	0.063	0.250E-02
0.240	0.852E-05	0.061	0.253E-02
0.260	0.537E-05	0.066	0.281E-02
0.280	0.100E-04	0.067	0.298E-02
0.300	0.461E-05	0.067	0.326E-02
0.325	0.179E-04	0.067	0.380E-02
0.360	0.756E-04	0.078	0.354E-02
0.370	0.147E-03	0.079	0.342E-02
0.400	0.277E-03	0.083	0.344E-02
0.433	0.858E-03	0.094	0.324E-02
0.466	0.175E-02	0.102	0.315E-02
0.500	0.345E-02	0.110	0.302E-02
0.533	0.608E-02	0.124	0.290E-02
0.566	0.100E-01	0.136	0.277E-02
0.600	0.129E-01	0.145	0.278E-02
0.633	0.167E-01	0.156	0.275E-02
0.666	0.235E-01	0.150	0.265E-02
0.700	0.306E-01	0.166	0.259E-02
0.817	0.472E-01	0.183	0.264E-02
0.907	0.645E-01	0.193	0.261E-02
1.000	0.716E-01	0.201	0.277E-02
1.105	0.846E-01	0.184	0.287E-02

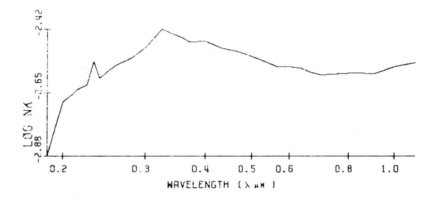

Table A.13 Optical Constants for Albite. nk Determined from KM Theory.
         Thickness = 0.349 cm

LAMBDA	T	R	NK	N
0.185	0.691E-03	0.169	0.815E-05	1.569
0.190	0.284E-02	0.231	0.474E-05	1.567
0.200	0.349E-02	0.227	0.492E-05	1.564
0.210	0.346E-02	0.206	0.580E-05	1.562
0.215	0.311E-02	0.235	0.520E-05	1.560
0.220	0.449E-02	0.246	0.473E-05	1.558
0.225	0.546E-02	0.226	0.517E-05	1.555
0.233	0.615E-02	0.238	0.493E-05	1.553
0.240	0.677E-02	0.240	0.494E-05	1.551
0.260	0.141E-01	0.282	0.377E-05	1.549
0.280	0.343E-01	0.295	0.304E-05	1.546
0.300	0.201E-01	0.328	0.330E-05	1.544
0.325	0.299E-01	0.347	0.297E-05	1.542
0.360	0.414E-01	0.369	0.272E-05	1.540
0.370	0.402E-01	0.377	0.275E-05	1.537
0.400	0.453E-01	0.391	0.271E-05	1.535
0.433	0.542E-01	0.402	0.264E-05	1.530
0.466	0.608E-01	0.410	0.265E-05	1.526
0.500	0.682E-01	0.423	0.260E-05	1.520
0.533	0.714E-01	0.423	0.272E-05	1.519
0.566	0.713E-01	0.414	0.300E-05	1.517
0.600	0.745E-01	0.411	0.317E-05	1.515
0.633	0.731E-01	0.406	0.343E-05	1.516
0.666	0.809E-01	0.417	0.331E-05	1.516
0.700	0.868E-01	0.405	0.353E-05	1.517
0.817	0.101E+00	0.414	0.370E-05	1.517
0.907	0.111E+00	0.452	0.336E-05	1.511
1.000	0.117E+00	0.436	0.388E-05	1.504
1.105	0.109E+00	0.434	0.449E-05	1.504

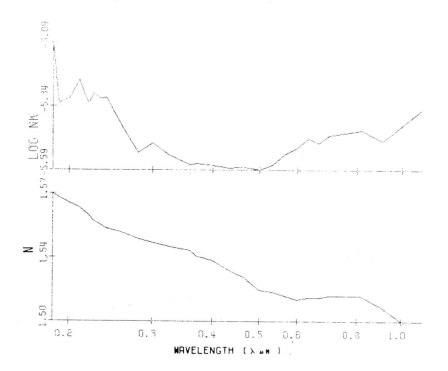

Table A.14 Optical Constants for Ambient Air, Ft. Collins, Colo.
nk Determined from MKM Theory. Effective Sample Thickness =
0.00000737 cm. Colorado State University Sample 8

LAMBDA	T	R	NK	N
0.185	0.156E+00	0.121	0.148E+00	1.390
0.190	0.178E+00	0.141	0.133E+00	1.387
0.200	0.171E+00	0.134	0.150E+00	1.383
0.210	0.163E+00	0.122	0.176E+00	1.380
0.215	0.171E+00	0.127	0.170E+00	1.377
0.220	0.174E+00	0.133	0.166E+00	1.374
0.225	0.217E+00	0.133	0.150E+00	1.370
0.233	0.230E+00	0.128	0.148E+00	1.367
0.240	0.203E+00	0.130	0.170E+00	1.364
0.260	0.238E+00	0.133	0.164E+00	1.361
0.280	0.249E+00	0.138	0.173E+00	1.357
0.300	0.260E+00	0.148	0.163E+00	1.354
0.325	0.283E+00	0.152	0.162E+00	1.351
0.360	0.323E+00	0.140	0.174E+00	1.347
0.370	0.340E+00	0.136	0.175E+00	1.344
0.400	0.375E+00	0.133	0.175E+00	1.341
0.433	0.411E+00	0.130	0.175E+00	1.340
0.466	0.438E+00	0.123	0.183E+00	1.339
0.500	0.443E+00	0.137	0.177E+00	1.338
0.533	0.463E+00	0.118	0.202E+00	1.337
0.566	0.481E+00	0.113	0.210E+00	1.336
0.600	0.490E+00	0.113	0.217E+00	1.335
0.633	0.515E+00	0.106	0.223E+00	1.334
0.666	0.526E+00	0.102	0.233E+00	1.333
0.700	0.530E+00	0.106	0.235E+00	1.333
0.817	0.574E+00	0.097	0.273E+00	1.332
0.907	0.599E+00	0.087	0.277E+00	1.331
1.000	0.615E+00	0.054	0.273E+00	1.330
1.105	0.637E+00	0.065	0.299E+00	1.329

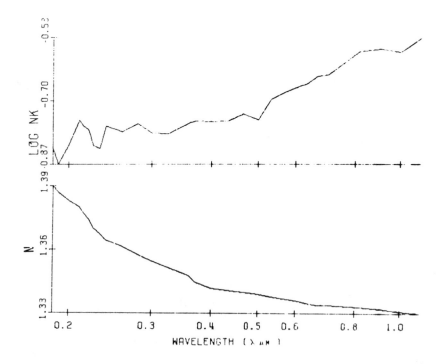

Table A.15 Optical Constants for Ammonium Sulfate [$(NH_4)_2SO_4$].
nk Determined from KM Theory. Thickness = 0.0475 cm

LAMBDA	T	R	NK	N
0.185	0.152E+00	0.302	0.794E-05	1.537
0.190	0.157E+00	0.321	0.736E-05	1.532
0.200	0.168E+00	0.327	0.728E-05	1.526
0.210	0.150E+00	0.318	0.857E-05	1.520
0.215	0.182E+00	0.332	0.737E-05	1.514
0.220	0.185E+00	0.337	0.733E-05	1.509
0.225	0.161E+00	0.326	0.866E-05	1.503
0.233	0.154E+00	0.335	0.885E-05	1.497
0.240	0.153E+00	0.315	0.101E-04	1.491
0.260	0.188E+00	0.348	0.836E-05	1.486
0.280	0.205E+00	0.344	0.870E-05	1.480
0.300	0.286E+00	0.355	0.530E-05	1.474
0.325	0.246E+00	0.355	0.676E-05	1.468
0.360	0.294E+00	0.355	0.624E-05	1.463
0.370	0.227E+00	0.350	0.877E-05	1.457
0.400	0.236E+00	0.352	0.773E-05	1.451
0.433	0.347E+00	0.358	0.612E-05	1.450
0.466	0.337E+00	0.351	0.719E-05	1.449
0.500	0.345E+00	0.355	0.774E-05	1.448
0.533	0.353E+00	0.350	0.772E-05	1.448
0.566	0.357E+00	0.355	0.810E-05	1.448
0.600	0.355E+00	0.353	0.846E-05	1.448
0.633	0.360E+00	0.355	0.862E-05	1.447
0.666	0.357E+00	0.344	0.125E-04	1.445
0.700	0.348E+00	0.370	0.118E-04	1.444
0.817	0.345E+00	0.363	0.145E-04	1.442
0.907	0.360E+00	0.356	0.158E-04	1.441
1.000	0.364E+00	0.360	0.168E-04	1.439
1.105	0.367E+00	0.326	0.224E-04	1.438

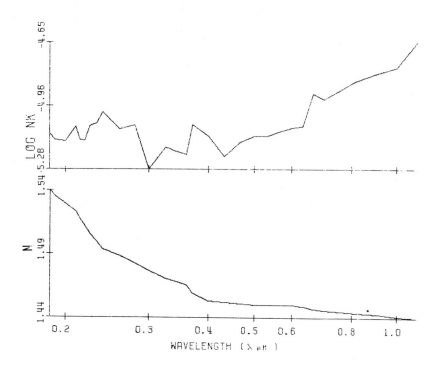

Table A.16 Optical Constants for Anatase (TiO$_2$) octahedrite.
nk Determined from KM Theory. Effective Sample Thickness = 0.00422 cm

LAMBDA	T	R	NK	N
0.185	0.821E-04	0.100	0.156E-02	1.502
0.190	0.102E-02	0.122	0.958E-03	1.516
0.200	0.933E-04	0.106	0.157E-02	1.529
0.210	0.250E-04	0.101	0.199E-02	1.542
0.215	0.235E-04	0.103	0.201E-02	1.556
0.220	0.179E-04	0.104	0.210E-02	1.569
0.233	0.226E-05	0.104	0.265E-02	1.596
0.240	0.135E-05	0.102	0.298E-02	1.610
0.260	0.525E-06	0.100	0.358E-02	1.623
0.280	0.225E-06	0.099	0.417E-02	1.637
0.300	0.228E-05	0.096	0.390E-02	1.650
0.325	0.391E-05	0.104	0.370E-02	1.605
0.360	0.675E-04	0.254	0.105E-02	1.560
0.370	0.105E-02	0.405	0.462E-03	1.515
0.400	0.279E-02	0.475	0.346E-03	1.470
0.433	0.552E-02	0.523	0.231E-03	1.476
0.466	0.171E-01	0.553	0.199E-03	1.483
0.500	0.202E-01	0.556	0.200E-03	1.489
0.533	0.376E-01	0.554	0.146E-03	1.507
0.566	0.409E-01	0.608	0.129E-03	1.525
0.600	0.424E-01	0.554	0.180E-03	1.543
0.633	0.546E-01	0.551	0.148E-03	1.577
0.666	0.807E-01	0.619	0.989E-04	1.600
0.700	0.933E-01	0.627	0.951E-04	1.623
0.800	0.122E+00	0.621	0.902E-04	1.654
0.907	0.147E+00	0.603	0.952E-04	1.669
1.000	0.152E+00	0.562	0.125E-03	1.684
1.105	0.170E+00	0.548	0.131E-03	1.706
1.200	0.232E+00	0.516	0.122E-03	1.706
1.300	0.241E+00	0.507	0.133E-03	1.706
1.400	0.320E+00	0.457	0.130E-03	1.706
1.500	0.151E+00	0.493	0.257E-03	1.706
1.700	0.322E+00	0.391	0.235E-03	1.706
2.000	0.181E+00	0.446	0.371E-03	1.706
2.300	0.318E+00	0.310	0.524E-03	1.706
2.500	0.957E-01	0.537	0.472E-03	1.706

LOG NK

N

WAVELENGTH [λ µM]

Table A.17 Optical Constants for Anatase (TiO$_2$) octahedrite.
nk Determined from MKM Theory. Effective Sample Thickness = 0.00422 cm

LAMBDA	T	R	NK	N
0.185	0.821E-04	0.100	0.314E-02	2.148
0.190	0.102E-02	0.122	0.247E-02	2.183
0.200	0.922E-04	0.106	0.338E-02	2.217
0.210	0.250E-04	0.101	0.438E-02	2.250
0.215	0.235E-04	0.103	0.391E-02	2.286
0.220	0.175E-04	0.104	0.403E-02	2.319
0.225	0.100E-04	0.106	0.427E-02	2.356
0.233	0.226E-05	0.104	0.455E-02	2.390
0.240	0.138E-05	0.102	0.463E-02	2.426
0.260	0.525E-04	0.100	0.536E-02	2.460
0.280	0.225E-04	0.095	0.534E-02	2.497
0.300	0.228E-05	0.098	0.578E-02	2.532
0.325	0.251E-05	0.104	0.615E-02	2.413
0.360	0.675E-04	0.254	0.622E-02	2.296
0.370	0.105E-02	0.405	0.466E-02	2.181
0.400	0.275E-02	0.475	0.424E-02	2.067
0.433	0.952E-02	0.533	0.349E-02	2.082
0.466	0.171E-01	0.553	0.325E-02	2.100
0.500	0.202E-01	0.556	0.333E-02	2.115
0.533	0.376E-01	0.594	0.286E-02	2.161
0.566	0.409E-01	0.608	0.271E-02	2.206
0.600	0.424E-01	0.554	0.316E-02	2.252
0.633	0.546E-01	0.581	0.298E-02	2.340
0.666	0.863E-01	0.615	0.241E-02	2.400
0.700	0.932E-01	0.627	0.246E-02	2.460
0.800	0.122E+00	0.621	0.241E-02	2.542
0.907	0.147E+00	0.603	0.247E-02	2.582
1.000	0.152E+00	0.562	0.281E-02	2.622
1.105	0.170E+00	0.548	0.291E-02	2.681
1.200	0.223E+00	0.516	0.254E-02	2.681
1.300	0.241E+00	0.507	0.268E-02	2.681
1.400	0.320E+00	0.457	0.228E-02	2.681
1.500	0.151E+00	0.453	0.453E-02	2.681
1.700	0.322E+00	0.391	0.302E-02	2.681
2.000	0.191E+00	0.446	0.557E-02	2.681
2.300	0.318E+00	0.310	0.447E-02	2.681
2.500	0.557E-01	0.537	0.943E-02	2.681

Table A.18 Optical Constants for Anorthosite. nk Determined from KM Theory.
Thickness = 0.0737 cm

LAMBDA	T	R	NK	N
0.185	0.150E-04	0.064	0.219E-03	1.603
0.190	0.866E-04	0.048	0.185E-03	1.600
0.200	0.800E-04	0.054	0.195E-03	1.598
0.210	0.624E-05	0.053	0.231E-03	1.595
0.215	0.758E-05	0.060	0.254E-03	1.593
0.220	0.246E-04	0.060	0.229E-03	1.590
0.225	0.529E-05	0.051	0.244E-03	1.588
0.233	0.644E-05	0.050	0.257E-03	1.585
0.240	0.156E-04	0.051	0.257E-03	1.583
0.260	0.577E-03	0.045	0.208E-03	1.580
0.280	0.146E-02	0.052	0.197E-03	1.578
0.300	0.280E-02	0.059	0.190E-03	1.575
0.325	0.792E-02	0.067	0.169E-03	1.573
0.360	0.143E-01	0.072	0.165E-03	1.570
0.370	0.166E-01	0.072	0.164E-03	1.568
0.400	0.221E-01	0.074	0.164E-03	1.565
0.433	0.309E-01	0.074	0.162E-03	1.562
0.466	0.375E-01	0.078	0.165E-03	1.559
0.500	0.422E-01	0.081	0.170E-03	1.556
0.533	0.467E-01	0.078	0.176E-03	1.553
0.566	0.485E-01	0.076	0.185E-03	1.550
0.600	0.509E-01	0.077	0.193E-03	1.547
0.633	0.525E-01	0.076	0.201E-03	1.547
0.666	0.554E-01	0.072	0.208E-03	1.546
0.700	0.617E-01	0.071	0.210E-03	1.546
0.817	0.682E-01	0.076	0.236E-03	1.545
0.907	0.784E-01	0.082	0.249E-03	1.545
1.000	0.826E-01	0.077	0.269E-03	1.545
1.105	0.765E-01	0.068	0.306E-03	1.545

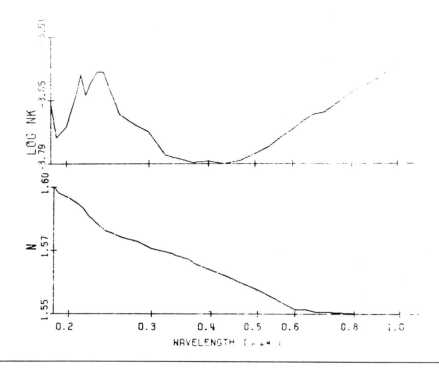

Table A.19 Optical Constants for Arizona Road Dust. nk Determined from MKM Theory. Effective Sample Thickness = 0.000149 cm. Colorado State University Sample 1

LAMBDA	T	R	NK	N
0.185	0.165E+00	0.125	0.742E-02	1.390
0.190	0.157E+00	0.134	0.652E-02	1.387
0.200	0.155E+00	0.125	0.715E-02	1.383
0.210	0.202E+00	0.126	0.752E-02	1.380
0.215	0.210E+00	0.131	0.728E-02	1.377
0.220	0.215E+00	0.136	0.712E-02	1.374
0.225	0.217E+00	0.134	0.736E-02	1.370
0.233	0.224E+00	0.137	0.734E-02	1.367
0.240	0.223E+00	0.136	0.764E-02	1.364
0.260	0.285E+00	0.152	0.632E-02	1.361
0.280	0.300E+00	0.153	0.652E-02	1.357
0.300	0.344E+00	0.155	0.598E-02	1.354
0.325	0.358E+00	0.153	0.648E-02	1.351
0.360	0.410E+00	0.149	0.637E-02	1.347
0.370	0.419E+00	0.151	0.631E-02	1.344
0.400	0.433E+00	0.152	0.652E-02	1.341
0.433	0.459E+00	0.145	0.683E-02	1.340
0.466	0.471E+00	0.144	0.714E-02	1.339
0.500	0.490E+00	0.139	0.746E-02	1.338
0.533	0.503E+00	0.137	0.774E-02	1.337
0.566	0.509E+00	0.137	0.806E-02	1.336
0.600	0.525E+00	0.133	0.833E-02	1.335
0.633	0.529E+00	0.133	0.868E-02	1.334
0.666	0.556E+00	0.123	0.893E-02	1.333
0.700	0.530E+00	0.124	0.102E-01	1.333
0.817	0.574E+00	0.111	0.112E-01	1.332
0.907	0.599E+00	0.102	0.122E-01	1.331
1.000	0.609E+00	0.106	0.125E-01	1.330
1.105	0.637E+00	0.112	0.118E-01	1.329

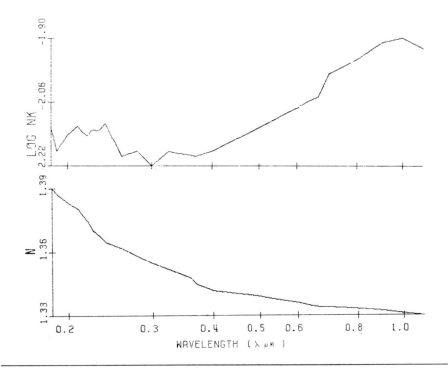

Table A.20 Optical Constants for Arizona Road Dust. nk Determined from MKM Theory.
Effective Sample Thickness = 0.00263 cm. Colorado State University Sample 1A

LAMBDA	T	R	NK	N
0.185	0.981E-02	0.073	0.156E-02	1.390
0.190	0.105E-01	0.069	0.164E-02	1.387
0.200	0.827E-02	0.062	0.193E-02	1.383
0.210	0.842E-02	0.060	0.206E-02	1.380
0.215	0.101E-01	0.062	0.200E-02	1.377
0.220	0.112E-01	0.067	0.191E-02	1.374
0.225	0.111E-01	0.064	0.200E-02	1.370
0.233	0.121E-01	0.066	0.199E-02	1.367
0.240	0.136E-01	0.066	0.201E-02	1.364
0.260	0.259E-01	0.075	0.171E-02	1.361
0.280	0.338E-01	0.075	0.171E-02	1.357
0.300	0.394E-01	0.080	0.168E-02	1.354
0.325	0.481E-01	0.084	0.166E-02	1.351
0.360	0.705E-01	0.087	0.157E-02	1.347
0.370	0.791E-01	0.092	0.149E-02	1.344
0.400	0.958E-01	0.094	0.148E-02	1.341
0.433	0.121E+00	0.101	0.138E-02	1.340
0.466	0.136E+00	0.104	0.138E-02	1.339
0.500	0.156E+00	0.113	0.130E-02	1.338
0.533	0.177E+00	0.122	0.122E-02	1.337
0.566	0.207E+00	0.141	0.106E-02	1.336
0.600	0.224E+00	0.149	0.102E-02	1.335
0.633	0.241E+00	0.152	0.100E-02	1.334
0.666	0.267E+00	0.150	0.997E-03	1.333
0.700	0.270E+00	0.156	0.100E-02	1.333
0.817	0.302E+00	0.151	0.111E-02	1.332
0.907	0.317E+00	0.155	0.115E-02	1.331
1.000	0.340E+00	0.143	0.128E-02	1.330
1.105	0.363E+00	0.146	0.129E-02	1.329

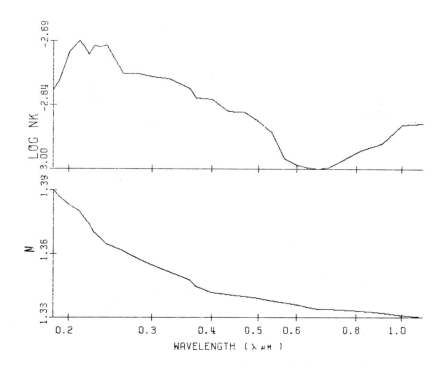

Table A.21 Optical Constants for Arizona Road Dust. nk Determined from MKM Theory. Effective Sample Thickness = 0.00488 cm. Colorado State University Sample 1B

LAMBDA	T	R	NK	N
0.185	0.242E-02	0.061	0.134E-02	1.390
0.190	0.877E-03	0.065	0.148E-02	1.387
0.200	0.114E-02	0.068	0.151E-02	1.383
0.210	0.118E-02	0.059	0.172E-02	1.380
0.215	0.157E-02	0.067	0.156E-02	1.377
0.220	0.164E-02	0.071	0.153E-02	1.374
0.225	0.171E-02	0.070	0.157E-02	1.370
0.233	0.187E-02	0.071	0.158E-02	1.367
0.240	0.215E-02	0.070	0.160E-02	1.364
0.260	0.518E-02	0.081	0.135E-02	1.361
0.280	0.648E-02	0.081	0.140E-02	1.357
0.300	0.975E-02	0.087	0.131E-02	1.354
0.325	0.126E-01	0.091	0.131E-02	1.351
0.360	0.191E-01	0.095	0.128E-02	1.347
0.370	0.220E-01	0.099	0.123E-02	1.344
0.400	0.263E-01	0.104	0.123E-02	1.341
0.433	0.372E-01	0.115	0.112E-02	1.340
0.466	0.449E-01	0.123	0.108E-02	1.339
0.500	0.543E-01	0.115	0.112E-02	1.338
0.533	0.670E-01	0.143	0.958E-03	1.337
0.566	0.853E-01	0.169	0.802E-03	1.336
0.600	0.960E-01	0.182	0.757E-03	1.335
0.633	0.108E+00	0.185	0.750E-03	1.334
0.666	0.122E+00	0.185	0.749E-03	1.333
0.700	0.116E+00	0.192	0.776E-03	1.333
0.817	0.155E+00	0.198	0.766E-03	1.332
0.907	0.165E+00	0.195	0.837E-03	1.331
1.000	0.187E+00	0.192	0.876E-03	1.330
1.105	0.204E+00	0.191	0.926E-03	1.329

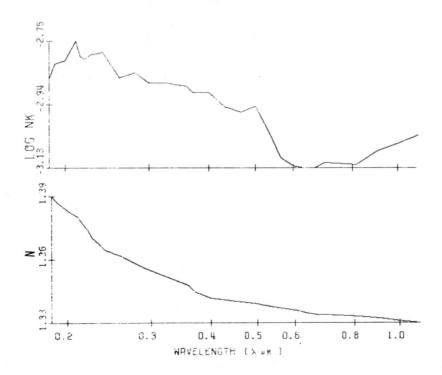

Table A.22 Optical Constants for Barium Sulfate ($BaSO_4$).
nk Determined from MKM Theory. Effective Sample Thickness = 0.00581 cm

LAMBDA	T	R	NK	N
0.185	0.206E-01	0.282	0.147E-03	1.546
0.190	0.291E-01	0.312	0.121E-03	1.543
0.200	0.263E-01	0.306	0.135E-03	1.540
0.210	0.275E-01	0.304	0.141E-03	1.537
0.215	0.188E-01	0.286	0.174E-03	1.534
0.220	0.215E-01	0.296	0.165E-03	1.531
0.225	0.337E-01	0.330	0.129E-03	1.528
0.233	0.359E-01	0.332	0.130E-03	1.526
0.240	0.376E-01	0.343	0.127E-03	1.523
0.260	0.486E-01	0.371	0.113E-03	1.520
0.280	0.679E-01	0.408	0.931E-04	1.517
0.300	0.788E-01	0.433	0.851E-04	1.514
0.325	0.948E-01	0.459	0.768E-04	1.505
0.360	0.122E+00	0.482	0.679E-04	1.497
0.370	0.132E+00	0.493	0.640E-04	1.488
0.400	0.155E+00	0.509	0.583E-04	1.479
0.433	0.163E+00	0.512	0.606E-04	1.471
0.466	0.170E+00	0.498	0.678E-04	1.462
0.500	0.180E+00	0.501	0.695E-04	1.454
0.533	0.182E+00	0.493	0.765E-04	1.450
0.566	0.180E+00	0.473	0.908E-04	1.445
0.600	0.188E+00	0.462	0.984E-04	1.441
0.633	0.192E+00	0.453	0.106E-03	1.441
0.666	0.203E+00	0.457	0.106E-03	1.430
0.700	0.208E+00	0.462	0.108E-03	1.419
0.800	0.226E+00	0.449	0.123E-03	1.415
0.907	0.240E+00	0.440	0.137E-03	1.418
1.000	0.252E+00	0.414	0.164E-03	1.420
1.105	0.261E+00	0.412	0.177E-03	1.419
1.200	0.343E+00	0.462	0.969E-04	1.419
1.300	0.281E+00	0.301	0.341E-03	1.419
1.400	0.327E+00	0.283	0.348E-03	1.419
1.500	0.253E+00	0.265	0.517E-03	1.419
1.700	0.299E+00	0.469	0.167E-03	1.419
2.000	0.274E+00	0.280	0.598E-03	1.419
2.300	0.392E+00	0.096	0.155E-02	1.419
2.500	0.367E+00	0.373	0.321E-03	1.419

Table A.23 Optical Constants for Coral (Florida). nk Determined from MKM Theory.
Effective Sample Thickness = 0.00414 cm

LAMBDA	T	R	NK	N
0.185	0.609E-02	0.224	0.365E-03	1.509
0.190	0.338E-02	0.224	0.417E-03	1.509
0.200	0.112E-01	0.231	0.337E-03	1.508
0.210	0.113E-01	0.233	0.350E-03	1.507
0.215	0.118E-01	0.252	0.325E-03	1.506
0.220	0.132E-01	0.263	0.308E-03	1.506
0.225	0.148E-01	0.274	0.292E-03	1.505
0.233	0.163E-01	0.239	0.348E-03	1.504
0.240	0.162E-01	0.251	0.339E-03	1.503
0.260	0.276E-01	0.263	0.303E-03	1.503
0.280	0.237E-01	0.297	0.265E-03	1.502
0.300	0.357E-01	0.324	0.250E-03	1.501
0.325	0.542E-01	0.356	0.207E-03	1.500
0.360	0.820E-01	0.364	0.174E-03	1.500
0.370	0.831E-01	0.368	0.175E-03	1.499
0.400	0.953E-01	0.393	0.175E-03	1.498
0.433	0.110E+00	0.407	0.167E-03	1.494
0.466	0.125E+00	0.404	0.171E-03	1.491
0.500	0.164E+00	0.403	0.157E-03	1.487
0.533	0.143E+00	0.410	0.178E-03	1.484
0.566	0.183E+00	0.404	0.165E-03	1.481
0.600	0.156E+00	0.408	0.193E-03	1.478
0.633	0.164E+00	0.399	0.209E-03	1.477
0.666	0.166E+00	0.386	0.227E-03	1.476
0.700	0.169E+00	0.396	0.226E-03	1.475
0.817	0.187E+00	0.339	0.320E-03	1.474
0.907	0.206E+00	0.368	0.288E-03	1.478
1.000	0.222E+00	0.365	0.303E-03	1.482
1.105	0.209E+00	0.346	0.384E-03	1.482

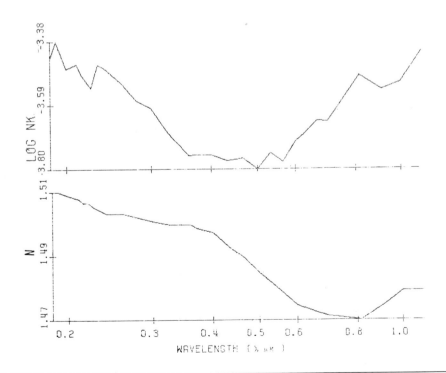

Table A.24 Optical Constants for Farm Soil (Long Island).
nk Determined from MKM Theory. Effective Sample Thickness = 0.00372 cm

LAMBDA	T	R	NK	N
0.185	0.443E-03	0.066	0.188E-02	1.304
0.190	0.570E-03	0.051	0.218E-02	1.317
0.200	0.708E-03	0.055	0.216E-02	1.330
0.210	0.760E-03	0.050	0.236E-02	1.344
0.215	0.952E-03	0.062	0.210E-02	1.357
0.220	0.103E-02	0.057	0.224E-02	1.370
0.225	0.131E-02	0.051	0.237E-02	1.383
0.233	0.149E-02	0.052	0.241E-02	1.396
0.240	0.146E-02	0.052	0.253E-02	1.409
0.260	0.200E-02	0.036	0.316E-02	1.422
0.280	0.262E-02	0.050	0.282E-02	1.435
0.300	0.234E-02	0.049	0.316E-02	1.449
0.325	0.364E-02	0.056	0.295E-02	1.462
0.360	0.577E-02	0.051	0.324E-02	1.475
0.370	0.106E-01	0.050	0.301E-02	1.488
0.400	0.258E-01	0.069	0.211E-02	1.501
0.433	0.481E-01	0.070	0.184E-02	1.492
0.466	0.686E-01	0.060	0.197E-02	1.483
0.500	0.102E+00	0.051	0.130E-02	1.474
0.533	0.129E+00	0.107	0.108E-02	1.473
0.566	0.163E+00	0.130	0.855E-03	1.473
0.600	0.198E+00	0.156	0.678E-03	1.472
0.633	0.212E+00	0.140	0.768E-03	1.471
0.666	0.241E+00	0.142	0.734E-03	1.471
0.700	0.258E+00	0.145	0.721E-03	1.470
0.817	0.304E+00	0.172	0.620E-03	1.468
0.907	0.323E+00	0.163	0.654E-03	1.466
1.000	0.350E+00	0.155	0.732E-03	1.464
1.105	0.391E+00	0.160	0.718E-03	1.462

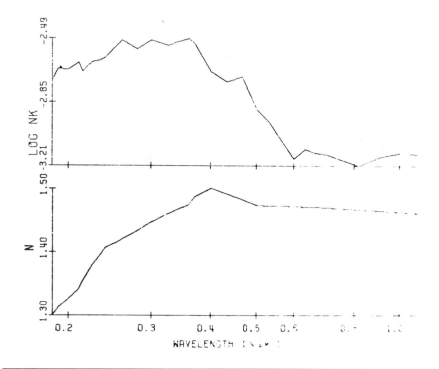

Table A.25 Optical Constants for FeO (Ferrous Oxide). nk Determined from KM Theory.
Effective Sample Thickness = 0.00484 cm

LAMBDA	T	R	NK	N
0.185	0.112E-02	0.063	0.206E-02	1.627
0.190	0.121E-02	0.067	0.205E-02	1.627
0.200	0.141E-02	0.071	0.216E-02	1.627
0.210	0.135E-02	0.070	0.226E-02	1.627
0.215	0.144E-02	0.075	0.231E-02	1.627
0.220	0.151E-02	0.075	0.234E-02	1.627
0.225	0.158E-02	0.067	0.238E-02	1.627
0.233	0.163E-02	0.067	0.246E-02	1.627
0.240	0.167E-02	0.067	0.253E-02	1.683
0.260	0.178E-02	0.068	0.269E-02	1.740
0.280	0.193E-02	0.053	0.287E-02	1.796
0.300	0.256E-02	0.062	0.294E-02	1.853
0.325	0.197E-02	0.067	0.332E-02	1.909
0.360	0.328E-02	0.062	0.338E-02	1.965
0.370	0.437E-02	0.062	0.330E-02	2.022
0.400	0.488E-02	0.080	0.349E-02	2.078
0.433	0.673E-02	0.089	0.356E-02	2.076
0.466	0.825E-02	0.101	0.367E-02	2.075
0.500	0.983E-02	0.115	0.379E-02	2.073
0.533	0.129E-01	0.136	0.380E-02	2.052
0.566	0.171E-01	0.160	0.376E-02	2.032
0.600	0.197E-01	0.176	0.384E-02	2.011
0.633	0.245E-01	0.195	0.382E-02	1.992
0.666	0.298E-01	0.215	0.379E-02	1.972
0.700	0.342E-01	0.234	0.382E-02	1.953
0.817	0.335E-01	0.243	0.493E-02	1.933
0.907	0.431E-01	0.236	0.460E-02	1.888
1.000	0.522E-01	0.251	0.475E-02	1.843
1.105	0.627E-01	0.265	0.489E-02	1.798

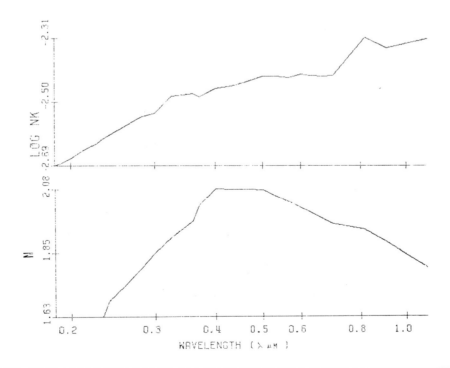

Table A.26 Optical Constants for Fe_2O_3 (Hematite). nk Determined from KM Theory.
Effective Sample Thickness = 0.0002 cm. n determined using bulk sample

LAMBDA	T	R	NK	N
0.185	0.785E-02	0.090	0.356E-01	1.051
0.225	0.103E-01	0.062	0.400E-01	1.552
0.233	0.107E-01	0.063	0.420E-01	1.523
0.240	0.109E-01	0.063	0.431E-01	1.555
0.260	0.113E-01	0.049	0.464E-01	1.509
0.280	0.120E-01	0.060	0.492E-01	1.461
0.300	0.121E-01	0.057	0.527E-01	1.414
0.325	0.136E-01	0.062	0.555E-01	1.367
0.360	0.115E-01	0.057	0.639E-01	1.323
0.370	0.125E-01	0.056	0.644E-01	1.277
0.400	0.168E-01	0.072	0.649E-01	1.232
0.433	0.173E-01	0.070	0.698E-01	1.256
0.466	0.196E-01	0.067	0.728E-01	1.270
0.500	0.211E-01	0.068	0.767E-01	1.303
0.533	0.249E-01	0.070	0.782E-01	1.327
0.566	0.500E-01	0.100	0.672E-01	1.352
0.600	0.779E-01	0.137	0.605E-01	1.376
0.633	0.103E+00	0.155	0.566E-01	1.500
0.666	0.137E+00	0.170	0.519E-01	1.624
0.700	0.165E+00	0.195	0.491E-01	1.754
0.817	0.202E+00	0.175	0.500E-01	1.754
0.907	0.230E+00	0.170	0.510E-01	1.872
1.000	0.298E+00	0.220	0.460E-01	1.990
1.105	0.340E+00	0.235	0.445E-01	2.110

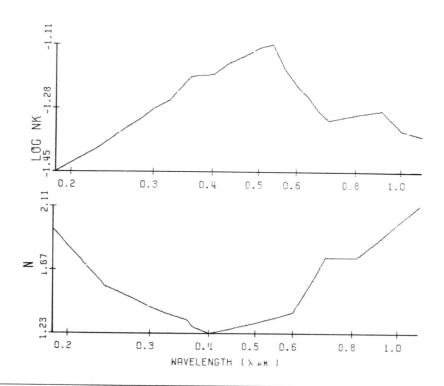

Table A.27 Optical Constants for Fe_2O_3 (Hematite). nk Determined from MKM Theory. Effective Sample Thickness = 0.0002 cm. n determined using pressed powder sample

LAMBDA	T	R	NK	N
0.185	0.785E-02	0.050	0.185E-01	1.423
0.190	0.875E-02	0.062	0.241E-01	1.403
0.200	0.895E-02	0.066	0.236E-01	1.382
0.210	0.893E-02	0.066	0.252E-01	1.361
0.215	0.932E-02	0.073	0.239E-01	1.340
0.220	0.986E-02	0.073	0.241E-01	1.320
0.225	0.103E-01	0.062	0.268E-01	1.299
0.233	0.107E-01	0.063	0.273E-01	1.278
0.240	0.105E-01	0.063	0.281E-01	1.257
0.260	0.113E-01	0.049	0.397E-01	1.237
0.280	0.120E-01	0.060	0.330E-01	1.216
0.300	0.121E-01	0.057	0.361E-01	1.195
0.325	0.136E-01	0.062	0.369E-01	1.174
0.360	0.115E-01	0.057	0.444E-01	1.154
0.370	0.125E-01	0.056	0.454E-01	1.133
0.400	0.168E-01	0.072	0.420E-01	1.112
0.433	0.173E-01	0.070	0.454E-01	1.123
0.466	0.196E-01	0.067	0.482E-01	1.134
0.500	0.211E-01	0.066	0.500E-01	1.145
0.533	0.249E-01	0.070	0.504E-01	1.156
0.566	0.500E-01	0.100	0.365E-01	1.167
0.600	0.779E-01	0.137	0.270E-01	1.178
0.633	0.103E+00	0.155	0.221E-01	1.233
0.666	0.137E+00	0.170	0.180E-01	1.287
0.700	0.165E+00	0.195	0.143E-01	1.342
0.817	0.202E+00	0.175	0.166E-01	1.342
0.907	0.230E+00	0.170	0.168E-01	1.391
1.000	0.295E+00	0.220	0.110E-01	1.439
1.105	0.340E+00	0.235	0.917E-02	1.487

Table A.28 Optical Constants for Fe_3O_4. nk Determined from KM Theory.
Effective Sample Thickness = 0.000246 cm

LAMBDA	T	R	NK	N
0.185	0.357E-03	0.065	0.478E-01	1.145
0.190	0.188E-03	0.065	0.530E-01	1.145
0.200	0.534E-03	0.070	0.482E-01	1.145
0.210	0.555E-03	0.067	0.506E-01	1.145
0.215	0.538E-03	0.074	0.529E-01	1.145
0.220	0.516E-03	0.077	0.535E-01	1.145
0.225	0.520E-03	0.065	0.548E-01	1.145
0.233	0.543E-03	0.066	0.573E-01	1.145
0.240	0.542E-03	0.065	0.588E-01	1.280
0.260	0.593E-03	0.052	0.619E-01	1.415
0.280	0.607E-03	0.062	0.665E-01	1.550
0.300	0.706E-03	0.062	0.701E-01	1.685
0.325	0.551E-03	0.065	0.786E-01	1.820
0.360	0.635E-03	0.060	0.853E-01	1.955
0.370	0.722E-03	0.057	0.863E-01	2.090
0.400	0.755E-03	0.070	0.925E-01	2.225
0.433	0.654E-03	0.065	0.102E+00	2.224
0.466	0.792E-03	0.063	0.107E+00	2.222
0.500	0.886E-03	0.062	0.113E+00	2.221
0.533	0.965E-03	0.061	0.119E+00	2.219
0.566	0.107E-02	0.060	0.125E+00	2.217
0.600	0.118E-02	0.061	0.130E+00	2.215
0.633	0.131E-02	0.062	0.136E+00	2.230
0.666	0.142E-02	0.060	0.141E+00	2.246
0.700	0.295E-02	0.055	0.132E+00	2.261
0.817	0.236E-02	0.055	0.160E+00	2.169
0.907	0.552E-02	0.060	0.150E+00	2.133
1.000	0.539E-02	0.060	0.165E+00	2.098
1.105	0.353E-02	0.065	0.201E+00	2.090

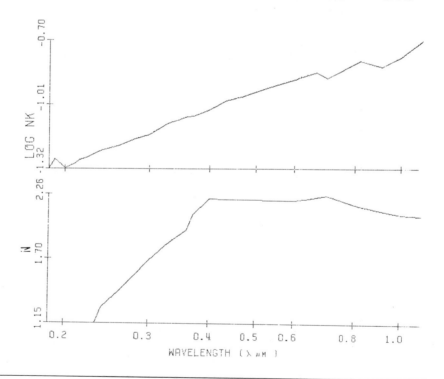

Table A.29 Absorption Properties for Graphite. nk Determined from KM Theory.
Effective Sample Thickness = 0.00208 cm

LAMBDA	T	R	NK	N
0.185	0.508E-02	0.181	0.372E-02	0.879
0.190	0.183E-02	0.159	0.458E-02	1.797
0.200	0.165E-02	0.163	0.487E-02	1.767
0.210	0.240E-02	0.155	0.482E-02	1.856
0.215	0.198E-02	0.161	0.511E-02	1.867
0.220	0.234E-02	0.156	0.506E-02	1.947
0.225	0.368E-02	0.150	0.480E-02	1.939
0.233	0.401E-02	0.149	0.490E-02	1.951
0.240	0.389E-02	0.138	0.508E-02	1.975
0.260	0.398E-02	0.131	0.547E-02	2.031
0.280	0.441E-02	0.124	0.579E-02	2.059
0.300	0.397E-02	0.094	0.633E-02	2.062
0.325	0.379E-02	0.113	0.691E-02	2.065
0.360	0.593E-02	0.109	0.704E-02	2.035
0.370	0.598E-02	0.102	0.723E-02	2.026
0.400	0.567E-02	0.104	0.790E-02	2.000
0.433	0.664E-02	0.104	0.829E-02	2.010
0.466	0.650E-02	0.077	0.897E-02	2.020
0.500	0.662E-02	0.099	0.958E-02	2.030
0.533	0.728E-02	0.101	0.100E-01	2.038
0.566	0.761E-02	0.101	0.105E-01	2.045
0.600	0.800E-02	0.106	0.111E-01	2.053
0.633	0.831E-02	0.101	0.116E-01	2.054
0.666	0.911E-02	0.096	0.119E-01	2.055
0.700	0.926E-02	0.095	0.125E-01	2.057
0.817	0.849E-02	0.108	0.149E-01	2.060
0.907	0.855E-02	0.114	0.165E-01	2.064
1.000	0.891E-02	0.117	0.180E-01	2.067
1.105	0.932E-02	0.111	0.197E-01	2.071

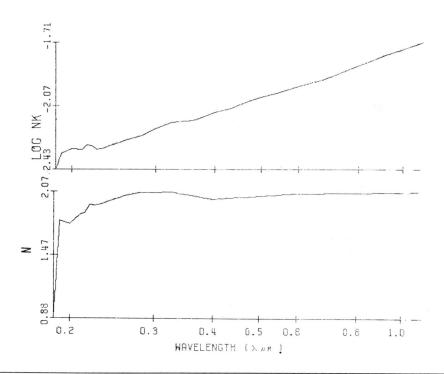

Table A.30 Graphite I (Ellipsometry)

LAMBDA	N	NK
0.185	0.472E+00	0.338
0.190	0.133E+01	0.834
0.200	0.141E+01	0.616
0.210	0.137E+01	0.897
0.215	0.151E+01	0.659
0.220	0.149E+01	0.881
0.225	0.163E+01	0.550
0.233	0.166E+01	0.511
0.240	0.168E+01	0.519
0.260	0.168E+01	0.711
0.280	0.168E+01	0.771
0.300	0.174E+01	0.653
0.325	0.170E+01	0.758
0.360	0.167E+01	0.731
0.370	0.166E+01	0.728
0.400	0.167E+01	0.649
0.433	0.168E+01	0.658
0.466	0.169E+01	0.656
0.500	0.170E+01	0.652
0.533	0.169E+01	0.703
0.566	0.173E+01	0.616
0.600	0.172E+01	0.680
0.633	0.173E+01	0.650
0.666	0.177E+01	0.508
0.700	0.175E+01	0.591
0.817	0.176E+01	0.576
0.907	0.175E+01	0.623
1.000	0.175E+01	0.624
1.105	0.176E+01	0.626

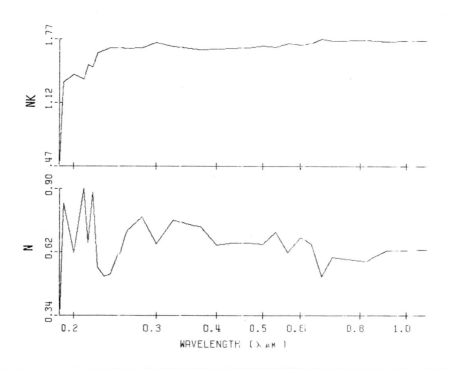

WAVELENGTH (λ μM)

Table A.31 Optical Constants for Hematite (Gamma Fe$_2$O$_3$). nk Determined from KM Theory. Effective Sample Thickness = 0.00324 cm. n determined using pressed powder sample

LAMBDA	T	R	NK	N
0.185	0.110E-01	0.112	0.204E-01	1.146
0.190	0.106E-01	0.072	0.212E-01	1.198
0.200	0.890E-02	0.068	0.232E-01	1.250
0.210	0.660E-02	0.069	0.259E-01	1.302
0.215	0.670E-02	0.069	0.264E-01	1.354
0.220	0.780E-02	0.065	0.262E-01	1.406
0.225	0.640E-02	0.066	0.279E-01	1.458
0.233	0.660E-02	0.067	0.287E-01	1.510
0.240	0.650E-02	0.067	0.297E-01	1.562
0.260	0.760E-02	0.062	0.311E-01	1.614
0.280	0.760E-02	0.062	0.335E-01	1.666
0.300	0.710E-02	0.063	0.364E-01	1.718
0.325	0.720E-02	0.067	0.394E-01	1.770
0.360	0.720E-02	0.070	0.436E-01	1.822
0.370	0.710E-02	0.074	0.450E-01	1.874
0.400	0.102E-01	0.074	0.450E-01	1.926
0.433	0.106E-01	0.074	0.483E-01	1.951
0.466	0.142E-01	0.076	0.486E-01	1.976
0.500	0.193E-01	0.078	0.484E-01	2.001
0.533	0.440E-01	0.089	0.408E-01	1.998
0.566	0.680E-01	0.107	0.372E-01	1.996
0.600	0.900E-01	0.114	0.353E-01	1.993
0.633	0.120E+00	0.134	0.327E-01	1.984
0.666	0.319E+00	0.151	0.183E-01	1.975
0.700	0.205E+00	0.175	0.267E-01	1.966
0.750	0.214E+00	0.190	0.277E-01	1.958
0.817	0.280E+00	0.194	0.247E-01	1.949
0.907	0.318E+00	0.173	0.248E-01	1.940
1.000	0.348E+00	0.169	0.251E-01	1.931
1.105	0.385E+00	0.150	0.252E-01	1.922

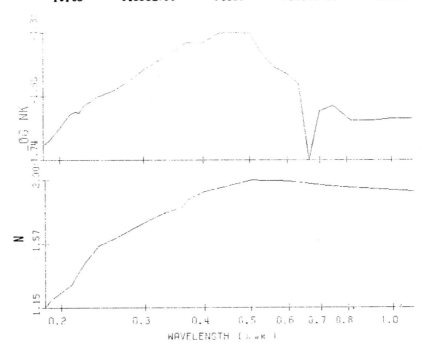

Table A.32 Optical Constants for Halon (Teflon). nk Determined from MKM Theory. Effective Sample Thickness = 0.00662 cm

LAMBDA	T	R	NK	N
0.185	0.745E-01	0.451	0.406E-04	1.441
0.190	0.873E-01	0.473	0.419E-04	1.437
0.200	0.937E-01	0.451	0.414E-04	1.433
0.210	0.959E-01	0.484	0.426E-04	1.430
0.215	0.105E+00	0.401	0.590E-04	1.426
0.220	0.123E+00	0.394	0.576E-04	1.422
0.225	0.128E+00	0.477	0.407E-04	1.418
0.233	0.110E+00	0.472	0.472E-04	1.414
0.240	0.112E+00	0.478	0.472E-04	1.410
0.260	0.136E+00	0.528	0.366E-04	1.406
0.280	0.147E+00	0.467	0.495E-04	1.402
0.300	0.162E+00	0.502	0.424E-04	1.399
0.325	0.137E+00	0.480	0.573E-04	1.395
0.360	0.179E+00	0.464	0.570E-04	1.391
0.370	0.215E+00	0.449	0.544E-04	1.387
0.400	0.224E+00	0.447	0.575E-04	1.383
0.433	0.231E+00	0.424	0.678E-04	1.382
0.466	0.234E+00	0.416	0.751E-04	1.381
0.500	0.243E+00	0.404	0.827E-04	1.380
0.533	0.249E+00	0.397	0.893E-04	1.379
0.566	0.254E+00	0.382	0.100E-03	1.377
0.600	0.255E+00	0.386	0.104E-03	1.376
0.633	0.259E+00	0.384	0.109E-03	1.378
0.666	0.266E+00	0.402	0.102E-03	1.379
0.700	0.268E+00	0.408	0.103E-03	1.381
0.817	0.269E+00	0.387	0.134E-03	1.377
0.907	0.285E+00	0.374	0.151E-03	1.371
1.000	0.286E+00	0.278	0.269E-03	1.373
1.105	0.297E+00	0.366	0.183E-03	1.363

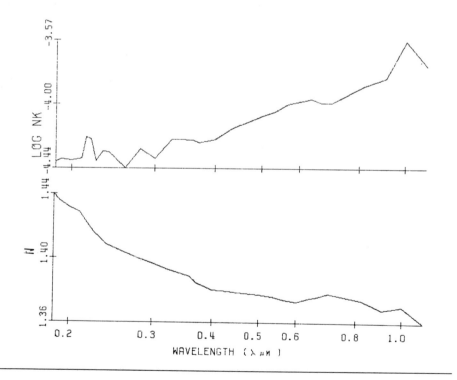

Table A.33 Optical Constants for Israel, No. 1. nk Determined from KM Theory. Effective Sample Thickness = 0.0165 cm

LAMBDA	T	R	NK	N
0.185	0.100E-01	0.071	0.410E-03	1.500
0.190	0.641E-03	0.064	0.670E-03	1.501
0.200	0.180E-03	0.054	0.823E-03	1.501
0.210	0.939E-05	0.044	0.998E-03	1.501
0.215	0.482E-04	0.040	0.101E-02	1.502
0.220	0.156E-04	0.045	0.105E-02	1.502
0.225	0.562E-05	0.044	0.108E-02	1.503
0.233	0.262E-04	0.048	0.111E-02	1.503
0.240	0.558E-04	0.047	0.109E-02	1.504
0.260	0.932E-04	0.047	0.112E-02	1.504
0.280	0.992E-04	0.045	0.121E-02	1.525
0.300	0.195E-03	0.049	0.122E-02	1.511
0.325	0.305E-03	0.048	0.125E-02	1.501
0.360	0.490E-03	0.051	0.131E-02	1.491
0.370	0.973E-03	0.056	0.123E-02	1.482
0.400	0.139E-02	0.058	0.127E-02	1.473
0.433	0.289E-02	0.070	0.122E-02	1.472
0.466	0.475E-02	0.075	0.120E-02	1.470
0.500	0.667E-02	0.072	0.121E-02	1.469
0.533	0.115E-01	0.076	0.115E-02	1.475
0.566	0.178E-01	0.087	0.110E-02	1.481
0.600	0.215E-01	0.091	0.111E-02	1.487
0.633	0.268E-01	0.094	0.110E-02	1.483
0.666	0.311E-01	0.106	0.111E-02	1.478
0.700	0.388E-01	0.096	0.109E-02	1.474
0.817	0.625E-01	0.127	0.109E-02	1.470
0.907	0.800E-01	0.136	0.110E-02	1.465
1.000	0.100E+00	0.135	0.110E-02	1.461
1.105	0.124E+00	0.137	0.110E-02	1.457

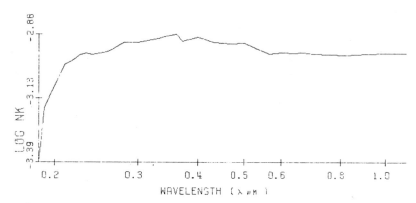

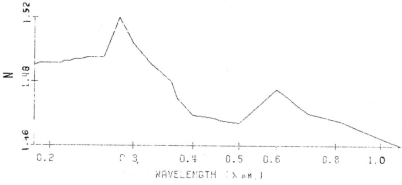

Table A.34 Optical Constants for Israel, No. 2. nk Determined from KM Theory.
Effective Sample Thickness = 0.0157 cm

LAMBDA	T	R	NK	N
0.185	0.864E-03	0.143	0.663E-03	1.515
0.190	0.386E-03	0.082	0.768E-03	1.512
0.200	0.351E-03	0.070	0.802E-03	1.510
0.210	0.223E-03	0.068	0.881E-03	1.508
0.215	0.549E-03	0.069	0.811E-03	1.505
0.220	0.482E-03	0.072	0.852E-03	1.503
0.225	0.588E-03	0.070	0.843E-03	1.500
0.233	0.577E-03	0.073	0.888E-03	1.598
0.240	0.640E-03	0.072	0.893E-03	1.595
0.260	0.167E-02	0.078	0.842E-03	1.493
0.280	0.196E-02	0.082	0.882E-03	1.494
0.300	0.246E-02	0.094	0.911E-03	1.453
0.325	0.372E-02	0.095	0.919E-03	1.453
0.360	0.500E-02	0.181	0.961E-03	1.452
0.370	0.875E-02	0.122	0.886E-03	1.449
0.400	0.130E-01	0.137	0.877E-03	1.446
0.433	0.242E-01	0.188	0.809E-03	1.439
0.466	0.338E-01	0.188	0.792E-03	1.432
0.500	0.507E-01	0.216	0.744E-03	1.425
0.533	0.769E-01	0.266	0.673E-03	1.427
0.566	0.110E+00	0.321	0.602E-03	1.430
0.600	0.125E+00	0.337	0.595E-03	1.432
0.633	0.134E+00	0.336	0.606E-03	1.431
0.666	0.142E+00	0.353	0.613E-03	1.430
0.700	0.152E+00	0.350	0.621E-03	1.430
0.817	0.189E+00	0.398	0.615E-03	1.429
0.907	0.195E+00	0.358	0.685E-03	1.428
1.000	0.212E+00	0.367	0.708E-03	1.427
1.105	0.234E+00	0.352	0.734E-03	1.426

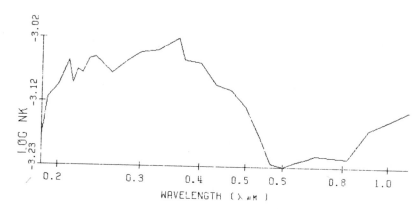

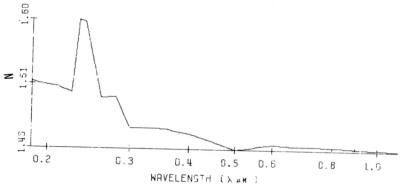

Table A.35 Optical Constants for Israel, No. 3. nk Determined from KM Theory.
Effective Sample Thickness = 0.0127 cm

LAMBDA	T	R	NK	N
0.185	0.411E-02	0.121	0.635E-03	1.583
0.190	0.256E-02	0.100	0.709E-03	1.572
0.200	0.244E-02	0.109	0.752E-03	1.562
0.210	0.247E-02	0.101	0.788E-03	1.552
0.215	0.332E-02	0.100	0.767E-03	1.541
0.220	0.344E-02	0.106	0.779E-03	1.531
0.225	0.397E-02	0.115	0.777E-03	1.520
0.233	0.385E-02	0.111	0.809E-03	1.510
0.240	0.470E-02	0.119	0.803E-03	1.499
0.260	0.960E-02	0.134	0.754E-03	1.489
0.280	0.990E-02	0.139	0.806E-03	1.585
0.300	0.106E-01	0.092	0.853E-03	1.425
0.325	0.146E-01	0.094	0.859E-03	1.434
0.360	0.178E-01	0.106	0.906E-03	1.443
0.370	0.255E-01	0.215	0.840E-03	1.423
0.400	0.347E-01	0.241	0.828E-03	1.403
0.433	0.531E-01	0.306	0.770E-03	1.417
0.466	0.674E-01	0.304	0.759E-03	1.432
0.500	0.848E-01	0.322	0.738E-03	1.446
0.533	0.109E+00	0.366	0.692E-03	1.453
0.566	0.134E+00	0.403	0.648E-03	1.459
0.600	0.142E+00	0.409	0.663E-03	1.466
0.633	0.149E+00	0.409	0.680E-03	1.460
0.666	0.156E+00	0.409	0.696E-03	1.453
0.700	0.164E+00	0.398	0.715E-03	1.447
0.817	0.178E+00	0.443	0.766E-03	1.441
0.907	0.190E+00	0.401	0.839E-03	1.434
1.000	0.200E+00	0.413	0.885E-03	1.428
1.105	0.206E+00	0.410	0.959E-03	1.422

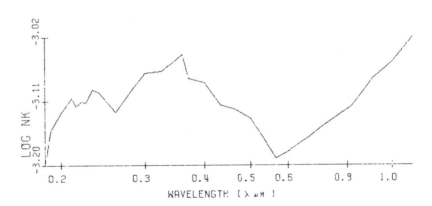

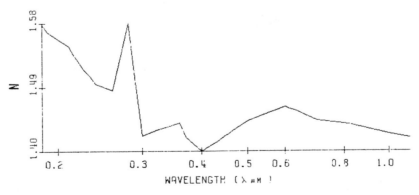

Table A.36 Optical Constants for Israel, No. 4. nk Determined from KM Theory.
            Effective Sample Thickness = 0.00521 cm

LAMBDA	T	R	NK	N
0.185	0.137E-01	0.071	0.121E-02	1.533
0.190	0.140E-02	0.045	0.191E-02	1.533
0.200	0.116E-02	0.050	0.206E-02	1.533
0.210	0.172E-02	0.047	0.203E-02	1.533
0.215	0.144E-02	0.043	0.215E-02	1.533
0.220	0.225E-02	0.047	0.205E-02	1.533
0.225	0.224E-02	0.046	0.209E-02	1.534
0.233	0.250E-02	0.048	0.213E-02	1.534
0.240	0.260E-02	0.047	0.218E-02	1.534
0.260	0.332E-02	0.045	0.226E-02	1.534
0.280	0.374E-02	0.045	0.239E-02	1.547
0.300	0.332E-02	0.045	0.261E-02	1.537
0.325	0.396E-02	0.044	0.275E-02	1.537
0.360	0.431E-02	0.046	0.299E-02	1.536
0.370	0.550E-02	0.048	0.294E-02	1.541
0.400	0.657E-02	0.051	0.307E-02	1.545
0.433	0.113E-01	0.055	0.296E-02	1.546
0.466	0.178E-01	0.057	0.286E-02	1.548
0.500	0.268E-01	0.058	0.276E-02	1.549
0.533	0.436E-01	0.062	0.255E-02	1.549
0.566	0.844E-01	0.082	0.213E-02	1.549
0.600	0.113E+00	0.094	0.199E-02	1.549
0.633	0.142E+00	0.108	0.188E-02	1.546
0.666	0.191E+00	0.116	0.167E-02	1.542
0.700	0.218E+00	0.110	0.161E-02	1.539
0.817	0.244E+00	0.124	0.174E-02	1.536
0.907	0.288E+00	0.114	0.171E-02	1.532
1.000	0.321E+00	0.116	0.171E-02	1.529
1.105	0.348E+00	0.111	0.176E-02	1.526

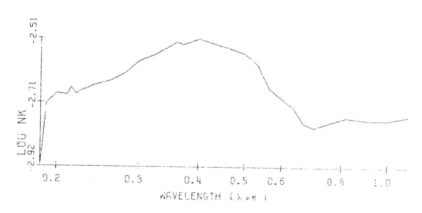

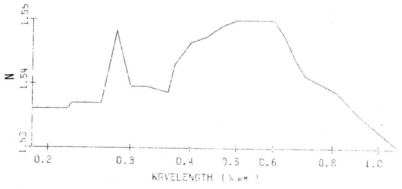

Table A.37 Lampblack (Fisher)

LAMBDA	N	NK
0.185	0.711E+00	0.404
0.190	0.723E+00	0.383
0.200	0.718E+00	0.392
0.210	0.725E+00	0.379
0.215	0.728E+00	0.373
0.220	0.732E+00	0.367
0.225	0.750E+00	0.328
0.233	0.792E+00	0.206
0.240	0.791E+00	0.211
0.260	0.792E+00	0.206
0.280	0.784E+00	0.234
0.300	0.780E+00	0.247
0.325	0.780E+00	0.247
0.360	0.761E+00	0.301
0.370	0.725E+00	0.379
0.400	0.753E+00	0.319
0.433	0.763E+00	0.333
0.466	0.773E+00	0.344
0.500	0.763E+00	0.399
0.533	0.786E+00	0.394
0.566	0.812E+00	0.383
0.600	0.857E+00	0.314
0.633	0.876E+00	0.320
0.666	0.907E+00	0.288
0.700	0.926E+00	0.294
0.807	0.935E+00	0.333
0.917	0.952E+00	0.321
1.000	0.974E+00	0.319
1.105	0.994E+00	0.315

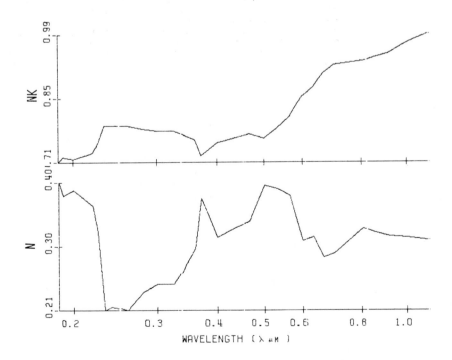

Table A.38 Optical Constants for Magnesium Carbonate (MgCO₃). nk Determined from MKM Theory. Effective Sample Thickness = 0.00439 cm

LAMBDA	LOG T	R	LOG NK	N
0.185	0.445E-02	0.183	0.455E-03	1.52
0.190	0.103E-01	0.192	0.378E-03	1.517
0.200	0.751E-02	0.189	0.432E-03	1.515
0.210	0.789E-02	0.187	0.455E-03	1.513
0.215	0.656E-02	0.186	0.487E-03	1.510
0.220	0.741E-02	0.178	0.512E-03	1.508
0.225	0.121E-01	0.205	0.405E-03	1.505
0.233	0.133E-01	0.207	0.406E-03	1.503
0.240	0.147E-01	0.212	0.399E-03	1.500
0.260	0.260E-01	0.226	0.350E-03	1.498
0.280	0.382E-01	0.235	0.324E-03	1.495
0.300	0.616E-01	0.242	0.288E-03	1.493
0.325	0.869E-01	0.273	0.237E-03	1.487
0.360	0.121E+00	0.315	0.188E-03	1.480
0.370	0.162E+00	0.333	0.154E-03	1.474
0.400	0.180E+00	0.356	0.141E-03	1.467
0.433	0.216E+00	0.363	0.129E-03	1.469
0.466	0.243E+00	0.359	0.129E-03	1.471
0.500	0.261E+00	0.354	0.133E-03	1.473
0.533	0.285E+00	0.354	0.130E-03	1.472
0.566	0.299E+00	0.341	0.141E-03	1.471
0.600	0.306E+00	0.328	0.156E-03	1.470
0.633	0.328E+00	0.319	0.161E-03	1.470
0.666	0.338E+00	0.310	0.172E-03	1.469
0.700	0.349E+00	0.295	0.190E-03	1.468
0.800	0.384E+00	0.269	0.223E-03	1.470
0.907	0.403E+00	0.245	0.275E-03	1.470
1.000	0.461E+00	0.233	0.268E-03	1.469
1.105	0.448E+00	0.217	0.348E-03	1.461
1.200	0.482E+00	0.225	0.316E-03	1.461
1.300	0.544E+00	0.104	0.706E-03	1.461
1.400	0.403E+00	0.075	0.148E-02	1.461
1.500	0.288E+00	0.051	0.273E-02	1.461
1.700	0.473E+00	0.044	0.209E-02	1.461
2.000	0.347E+00	0.056	0.297E-02	1.461
2.300	0.437E+00	0.133	0.136E-02	1.461
2.500	0.327E+00	0.292	0.747E-03	1.461

Table A.39 Optical Constants for Magnetite. nk Determined from KM Theory.
Effective Sample Thickness = 0.000381 cm

LAMBDA	T	R	NK	N
0.185	0.460E-01	0.078	0.115E-01	1.145
0.190	0.518E-01	0.064	0.117E-01	1.145
0.200	0.500E-01	0.066	0.125E-01	1.145
0.210	0.555E-01	0.064	0.127E-01	1.145
0.215	0.588E-01	0.068	0.127E-01	1.145
0.220	0.589E-01	0.078	0.130E-01	1.145
0.225	0.584E-01	0.060	0.133E-01	1.145
0.233	0.623E-01	0.062	0.135E-01	1.145
0.240	0.600E-01	0.060	0.141E-01	1.290
0.260	0.710E-01	0.062	0.143E-01	1.415
0.280	0.761E-01	0.045	0.150E-01	1.550
0.300	0.670E-01	0.056	0.169E-01	1.685
0.325	0.622E-01	0.060	0.188E-01	1.820
0.360	0.777E-01	0.053	0.192E-01	1.955
0.370	0.940E-01	0.050	0.182E-01	2.090
0.400	0.119E+00	0.060	0.177E-01	2.225
0.433	0.105E+00	0.056	0.203E-01	2.224
0.466	0.106E+00	0.056	0.218E-01	2.222
0.500	0.110E+00	0.056	0.230E-01	2.221
0.533	0.112E+00	0.055	0.243E-01	2.219
0.566	0.115E+00	0.054	0.255E-01	2.217
0.600	0.118E+00	0.055	0.267E-01	2.215
0.633	0.120E+00	0.056	0.280E-01	2.197
0.666	0.124E+00	0.052	0.290E-01	2.179
0.700	0.126E+00	0.056	0.302E-01	2.161
0.817	0.812E-01	0.051	0.428E-01	2.169
0.907	0.146E+00	0.051	0.364E-01	2.133
1.000	0.148E+00	0.045	0.399E-01	2.098
1.105	0.155E+00	0.045	0.430E-01	2.090

Table A.40 Optical Constants for Maple Leaf (Acer saccharum).
nk Determined from MKM Theory. Thickness = 0.0140 cm

LAMBDA	T	R	NK	N
0.185	0.492E-03	0.053	0.567E-03	1.338
0.190	0.104E-03	0.036	0.842E-03	1.345
0.200	0.475E-04	0.040	0.925E-03	1.352
0.210	0.108E-03	0.040	0.899E-03	1.359
0.215	0.300E-04	0.039	0.106E-02	1.366
0.220	0.397E-04	0.035	0.112E-02	1.373
0.225	0.313E-05	0.038	0.671E-03	1.380
0.233	0.677E-04	0.042	0.946E-03	1.370
0.240	0.123E-05	0.043	0.148E-02	1.360
0.260	0.461E-06	0.035	0.178E-02	1.350
0.280	0.376E-05	0.035	0.162E-02	1.339
0.300	0.599E-04	0.035	0.134E-02	1.328
0.325	0.650E-04	0.070	0.108E-02	1.318
0.360	0.316E-03	0.035	0.132E-02	1.307
0.370	0.113E-02	0.036	0.114E-02	1.296
0.400	0.125E-02	0.038	0.122E-02	1.296
0.433	0.249E-02	0.046	0.110E-02	1.291
0.466	0.657E-02	0.046	0.994E-03	1.285
0.500	0.135E-01	0.046	0.915E-03	1.280
0.533	0.466E-01	0.072	0.570E-03	1.300
0.566	0.374E-01	0.065	0.689E-03	1.219
0.600	0.288E-01	0.056	0.855E-03	1.139
0.633	0.200E-01	0.050	0.101E-02	1.194
0.666	0.305E-01	0.082	0.755E-03	1.248
0.700	0.132E+00	0.167	0.275E-03	1.303
0.817	0.330E+00	0.415	0.433E-04	1.358
0.907	0.346E+00	0.402	0.448E-04	1.412
1.000	0.326E+00	0.470	0.310E-04	1.467
1.105	0.335E+00	0.363	0.629E-04	1.522

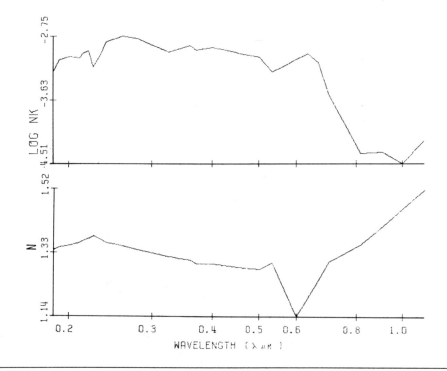

WAVELENGTH (λ μM)

Table A.41 Methylene Blue (Ellipsometry)

LAMBDA	N	NK
0.185	0.121E+01	0.431
0.190	0.120E+01	0.453
0.200	0.120E+01	0.474
0.210	0.120E+01	0.468
0.215	0.118E+01	0.517
0.220	0.120E+01	0.473
0.225	0.121E+01	0.459
0.233	0.121E+01	0.467
0.240	0.121E+01	0.456
0.260	0.127E+01	0.254
0.280	0.128E+01	0.205
0.300	0.129E+01	0.163
0.325	0.130E+01	0.154
0.360	0.130E+01	0.116
0.370	0.126E+01	0.108
0.400	0.121E+01	0.118
0.433	0.107E+01	0.201
0.466	0.946E+00	0.204
0.500	0.756E+00	0.512
0.533	0.105E+01	0.586
0.566	0.143E+01	0.522
0.600	0.158E+01	0.426
0.633	0.156E+01	0.364
0.666	0.158E+01	0.072
0.700	0.155E+01	0.001
0.817	0.152E+01	0.016
0.907	0.143E+01	0.015
1.000	0.147E+01	0.001
1.105	0.147E+01	0.001

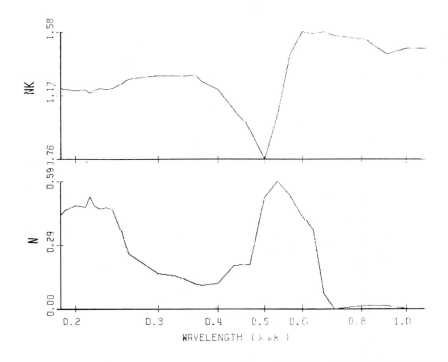

Table A.42 Optical Constants for Methylene Blue (KBr Pellet).
nk Determined from KM Theory. Effective Sample Thickness = 0.00012 cm.
Colorado State University Sample 6

LAMBDA	T	R	NK	N
0.185	0.576E-02	0.080	0.632E-01	1.535
0.190	0.100E-01	0.072	0.580E-01	1.536
0.200	0.807E-02	0.067	0.639E-01	1.537
0.210	0.776E-02	0.064	0.676E-01	1.539
0.215	0.743E-02	0.067	0.698E-01	1.540
0.220	0.650E-02	0.073	0.726E-01	1.541
0.225	0.675E-02	0.074	0.745E-01	1.543
0.233	0.678E-02	0.075	0.771E-01	1.544
0.240	0.664E-02	0.074	0.792E-01	1.545
0.260	0.124E-01	0.076	0.756E-01	1.547
0.280	0.150E-01	0.072	0.775E-01	1.548
0.300	0.196E-01	0.072	0.781E-01	1.549
0.325	0.347E-01	0.078	0.723E-01	1.551
0.360	0.683E-01	0.054	0.639E-01	1.552
0.370	0.653E-01	0.052	0.653E-01	1.510
0.400	0.956E-01	0.110	0.637E-01	1.467
0.433	0.731E-01	0.085	0.749E-01	1.356
0.466	0.416E-01	0.057	0.982E-01	1.243
0.500	0.200E-01	0.061	0.130E+00	1.190
0.533	0.195E-01	0.077	0.146E+00	1.451
0.566	0.168E-01	0.085	0.153E+00	1.750
0.600	0.185E-01	0.085	0.158E+00	1.854
0.633	0.318E-01	0.082	0.144E+00	1.829
0.666	0.478E-01	0.084	0.134E+00	1.803
0.700	0.665E-01	0.092	0.125E+00	1.778
0.817	0.194E+00	0.232	0.857E-01	1.752
0.907	0.250E+00	0.302	0.772E-01	1.727
1.000	0.265E+00	0.300	0.813E-01	1.701
1.105	0.283E+00	0.266	0.856E-01	1.676

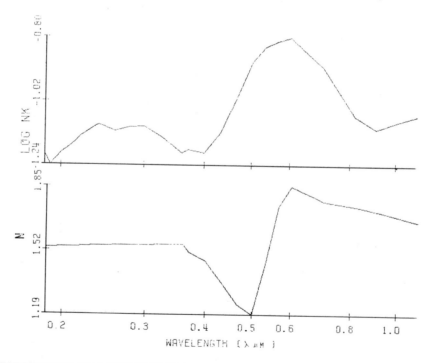

Table A.43 Optical Constants for Methylene Blue (KBr Pellet).
nk Determined from KM Theory. Effective Sample Thickness = 0.000536 cm.
Colorado State University Sample 6A

LAMBDA	T	R	NK	N
0.185	0.443E-01	0.070	0.855E-02	1.535
0.190	0.500E-01	0.076	0.843E-02	1.536
0.200	0.488E-01	0.070	0.855E-02	1.537
0.210	0.500E-01	0.068	0.933E-02	1.539
0.215	0.525E-01	0.071	0.937E-02	1.540
0.220	0.524E-01	0.075	0.961E-02	1.541
0.225	0.505E-01	0.075	0.966E-02	1.543
0.233	0.367E-01	0.076	0.114E-01	1.544
0.240	0.503E-01	0.075	0.106E-01	1.545
0.260	0.684E-01	0.075	0.103E-01	1.547
0.280	0.751E-01	0.071	0.107E-01	1.548
0.300	0.853E-01	0.071	0.105E-01	1.549
0.325	0.104E+00	0.060	0.106E-01	1.551
0.360	0.143E+00	0.065	0.104E-01	1.552
0.370	0.150E+00	0.064	0.104E-01	1.510
0.400	0.170E+00	0.065	0.105E-01	1.467
0.433	0.153E+00	0.061	0.120E-01	1.356
0.466	0.132E+00	0.057	0.140E-01	1.243
0.500	0.115E+00	0.061	0.160E-01	1.190
0.533	0.107E+00	0.075	0.176E-01	1.451
0.566	0.104E+00	0.075	0.190E-01	1.750
0.600	0.112E+00	0.080	0.194E-01	1.854
0.633	0.126E+00	0.076	0.194E-01	1.829
0.666	0.148E+00	0.068	0.188E-01	1.803
0.700	0.154E+00	0.070	0.194E-01	1.778
0.817	0.268E+00	0.057	0.155E-01	1.752
0.907	0.338E+00	0.158	0.142E-01	1.727
1.000	0.365E+00	0.163	0.145E-01	1.701
1.105	0.358E+00	0.176	0.145E-01	1.676

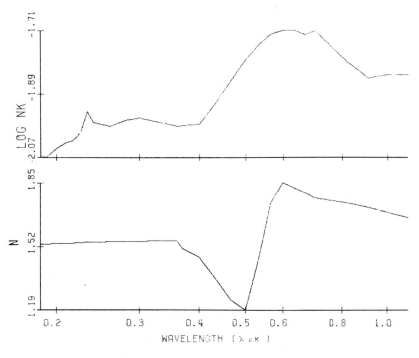

Table A.44 Optical Constants for Mica. nk Determined from KM Theory.
Effective Sample Thickness = 0.0297 cm

LAMBDA	T	R	NK	N
0.185	0.105E-02	0.055	0.338E-03	1.705
0.190	0.109E-01	0.066	0.230E-03	1.694
0.200	0.235E-02	0.060	0.324E-03	1.682
0.210	0.134E-03	0.059	0.487E-03	1.678
0.215	0.196E-03	0.061	0.487E-03	1.674
0.220	0.281E-03	0.061	0.477E-03	1.671
0.225	0.155E-03	0.060	0.523E-03	1.669
0.233	0.193E-03	0.058	0.525E-03	1.663
0.240	0.229E-03	0.059	0.532E-03	1.661
0.260	0.119E-02	0.058	0.468E-03	1.659
0.280	0.177E-01	0.056	0.302E-03	1.657
0.300	0.762E-01	0.055	0.207E-03	1.655
0.325	0.162E+00	0.056	0.158E-03	1.662
0.360	0.336E+00	0.061	0.105E-03	1.669
0.370	0.427E+00	0.066	0.838E-04	1.676
0.400	0.517E+00	0.073	0.699E-04	1.684
0.433	0.620E+00	0.081	0.542E-04	1.691
0.466	0.666E+00	0.086	0.490E-04	1.698
0.500	0.673E+00	0.085	0.511E-04	1.705
0.533	0.691E+00	0.085	0.508E-04	1.694
0.566	0.703E+00	0.087	0.511E-04	1.683
0.600	0.759E+00	0.094	0.408E-04	1.673
0.633	0.768E+00	0.092	0.411E-04	1.662
0.666	0.819E+00	0.092	0.307E-04	1.651
0.700	0.810E+00	0.086	0.352E-04	1.640
0.817	0.803E+00	0.093	0.423E-04	1.630
0.907	0.818E+00	0.103	0.403E-04	1.619
1.000	0.833E+00	0.097	0.398E-04	1.608
1.105	0.842E+00	0.083	0.433E-04	1.597

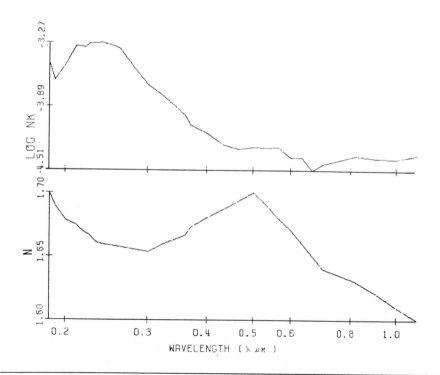

Table A.45 Optical Constants for Nextel, Black. nk Determined from MKM Theory.
Effective Sample Thickness = 0.00368 cm

LAMBDA	T	R	NK	N
0.185	0.555E-04	0.066	0.240E-02	1.400
0.190	0.170E-03	0.044	0.291E-02	1.401
0.200	0.141E-03	0.046	0.300E-02	1.401
0.210	0.131E-03	0.046	0.324E-02	1.402
0.215	0.191E-03	0.066	0.255E-02	1.403
0.220	0.243E-03	0.064	0.260E-02	1.404
0.225	0.156E-03	0.045	0.347E-02	1.404
0.233	0.160E-03	0.044	0.361E-02	1.405
0.240	0.166E-03	0.043	0.374E-02	1.406
0.260	0.123E-03	0.039	0.440E-02	1.407
0.280	0.178E-03	0.042	0.439E-02	1.407
0.300	0.233E-03	0.043	0.453E-02	1.408
0.325	0.280E-03	0.047	0.459E-02	1.409
0.360	0.598E-03	0.036	0.508E-02	1.410
0.370	0.475E-03	0.036	0.540E-02	1.410
0.400	0.357E-03	0.072	0.415E-02	1.411
0.433	0.341E-03	0.060	0.513E-02	1.411
0.466	0.411E-03	0.060	0.542E-02	1.411
0.500	0.475E-03	0.055	0.577E-02	1.411
0.533	0.503E-03	0.055	0.612E-02	1.409
0.566	0.550E-03	0.056	0.648E-02	1.408
0.600	0.583E-03	0.055	0.700E-02	1.406
0.633	0.667E-03	0.056	0.719E-02	1.405
0.666	0.138E-02	0.055	0.688E-02	1.405
0.700	0.754E-03	0.054	0.795E-02	1.405
0.800	0.118E-02	0.056	0.840E-02	1.405
0.907	0.101E-02	0.058	0.947E-02	1.405
1.000	0.156E-02	0.053	0.104E-01	1.405
1.105	0.591E-02	0.049	0.946E-02	1.405
1.200	0.190E-01	0.076	0.608E-02	1.405
1.300	0.720E-02	0.042	0.117E-01	1.405
1.400	0.350E-01	0.066	0.552E-02	1.405
1.500	0.166E-01	0.196	0.339E-02	1.405
1.700	0.210E-01	0.581	0.765E-03	1.405
2.000	0.257E-01	0.505	0.114E-02	1.405
2.300	0.935E-01	0.224	0.268E-02	1.405
2.500	0.319E-01	0.100	0.900E-02	1.405

LOG NK (-1.93 -2.52 -3.12)

N (X10⁻¹) (14.1 14.05 14.00)

WAVELENGTH (λ μM)   0.2   0.3   0.4   0.5 0.6   0.8   1.0   1.3   1.6   2.0   2.5

Table A.46 Optical Constants for Nextel, Blue. nk Determined from MKM Theory.
Effective Sample Thickness = 0.00564 cm

LAMBDA	T	R	NK	N
0.185	0.235E-04	0.066	0.173E-02	1.314
0.190	0.462E-03	0.054	0.144E-02	1.320
0.200	0.348E-04	0.056	0.156E-02	1.325
0.210	0.142E-04	0.054	0.231E-02	1.331
0.215	0.133E-04	0.066	0.208E-02	1.337
0.220	0.630E-05	0.066	0.273E-02	1.343
0.225	0.467E-05	0.052	0.275E-02	1.348
0.233	0.454E-05	0.056	0.281E-02	1.354
0.240	0.266E-05	0.054	0.304E-02	1.360
0.260	0.678E-05	0.053	0.315E-02	1.366
0.280	0.935E-05	0.048	0.220E-02	1.371
0.300	0.120E-04	0.057	0.333E-02	1.377
0.325	0.105E-04	0.063	0.342E-02	1.383
0.360	0.705E-04	0.060	0.292E-02	1.389
0.370	0.873E-04	0.110	0.209E-03	1.394
0.400	0.357E-03	0.186	0.116E-02	1.400
0.433	0.176E-02	0.252	0.743E-03	1.394
0.466	0.305E-02	0.276	0.665E-03	1.389
0.500	0.550E-03	0.193	0.124E-02	1.383
0.533	0.136E-03	0.105	0.292E-02	1.376
0.566	0.254E-04	0.067	0.528E-02	1.370
0.600	0.347E-04	0.061	0.576E-02	1.363
0.633	0.511E-03	0.067	0.426E-02	1.369
0.666	0.566E-03	0.073	0.389E-02	1.369
0.700	0.138E-02	0.070	0.365E-02	1.369
0.800	0.166E-01	0.307	0.736E-03	1.369
0.907	0.743E-01	0.565	0.184E-03	1.369
1.000	0.947E-01	0.577	0.172E-03	1.369
1.105	0.902E-01	0.565	0.203E-03	1.369
1.200	0.105E+00	0.660	0.123E-03	1.369
1.300	0.775E-01	0.534	0.300E-03	1.369
1.400	0.781E-01	0.340	0.706E-03	1.369
1.500	0.833E-01	0.197	0.142E-02	1.369
1.700	0.568E-01	0.678	0.170E-03	1.369
2.000	0.113E+00	0.633	0.232E-03	1.369
2.300	0.422E-01	0.255	0.206E-02	1.369
2.500	0.479E-02	0.561	0.118E-02	1.369

Table A.47 Optical Constants for Nextel, Red. nk Determined from MKM Theory.
Effective Sample Thickness = 0.00471 cm

LAMBDA	T	R	NK	N
0.185	0.528E-04	0.064	0.168E-02	1.453
0.190	0.103E-03	0.064	0.200E-02	1.453
0.200	0.119E-03	0.066	0.202E-02	1.452
0.210	0.876E-04	0.065	0.222E-02	1.452
0.215	0.605E-04	0.071	0.221E-02	1.452
0.220	0.411E-04	0.070	0.237E-02	1.452
0.225	0.917E-04	0.066	0.235E-02	1.451
0.233	0.100E-03	0.064	0.246E-02	1.451
0.240	0.974E-04	0.064	0.252E-02	1.451
0.260	0.963E-04	0.062	0.281E-02	1.451
0.280	0.125E-03	0.064	0.287E-02	1.450
0.300	0.176E-03	0.066	0.283E-02	1.450
0.325	0.243E-03	0.071	0.288E-02	1.450
0.360	0.404E-03	0.065	0.306E-02	1.450
0.370	0.418E-03	0.067	0.320E-02	1.445
0.400	0.321E-03	0.065	0.286E-02	1.440
0.433	0.421E-03	0.076	0.336E-02	1.443
0.466	0.411E-03	0.075	0.366E-02	1.438
0.500	0.376E-03	0.074	0.401E-02	1.432
0.533	0.472E-03	0.076	0.400E-02	1.440
0.566	0.113E-01	0.127	0.164E-02	1.465
0.600	0.704E-02	0.246	0.929E-03	1.482
0.633	0.357E-01	0.477	0.250E-03	1.491
0.666	0.641E-01	0.564	0.153E-03	1.491
0.700	0.708E-01	0.557	0.132E-03	1.491
0.800	0.935E-01	0.621	0.116E-03	1.491
0.907	0.103E+00	0.625	0.121E-03	1.491
1.000	0.113E+00	0.608	0.137E-03	1.491
1.105	0.114E+00	0.601	0.156E-03	1.491
1.200	0.199E+00	0.661	0.671E-04	1.491
1.300	0.136E+00	0.578	0.181E-03	1.491
1.400	0.315E-01	0.535	0.476E-03	1.491
1.500	0.234E+00	0.550	0.105E-03	1.491
1.700	0.164E+00	0.625	0.154E-03	1.491
2.000	0.120E+00	0.511	0.417E-03	1.491
2.300	0.172E+00	0.294	0.101E-03	1.491
2.500	0.981E-01	0.296	0.143E-02	1.491

LOG NK

N

WAVELENGTH (λ μM)

Table A.48 Optical Constants for Nextel, White. nk Determined from MKM Theory.
Effective Sample Thickness = 0.00484 cm

LAMBDA	T	R	NK	N
0.185	0.192E-02	0.076	0.109E-02	1.405
0.190	0.160E-02	0.091	0.103E-02	1.408
0.200	0.475E-03	0.077	0.146E-02	1.411
0.210	0.112E-03	0.075	0.185E-02	1.414
0.215	0.135E-03	0.066	0.204E-02	1.417
0.220	0.754E-04	0.066	0.221E-02	1.419
0.225	0.114E-03	0.074	0.200E-02	1.422
0.233	0.550E-04	0.074	0.210E-02	1.425
0.240	0.586E-04	0.073	0.220E-02	1.428
0.260	0.119E-03	0.073	0.234E-02	1.430
0.280	0.145E-02	0.073	0.183E-02	1.433
0.300	0.160E-03	0.075	0.258E-02	1.436
0.325	0.179E-03	0.073	0.284E-02	1.457
0.360	0.263E-03	0.054	0.241E-02	1.478
0.370	0.570E-03	0.165	0.118E-02	1.499
0.400	0.107E-01	0.367	0.302E-03	1.520
0.433	0.342E-01	0.573	0.115E-03	1.524
0.466	0.382E-01	0.570	0.120E-03	1.528
0.500	0.436E-01	0.552	0.111E-03	1.532
0.533	0.650E-01	0.614	0.901E-04	1.532
0.566	0.703E-01	0.615	0.502E-04	1.531
0.600	0.704E-01	0.605	0.100E-03	1.531
0.633	0.774E-01	0.621	0.965E-04	1.517
0.666	0.855E-01	0.615	0.851E-04	1.535
0.700	0.861E-01	0.621	0.103E-03	1.497
0.800	0.987E-01	0.611	0.119E-03	1.459
0.907	0.108E+00	0.606	0.135E-03	1.430
1.000	0.114E+00	0.566	0.165E-03	1.400
1.105	0.120E+00	0.560	0.165E-03	1.479
1.200	0.137E+00	0.675	0.525E-04	1.479
1.300	0.148E+00	0.457	0.248E-03	1.479
1.400	0.667E-01	0.455	0.488E-03	1.479
1.500	0.522E-01	0.455	0.484E-03	1.479
1.700	0.232E+00	0.750	0.586E-05	1.479
2.000	0.156E+00	0.603	0.215E-03	1.479
2.300	0.111E+00	0.132	0.315E-02	1.479
2.500	0.270E+00	0.475	0.303E-03	1.479

Table A.49 Optical Constants for Nontronite. nk Determined from MKM Theory.
Effective Sample Thickness = 0.00186 cm

LAMBDA	T	R	NK	N
0.185	0.103E-01	0.084	0.198E-02	1.397
0.190	0.101E-01	0.059	0.258E-02	1.398
0.200	0.970E-02	0.052	0.295E-02	1.399
0.210	0.960E-02	0.056	0.301E-02	1.400
0.215	0.103E-01	0.054	0.307E-02	1.401
0.220	0.107E-01	0.053	0.317E-02	1.402
0.225	0.110E-01	0.053	0.323E-02	1.403
0.233	0.114E-01	0.054	0.328E-02	1.404
0.240	0.104E-01	0.053	0.348E-02	1.405
0.260	0.120E-01	0.051	0.374E-02	1.406
0.280	0.130E-01	0.049	0.404E-02	1.407
0.300	0.148E-01	0.047	0.430E-02	1.408
0.325	0.157E-01	0.052	0.434E-02	1.409
0.360	0.184E-01	0.049	0.478E-02	1.410
0.370	0.210E-01	0.051	0.467E-02	1.411
0.400	0.275E-01	0.051	0.470E-02	1.412
0.433	0.415E-01	0.052	0.444E-02	1.409
0.466	0.525E-01	0.055	0.430E-02	1.406
0.500	0.730E-01	0.054	0.410E-02	1.403
0.533	0.100E+00	0.061	0.362E-02	1.401
0.566	0.145E+00	0.076	0.283E-02	1.398
0.600	0.175E+00	0.087	0.247E-02	1.396
0.633	0.210E+00	0.101	0.211E-02	1.395
0.666	0.250E+00	0.126	0.166E-02	1.393
0.700	0.269E+00	0.111	0.184E-02	1.392
0.750	0.288E+00	0.134	0.160E-02	1.391
0.817	0.325E+00	0.123	0.170E-02	1.390
0.907	0.340E+00	0.112	0.196E-02	1.388
1.000	0.326E+00	0.110	0.227E-02	1.387
1.105	0.361E+00	0.097	0.251E-02	1.386

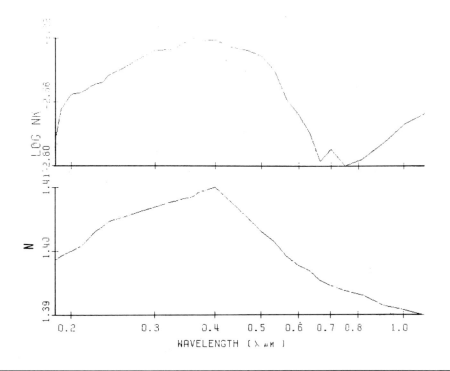

Table A.50 Optical Constants for Olivine. nk Determined from KM Theory.
Thickness = 0.310 cm

LAMBDA	T	R	NK	N
0.185	0.472E-04	0.063	0.454E-04	1.741
0.190	0.228E-03	0.061	0.410E-04	1.736
0.200	0.114E-03	0.062	0.445E-04	1.730
0.210	0.402E-05	0.067	0.563E-04	1.725
0.215	0.112E-04	0.072	0.545E-04	1.720
0.220	0.163E-04	0.072	0.556E-04	1.715
0.225	0.660E-05	0.063	0.575E-04	1.709
0.233	0.172E-04	0.062	0.582E-04	1.704
0.240	0.911E-05	0.063	0.640E-04	1.699
0.260	0.195E-04	0.053	0.701E-04	1.694
0.290	0.106E-03	0.056	0.648E-04	1.688
0.300	0.175E-02	0.066	0.487E-04	1.683
0.325	0.133E-01	0.063	0.360E-04	1.678
0.360	0.403E-01	0.102	0.296E-04	1.673
0.370	0.457E-01	0.106	0.293E-04	1.667
0.400	0.395E-01	0.115	0.330E-04	1.662
0.433	0.466E-01	0.123	0.339E-04	1.657
0.466	0.558E-01	0.132	0.337E-04	1.651
0.500	0.827E-01	0.161	0.316E-04	1.646
0.533	0.106E+00	0.167	0.302E-04	1.644
0.566	0.109E+00	0.162	0.316E-04	1.641
0.600	0.106E+00	0.165	0.340E-04	1.639
0.633	0.946E-01	0.166	0.378E-04	1.636
0.666	0.965E-01	0.167	0.395E-04	1.634
0.700	0.854E-01	0.135	0.430E-04	1.631
0.817	0.322E-01	0.066	0.719E-04	1.628
0.907	0.168E-01	0.080	0.950E-04	1.626
1.000	0.983E-02	0.066	0.119E-03	1.623
1.105	0.105E-01	0.083	0.129E-03	1.620

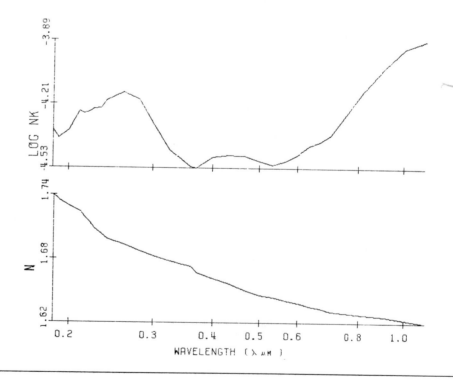

Table A.51 Optical Constants for Propane Soot. nk Determined from KM Theory.
Effective Sample Thickness = 0.0000831 cm. Colorado State University Sample 2

LAMBDA	T	R	NK	N
0.185	0.859E-02	0.083	0.841E-01	0.472
0.190	0.174E-01	0.080	0.736E-01	1.331
0.200	0.159E-01	0.081	0.792E-01	1.409
0.210	0.161E-01	0.078	0.829E-01	1.367
0.215	0.172E-01	0.080	0.835E-01	1.508
0.220	0.176E-01	0.084	0.850E-01	1.491
0.225	0.176E-01	0.081	0.869E-01	1.633
0.233	0.182E-01	0.081	0.892E-01	1.659
0.240	0.187E-01	0.079	0.913E-01	1.684
0.260	0.255E-01	0.080	0.912E-01	1.678
0.280	0.271E-01	0.078	0.966E-01	1.684
0.300	0.305E-01	0.078	0.100E+00	1.737
0.325	0.334E-01	0.074	0.106E+00	1.697
0.360	0.412E-01	0.072	0.110E+00	1.674
0.370	0.422E-01	0.073	0.112E+00	1.665
0.400	0.471E-01	0.075	0.117E+00	1.668
0.433	0.529E-01	0.075	0.122E+00	1.675
0.466	0.570E-01	0.075	0.128E+00	1.687
0.500	0.641E-01	0.072	0.131E+00	1.701
0.533	0.691E-01	0.073	0.136E+00	1.689
0.566	0.741E-01	0.074	0.141E+00	1.731
0.600	0.816E-01	0.076	0.144E+00	1.716
0.633	0.861E-01	0.076	0.148E+00	1.729
0.666	0.963E-01	0.069	0.149E+00	1.775
0.700	0.964E-01	0.071	0.156E+00	1.753
0.817	0.117E+00	0.069	0.167E+00	1.762
0.907	0.130E+00	0.078	0.177E+00	1.750
1.000	0.139E+00	0.076	0.188E+00	1.753
1.105	0.162E+00	0.075	0.192E+00	1.757

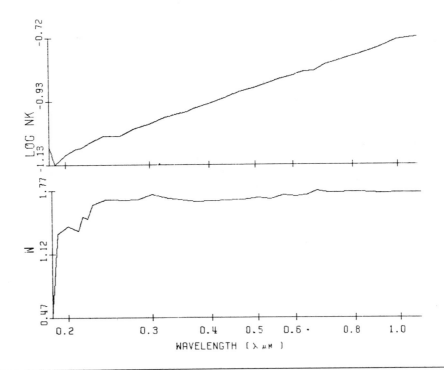

WAVELENGTH (λ μM)

Table A.52 Optical Constants for Propane Soot $+ (NH_4)_2 SO_4$.
nk Determined from KM Theory. Effective Sample Thickness = 0.000229 cm.
Colorado State University Sample 2A

LAMBDA	T	R	NK	N
0.185	0.615E-02	0.074	0.327E-01	0.472
0.190	0.216E-02	0.076	0.404E-01	1.331
0.200	0.299E-02	0.076	0.404E-01	1.409
0.210	0.355E-02	0.072	0.411E-01	1.367
0.215	0.421E-02	0.076	0.408E-01	1.508
0.220	0.421E-02	0.077	0.417E-01	1.491
0.225	0.326E-02	0.075	0.446E-01	1.633
0.233	0.345E-02	0.076	0.457E-01	1.659
0.240	0.322E-02	0.075	0.477E-01	1.684
0.260	0.510E-02	0.076	0.476E-01	1.678
0.280	0.568E-02	0.074	0.502E-01	1.684
0.300	0.597E-02	0.075	0.533E-01	1.737
0.325	0.691E-02	0.072	0.562E-01	1.697
0.360	0.567E-02	0.070	0.580E-01	1.674
0.370	0.587E-02	0.071	0.593E-01	1.665
0.400	0.131E-01	0.073	0.602E-01	1.668
0.433	0.157E-01	0.073	0.624E-01	1.675
0.466	0.192E-01	0.073	0.639E-01	1.687
0.500	0.200E-01	0.070	0.679E-01	1.701
0.533	0.218E-01	0.071	0.708E-01	1.689
0.566	0.246E-01	0.074	0.727E-01	1.731
0.600	0.256E-01	0.073	0.763E-01	1.716
0.633	0.290E-01	0.074	0.778E-01	1.729
0.666	0.297E-01	0.069	0.813E-01	1.775
0.700	0.314E-01	0.071	0.841E-01	1.753
0.817	0.383E-01	0.069	0.925E-01	1.762
0.907	0.458E-01	0.075	0.970E-01	1.750
1.000	0.478E-01	0.073	0.105E+00	1.753
1.105	0.503E-01	0.078	0.115E+00	1.757

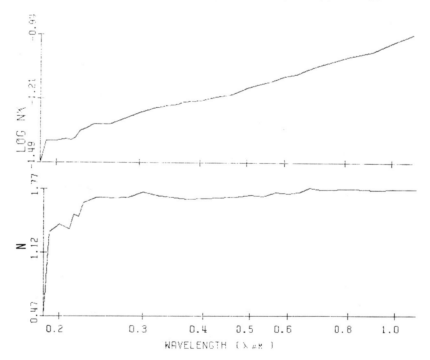

Table A.53 Optical Constants for Propane Soot + $(NH_4)_2SO_4$.
nk Determined from KM Theory. Effective Sample Thickness = 0.000112 cm.
Colorado State University Sample 5

LAMBDA	T	R	NK	N
0.185	0.385E-01	0.093	0.426E-01	1.005
0.190	0.444E-01	0.072	0.420E-01	1.432
0.200	0.445E-01	0.074	0.440E-01	1.468
0.210	0.437E-01	0.071	0.466E-01	1.444
0.215	0.455E-01	0.073	0.471E-01	1.511
0.220	0.465E-01	0.075	0.477E-01	1.500
0.225	0.508E-01	0.072	0.476E-01	1.568
0.233	0.516E-01	0.072	0.490E-01	1.578
0.240	0.555E-01	0.070	0.492E-01	1.589
0.260	0.652E-01	0.071	0.503E-01	1.582
0.280	0.692E-01	0.069	0.530E-01	1.582
0.300	0.845E-01	0.070	0.525E-01	1.606
0.325	0.868E-01	0.066	0.563E-01	1.583
0.360	0.105E+00	0.064	0.575E-01	1.569
0.370	0.111E+00	0.064	0.577E-01	1.561
0.400	0.108E+00	0.067	0.631E-01	1.560
0.433	0.114E+00	0.065	0.667E-01	1.563
0.466	0.123E+00	0.066	0.692E-01	1.568
0.500	0.132E+00	0.056	0.718E-01	1.575
0.533	0.172E+00	0.071	0.665E-01	1.569
0.566	0.145E+00	0.066	0.775E-01	1.590
0.600	0.154E+00	0.066	0.796E-01	1.582
0.633	0.162E+00	0.067	0.817E-01	1.588
0.666	0.170E+00	0.060	0.837E-01	1.610
0.700	0.173E+00	0.063	0.871E-01	1.599
0.817	0.200E+00	0.056	0.932E-01	1.602
0.907	0.222E+00	0.065	0.967E-01	1.596
1.000	0.187E+00	0.057	0.119E+00	1.596
1.105	0.248E+00	0.063	0.109E+00	1.598

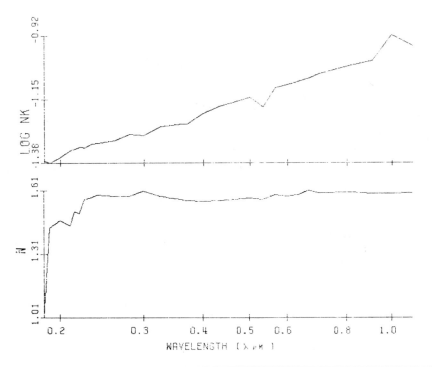

Table A.54 Optical Constants for Propane Soot + $(NH_4)_2SO_4$.
nk Determined from KM Theory. Effective Sample Thickness = 0.0000850 cm.
Colorado State University Sample 5A

LAMBDA	T	R	NK	N
0.185	0.608E-02	0.072	0.883E-01	1.005
0.190	0.517E-02	0.065	0.935E-01	1.432
0.200	0.461E-02	0.065	0.101E+00	1.468
0.210	0.408E-02	0.063	0.108E+00	1.444
0.215	0.446E-02	0.067	0.109E+00	1.511
0.220	0.448E-02	0.065	0.111E+00	1.500
0.225	0.451E-02	0.067	0.114E+00	1.568
0.233	0.464E-02	0.067	0.117E+00	1.578
0.240	0.503E-02	0.066	0.119E+00	1.588
0.260	0.636E-02	0.066	0.123E+00	1.582
0.280	0.703E-02	0.064	0.130E+00	1.582
0.300	0.853E-02	0.065	0.132E+00	1.606
0.325	0.106E-01	0.063	0.138E+00	1.583
0.360	0.140E-01	0.060	0.144E+00	1.569
0.370	0.149E-01	0.060	0.146E+00	1.561
0.400	0.201E-01	0.057	0.146E+00	1.560
0.433	0.225E-01	0.056	0.152E+00	1.563
0.466	0.274E-01	0.060	0.157E+00	1.568
0.500	0.312E-01	0.057	0.162E+00	1.575
0.533	0.335E-01	0.055	0.169E+00	1.569
0.566	0.375E-01	0.055	0.174E+00	1.590
0.600	0.403E-01	0.057	0.180E+00	1.582
0.633	0.438E-01	0.056	0.185E+00	1.588
0.666	0.479E-01	0.053	0.189E+00	1.610
0.700	0.486E-01	0.051	0.198E+00	1.599
0.817	0.638E-01	0.052	0.210E+00	1.602
0.907	0.735E-01	0.056	0.221E+00	1.596
1.000	0.826E-01	0.053	0.233E+00	1.596
1.105	0.912E-01	0.050	0.248E+00	1.598

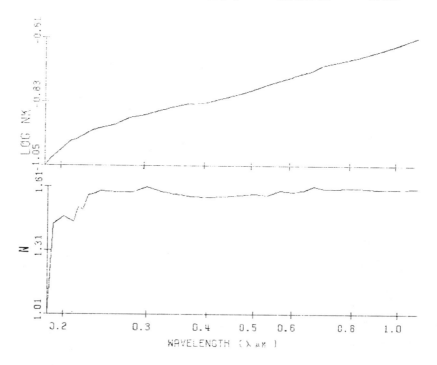

Table A.55 Optical Constants for Rutile (TiO$_2$) ortho.
nk Determined from MKM Theory. Effective Sample Thickness = 0.00366 cm

LAMBDA	T	R	NK	N
0.195	0.842E-04	0.109	0.366E-02	2.276
0.190	0.355E-03	0.126	0.325E-02	2.400
0.200	0.140E-03	0.115	0.776E-02	2.524
0.210	0.899E-05	0.117	0.458E-02	2.640
0.215	0.278E-04	0.122	0.459E-02	2.779
0.220	0.194E-04	0.126	0.467E-02	2.907
0.225	0.188E-04	0.124	0.598E-02	3.036
0.233	0.200E-05	0.123	0.524E-02	3.170
0.240	0.148E-04	0.126	0.513E-02	3.303
0.260	0.417E-06	0.132	0.595E-02	3.437
0.290	0.840E-06	0.136	0.647E-02	3.575
0.300	0.325E-05	0.121	0.657E-02	3.711
0.325	0.489E-05	0.111	0.725E-02	3.477
0.360	0.755E-05	0.111	0.918E-02	3.246
0.370	0.651E-04	0.163	0.735E-02	3.020
0.400	0.450E-02	0.378	0.456E-02	2.797
0.433	0.305E-01	0.613	0.284E-02	2.727
0.466	0.378E-01	0.631	0.290E-02	2.654
0.500	0.398E-01	0.636	0.294E-02	2.595
0.533	0.497E-01	0.650	0.284E-02	2.550
0.566	0.533E-01	0.650	0.293E-02	2.513
0.600	0.521E-01	0.639	0.317E-02	2.470
0.633	0.551E-01	0.640	0.326E-02	2.473
0.666	0.635E-01	0.653	0.318E-02	2.473
0.700	0.622E-01	0.652	0.338E-02	2.473
0.800	0.701E-01	0.641	0.369E-02	2.484
0.907	0.730E-01	0.635	0.413E-02	2.526
1.000	0.750E-01	0.617	0.458E-02	2.566
1.105	0.849E-01	0.622	0.473E-02	2.617
1.200	0.840E-01	0.663	0.493E-02	2.617
1.300	0.107E+00	0.556	0.525E-02	2.617
1.400	0.114E+00	0.525	0.561E-02	2.617
1.500	0.511E-01	0.494	0.878E-02	2.617
1.700	0.164E+00	0.716	0.377E-02	2.617
2.000	0.111E+00	0.748	0.582E-02	2.617
2.300	0.332E-01	0.457	0.159E-01	2.617
2.500	0.450E-01	0.489	0.154E-01	2.617

WAVELENGTH (λ μM)

Table A.56 Optical Constants for Rutile (TiO$_2$) ortho. nk Determined from KM Theory. Effective Sample Thickness = 0.00366 cm

LAMBDA	T	R	NK	N
0.185	0.842E-04	0.109	0.165E-02	1.552
0.190	0.355E-03	0.129	0.120E-02	1.600
0.200	0.140E-03	0.119	0.159E-02	1.647
0.210	0.899E-05	0.117	0.230E-02	1.694
0.215	0.278E-04	0.122	0.207E-02	1.742
0.220	0.194E-04	0.126	0.219E-02	1.789
0.225	0.188E-04	0.124	0.247E-02	1.836
0.233	0.200E-05	0.123	0.341E-02	1.884
0.240	0.148E-04	0.126	0.310E-02	1.931
0.260	0.417E-06	0.132	0.439E-02	1.978
0.280	0.838E-06	0.136	0.467E-02	2.026
0.300	0.325E-05	0.121	0.806E-02	2.070
0.325	0.489E-05	0.111	0.820E-02	1.992
0.360	0.755E-05	0.111	0.649E-02	1.911
0.370	0.651E-04	0.163	0.215E-02	1.830
0.400	0.450E-02	0.378	0.413E-03	1.749
0.433	0.305E-01	0.613	0.110E-03	1.723
0.466	0.378E-01	0.631	0.103E-03	1.696
0.500	0.398E-01	0.636	0.109E-03	1.670
0.533	0.497E-01	0.650	0.100E-03	1.657
0.566	0.533E-01	0.650	0.104E-03	1.643
0.600	0.521E-01	0.639	0.119E-03	1.630
0.633	0.551E-01	0.640	0.122E-03	1.628
0.666	0.635E-01	0.653	0.113E-03	1.628
0.700	0.622E-01	0.652	0.120E-03	1.628
0.800	0.701E-01	0.641	0.136E-03	1.632
0.907	0.730E-01	0.635	0.153E-03	1.648
1.000	0.750E-01	0.617	0.179E-03	1.663
1.105	0.849E-01	0.622	0.177E-03	1.682
1.200	0.840E-01	0.663	0.156E-03	1.682
1.300	0.107E+00	0.556	0.249E-03	1.682
1.400	0.114E+00	0.525	0.296E-03	1.682
1.500	0.511E-01	0.494	0.537E-03	1.682
1.700	0.164E+00	0.716	0.793E-04	1.682
2.000	0.111E+00	0.748	0.116E-03	1.682
2.300	0.332E-01	0.457	0.111E-02	1.682
2.500	0.450E-01	0.489	0.959E-03	1.682

Table A.57 Optical Constants for Sahara Sand 1. nk Determined from MKM Theory.
Effective Sample Thickness = 0.00562 cm

LAMBDA	T	R	NK	N
0.185	0.196E-02	0.045	0.141E-02	1.458
0.190	0.197E-03	0.052	0.182E-02	1.461
0.200	0.243E-03	0.050	0.192E-02	1.463
0.210	0.151E-03	0.049	0.215E-02	1.466
0.215	0.217E-03	0.050	0.209E-02	1.468
0.220	0.194E-03	0.051	0.215E-02	1.470
0.225	0.273E-03	0.050	0.215E-02	1.473
0.233	0.290E-03	0.050	0.222E-02	1.475
0.240	0.336E-03	0.051	0.222E-02	1.477
0.260	0.913E-03	0.052	0.205E-02	1.480
0.290	0.121E-02	0.052	0.212E-02	1.482
0.300	0.163E-02	0.055	0.203E-02	1.485
0.325	0.129E-02	0.055	0.229E-02	1.487
0.360	0.328E-02	0.056	0.201E-02	1.489
0.370	0.461E-02	0.052	0.233E-02	1.492
0.400	0.122E-01	0.064	0.177E-02	1.494
0.433	0.173E-01	0.061	0.181E-02	1.491
0.466	0.249E-01	0.070	0.160E-02	1.487
0.500	0.312E-01	0.073	0.156E-02	1.484
0.533	0.454E-01	0.075	0.138E-02	1.483
0.566	0.764E-01	0.103	0.980E-03	1.483
0.600	0.962E-01	0.125	0.792E-03	1.482
0.633	0.118E+00	0.138	0.696E-03	1.477
0.666	0.135E+00	0.145	0.657E-03	1.471
0.700	0.163E+00	0.151	0.605E-03	1.466
0.817	0.144E+00	0.176	0.634E-03	1.460
0.907	0.161E+00	0.164	0.720E-03	1.455
1.000	0.197E+00	0.173	0.682E-03	1.440
1.105	0.243E+00	0.182	0.627E-03	1.444

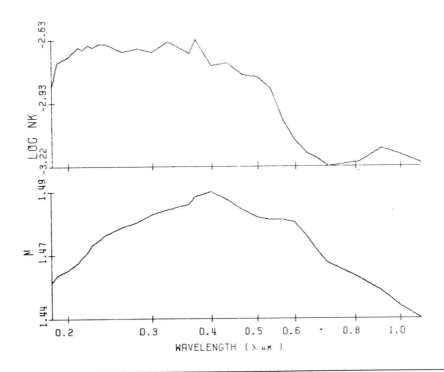

Table A.58 Optical Constants for Sahara Sand 2A. nk Determined from MKM Theory.
Effective Sample Thickness = 0.0153 cm

LAMBDA	T	R	NK	N
0.185	0.142E-02	0.174	0.159E-03	1.649
0.190	0.109E-03	0.187	0.206E-03	1.639
0.200	0.117E-03	0.189	0.213E-03	1.629
0.210	0.580E-04	0.188	0.243E-03	1.620
0.215	0.796E-04	0.191	0.237E-03	1.610
0.220	0.449E-04	0.195	0.252E-03	1.600
0.225	0.114E-03	0.198	0.231E-03	1.591
0.233	0.120E-03	0.199	0.238E-03	1.581
0.240	0.134E-03	0.203	0.224E-03	1.571
0.260	0.285E-03	0.228	0.208E-03	1.562
0.280	0.366E-03	0.235	0.212E-03	1.552
0.300	0.518E-03	0.249	0.205E-03	1.542
0.325	0.832E-03	0.263	0.197E-03	1.533
0.360	0.134E-02	0.295	0.179E-03	1.523
0.370	0.240E-02	0.251	0.204E-03	1.513
0.400	0.339E-02	0.329	0.152E-03	1.504
0.433	0.652E-02	0.380	0.121E-03	1.494
0.466	0.963E-02	0.405	0.110E-03	1.484
0.500	0.148E-01	0.435	0.966E-04	1.473
0.533	0.207E-01	0.468	0.843E-04	1.463
0.566	0.306E-01	0.512	0.671E-04	1.467
0.600	0.354E-01	0.536	0.615E-04	1.470
0.633	0.408E-01	0.564	0.548E-04	1.474
0.666	0.505E-01	0.556	0.558E-04	1.461
0.700	0.568E-01	0.570	0.534E-04	1.448
0.817	0.535E-01	0.606	0.553E-04	1.435
0.907	0.602E-01	0.601	0.606E-04	1.422
1.000	0.727E-01	0.605	0.609E-04	1.409
1.105	0.836E-01	0.612	0.615E-04	1.396

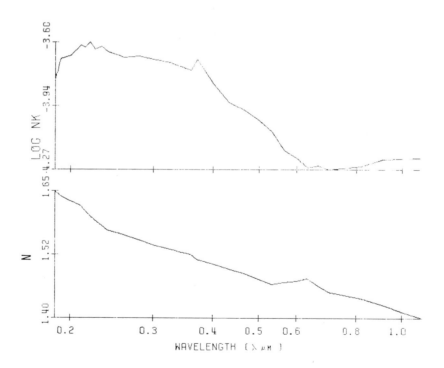

Table A.59 Optical Constants for Sahara Sand 2B. nk Determined from MKM Theory.
Effective Sample Thickness = 0.0132 cm

LAMBDA	T	R	NK	N
0.185	0.280E-02	0.115	0.284E-03	1.310
0.190	0.136E-03	0.124	0.404E-03	1.318
0.200	0.115E-03	0.120	0.444E-03	1.326
0.210	0.429E-03	0.190	0.262E-03	1.334
0.215	0.552E-04	0.121	0.504E-03	1.341
0.220	0.412E-04	0.125	0.517E-03	1.349
0.225	0.773E-04	0.127	0.488E-03	1.357
0.233	0.880E-04	0.128	0.454E-03	1.365
0.240	0.106E-03	0.131	0.497E-03	1.373
0.260	0.240E-03	0.148	0.429E-03	1.381
0.280	0.294E-03	0.151	0.440E-03	1.389
0.300	0.364E-03	0.165	0.420E-03	1.397
0.325	0.737E-03	0.174	0.393E-03	1.404
0.360	0.103E-02	0.193	0.373E-03	1.412
0.370	0.191E-02	0.195	0.400E-03	1.420
0.400	0.317E-02	0.230	0.287E-03	1.428
0.433	0.705E-02	0.277	0.215E-03	1.445
0.466	0.106E-01	0.307	0.185E-03	1.461
0.500	0.180E-01	0.343	0.150E-03	1.478
0.533	0.285E-01	0.390	0.118E-03	1.475
0.566	0.491E-01	0.454	0.830E-04	1.472
0.600	0.556E-01	0.461	0.732E-04	1.469
0.633	0.671E-01	0.526	0.619E-04	1.459
0.666	0.653E-01	0.512	0.706E-04	1.449
0.700	0.102E+00	0.526	0.570E-04	1.440
0.817	0.884E-01	0.554	0.643E-04	1.430
0.907	0.914E-01	0.557	0.700E-04	1.420
1.000	0.106E+00	0.555	0.725E-04	1.410
1.105	0.128E+00	0.556	0.717E-04	1.400

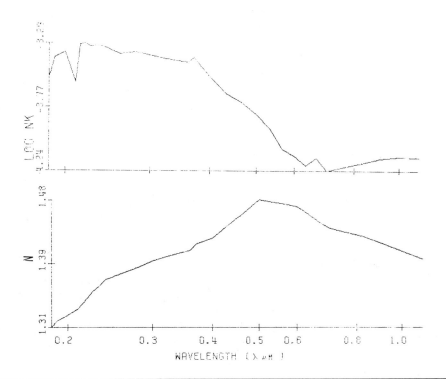

Table A.60 Optical Constants for Sahara Sand 3. nk Determined from MKM Theory.
Effective Sample Thickness = 0.0128 cm

LAMBDA	T	R	NK	N
0.185	0.725E-03	0.123	0.330E-03	1.355
0.190	0.340E-03	0.134	0.345E-03	1.358
0.200	0.550E-04	0.131	0.426E-03	1.360
0.210	0.584E-04	0.131	0.469E-03	1.362
0.215	0.595E-04	0.133	0.448E-03	1.364
0.220	0.741E-04	0.137	0.455E-03	1.367
0.225	0.105E-03	0.140	0.442E-03	1.369
0.233	0.120E-03	0.141	0.450E-03	1.371
0.240	0.134E-03	0.145	0.445E-03	1.373
0.260	0.361E-03	0.171	0.367E-03	1.376
0.280	0.453E-03	0.172	0.382E-03	1.378
0.300	0.633E-03	0.196	0.341E-03	1.380
0.325	0.117E-02	0.208	0.323E-03	1.382
0.360	0.180E-02	0.236	0.254E-03	1.385
0.370	0.346E-02	0.207	0.312E-03	1.387
0.400	0.535E-02	0.275	0.229E-03	1.389
0.433	0.121E-01	0.332	0.166E-03	1.400
0.466	0.176E-01	0.355	0.145E-03	1.412
0.500	0.280E-01	0.393	0.120E-03	1.423
0.533	0.430E-01	0.448	0.902E-04	1.427
0.566	0.670E-01	0.506	0.640E-04	1.432
0.600	0.760E-01	0.522	0.598E-04	1.436
0.633	0.846E-01	0.558	0.498E-04	1.460
0.666	0.884E-01	0.545	0.528E-04	1.484
0.700	0.115E+00	0.548	0.457E-04	1.509
0.817	0.884E-01	0.554	0.585E-04	1.533
0.907	0.925E-01	0.543	0.652E-04	1.557
1.000	0.103E+00	0.530	0.699E-04	1.581
1.105	0.126E+00	0.536	0.646E-04	1.605

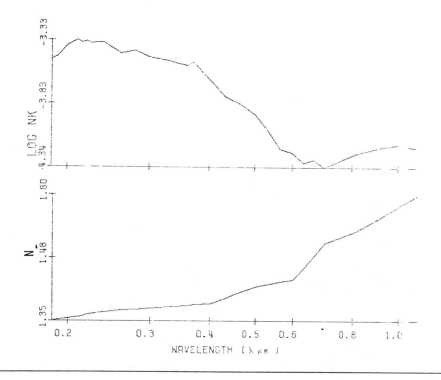

Table A.61 Optical Constants for Sahara Sand 4. nk Determined from MKM Theory.
Effective Sample Thickness = 0.0118 cm

LAMBDA	LOG T	R	LOG NK	N
0.185	0.266E-02	0.094	0.115E-03	1.456
0.190	0.677E-04	0.114	0.505E-03	1.457
0.200	0.224E-04	0.108	0.321E-02	1.457
0.210	0.185E-04	0.108	0.381E-02	1.458
0.215	0.285E-04	0.107	0.597E-02	1.458
0.220	0.338E-04	0.111	0.188E-02	1.458
0.225	0.455E-04	0.113	0.150E-02	1.459
0.233	0.520E-04	0.113	0.154E-02	1.459
0.240	0.688E-04	0.117	0.621E-03	1.459
0.260	0.171E-03	0.139	0.513E-03	1.460
0.280	0.208E-03	0.144	0.521E-03	1.460
0.300	0.441E-03	0.161	0.455E-03	1.461
0.325	0.605E-03	0.170	0.449E-03	1.461
0.360	0.102E-02	0.197	0.396E-03	1.461
0.370	0.220E-02	0.173	0.417E-03	1.462
0.400	0.313E-02	0.230	0.313E-03	1.462
0.433	0.786E-02	0.267	0.243E-03	1.458
0.466	0.137E-01	0.295	0.207E-03	1.453
0.500	0.200E-01	0.325	0.180E-03	1.449
0.533	0.322E-01	0.357	0.149E-03	1.447
0.566	0.475E-01	0.404	0.117E-03	1.444
0.600	0.510E-01	0.411	0.118E-03	1.442
0.633	0.630E-01	0.429	0.107E-03	1.440
0.666	0.800E-01	0.439	0.984E-04	1.437
0.700	0.938E-01	0.445	0.940E-04	1.435
0.817	0.894E-01	0.476	0.595E-04	1.432
0.907	0.103E+00	0.462	0.109E-03	1.430
1.000	0.130E+00	0.447	0.113E-03	1.427
1.105	0.153E+00	0.444	0.115E-03	1.425

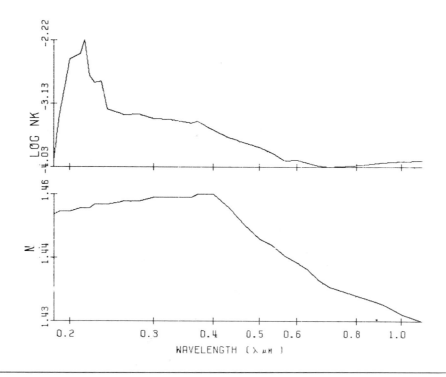

Table A.62 Optical Constants for Sahara Sand 5. nk Determined from MKM Theory.
Effective Sample Thickness = 0.00726 cm

LAMBDA	T	R	NK	N
0.185	0.204E-02	0.053	0.657E-03	1.222
0.190	0.856E-03	0.104	0.707E-03	1.225
0.200	0.104E-02	0.100	0.748E-03	1.228
0.210	0.111E-02	0.058	0.786E-03	1.231
0.215	0.121E-02	0.058	0.796E-03	1.234
0.220	0.142E-02	0.055	0.789E-03	1.237
0.225	0.173E-02	0.100	0.776E-03	1.240
0.233	0.152E-02	0.104	0.796E-03	1.243
0.240	0.169E-02	0.101	0.823E-03	1.246
0.260	0.284E-02	0.108	0.784E-03	1.249
0.280	0.321E-02	0.110	0.815E-03	1.252
0.300	0.199E-02	0.117	0.899E-03	1.255
0.325	0.102E-02	0.115	0.106E-02	1.258
0.360	0.546E-02	0.134	0.820E-03	1.261
0.370	0.840E-02	0.117	0.861E-03	1.264
0.400	0.152E-01	0.155	0.656E-03	1.267
0.433	0.261E-01	0.150	0.520E-03	1.268
0.466	0.370E-01	0.214	0.455E-03	1.268
0.500	0.530E-01	0.246	0.378E-03	1.269
0.533	0.787E-01	0.295	0.285E-03	1.268
0.566	0.125E+00	0.358	0.162E-03	1.266
0.600	0.143E+00	0.426	0.142E-03	1.265
0.633	0.160E+00	0.455	0.124E-03	1.259
0.666	0.177E+00	0.445	0.129E-03	1.254
0.700	0.193E+00	0.452	0.124E-03	1.248
0.817	0.144E+00	0.467	0.167E-03	1.242
0.907	0.155E+00	0.440	0.201E-03	1.237
1.000	0.182E+00	0.430	0.210E-03	1.231
1.105	0.205E+00	0.425	0.219E-03	1.225

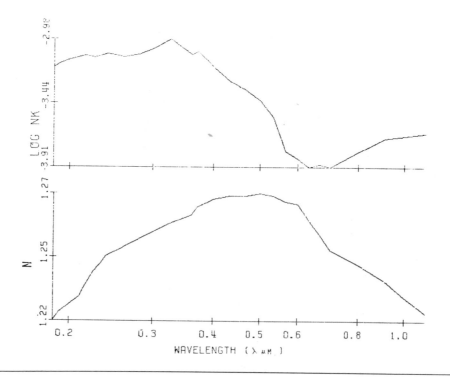

Table A.63 Optical Constants for Sand, Cape Fear, North Carolina.
nk Determined MKM Theory. Effective Sample Thickness = 0.0108 cm

LAMBDA	T	R	NK	N
0.185	0.285E-02	0.210	0.139E-03	1.499
0.190	0.396E-02	0.201	0.178E-03	1.487
0.200	0.358E-02	0.211	0.182E-03	1.476
0.210	0.407E-02	0.215	0.185E-03	1.465
0.215	0.436E-02	0.214	0.190E-03	1.453
0.220	0.514E-02	0.227	0.178E-03	1.442
0.225	0.533E-02	0.226	0.184E-03	1.430
0.233	0.605E-02	0.229	0.185E-03	1.419
0.240	0.583E-02	0.232	0.191E-03	1.408
0.260	0.890E-02	0.227	0.157E-03	1.396
0.280	0.940E-02	0.230	0.200E-03	1.385
0.300	0.109E-01	0.270	0.184E-03	1.374
0.325	0.155E-01	0.296	0.167E-03	1.362
0.360	0.249E-01	0.319	0.152E-03	1.351
0.370	0.263E-01	0.332	0.148E-03	1.339
0.400	0.314E-01	0.364	0.137E-03	1.328
0.433	0.435E-01	0.399	0.116E-03	1.335
0.466	0.526E-01	0.414	0.110E-03	1.343
0.500	0.564E-01	0.433	0.106E-03	1.350
0.533	0.691E-01	0.450	0.582E-04	1.344
0.566	0.766E-01	0.467	0.942E-04	1.337
0.600	0.799E-01	0.468	0.984E-04	1.331
0.633	0.862E-01	0.479	0.959E-04	1.331
0.666	0.954E-01	0.471	0.994E-04	1.331
0.700	0.945E-01	0.453	0.113E-03	1.331
0.817	0.994E-01	0.423	0.145E-03	1.331
0.907	0.109E+00	0.455	0.114E-03	1.331
1.000	0.119E+00	0.458	0.119E-03	1.331
1.105	0.121E+00	0.444	0.163E-03	1.331

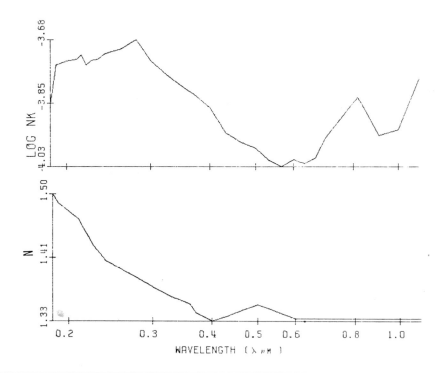

Table A.64 Optical Constants for Sand, Cape Hatteras, North Carolina.
nk Determined from MKM Theory. Effective Sample Thickness = 0.0144 cm

LAMBDA	T	R	NK	N
0.185	0.495E-03	0.190	0.192E-03	1.449
0.190	0.112E-02	0.193	0.174E-03	1.451
0.200	0.832E-03	0.200	0.183E-03	1.454
0.210	0.807E-03	0.204	0.189E-03	1.457
0.215	0.806E-03	0.214	0.183E-03	1.459
0.220	0.861E-03	0.210	0.189E-03	1.462
0.225	0.108E-02	0.218	0.180E-03	1.464
0.233	0.122E-02	0.220	0.181E-03	1.467
0.240	0.126E-02	0.221	0.184E-03	1.470
0.260	0.197E-02	0.227	0.173E-03	1.472
0.280	0.248E-02	0.247	0.171E-03	1.475
0.300	0.422E-02	0.236	0.176E-03	1.478
0.325	0.488E-02	0.265	0.162E-03	1.480
0.360	0.826E-02	0.296	0.141E-03	1.483
0.370	0.110E-01	0.317	0.126E-03	1.486
0.400	0.135E-01	0.348	0.115E-03	1.488
0.433	0.232E-01	0.330	0.116E-03	1.491
0.466	0.299E-01	0.436	0.769E-04	1.493
0.500	0.399E-01	0.470	0.658E-04	1.496
0.533	0.484E-01	0.490	0.604E-04	1.499
0.566	0.552E-01	0.601	0.381E-04	1.501
0.600	0.597E-01	0.558	0.471E-04	1.504
0.633	0.614E-01	0.554	0.497E-04	1.507
0.666	0.821E-01	0.578	0.409E-04	1.509
0.700	0.800E-01	0.578	0.434E-04	1.512
0.817	0.798E-01	0.525	0.627E-04	1.515
0.907	0.858E-01	0.571	0.557E-04	1.517
1.000	0.906E-01	0.563	0.616E-04	1.520
1.105	0.870E-01	0.567	0.622E-04	1.522

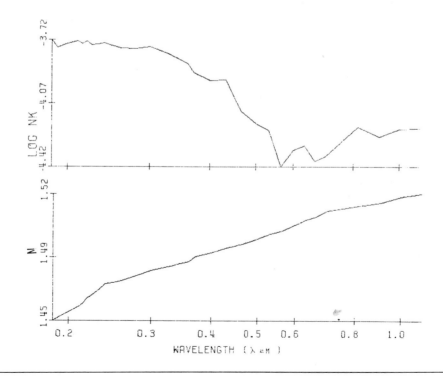

Table A.65 Optical Constants for Sand, Jones Beach, New York.
nk Determined from MKM Theory. Effective Sample Thickness = 0.0128 cm

LAMBDA	T	R	NK	N
0.185	0.760E-03	0.354	0.113E-03	1.340
0.190	0.131E-02	0.355	0.104E-03	1.352
0.200	0.553E-03	0.410	0.117E-03	1.363
0.210	0.136E-02	0.418	0.105E-03	1.374
0.215	0.142E-02	0.403	0.111E-03	1.385
0.220	0.156E-02	0.412	0.107E-03	1.397
0.225	0.199E-02	0.462	0.872E-04	1.408
0.233	0.226E-02	0.452	0.906E-04	1.419
0.240	0.240E-02	0.455	0.851E-04	1.430
0.260	0.479E-02	0.447	0.877E-04	1.442
0.280	0.670E-02	0.478	0.780E-04	1.453
0.300	0.766E-02	0.535	0.640E-04	1.464
0.325	0.111E-01	0.550	0.535E-04	1.475
0.360	0.204E-01	0.606	0.446E-04	1.487
0.370	0.248E-01	0.625	0.392E-04	1.498
0.400	0.310E-01	0.636	0.370E-04	1.509
0.433	0.414E-01	0.658	0.322E-04	1.512
0.466	0.465E-01	0.660	0.326E-04	1.516
0.500	0.437E-01	0.648	0.379E-04	1.519
0.533	0.472E-01	0.678	0.335E-04	1.522
0.566	0.499E-01	0.680	0.342E-04	1.526
0.600	0.524E-01	0.697	0.322E-04	1.529
0.633	0.535E-01	0.701	0.326E-04	1.532
0.666	0.667E-01	0.656	0.388E-04	1.536
0.700	0.658E-01	0.678	0.368E-04	1.539
0.817	0.640E-01	0.692	0.402E-04	1.542
0.907	0.692E-01	0.685	0.443E-04	1.546
1.000	0.747E-01	0.696	0.436E-04	1.549
1.105	0.800E-01	0.870	0.756E-05	1.552

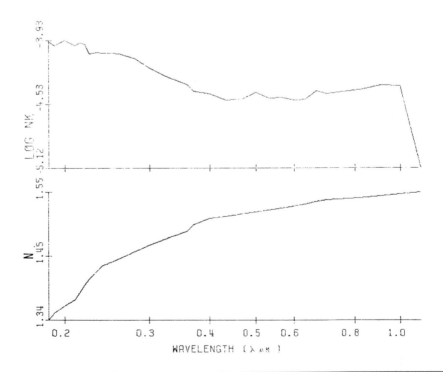

Table A.66 Optical Constants for Sand, Ocracoke, North Carolina.
nk Determined from MKM Theory. Effective Sample Thickness = 0.0134 cm

LAMBDA	T	R	NK	N
0.185	0.311E-03	0.265	0.140E-03	1.583
0.190	0.773E-03	0.271	0.124E-03	1.578
0.200	0.521E-03	0.275	0.136E-03	1.573
0.210	0.631E-03	0.277	0.139E-03	1.567
0.215	0.587E-03	0.284	0.140E-03	1.562
0.220	0.554E-03	0.287	0.142E-03	1.557
0.225	0.862E-03	0.297	0.132E-03	1.552
0.233	0.900E-03	0.304	0.135E-03	1.547
0.240	0.977E-03	0.303	0.136E-03	1.542
0.260	0.181E-02	0.321	0.126E-03	1.537
0.280	0.238E-02	0.335	0.123E-03	1.532
0.300	0.345E-02	0.320	0.131E-03	1.527
0.325	0.448E-02	0.350	0.175E-03	1.522
0.360	0.862E-02	0.390	0.102E-03	1.517
0.370	0.120E-01	0.405	0.911E-04	1.511
0.400	0.176E-01	0.455	0.758E-04	1.506
0.433	0.258E-01	0.508	0.579E-04	1.501
0.466	0.382E-01	0.537	0.514E-04	1.496
0.500	0.463E-01	0.556	0.459E-04	1.491
0.533	0.567E-01	0.572	0.441E-04	1.486
0.566	0.646E-01	0.656	0.240E-04	1.481
0.600	0.681E-01	0.636	0.343E-04	1.476
0.633	0.713E-01	0.635	0.357E-04	1.471
0.666	0.872E-01	0.661	0.294E-04	1.466
0.700	0.826E-01	0.651	0.339E-04	1.461
0.817	0.867E-01	0.574	0.567E-04	1.455
0.907	0.542E-01	0.640	0.436E-04	1.450
1.000	0.655E-01	0.624	0.512E-04	1.445
1.105	0.858E-01	0.567	0.740E-04	1.440

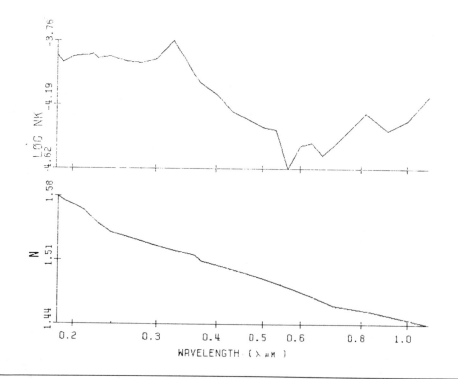

Table A.67 Optical Constants for Sand, Sinai. nk Determined from MKM Theory.
Effective Sample Thickness = 0.0101 cm

LAMBDA	T	R	NK	N
0.185	0.135E-02	0.218	0.439E-04	1.569
0.190	0.105E-02	0.206	0.218E-03	1.550
0.200	0.236E-02	0.213	0.194E-03	1.530
0.210	0.251E-02	0.215	0.208E-03	1.511
0.215	0.198E-02	0.214	0.225E-03	1.492
0.220	0.240E-02	0.227	0.212E-03	1.473
0.225	0.211E-02	0.234	0.217E-03	1.453
0.233	0.218E-02	0.237	0.224E-03	1.434
0.240	0.242E-02	0.246	0.221E-03	1.415
0.260	0.197E-02	0.239	0.259E-03	1.396
0.280	0.351E-02	0.253	0.244E-03	1.376
0.300	0.817E-02	0.255	0.222E-03	1.357
0.325	0.150E-01	0.282	0.196E-03	1.338
0.360	0.208E-01	0.266	0.191E-03	1.319
0.370	0.292E-01	0.332	0.162E-03	1.299
0.400	0.205E-01	0.355	0.179E-03	1.280
0.433	0.234E-01	0.409	0.138E-03	1.287
0.466	0.392E-01	0.433	0.128E-03	1.295
0.500	0.509E-01	0.452	0.116E-03	1.302
0.533	0.629E-01	0.477	0.106E-03	1.287
0.566	0.752E-01	0.459	0.975E-04	1.271
0.600	0.859E-01	0.494	0.102E-03	1.256
0.633	0.871E-01	0.511	0.902E-04	1.324
0.666	0.967E-01	0.484	0.923E-04	1.393
0.700	0.954E-01	0.510	0.807E-04	1.461
0.817	0.106E+00	0.472	0.966E-04	1.529
0.907	0.116E+00	0.487	0.887E-04	1.598
1.000	0.125E+00	0.498	0.829E-04	1.666
1.105	0.111E+00	0.466	0.104E-03	1.666

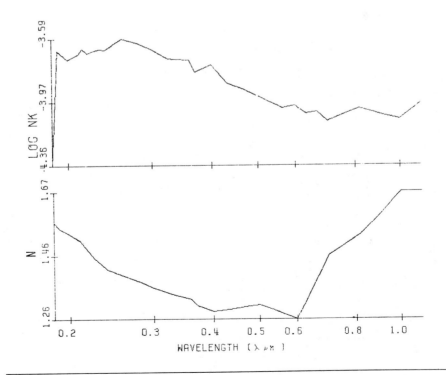

Table A.68 Optical Constants for Sand, Timber Island, Great South Bay,
Long Island, New York. nk Determined from MKM Theory. Effective Sample
Thickness = 0.0111 cm

LAMBDA	T	R	NK	N
0.185	0.202E-03	0.179	0.236E-03	1.647
0.190	0.400E-03	0.177	0.228E-03	1.628
0.200	0.350E-03	0.184	0.234E-03	1.609
0.210	0.346E-03	0.190	0.238E-03	1.591
0.215	0.416E-03	0.192	0.237E-03	1.572
0.220	0.143E-03	0.200	0.265E-03	1.553
0.225	0.537E-03	0.208	0.224E-03	1.534
0.233	0.585E-03	0.208	0.232E-03	1.515
0.240	0.625E-03	0.214	0.232E-03	1.496
0.260	0.108E-02	0.210	0.242E-03	1.477
0.280	0.142E-02	0.230	0.230E-03	1.458
0.300	0.128E-02	0.261	0.220E-03	1.440
0.325	0.236E-02	0.289	0.197E-03	1.421
0.360	0.424E-02	0.317	0.180E-03	1.402
0.370	0.469E-02	0.328	0.177E-03	1.383
0.400	0.756E-02	0.351	0.164E-03	1.364
0.433	0.122E-01	0.388	0.139E-03	1.366
0.466	0.165E-01	0.400	0.133E-03	1.368
0.500	0.200E-01	0.418	0.127E-03	1.370
0.533	0.290E-01	0.440	0.112E-03	1.370
0.566	0.321E-01	0.442	0.115E-03	1.369
0.600	0.341E-01	0.436	0.123E-03	1.369
0.633	0.398E-01	0.451	0.116E-03	1.369
0.666	0.406E-01	0.453	0.121E-03	1.369
0.700	0.421E-01	0.421	0.141E-03	1.369
0.817	0.467E-01	0.406	0.169E-03	1.369
0.907	0.492E-01	0.478	0.140E-03	1.369
1.000	0.568E-01	0.498	0.135E-03	1.369
1.105	0.539E-01	0.442	0.189E-03	1.369

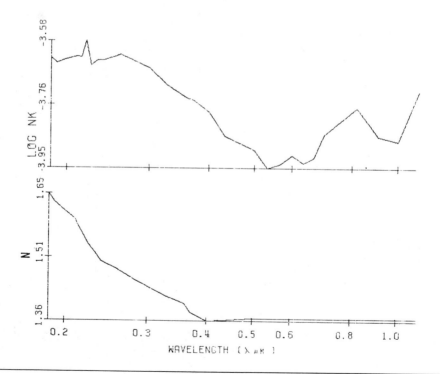

Table A.69 Optical Constants for Siderite. nk Determined from KM Theory.
Effective Sample Thickness = 0.000783 cm

LAMBDA	T	R	NK	N
0.185	0.149E-02	0.063	0.122E-01	1.836
0.190	0.190E-02	0.060	0.121E-01	1.834
0.200	0.188E-02	0.062	0.127E-01	1.831
0.210	0.208E-02	0.059	0.131E-01	1.828
0.215	0.203E-02	0.066	0.135E-01	1.825
0.220	0.206E-02	0.067	0.138E-01	1.823
0.225	0.243E-02	0.057	0.137E-01	1.820
0.233	0.257E-02	0.055	0.141E-01	1.817
0.240	0.217E-02	0.055	0.149E-01	1.814
0.260	0.237E-02	0.062	0.150E-01	1.812
0.280	0.406E-02	0.049	0.157E-01	1.809
0.300	0.453E-02	0.055	0.162E-01	1.806
0.325	0.473E-02	0.062	0.177E-01	1.803
0.360	0.680E-02	0.057	0.182E-01	1.801
0.370	0.512E-02	0.056	0.177E-01	1.798
0.400	0.858E-02	0.066	0.193E-01	1.795
0.433	0.142E-01	0.063	0.187E-01	1.789
0.466	0.185E-01	0.063	0.188E-01	1.784
0.500	0.264E-01	0.075	0.184E-01	1.778
0.533	0.390E-01	0.067	0.175E-01	1.772
0.566	0.643E-01	0.077	0.157E-01	1.765
0.600	0.779E-01	0.085	0.155E-01	1.759
0.633	0.948E-01	0.091	0.151E-01	1.754
0.666	0.111E+00	0.094	0.148E-01	1.750
0.700	0.125E+00	0.107	0.147E-01	1.746
0.817	0.955E-01	0.115	0.193E-01	1.740
0.907	0.180E+00	0.115	0.157E-01	1.743
1.000	0.199E+00	0.125	0.162E-01	1.742
1.105	0.221E+00	0.134	0.167E-01	1.734

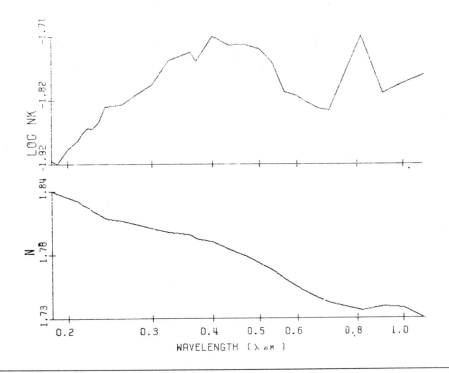

Table A.70 Optical Constants for Sulfur (Fisher).
nk Determined from KM Theory.
Effective Sample Thickness = 0.0137 cm

LAMBDA	T	R	NK	N
0.185	0.571E-03	0.075	0.797E-03	1.664
0.190	0.108E-02	0.081	0.750E-03	1.689
0.200	0.947E-03	0.075	0.821E-03	1.713
0.210	0.705E-03	0.075	0.883E-03	1.737
0.215	0.445E-03	0.076	0.957E-03	1.762
0.220	0.524E-03	0.076	0.965E-03	1.786
0.225	0.692E-03	0.075	0.947E-03	1.811
0.233	0.725E-03	0.076	0.973E-03	1.835
0.240	0.720E-03	0.075	0.100E-02	1.860
0.260	0.789E-03	0.072	0.108E-02	1.884
0.280	0.867E-03	0.071	0.114E-02	1.909
0.300	0.997E-03	0.067	0.120E-02	1.933
0.325	0.969E-03	0.064	0.130E-02	1.967
0.360	0.981E-03	0.063	0.144E-02	2.000
0.370	0.104E-02	0.071	0.148E-02	2.034
0.400	0.125E-02	0.115	0.154E-02	2.067
0.433	0.114E-01	0.353	0.109E-02	2.015
0.466	0.845E-01	0.558	0.567E-03	1.964
0.500	0.128E+00	0.620	0.451E-03	1.912
0.533	0.138E+00	0.623	0.455E-03	1.893
0.566	0.142E+00	0.606	0.485E-03	1.873
0.600	0.143E+00	0.585	0.523E-03	1.854
0.633	0.146E+00	0.592	0.542E-03	1.875
0.666	0.152E+00	0.584	0.560E-03	1.868
0.700	0.155E+00	0.576	0.582E-03	1.860
0.800	0.155E+00	0.562	0.675E-03	1.843
0.907	0.162E+00	0.554	0.756E-03	1.845
1.000	0.166E+00	0.543	0.830E-03	1.847
1.105	0.176E+00	0.536	0.883E-03	1.848
1.200	0.231E+00	0.636	0.607E-03	1.848
1.300	0.218E+00	0.512	0.901E-03	1.848
1.400	0.175E+00	0.526	0.112E-02	1.848
1.500	0.205E+00	0.650	0.847E-03	1.848
1.700	0.273E+00	0.641	0.637E-03	1.848
2.000	0.172E+00	0.778	0.775E-03	1.848
2.300	0.244E+00	0.214	0.182E-02	1.848
2.500	0.240E+00	0.607	0.131E-02	1.848

Table A.71 Optical Constants for Volcanic Soil, Haleakala, Hawaii.
nk Determined from KM Theory. Effective Sample Thickness = 0.00385 cm

LAMBDA	T	R	NK	N
0.185	0.85SE-02	0.061	0.265E-02	1.501
0.190	0.125E-02	0.045	0.261E-02	1.499
0.200	0.128E-02	0.050	0.275E-02	1.496
0.210	0.136E-02	0.048	0.286E-02	1.494
0.215	0.140E-02	0.055	0.291E-02	1.491
0.220	0.156E-02	0.056	0.293E-02	1.488
0.225	0.158E-02	0.048	0.300E-02	1.486
0.233	0.165E-02	0.045	0.307E-02	1.483
0.240	0.175E-02	0.045	0.314E-02	1.496
0.260	0.263E-02	0.034	0.319E-02	1.509
0.280	0.316E-02	0.045	0.332E-02	1.501
0.300	0.498E-02	0.050	0.329E-02	1.493
0.325	0.688E-02	0.055	0.335E-02	1.485
0.360	0.101E-01	0.055	0.342E-02	1.478
0.370	0.135E-01	0.052	0.327E-02	1.470
0.400	0.140E-01	0.060	0.352E-02	1.462
0.433	0.177E-01	0.055	0.361E-02	1.460
0.466	0.220E-01	0.054	0.367E-02	1.458
0.500	0.259E-01	0.060	0.377E-02	1.456
0.533	0.302E-01	0.064	0.385E-02	1.342
0.566	0.384E-01	0.076	0.381E-02	1.328
0.600	0.444E-01	0.083	0.386E-02	1.314
0.633	0.513E-01	0.093	0.388E-02	1.347
0.666	0.557E-01	0.095	0.387E-02	1.381
0.700	0.716E-01	0.110	0.380E-02	1.414
0.817	0.831E-01	0.117	0.418E-02	1.413
0.907	0.909E-01	0.117	0.447E-02	1.412
1.000	0.102E+00	0.122	0.465E-02	1.411
1.105	0.130E+00	0.128	0.462E-02	1.410

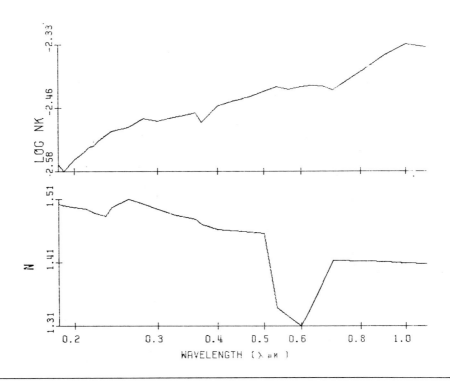

WAVELENGTH (λ μm)

Table A.72 Optical Constants for Volcanic Soil, Mt. St. Helens.
nk Determined from MKM Theory. Effective Sample Thickness = 0.0111 cm.
Collected at Kelso, Washington following eruption

LAMBDA	T	R	NK	N
0.185	0.206E-02	0.078	0.510E-03	1.536
0.190	0.159E-02	0.079	0.532E-03	1.530
0.200	0.140E-02	0.078	0.576E-03	1.525
0.210	0.146E-02	0.076	0.613E-03	1.520
0.215	0.196E-02	0.084	0.543E-03	1.514
0.220	0.214E-02	0.084	0.551E-03	1.509
0.225	0.304E-02	0.084	0.527E-03	1.503
0.233	0.322E-02	0.086	0.526E-03	1.498
0.240	0.359E-02	0.083	0.548E-03	1.493
0.260	0.701E-02	0.132	0.336E-03	1.487
0.280	0.878E-02	0.136	0.336E-03	1.482
0.300	0.137E-01	0.112	0.398E-03	1.477
0.325	0.146E-01	0.116	0.411E-03	1.471
0.360	0.251E-01	0.128	0.363E-03	1.466
0.370	0.302E-01	0.127	0.359E-03	1.460
0.400	0.366E-01	0.140	0.335E-03	1.455
0.433	0.463E-01	0.149	0.318E-03	1.450
0.466	0.530E-01	0.155	0.316E-03	1.446
0.500	0.605E-01	0.161	0.313E-03	1.441
0.533	0.699E-01	0.167	0.307E-03	1.437
0.566	0.792E-01	0.177	0.294E-03	1.433
0.600	0.809E-01	0.224	0.240E-03	1.429
0.633	0.875E-01	0.186	0.302E-03	1.424
0.666	0.902E-01	0.174	0.339E-03	1.418
0.700	0.955E-01	0.187	0.319E-03	1.413
0.817	0.111E+00	0.189	0.354E-03	1.407
0.907	0.115E+00	0.183	0.401E-03	1.402
1.000	0.123E+00	0.179	0.442E-03	1.396
1.105	0.131E+00	0.171	0.499E-03	1.391

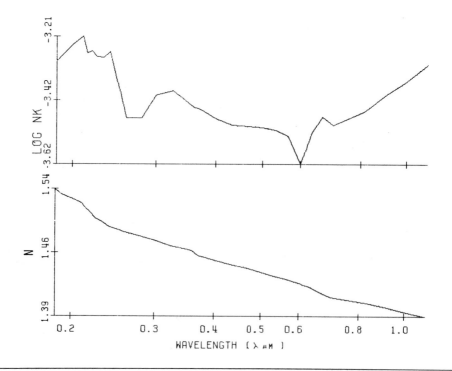

Table A.73 Optical Constants for Volcanic Soil, Mt. St. Helens.
nk Determined from MKM Theory. Effective Sample Thickness = 0.0136 cm.
Collected at Spokane Washington, following eruption

LAMBDA	T	R	NK	N
0.185	0.892E-03	0.061	0.453E-03	1.539
0.190	0.468E-03	0.093	0.440E-03	1.531
0.200	0.401E-03	0.085	0.495E-03	1.523
0.210	0.426E-03	0.086	0.530E-03	1.515
0.215	0.638E-03	0.097	0.453E-03	1.507
0.220	0.727E-03	0.096	0.460E-03	1.499
0.225	0.704E-03	0.101	0.448E-03	1.491
0.233	0.731E-03	0.103	0.452E-03	1.483
0.240	0.897E-03	0.102	0.457E-03	1.475
0.260	0.281E-02	0.175	0.241E-03	1.467
0.280	0.403E-02	0.143	0.303E-03	1.459
0.300	0.653E-02	0.166	0.257E-03	1.451
0.325	0.946E-02	0.178	0.242E-03	1.443
0.360	0.138E-01	0.198	0.222E-03	1.435
0.370	0.113E-01	0.198	0.240E-03	1.427
0.400	0.250E-01	0.213	0.200E-03	1.419
0.433	0.320E-01	0.236	0.181E-03	1.420
0.466	0.383E-01	0.247	0.175E-03	1.421
0.500	0.450E-01	0.251	0.176E-03	1.422
0.533	0.534E-01	0.270	0.163E-03	1.418
0.566	0.613E-01	0.282	0.157E-03	1.414
0.600	0.640E-01	0.291	0.159E-03	1.410
0.633	0.696E-01	0.295	0.160E-03	1.407
0.666	0.745E-01	0.273	0.182E-03	1.404
0.700	0.821E-01	0.293	0.169E-03	1.401
0.817	0.862E-01	0.293	0.194E-03	1.397
0.907	0.901E-01	0.280	0.225E-03	1.394
1.000	0.943E-01	0.281	0.243E-03	1.391
1.105	0.991E-01	0.266	0.280E-03	1.388

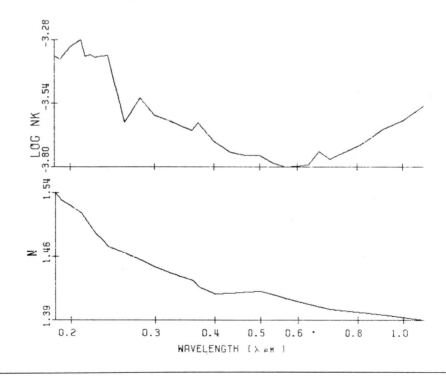

Table A.74 Optical Constants for Volcanic Soil, Mt. St. Helens 1.
nk Determined from MKM Theory. Effective Sample Thickness = 0.00420 cm.
Collected during eruption

LAMBDA	T	R	NK	N
0.185	0.205E-01	0.076	0.840E-03	1.530
0.190	0.200E-01	0.070	0.956E-03	1.526
0.200	0.213E-01	0.072	0.963E-03	1.523
0.210	0.221E-01	0.073	0.985E-03	1.519
0.215	0.256E-01	0.072	0.984E-03	1.516
0.220	0.264E-01	0.071	0.100E-02	1.512
0.225	0.326E-01	0.072	0.956E-03	1.509
0.233	0.333E-01	0.073	0.968E-03	1.505
0.240	0.359E-01	0.070	0.101E-02	1.502
0.260	0.527E-01	0.102	0.694E-03	1.498
0.280	0.640E-01	0.090	0.785E-03	1.495
0.300	0.751E-01	0.082	0.851E-03	1.491
0.325	0.839E-01	0.084	0.869E-03	1.488
0.360	0.115E+00	0.085	0.801E-03	1.484
0.370	0.124E+00	0.085	0.797E-03	1.481
0.400	0.145E+00	0.096	0.750E-03	1.477
0.433	0.164E+00	0.101	0.728E-03	1.468
0.466	0.176E+00	0.103	0.742E-03	1.459
0.500	0.192E+00	0.107	0.734E-03	1.450
0.533	0.206E+00	0.110	0.734E-03	1.446
0.566	0.215E+00	0.112	0.748E-03	1.442
0.600	0.222E+00	0.142	0.625E-03	1.438
0.633	0.232E+00	0.120	0.754E-03	1.429
0.666	0.243E+00	0.109	0.837E-03	1.420
0.700	0.245E+00	0.110	0.860E-03	1.411
0.817	0.263E+00	0.112	0.955E-03	1.402
0.907	0.285E+00	0.117	0.967E-03	1.393
1.000	0.300E+00	0.117	0.103E-02	1.384
1.105	0.318E+00	0.109	0.115E-02	1.375

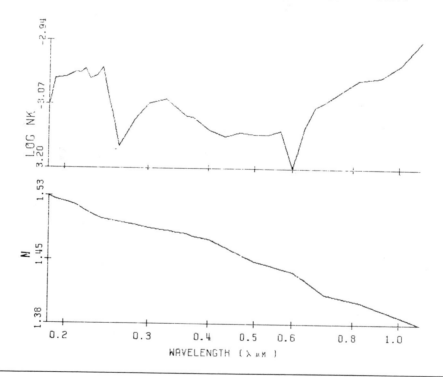

Table A.75 Optical Constants for Volcanic Soil, Mt. St. Helens 2.
nk Determined from MKM Theory. Effective Sample Thickness = 0.0129 cm.
Collected during eruption

LAMBDA	T	R	NK	N
0.185	0.403E-02	0.065	0.618E-03	1.530
0.190	0.375E-03	0.075	0.557E-03	1.526
0.200	0.265E-03	0.076	0.630E-03	1.523
0.210	0.151E-03	0.074	0.713E-03	1.519
0.215	0.143E-03	0.075	0.702E-03	1.516
0.220	0.235E-03	0.075	0.677E-03	1.512
0.225	0.311E-03	0.080	0.661E-03	1.509
0.233	0.351E-03	0.081	0.665E-03	1.505
0.240	0.355E-03	0.080	0.691E-03	1.502
0.260	0.765E-03	0.118	0.134E-02	1.498
0.280	0.112E-02	0.120	0.464E-03	1.495
0.300	0.171E-02	0.093	0.598E-03	1.491
0.325	0.146E-02	0.094	0.657E-03	1.488
0.360	0.406E-02	0.103	0.567E-03	1.484
0.370	0.510E-02	0.103	0.559E-03	1.481
0.400	0.529E-02	0.142	0.436E-03	1.477
0.433	0.790E-02	0.151	0.412E-03	1.468
0.466	0.968E-02	0.156	0.413E-03	1.459
0.500	0.127E-01	0.162	0.405E-03	1.450
0.533	0.147E-01	0.167	0.406E-03	1.446
0.566	0.178E-01	0.173	0.399E-03	1.442
0.600	0.198E-01	0.172	0.415E-03	1.438
0.633	0.217E-01	0.176	0.421E-03	1.429
0.666	0.333E-01	0.171	0.411E-03	1.420
0.700	0.361E-01	0.177	0.410E-03	1.411
0.817	0.365E-01	0.156	0.538E-03	1.402
0.907	0.408E-01	0.178	0.535E-03	1.393
1.000	0.481E-01	0.133	0.720E-03	1.384
1.105	0.511E-01	0.157	0.674E-03	1.375

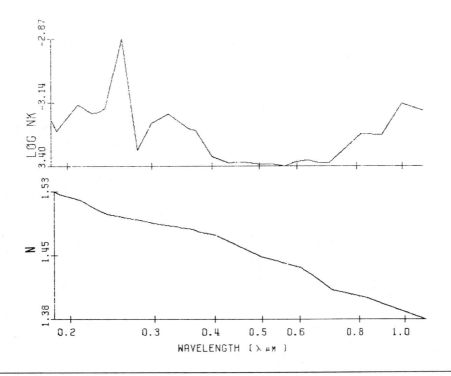

Table A.76 Optical Constants for Volcanic Soil, Vesuvius.
nk Determined from MKM Theory. Effective Sample Thickness = 0.00370 cm

LAMBDA	T	R	NK	N
0.185	0.736E-03	0.062	0.219E-02	1.541
0.190	0.545E-03	0.049	0.283E-02	1.536
0.200	0.129E-02	0.053	0.248E-02	1.531
0.210	0.132E-02	0.047	0.280E-02	1.526
0.215	0.136E-02	0.065	0.220E-02	1.520
0.220	0.161E-02	0.063	0.224E-02	1.515
0.225	0.164E-02	0.050	0.270E-02	1.510
0.233	0.179E-02	0.051	0.239E-02	1.505
0.240	0.192E-02	0.051	0.274E-02	1.500
0.260	0.331E-02	0.040	0.310E-02	1.495
0.280	0.427E-02	0.056	0.250E-02	1.490
0.300	0.670E-02	0.059	0.240E-02	1.485
0.325	0.100E-01	0.072	0.206E-02	1.479
0.360	0.128E-01	0.065	0.221E-02	1.474
0.370	0.171E-01	0.067	0.217E-02	1.469
0.400	0.212E-01	0.066	0.181E-02	1.464
0.433	0.260E-01	0.064	0.190E-02	1.458
0.466	0.322E-01	0.066	0.185E-02	1.453
0.500	0.377E-01	0.091	0.184E-02	1.447
0.533	0.440E-01	0.097	0.178E-02	1.440
0.566	0.547E-01	0.110	0.158E-02	1.432
0.600	0.626E-01	0.112	0.158E-02	1.425
0.633	0.690E-01	0.119	0.153E-02	1.433
0.666	0.808E-01	0.110	0.162E-02	1.441
0.700	0.845E-01	0.118	0.157E-02	1.449
0.817	0.104E+00	0.135	0.150E-02	1.430
0.907	0.111E+00	0.123	0.178E-02	1.412
1.000	0.120E+00	0.116	0.200E-02	1.393
1.105	0.135E+00	0.118	0.209E-02	1.374

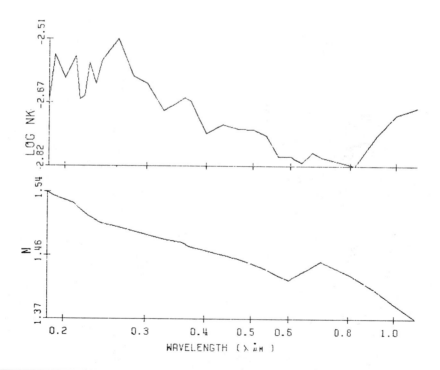

Bibliography

Adams, J. B. (1968) Lunar and Martian surfaces: Petrologic significance of absorption bands in the near-infrared. *Science* 159:1453–1455.

Adams, J. B., and Filice, A. L. (1967) Spectral reflectance 0.4–20 μm of silicate rock powders. *J. Geophys. Res.* 72:5705–5715.

Adams, J. B., and McCord, T. B. (1971) Alteration of lunar optical properties: Age and composition effects. *Science* 171:567–571.

Adams, J. B., and McCord, T. B. (1972) Optical evidence for regional cross-contamination of highland and mare soils, 3rd Lunar Sci. Conf. (Abst.), Lunar Science Institute and NASA Manned Spacecraft Center, Houston, TX.

Ahern, F. J., Goodenough, D. G., Jain, S. C., Rao, V. R., and Rochon, G. (1977) Use of clear lakes as standard reflectors for atmospheric measurements, Proc. 4th Can. Symp. Remote Sensing.

Aldrich, R. C. (1968) Remote sensing and the forest survey—present applications research, and a look at the future. Proc. 5th Symp. Remote Sensing, April 1968, Univ. Michigan, 357–372.

Allen, C. W. (1963) *Astrophysical Quantities*, 2nd ed. London: Athlone Press.

Allen, T. (1974) *Particle Size Measurement*. London: Chapman and Hall.

Allen, W. A., and Richardson, A. J. (1968) Interaction of light with a plant canopy, Proc. 5th Symp. Remote Sensing, April 1968, Univ. Michigan, 219–232.

Anders, E. (1964) Origin, age and composition of meteorites. *Space Sci. Rev.* 3:583–714.

Anders, E. (1971) Interrelations of meteorites, asteroids and comets. Physical Studies of Minor Planets, T. Gehrels (ed.). NASA Rep. SP-267.

Anderson, R. R., Brown, R. G., and Rappleye, R. D. (1968) *Chesapeake Sci.* 9 (3):145–146.

Arago, M. (1824) Extrait des Seances de l'Académie Royale des Sciences Séanc de Lundi 14 Juin, *Ann. Chem. Phys. Ser.* (2) 27:89.

Ashburn, E. V., and Weldon, R. G. (1956) Spectral diffuse reflectance of desert surfaces. *J. Opt. Soc. Am.* 46:583–586.

Ashok, N. M., Bhatt, H. C., Chandrasekhar, T., and Desai, J. N. (1982) *Nature* 300:620–621.

Archer, R. J. (1968) *Manual of Ellipsometry*. Chicago: Gaertner Scientific Corp.

Bancroft, G. M., and Burns, R. G. (1967) Interpretation of the electronic spectra of iron in pyroxenes. *Am. Mineral.* 52:1278–1287.

Barabashev, N. P., and Chekirda, A. O. (1959) A study of the rocks most closely resembling the surface constituents of the moon. *Astron. Zh.* 36:851–855.

Barabashev, N. P., and Garazhon, V. I. (1962) The microstructure of the lunar surface. *Soviet Astronomy* VI.

Barrett, E. C., and Curtis, L. F. (1976) *Introduction to Environmental Remote Sensing.* London: Chapman and Hall.

Basener, R. F., and McCoyd, G. C. (1967) Polarization of infrared light emitted by the sea. Grumman Res. Dept. Memorandum, RM-360.

Bauer, K. G., and Dutton, J. A. (1962) Albedo variations measured from an airplane over several types of surfaces. *J. Geophys. Res.* 67:2367–2376.

Becher, W. D., and Peace, G. M. (1982) Advanced imaging development. Paper presented at 4th Ann. Honeywell CAD CAM Workshop, Minneapolis, Mn. 19–21 October.

Beckmann, P. (1963) The depolarization of electromagnetic waves scattered from rough surfaces. *Electromagnetic Theory and Antennas*, Part 2. New York: Pergamon Press.

Beckmann, P. (1965) Radar backscatter from the surface of the moon. *J. Geophys. Res.* 70:2345–2350.

Beckmann, P. (1968) *The Depolarization of Electromagnetic Waves.* Boulder, Co.: The Golem Press.

Beckmann, P., and Spizzichino, A. (1963) *The Scattering of Electromagnetic Waves from Rough Surfaces.* New York: Macmillan.

Berge, G. (1963) A recording transparency meter for oceanic plankton estimation. *Rep. Norweg. Fish. Mar. Invest.* 13:95–105.

Bertie, J. E., Labbé, H. J., and Whalley, E. (1969) Absorptivity of ice in the range 4000–30 cm^{-1}. *J. Chem. Phys.* 50:4501–4520.

Binder, A. B., and Cruikshank, D. P. (1964) Comparison of the infrared spectrum of Mars with spectra of selected terrestrial rocks and minerals. *Commun. Lunar Planet. Lab.* 2:193.

Binder, A. B., Cruikshank, D. P., and Hartmann, W. K. (1965) Observations of the Moon and of the terrestrial rocks in the infrared. *Icarus* 4:415–420.

Binder, A. B., and Cruikshank, D. P. (1966a) Lithological and mineralogical investigation of the surface of Mars. *Icarus* 5:521–525.

Binder, A. B., and Cruikshank, D. P. (1966b) The composition of the surface layer of Mars. *Commun. Lunar Planet. Lab.* 4:111–120.

Birge, R. T. (1934) The velocity of light. *Nature* 134:771–772.

Biscaye, P. E. (1965) Mineralogy and sedimentation of recent deep-sea clays in the Atlantic and adjacent seas and oceans. *Geol. Soc. Am. Bull.* 76:803–832.

Blackman. R. B., and Tukey, J. W. (1958) *The Measurement of Power Spectra.* New York: Dover Publications.

Bodhaine, B. A., and Harris, J. M. (eds.) (1981) Geophysical Monitoring for Climatic Change, No. 10, U.S. Dept. Commerce, NOAA, ERL, Boulder, Co.

Born, M., and Wolf, E. (1959) *Principles of Optics.* New York: Macmillan.

Braslau, N., and Dave, J. V. (1972) Effect of aerosols on the transfer of solar energy through realistic model atmospheres. Part I: Non-absorbing aerosols. IBM Res. Rep. RC4114 (No. 18435), T. J. Watson Research Center, Yorktown Heights, N.Y.

Braslau, N., and Dave, J. V. (1973) Effect of aerosols on the transfer of solar energy through realistic model atmospheres. *J. Appl. Meterology* 12:601–615.

Brezenski, F. T. (1972) Laboratory Results—U.S. Virgin Islands. Rep. 2-SA-TEC; to Chief Surveillance Branch E.P.A.

Bryan, G. M. (1970) Hydrodynamic model of the Blake outer ridge. *J. Geophys. Res.* 75:4530–4537.

Buriez, J. C., Fouquart, Y., and Fymat, A. L. (1979a) Spectropolarimetry of Venus and Jupiter clouds: Information content of equilvalent widths. *Astron. Astrophys.* 79:287–298.

Buriez, J. C., Fouquart, Y., and Fymat, A. L. (1979b) Equivalent widths in spectropolarimetry can provide additional information on the atmosphere of Venus. *Astrophysical J.* 230:590–596.

Burns, R. G. (1970) *Mineralogical Applications of Crystal Field Theory*. Cambridge: Cambridge University Press.

Byrne, L. J. (1963) Lunar photographic orbiter: Lighting and viewing conditions. Bellcomm Rep. TM 6 3111 21.

Cabannes, J. (1929) *La Diffusion Moléculaire de la Lumière*. Paris: Les Presses Universitaires de France.

Carneggie, D. M. (1968) Applying remote sensing technology for range resource inventories. Proc. 5th Symp. Remote Sensing, April 1968, Univ. Michigan, 373–385.

Carrier, L. W., and Nugent, L. J. (1965) Comparison of some recent experimental results of coherent and incoherent light scattering with theory. *Appl. Opt.* 4:1457–1462.

Chandrasekhar, S. (1960) *Radiative Transfer*. New York: Dover.

Chanu, J. (1959) Extraction de la substance jaune dans les eaux côtières. *Rev. Opt.* 38:569.

Chapman, C. R. (1972) Surface properties of asteroids. Ph.D. Thesis, MIT, Cambridge, Ma.

Charlson, R. J., and Pilat, M. J. (1969) Climate: The influence of aerosols. *J. Appl. Meteor.* 8:1001–1002.

Chylek, P., and Grams, G. W. (1978) Scattering by nonspherical particles and optical properties of Martian dust. *Icarus* 36:198–203.

Clarke, D. (1965) Studies in astronomical polarimetry, III, the wavelength dependence of the polarization of light reflected by the Moon and Mars. *Mont. Not. R. Astron. Soc.* 130:83–94.

Clark, J. R., and Frank, J. L. (1963) Infrared measurement of sea surface temperatures. *Undersea Technol.* 4(10):20–23.

Cobb, W. E., and Wells, H. J. (1970) The electric conductivity of oceanic air and its correlation to global atmospheric pollution. *J. Atmos. Sci.* 27:814–819.

Coffeen, D. L. (1965) Wavelength dependence of polarization, IV Volcanic cinders and particles. *Astronomical J.* 70:403–413.

Cohen, A. J. (1966) Seasonal color changes on Mars. *Astron. J.* 71(9):849–850.

Colwell, J. E. (1981) Landsat feature enhancement. Proc. 15th Int. Symp. Remote Sensing Env., 11–15 May.

Connes, J., Connes, P., and Maillard, J. P. (1969) Atlas, near infrared spectra of Venus, Mars, Jupiter, and Saturn. Editions du Centre National de la Recherche Scientifique, 15, quai Anatole France, Paris VII.

Coulson, K. L. (1959) Characteristics of the radiation emerging from the top of a Rayleigh atmosphere—II, Total upward flux and albedo. *Planet. Space Sci.* 1:277–284.

Coulson, K. L. (1966) Effects of reflection properties of natural surfaces in aerial reconnaissance. *Appl. Opt.* 5:905–917.

Coulson, K. L. (1978) Zenith skylight characteristics in the sunrise period at Mauna Loa. *Mauna Loa Observatory*, A 20th Anniversary Report, U.S. Department of Commerce, NOAA Environmental Research Laboratories, U.S.G.P.O. 1978-677-034/17.

Coulson, K. L., Bouricius, G. M. B., and Gray, E. L. (1964) Technical Information Ser. Rep. R64SD74, General Electric Company, Missile and Space Division, Space Sciences Laboratory.

Coulson, K. L., Gray, E. L., and Bouricius, G. M. (1966) Effect of surface reflection on planetary albedo. *Icarus* 5:139–148.

Cox, C., and Munk, W. (1954) Measurement of the roughness of the sea surface from photographs of the sun's glitter. *J. Opt. Sci. Am.* 44:838–850.

Cox, C., and Munk, W. (1956) Slopes of the sea surface deduced from photographs of sun glitter. Bulletin Scripps Institution of Oceanography, University of California, LaJolla.

Craig, J. W. (1960) Optimum Approximation to a Matched Filter Response. IRE Transaction on Information Theory, Vol. IT6, No. 3.

Crittenden, E. C., Jr., Cooper, A. W., Milne, E. A., Rodeback, G. W., Armstead, R. L., Kalmbach, S. H., Land, D., and Katz, B. (1978) Optical resolution in the turbulent atmosphere of the marine boundary layer. Naval Postgraduate School Rep. NPS61-78-0003.

Cuny, J. J., Johnson, C. B., and Lynch, T. F. (1979a) Proximity focused image tube intensified charge injection device (CID) camera for low light level television. SPIE, Vol. 203, *Recent Advances in TV Sensors and Systems.*

Cuny, J. J., Lynch, T. F., and Johnson, C. B. (1979b) Small intensified charge injection device cameras for low light level television. MEDE '79 Conf. Proc. Interavia, S. A. Geneva, Switzerland.

Curran, P. J. (1978) A photographic method for the recording of polarized visible light for soil surface moisture indications. *Remote Sensing Environ.* 7:305–322.

Curran, P. J. (1979) The use of polarized panchromatic and false-color infrared film in the monitoring of soil surface moisture. *Remote Sensing Environ.* 8:249–266.

Currie, R. I. (1962) Pigments in zooplankton faeces. *Nature* 193:956–957.

Dana, R. W. (1975) Solar and atmospheric effects on satellite imagery derived from aircraft reflectance measurements. Proc. 10th Inst. Symp. Remote Sensing Environ.

Darss (1983) *Optical Multichannel Analysis System*, Tracor Northern, Middletown, Wisc.

Dave, J. V. (1978) Extensive data sets of the diffuse radiation in realistic atmospheric models with aerosols and common absorbing gases. *Solar Energy* 21:361–369.

Dave, J. V. (1970) Coefficients of the Legendre and Fourier series for the scattering functions of spherical particles. *Appl. Opt.* 9:1888–1896.

Dave, J. V. (1972) Development of programs for computing characteristics of ultraviolet radiation technical reports—Scalar case, Programs I–IV. International Business Machines Corp., Federal Systems Division, Gaithersburg, Md.

Dave, J. V. (1972a) Development of programs for computing characteristics of ultraviolet radiation, NASA Rep. No. S80-RADTMO, Goddard Space Flight Center, Greenbelt, Md.

Dave, J. V. (1972b) Development of programs for computing characteristics of ultraviolet radiation, Tech. Rep.—Vector Case, Program V. NASA Contract NAS 5-21680, Goddard Space Flight Center, Greenbelt, Md.

Dave, J. V., and Gazdag, J. (1970) A modified Fourier transform method for multiple scattering calculations in a plane parallel atmosphere. *Appl. Opt.* 9:1457–1466.

Dave, J. V., and Mateer, C. L. (1968) The effect of stratospheric dust on the color of the twilight sky. *J. Geophys. Res.* 73:6897–6913.

Davies, C. N. (ed.) (1966) *Aerosol Science*. New York: Academic Press.

Deirmendjian, D. (1969) *Electromagnetic Scattering on Spherical Polydispersions*, New York: American Elsevier.

Deirmendjian, D., Clasen, R., and Viezee, W. (1961) Mie scattering with complex index of refraction. *J. Opt. Soc. Am.* 51:620–633.

Deluisi, J. J. (ed.) (1980) Geophysical monitoring for climatic change, No. 9, U.S. Dept. Commerce, NOAA, ERL, Boulder, Co.

Dollfus, A. (1955) Etude de planètes par la polarization de leur lumière, Thesis, University of Paris; also in *Ann. Astrophys. Suppl. 4* (1957); English translation in NASA Rep. TTF-188 (1964).

Dollfus, A. (1956) Polarization de la lumière renvoyée par les corps solides et les nuages naturels. *Ann. d'Astrophys.* 19:83–113.

Dollfus, A. (1957) Etude des planets par la polarization de leur lumière. *Ann. d'Astrophys. Suppl. 4.*

Dollfus, A. (1961) Polarization studies of the planets. *Planets and Satellites.* (G. P. Kuiper and B. M. Middlehurst, eds.) Chicago: University of Chicago Press.

Dollfus, A. (1967) Polarization measurements. Unpublished talk, Symp. Surface of Mars, Goddard Institute for Space Studies, February 3–4.

Dollfus A., and Focas, J. H. (1966a) On the purity of the atmosphere of the planet Mars. *C. R. Acad. Sci. Ser. B* 262:1024–1027.

Dollfus, A., and Focas, J. H. (1966b) Polarimetric study of the planet Mars. *Final Scientific Report under Contract AF-61(052)-508*, through the European Office of Aerospace Research and U.S. Air Force Cambridge Research Laboratories.

Dollfus, A., and Geake, J. E. (1965) L'altération des Propriétés Polarimetriques de Sol Lunaire par l'action des Protons Solaires. *C. R. Acad. Sci.* 260:4921–4923.

Draper, A. L., Adamcek, J. A., and Gibson, E. K. (1964) Comparison of the spectra of Mars and a goethite–hematite mixture in the 1 to 2 micron region. *Icarus* 3:63–65.

Driscoll, W. G., and Vaughn, W. (1978) *Handbook of Optics*. New York: McGraw-Hill.

Duddek, M. (1967) *Photogrammetric Eng.* 33(10):1117–1125.

Duntley, S. Q. (1954) Measurement of the distribution of water wave slopes. *J. Opt. Soc. Am.* 44:574–575.

Dyck, H. M., and Jones, T. J. (1978) Near-infrared observations of interstellar polarization. *Astron. J.* 83:594–597.

Eastman Kodak Company, Rochester, N.Y. (1967) Kodak Neutral Density Filters. Pamphlet No. P-114.

Eastman Kodak Company, Rochester, NY (1968) Kodak Wratten Filters for Scientific and Technical Use. Publication B-3, 22nd Ed.

Eberhardt, E. H. (1969) Threshold sensitivity and noise ratings of multiplier phototubes. *Appl. Opt.* 6:251–255.

Egan, W. G. (1965a) Polarimetry and the lunar surface (a literature survey). Grumman Res. Dept. Memorandum, RM-271.

Egan, W. G. (1965b) Search for a lunar surface model (Volume II-Polarimetry). Grumman Res. Dept. Memorandum, RM-276.

Egan, W.G. (1967a) Polarimetry of a limonite surface as a function of wavelength and particle size. Grumman Res. Dept. Memorandum, RM-371J; *Trans A.G.U.* 48:149.

Egan, W. G. (1967b) Polarimetric measurements of simulated lunar surfaces. *J. Geophys. Res.* 72:3233–3245.

Egan, W. G. (1968) Aircraft polarimetric and photometric observations. Proc. 5th Symp. Remote Sensing, Univ. Mich., 169–189.

Egan, W. G. (1969) Practical calibration and control technique for type 8443 and Ektachrome films; aerial color photography in the plant sciences. Proc. Workshop, Univ. Florida, Gainesville, March 5–7.

Egan, W. G. (1969) Polarimetric and photometric simulation of the Martian surface. *Icarus* 10:223–227.

Egan, W. G. (1969) Manned submersible optical remote sensing within the Gulf Stream. Proc. 6th Int. Symp. Remote Sensing, Univ. Michigan, 721–735.

Egan, W. G. (1970) Nonimaging optical differentiation of forest foliage. *Forest Sci.* 16:79–94.

Egan, W. G. (1970) Optical Stokes parameters for farm crop identification. *Remote Sensing Environ.* 1:165–180.

Egan, W. G. (1971) Automated delineation of wetlands in photographic remote sensing. Proc. 7th Int. Symp. Remote Sensing, Univ. Michigan, 2231–2251.

Egan, W. G. (1971) Size classification of Mars simulation samples. *J. Geophys. Res.* 76:6213–6219.

Egan W. G. (1972) Water quality determinations in the Virgin Islands from ERTS-A data. Proc. 8th Int. Symp. Remote Sensing, Univ. Michigan, October, 685–708.

Egan, W. G. (1974a) Measurement of the fluorescence of Gulf Stream water with submerged *in situ* sensors. *Marine Technol. Soc. J.* 8:40–47.

Egan, W. G. (1974b) Boundaries of ERTS and aircraft data within which useful water quality information can be obtained. Proc. 9th Int. Symp. Remote Sensing, ERIM, 1319–1343.

Egan, W. G. (1974c) Boundaries of ERTS and aircraft data within which useful water quality information can be obtained. *Environ. Lett.* 6:95–101.

Egan, W. G. (1980) Optical remote sensing of the sea—A Caribbean example. Proc. 14th Int. Symp. Remote Sensing, ERIM.

Egan, W. G. (1982) Volumetric scattering and absorption by aerosols: Parametric sensitivity in Mie modeling and comparisons to observations. *Appl. Opt.* 21:1445–1453.

Egan, W. G. and Aspnes, D. E. (1982) Finite-wavelength effects in composite media. *Physical Rev. B*, 26:5313–5320.

Egan, W. G., and Cassin, J. M. (1973) Correlation of *in situ* fluorescence and bioluminescence with biota in the New York Bight. *Biol. Bull.* 144:262–275.

Egan, W. G., Cassin, J. M., and Hair, M. E. (1972) Interdisciplinary monitoring of the New York Bight. *Environ. Lett.* 2:205–215.

Egan, W. G., and Conrad, J. E. (1975) Summer abundance of zooplankton in the Gulf Stream. *Biol. Bull.* 149:492–505.

Egan, W. G., and Fischbein, W. L. (1975) Optical atmospheric scattering and absorption limitations on offset pointing from Earth Observatory Satellite sensors (EOS). Proc. 10th Int. Symp. Remote Sensing, 6–10 October, ERIM.

Egan, W. G., Fischbein, W. L., Hilgeman, T., and Smith, L. L. (1977) Martian atmosphere modeling between 0.4 and 3.5 μm: Comparison of theory and experiment. Proc. Symp. Planetary Atmospheres, A V. Jones (ed.): Pub. Ottawa: Royal Society of Canada, 57–60.

Egan, W. G., Fischbein, W. L., Smith, L. L., and Hilgeman, T. (1980) High resolution Martian atmosphere modeling. *Icarcus* 41:166–174.

Egan, W. G., and Foreman, K. M. (1967a) A new perspective on Martian polarimetric measurements. Grumman Res. Dept. Memorandum, RM-366J; also presented at the 1967 Natl. Tech. Symp. Am. Astronaut. Soc., Huntsville, Ala., June 11–14.

Egan, W. G., and Foreman, K. M. (1967b) Proc. 1967 Nat. Tech. Symp. Am. Astronaut. Soc., Huntsville, Ala.; also Grumman Res. Dept. Tech. Memorandum, RM-366J.

Egan, W. G., and Foreman, K. M. (1971) Mie Scattering and the Martian Atmosphere, in *Planetary Atmosphere*, C. Sagan (ed.) International Astronomical Union.

Egan, W. G., Foreman, K. M., and Nowatzki, E. A. (1967) Interpretation and application of photometric–polarimetric observations of planetary surfaces. *Use of Space Systems for Planetary Geology and Geophysics*, Vol. 17., (R. D. Enzmann ed.), American Astronautical Society, Washington.

Egan, W. G., Grusauskas, J., and Hallock, H. B. (1967) Optical depolarization properties of surfaces illuminated by coherent light. *J. Opt. Soc. Am.* 57:1430–1431.

Egan, W. G., Grusauskas, J., and Hallock, H. B. (1968) Optical depolarization properties of surfaces illuminated by coherent light. *Appl. Opt.* 7:1529–1534.

Egan, W. G., Hair, M. E., and Cok, A. E. (1969) Visible Region Sensors Workshop. U.S. Naval Oceanographic Office, Washington, D.C., December 17.

Egan, W. G., and Hallock, H. B. (1966) Polarimetry signature of terrestrial and planetary materials. Proc. 4th Symp. Remote Sensing, Univ. Michigan, April 12–14.

Egan, W. G., and Hilgeman, T. (1967) Polarimeter measures sea state characteristics using emitted infrared radiation. Proc. 11th Int. Symp. Remote Sensing, ERIM.

Egan, W. G., and Hilgeman, T. (1971) Optical properties of naturally occurring rhodochrosite between 0.33 μm and 2.5 μm. *Appl. Opt.* 10:2132–2136.

Egan, W. G., and Hilgeman, T. (1972) Fundamental optical properties of the Bruderheim meteorite. *Bull. A.A.S.* 4:425.

Egan, W. G., and Hilgeman, T. (1973) Optical constants for terrestrial analogs of lunar materials. *Astronom. J.* 78:799–804.

Egan, W. G., and Hilgeman, T. (1975) The interstellar medium: UV complex index of refraction of several candidate materials. *Astron. J.* 80:587–594.

Egan, W. G., and Hilgeman, T. (1976) Retroreflectance measurements of photometric standards and coatings. *Appl. Opt.* 15:1845–1849.

Egan, W. G., and Hilgeman, T. (1978) Spectral reflectance of particulate materials: A Monte Carlo model including asperity scattering. *Appl. Opt.* 17:245–252.

Egan, W. G., and Hilgeman, T. (1979) *Optical Properties of Inhomogeneous Materials*. New York: Academic Press.

Egan, W. G., and Hilgeman, T. (1980) Anomalous refractive index of submicron-sized particulates. *Appl. Opt.* 19:3724–3727.

Egan, W. G., Hilgeman, T., and Pang, K. (1975) Ultraviolet complex refractive index of Martian dust: Laboratory measurements of terrestrial analogs. *Icarus* 25:344–355.

Egan, W. G., Hilgeman, T., and Reichman, J. (1973) Determination of absorption and scattering coefficients for nonhomogeneous media. 2: Experiment. *Appl. Opt.* 12:1816–1823.

Egan, W. G., Hilgeman, T., and Smith, L. L. (1978) Comparison between infrared Martian disk spectra and optical properties of terrestrial analogs. *Icarus* 35:209–226.

Egan, W. G., and Nowatzki, E. A. (1966) A porosity polarization relationship applicable to the lunar surface. Paper presented to the 123rd Meeting Am. Astronom. Soc., University of California, December 27.

Egan, W. G., and Nowatzki, E. A. (1967) An investigation of the relationship between polarization and surface structure of rhyolytic pumice. Paper presented at 48th Ann. Meeting Am. Geophys. Union, April 18.

Egan, W. G., and Nowatzki, E. A. (1967) An investigation of the relationship between polarization and surface structure of rhyolytic pumice. Grumman Res. Dept. Memorandum, RM-370J; *Trans. A.G.U.* 48:153.

Egan, W. G., and Nowatzki, E. A. (1968) A porosity–polarization relationship applicable to the lunar surface. *Astron. J.* 73:57–59.

Egan, W. G., and Selby, J. E. A. (1980) Atmospheric radiative transfer effects from Mt. St. Helens eruption ash. Presented at 1980 Int. Rad. Symp. August 11–16, Colorado State University, Extended Abstracts.

Egan, W. G., and Shaw, G. E. (1981) Effect of aerosols on optical remotely sensed data. Proc. 15th Int. Symp. Remote Sensing, ERIM, 959–973.

Egan, W. G., and Smith, L. L., (1965) Polarimetric measurements of simulated lunar surfaces. Grumman Res. Dept. Memorandum, RM-304.

Egan, W. G., and Smith, L. L. (1966) Polarimetric measurements of simulated lunar surfaces. *Astronom. J.* 71:383.

Egan, W. G., and Smith, L. L. (1966) Polarimetric measurements of simulated lunar surfaces (Phase II). Grumman Res. Dept. Memorandum, RM-312.

Egan, W. G., and Smith, L. L. (1966) Polarimetric measurements of simulated lunar surfaces (Phase III). Grumman Res. Dept. Memorandum, RM-315.

Egan, W. G., Smith, L. L., and McCoyd, G. C. (1966) Polarimetric measurements of simulated lunar surfaces. Grumman Res. Dept. Rep., RE-250.

Egan, W. G., Veverka, J., Noland, M., and Hilgeman, T. (1973) Photometric and polarimetric properties of the Bruderheim chondritic meteorite. *Icarus* 19:358–371.

Eldridge, R. G. (1966) Climatic visibilities of the United States. *J. Appl. Meteorol.* 5:277–282.

Electronuclear Laboratories (1968) *Bull.* 1078.

Ellis, H. T., and Pueschel, R. F. (1971) Solar radiation: absence of air pollution and trends at Mauna Loa. *Science* 172:845–846.

Engel, J. L. (1980) Thematic mapper—an interim report on anticipated performance. Santa Barbara Research Center Rep., Goleta, California.

Engel, J. L., and Weinstein, O. (1982) The thematic mapper—an overview. *Digest*, 1982 Int. Geosci. Remote Sensing Symp. (IGARSS '82), June 1–4, Munich.

Engstrom, R. W. (1980) RCA Photomultiplier Handbook PMT-62. RCA Solid State Division, Electro Optics and Devices, Lancaster, Pa.

Estes, J. E. (1974) *Imaging with Photographic and Nonphotographic Sensor Systems in Remote Sensing: Techniques for Environmental Analysis* (J. E. Estes and L. W. Senger, eds.). Santa Barbara: Hamilton Publishing.

Farmer, C. B., Davies, D. W., and La Porte, D. D. (1976) Mars: northern summer ice cap water vapor observations from Viking 2. *Science* 194:1339–1341.

Fassett, N. C. (1957) *A Manual of Aquatic Plants*. Madison, Wi.: University of Wisconsin Press.

Felgett, P. B. (1949) Threshold sensitivity and noise ratings of multiplier phototubes. *Appl. Opt.* 6:251–255.

Fernald, M. L. (1950) *Gray's Manual of Botany*, 8th Centennial Ed. New York: American Book Company.

Fesenkov, V. G. (1961) The twilight method in the study of the optical properties of the atmosphere, in scattering and polarization of light in the atmosphere; translated from Russian. Israel Program for Scientific Translations, Jerusalem, 1965.

Fesenkov, V. G. (1962) Photometry of the Moon. *Physics and Astronomy of the Moon* (Z. Kopal, ed.). New York: Academic Press.

Fish, F. F., Jr. (1966) The stability of goethite on Mars. *J. Geophys. Res.* 71:3063–3068.

Fitch, B. W. (1981) Effects of reflection by natural surfaces on the radiation emerging from the top of the earth's atmosphere. *J. Atmos. Sci.* 38:2717–2729.

Fitch, B. W., and Coulson, K. L. (1983) Effects on skylight at South Pole Station, Antarctica, by ice crystal precipitation in the atmosphere. *Appl. Opt.* 22:71–74.

Flowers, E. C., McCormick, R. A., and Kurfis, K. R. (1969) Atmospheric turbidity over the United States, 1961–1966. *J. Appl. Meteor.* 8:955–962.

Fournier, R. O. (1966) North Atlantic deep sea fertility. *Science* 153:1250–1252.

Fraiture, T. (1965) Scattering of electromagnetic waves from rough surfaces. Annales de la Societé Scientifique de Bruxelles 7911:144.

Fritz, N. L. (1967) Optimum methods for using infrared sensitive color film. Presented at 33rd Ann. Meeting, Am. Soc. Photogrammetry, March 10.

Fung, A. K. (1966) Scattering and depolarization of EM waves from a rough surface. *Proc. IEEE* 54:395–396.

Fymat, A. L. (1972) Interferometric spectropolarimetry-alternate experimental methods. *Applied Optics* 11:2255–2264.

Fymat, A. L. (1979) High resolution spectropolarimetry: A new atmospheric remote sensor. *JPL Atmos. Publ.* No. 78-08.

Fymat, A. L. (1981) High resolution interferometry: Spectrophotopolarimetry. *Optic. Eng.* 20:25-30.

Gambling, D. J. (1975) Sun glitter on the surface of the ocean in the infrared spectral region. *Infrared Phys.* 15:149–155.

Gausman, H. W., and Cardenas, R. (1968) Effect of pubescence on reflectance of light. Proc. 5th Symp. Remote Sensing, Univ. Michigan.

G. E. Co. Z-7803S One Inch FPS Epicon. Bulletin TPD-8482B, Microwave and Imaging Devices Products Section, Owensboro, Ky.

Gehrels, T. (ed.) (1974) *Planets, Stars, and Nebulae Studied With Photopolarimetry*. Tucson, Az.: Univ. of Arizona Press.

Gehrels, T., Coffeen, T., and Owings, D. (1964) Wavelength dependence of polarization, III, The lunar surface. *Astronom. J.* 69:826–852.

Gehrels, T., Roemer, E., Taylor, E. C., and Zellner, B. H. (1970) Minor planets and related objects. IV. Asteroid (1566) Icarus. *Astron. J.* 75:186–195.

Gehrels, T., and Teska, T. M. (1960) A Wollaston photometer. *Publ. Astronom. Soc. Pacific.* 72:115–122.

Geller, R., and Johanson, N. (1967) 1965 NASA–Goddard Round Robin Results. Grumman Aircraft Eng. Corp. Thermal Lab, Memorandum TLM-67-19, 22 August 1967.

Gerber, H. E., and Hindman, E. E. (1982) *Light Absorption by Aerosol Particles*. Hampton: Spectrum Press.

Gilbert, G. D., and Pernicka, J. C. (1967) Improvement of underwater visibility by reduction of backscatter with a circular polarization technique. *Appl. Opt.* 6:741–746.

Gillett, F. C., and Forrest, W. J. (1973) Spectra of the Becklin-Neugebauer point source and the Kleinmann-Low nebula from 2.8 to 13.5 microns. *Ap. J.* 179:483–491.

Gold, T. (1958) In Dust on the Moon. *Vistas in Astronautics*. New York: Pergamon Press.

Goldman, S. (1948) *Frequency Analysis, Modulation and Noise*. New York: McGraw-Hill.

Goldman, S. (1953) *Information Theory*. New York: Prentice-Hall.

Gordon, J. I., and Church, P. V. (1966) Overcast sky luminances and directional luminous reflectances of objects and backgrounds under overcast skies. *Appl. Opt.* 5:919–923.

Gordon, J. I., Harris, J. L., Jr., and Duntley, S. Q. (1973) Measuring earth-to-space contrast transmittance from ground stations. *Appl. Opt.* 12:1317–1324.

Greenberg, J. M. (1973a) On the origin of the Solar System. Proc. of Symposium, Nice, France, Centre National de la Recherche Scientifique, 1973, 135.

Greenberg, J. M. (1973b) *Molecules in the Galactic Environment.* (M. A. Gordon and L. E. Snyder, eds.). New York: Wiley.

Greenberg, J. M. (1973c) Some scattering problems of interstellar grains, in *Interstellar Dust and Related Topics*, IAU Symp. No. 52, (J. M. Greenberg and H. C. Van de Hulst, eds.). Boston: D. Reidel Publishing Co., 3–9.

Greenberg, J. M., and Hong, S. (1973) The chemical composition and distribution of interstellar grains. IAU Symp. No. 60, Maroochydore, Australia, September.

Greenberg, J. M., and Van de Hulst, H. C. (1973) Interstellar dust and related topics. IAU Symp. No. 52. Boston: D. Reidel Publishing Co.

Greenstein, J. L. (1938) The theory of interstellar absorption. *Harvard Obs. Circ. No. 442.*

Griggs, M. (1973) A method to measure the atmospheric aerosol content using ERTS-1 data. 3rd ERTS Symp. December 10–14.

Guinn, J. A., Jr., Plass, G. N., and Kattawar, G. W. (1979) Sunlight glitters on a wind-ruffled sea: further studies. *Appl. Opt.* 18:842–849.

Hair, M. E. (1970) An Experimental Survey of a Wetlands Study Site. Prepared for the State of Maryland, Department of Natural Resources by Grumman Ecosystems Corporation, December 29.

Halajian, J. D. (1964a) Photometric measurements of simulated lunar surfaces. Grumman Res. Dept. Memorandum, RM-250.

Halajian, J. D. (1964b) The case for a cohesive lunar surface model. Grumman Rep. ADR-04-04-62.2. Presented at the N. Y. Academy of Sciences Meeting on Geological Problems in Lunar Research.

Halajian, J. D. (1964c) Photometric measurements of simulated lunar surfaces. Grumman Res. Dept. Memorandum, RM-250.

Halajian, J. D. (1965) Photometric measurements of simulated lunar surfaces. Grumman Res. Dept. Rep., RE-219.

Halajian, J. D. (1965) Photometric measurements of simulated lunar surfaces. Grumman Res. Dept. Memorandum, RM-262.

Halajian, J. D. (1965) The case for a cohesive lunar surface model. *Ann. N.Y. Acad. Sci.* 123:671–710.

Halajian, J. D., and Spagnolo, F. A. (1966) Photometric measurements of simulated lunar surfaces. Grumman Res. Dept. Memorandum, RM-308.

Halajian, J. D., and Spagnolo, F. A. (1966) Photometric measurements of simulated lunar surfaces. Grumman Res. Dept. Rep., RE-245.

Halajian, J. D., and Reichman, J. (1966) Correlation of mechanical and thermal properties of extraterrestrial materials. Grumman Res. Dept. Memorandum, RM-309.

Halajian, J. D., and Hallock, H. B. (1968) Uses of temporal data in remote sensing. Advanced Systems Space Sciences Rep. No. SAS-425-6.

Hale, G. M, and Querry, M. R. (1973) Optical constants of water in the 200-nm to 200-μm wavelength range. *Appl. Opt.* 12:555–563.

Hall, F. F., Jr. (1964) The polarized emissivity of water in the infrared. *Appl. Opt.* 3:781–782.

Hall, J. S. (1949) Observations of the polarized light from stars. *Science* 109:166–167.

Handbook of Chemistry and Physics, 32nd Ed. (1950–1951) Cincinnati: Chemical Rubber Publ. Co.

Hanel, G. (1968) The real part of the mean complex index of refraction and the mean density of samples of atmospheric aerosol particles. *Tellus* 20:371–379.

Hansen, D. V. (1970) Gulf Stream meanders between Cape Hatteras and the Grand Banks. *Deep-Sea Research* 17:495–511.

Hanson, J. D. (1969) Survey of annosus root-rot-caused mortality in thinned short leaf pine plantations using infrared film. Proc. Symp. Aerial Color Photography Plant Sci., Univ. Florida.

Hapke, B. W. (1963) A theoretical photometric function for the lunar surface. Center for Radiophysics and Space Research.

Hapke, B. W. (1963) A theoretical photometric function for the lunar surface. *J. Geophys. Res.* 68:4571–4586.

Hapke, B. W. (1964) Effects of a simulated solar wind on the photometric properties of rocks and powders. Center for Radiophysics and Space Research, Cornell University, Rep. CRSR 169.

Hapke, B. W. (1965) Some comments on Gehrels' model of the lunar surface. Center for Radiophysics and Space Research, Cornell University, Rep. CRSR 199.

Hapke, B. W. (1966a) An Improved theoretical lunar photometric function. 121st Am. Astronom. Soc. Meeting, March 28–31.

Hapke, B. W. (1966b) An improved theoretical lunar photometric function. *Astron. J.* 73:333–339.

Hapke, B. W., and Van Horn, H. (1962) Photometric studies of complex surfaces with applications to the Moon. Center for Radiophysics and Space Research, Cornell University, Publication No. 127.

Hapke, B. W., and Van Horn, H. (1963) Photometric studies of complex surfaces, with applications to the Moon. *J. Geophys. Res.* 68:4545–4570.

Harder, J. W., Rosen, J. M., and Hoffman, D. J. (1983) Measurement of the sulfuric acid weight percent in the stratospheric aerosol from the El Chichón eruption. *Trans. AGU* 64:197.

Harlow, W. M., and Harrar, E. S. (1968) *Textbook of Dendrology*. New York: McGraw-Hill.

Harris, F. S., Jr., Sherman, G. C., and Morse, F. L. (1967) Experimental comparison of scattering of coherent and incoherent light. *IEEE Trans. Antennas Propag.* AP-15:141–147.

Hartwell, J. H. (1969) A remote sensing method to automatically classify and delineate plant communities. Symp. Aerial Color Photography Plant Sci. Proc., Univ. Florida.

Heller, R. C. (1968) Previsual detection of ponderosa pine trees dying from bark beetle attack. Proc. 5th Symp. Remote Sensing, Univ. Michigan, April 16–18.

Heller, R. C., Doverspike, G. E., and Aldrich, R. C. (1964) Identification of tree species on large-scale panchromatic and color aerial photographs. Handbook No. 261, U.S. Department of Agriculture. Washington, D.C.: U.S. Government Printing Office.

Hemphill, W. (1969) Fraunhofer line discriminator. Presented at Symp. Oceanogr. Space Using Visible Region Sensors, December 16. Office of Naval Research, Washington, D.C.

Henderson, S. T., and Hodgkiss, D. (1964) The spectral energy distribution of daylight, *Brit. J. Appl. Phys.* 15:947–952.

Herbert, G. A. (ed.) (1979) Geophysical monitoring for climatic change, No. 8. U.S. Dept. Commerce, NOAA, ERL, Boulder, Colo.

Herschel, F. J. W. (1847) *Results of Astronomical Observations Made at the Cape of Good Hope*. London: Proc. R. Soc.

Hiltner, W. A. (1949) Polarization of light from distant stars by Interstellar Medium. *Science* 109:165.

Hiltner, W. A. (ed.) (1962) *Astronomical Techniques*. Chicago: The University of Chicago Press.

Hodge, D. B., and Schultz, F. V. (1965) The Born approximation applied to electromagnetic scattering from a finite cone. Purdue University TR-EE64-18.

Hollinger, J. P. (1971) Remote passive microwave sensing of the ocean surface. Proc. 7th Int. Symp. Remote Sensing, Ann Arbor, Michigan.

Holm-Hansen, O., Lorenzen, C. J., Holmes, R. W., and Strickland, J. D. H. (1965). Fluorometer determination of chlorophyll. *J. Cons. Perm. Int. Explor. Mer.* 30:3–15.

Hopfield, J. J. (1966) Mechanism of lunar polarization. *Science* 151:1380–1381.

Hopfield, R. F. (1967) A comment on the scattering of coherent light. *Appl. Opt.* 6:170–171.

Hovis, W. A., Jr. (1965) Infrared reflectivity of $Fe_2O_3 \cdot nH_2O$: Influence on Martian reflection spectra. *Icarus* 4:41–42.

Huck, F. O., Jobson, D. J., Park, S. K., Wall, S. D., Arvidson, R. E., Patterson, W. R., and Benton, W. B. (1977) Spectrophotometric and color estimates of the Viking Lander sites. *J. Geophys. Res.* 28:4401–4411.

Huffman, P. J., Luers, J. K., Engler, N. A., and Felt, J. E. (1983) Proc. Workshop Standardize Atmospheric Measurements in Support of Electro-Optical Systems. Dayton: University of Dayton Research Institute.

Hufnagel, R. E., and Stanley, N. R. (1964) Modulation transfer function associated with image transmission through turbulent media. *J. Opt. Soc. Am.* 54:52–61.

Hunt, G. R., and Salisbury, J. W. (1970) Visible and near-infrared spectra of minerals and rocks: I. Silicate minerals. *Mod. Geol.* 1:283–300.

Hurlbut, S., Jr. (1966) *Dana's Manual of Mineralogy*, 17th Ed. New York: John Wiley.

Industrial Optics/3M (1978) 3M Polacoat Polarizing Filters. Industrial Optics Bulletin. St. Paul, Minn.

Infrared Components Book (1983) Edition 17. Santa Barbara Research Center, Goleta, California.

Irvine, W. M. (1966) The shadowing effect in diffuse reflection. *J. Geophys. Res.* 71:2931–2937.

Irvine, W. M., Pikoos, C., Charon, J., and Lecomte, G. (1964) Effect of high voltage on spectral sensitivity for two photomultipliers. *Astrophys. J.* 140:1629–1631.

Irvine, W. M., and Pollack, J. B. (1968) Infrared optical properties of water and ice spheres. *Icarus* 8:324–360.

Ivanov, A. P., and Toporets, A. S. (1957) Investigation of diffuse reflection by means of polarized light. *Sov. Phys.—Tech. Phys.* 1:598–605.

Jacobs, M. B., and Ewing, M. (1969) Mineral source and transport in waters of the Gulf of Mexico and Caribbean Sea. *Science* 163:805–809.

Jacobs, S. F., and Sargent, M. III. (1970) Photon noise limited D* for low temperature backgrounds and long wavelengths. *Infrared Phys.* 10:233–235.

Jenkins, F. A., and White, H. E. (1957) *Principles of Optics*. New York: McGraw-Hill.

Jerlov, N. G. (1968) *Optical Oceanography*. New York: Elsevier.

Johnson, C. B. (1980) Photoelectronic sensor development projections: 1980-2000. *Huntsville EO Tech. Symp. SPIE* 285:53–63.

Johnson, C. B. (1982) Intensified self-scanned array technology review. *IEEE Trans. Nucl. Sci.* NS-29:200–206.

Johnson, C. B., and Blank, R. E. (1980) Digital and low-light-level SSA instrument and TV camera sub-systems review. Military Electronics Defense Expo. '80:598–609. Interavia, S. A.

Johnson, C. B., and Blank, R. E. (1981a) An image intensified linear self-scanned array instrument. Los Alamos Conf. Opt. 288:447–453.

Johnson, C. B., and Blank, R. E. (1981b) Image tube intensified linear and area self-scanned array detectors for astronomy. *Solid State Imagers for Astronomy SPIE* 290-31.

Johnson, C. B., Lynch, T. F., and Cuny, J. J. (1979) Microchannel plate wafer image tube intensified charge-injection device cameras. 5th Int. Conf. Charge Coupled Devices, University of Edinburgh.

Johnson, C. B., Nevin, S., Bebris, J., and Abshire, J. B. (1980) Circular-scan streak tube with solid-state readout. *Appl. Opt.* 19:3491–3495.

Johnson, H. L., and Morgan, W. W. (1951) On the color magnitude diagram of the Pleiades. *Ap. J.* 114:522–543.

Johnson, H. L., and Morgan, W. W. (1953) Fundamental stellar photometry for standards of spectral type on the revised system of the Yerkes Spectral Atlas. *Ap. J.* 117:313–352.

Judd, D. B., and Kelley, K. L. (1939) Method of designating colors. *J. Res. Natl. Bur. Stand.* 23:355–385.

Junge, C. E. (1963) *Air Chemistry and Radioactivity*. New York: Academic Press.

Kalle, K. (1938) Zum Problem der Meereswasserfarbe. *Ann. Hydrol. Marine Mitt.* 66:1–13.

Kalle, K. (1962) Über die gelösten organischen Komponenten im Meerwasser. *Kieler Meeresforsch.* 18:128–131.

Kalle, K. (1966) The problem of the Gelbstoff in the sea. *Oceanogr. Marine Biol. Ann. Rev.* 4:91–104.

Kamijo, F. (1963a) A theoretical study on the long period variable star, Part III: Formation of solid or liquid particles in the circumstellar envelope. *Publ. Astron. Soc. Japan* 15:440–448.

Kamijo, F., and de Jong, T. (1963b) Infrared 10 μ emission from condensation nuclei of interstellar grains. *Astron. & Astrophysics* 25:371–377.

Kattawar, G. W., and Plass, G. N. (1967) Electromagnetic scattering from absorbing spheres. *Appl. Opt.* 6:1377–1382.

Keil, K., and Fredriksson, K. (1964) The iron, magnesium and calcium distribution in coexisting olivines and rhombic pyroxenes of chondrites, *J. Geophys. Res.* 69:3487–3515.

Kemp, J. C. (1973) Interstellar circular polarization of Upper Scorpius Stars, in *Interstellar Dust and Related Topics*, IAU Symp. No. 52, J. M. Greenberg and H. C. Van de Hulst (eds.). Boston: D. Reidel Publishing Co., 181–185.

Ken Knight, E. C., Rosenberg, D. L., and Wehner, G. K. (1967) Parameters of the optical properties of the lunar surface powder in relation to solar wind bombardment. *J. Geophys. Res.* 72:3105–3129.

Kerker, M. (1969) *The Scattering of Light*. New York: Academic Press.

Kislovskii, L. D. (1959) Optical characteristics of water and ice in the infrared and radiowave regions of the spectrum. *Opt. Spectrosc.* 7:201.

Kneizys, F. X., Shettle, E. P., Gallery, W. O., Chetwynd, J. H., Jr., Abreu, L. W., Selby, J. E. A., Fenn, R. W., and McClatchey, R. A. (1980) Atmospheric Transmittance/Radiance: Computer Code LOWTRAN 5. AFGL-TR-80-0067. Environmental Research Paper No. 697, Bedford, Ma.

Kodis, R. D. (1966) A note on the theory of scattering from an irregular surface. *IEEE Trans.* AP-14:77–82.

Kondratiev, K. I., and Manolova, M. P. (1955) The problem of incoming diffuse and total radiation on the surface of a slope. *Meteorol. Gidrol* 6:31–34.

Krinov, E. L. (1953) Spectral reflectance properties of natural surfaces. Tech. Transl. TT-439, 1947, Ottawa: National Research Council of Canada.

Krishen, K. (1971) Mathematical model for the relationship of radar backscattering cross sections with ocean scene and wind velocity. Proc. 7th Int. Symp. Remote Sensing, Ann Arbor, Mi.

Kruse, P. W., McGlaughlin, L. D., and McQuistan, R. B. (1963) Elements of infrared technology. New York: Wiley.

Kuiper, G. P. (1964) Infrared spectra of stars and planets IV: The spectrum of Mars, 1-2.5 microns, and the structure of its atmosphere. *Commun. Lunar Planet. Lab.* 2:79–112.

Landaberry, S. C., and Magalhaes, A. M. (1977) Magnetite and the interstellar medium. *An. Acad. Bras. Cienc.* (Brazil) 48(2):199–204.

Landsberg, H. E. (1970) Man made climate changes. *Science* 170: 1265–1274.

LARS (1967a) Laboratory for Agricultural Remote Sensing, Purdue University Agricultural Experiment Station, Lafayette, Ind. Res. Bull. No. 831.

LARS (1967b) Laboratory for Agricultural Remote Sensing, Purdue University Agricultural Experiment Station, Lafayette, Ind. Res. Bull. No. 832.

LARS (1968) Remote multispectral sensing in agriculture. Laboratory for Agricultural Remote Sensing, Purdue University Agricultural Experiment Station, Lafayette, Ind. Res. Bull. No. 844, Vol. 3, Ann. Rep.

Leovy, C. (1966) Note on thermal property of Mars. *Icarus* 5:1–6.

Lillesand, T. M., and Kiefer, R. W. (1978) *Remote Sensing and Image Interpretation*. New York: Wiley.

Lindberg, J. D., and Laude, L. S. (1974) Measurement of the absorption coefficient of atmospheric dust. *Appl. Opt.* 13:1923–1927.

Lindblad, B. (1935) A condensation theory of meteoritic matter and its cosmological significance. *Nature* 135:133–135.

Linz, J., Jr., and Simonett, D. S. (1976) *Remote Sensing of Environment.* Reading, Ma.: Addison-Wesley.

Lorenzen, C. J. (1966) A method for the continuous measurement of *in vivo* chlorophyll concentration. *Deep-Sea Res.* 13:223–227.

Lougheed, R. M., McCubbrey, D. L., and Sternberg, S. (1980) Cytocomputers™: Architectures for Parallel Image Processing. IEEE Workshop on Picture Data Description and Management.

Lowan, A. N. (1948) Tables of scattering functions for spherical particles. *Natl. Bur. Stand. (U.S.) Appl. Math. Ser.* 4.

Lowrance, J., and Zucchino, P. (1979) Intensified charge coupled device (CCD) single photoelectron response. *SPIE* 172: *Instrumentation in Astronomy.*

Ludwig, F. L., and Robinson, E. (1970) Observations of aerosols and droplets in California stratus. *Tellus* 22:94–105.

Lundgren, D. A., Harris, F. S., Jr., Marlow, W. H., Lippmann, M., Clark, W. E., and Durham, M. D. (1979) *Aerosol Measurement.* Gainesville: University Press of Florida.

Lutomirski, R. F., and Yura, H. T. (1971) Propagation of a Finite Optical Beam in an Inhomogeneous Medium. *Appl. Opt.* 10:1652–1658.

Lyot, B. (1929) Research on the polarization of light from planets and some terrestrial substances. *Annales de l'Observatoire de Paris,* Section de Meudon, VIII, No. 1. Also NASA Tech. Transl. TTF-187.

Mairs, R. L. (1970) Oceanographic interpretation of Apollo photographs. *Photogramm. Eng.* 36:1045–1058.

Majkowski, R. F. (1977) Infrared absorption coefficient of H_2SO_4 vapor from 1190 to 1260 cm^{-1}. *J. Opt. Soc. Am.* 67:624–627.

Manabe, S., and Weatherald, R. T. (1967) Thermal equilibrium in the atmosphere with a given distribution of relative humidity. *J. Atmos. Sci.* 24:241–259.

Markham, B. L., and Townshend, J. R. G. (1981) Land cover classification accuracy as a function of sensor spatial resolution. Proc. 15th Int. Symp. Remote Sensing, Ann Arbor, Mich.

Maryland State Planning Department (1970) Volume II Draft Report—Wetlands in Maryland. State Office Building, Baltimore, Md. U.S. Dept. Interior, Fish and Wildlife Service, Bur. Sport Fisheries and Wildlife and Bur. Commercial Fisheries. *National Estuary Study, Volume 3, Chesapeake Bay, MD and VA.*

Mason, B. (1962) *Meteorites,* New York: Wiley.

Matthews, W. H., Kellogg, W. W., and Robinson, G. D. (eds.) (1971) Report of Study of Critical Environmental Problems (SCEP). *Man's Impact on the Climate.* Cambridge, Ma.: MIT Press.

Martin, P. G. (1973a) Interstellar circular polarization and the composition of interstellar dust, in *Interstellar Dust and Related Topics,* IAU Symp. No. 52, (J. M. Greenberg and H. C. Van de Hulst, eds.). Boston: D. Reidel Publishing Co., 161–167.

Martin, P. G., Illing, R., and Angel, J. R. P. (1973b) Discovery of interstellar circular polarization in the direction of the Crab Nebula, in *Interstellar Dust and Related Topics,* IAU Symp. No. 52, (J. M. Greenberg and H. C. Van de Hulst, eds.). Boston: D. Reidel Publishing Co., 169–179.

Maxwell, J. R. (1978) Statistical analyses of selected terrain and water background measurements data. ERIM, Ann Arbor, Rep. No. 132300-1-F.

McCamy, C. S. (1968) A half century of photographic standardization in the United States. *Photogr. Sci. Eng.* 12:308–312.

McClatchey, R. A., Benedict, W. S., Clough, S. A., Burch, D. E., Calfee, R. F., Fox, K., Rothman, L. S., and Garing, J. S. (1973) Air Force Cambridge Research Laboratories

Atmospheric Absorption Line Parameters Compilation AFCRL-TR-73-0096. Available on magnetic tape from U.S. Dept. Commerce, National Climatic Center.

McClatchey, R. A., Fenn, R. W., Selby, J. E. A., Volz, F. E., and Garing, J. S. (1972) Optical properties of the atmosphere. Environ. Res. Papers No. 411, AFCRL-72-0497, 24 August. Air Force Cambridge Laboratories, Bedford, Mass.

McCord, T. B., Adams, J. B., and Johnson, T. V. (1970) Asteroid Vesta: spectral reflectivity and compositional implications. *Science* 168:1445–1447.

McCormick, R. A., and Ludwig, J. H. (1967) Climate modification by atmospheric aerosols. *Science* 156:1358–1359.

Meinel, A. P., and Meinel, A. B. (1963) Late twilight glow of the ash stratum from the eruption of Agung Volcano. *Science* 142:582–583.

Michan, G. J., and Burke, H. K. (1976) Charge injection devices for solid state imaging. *Solid State Imaging*. Boston: Noordhoff Publishing.

Middleton, D. (1960) *An Introduction to Statistical Communication Theory*. New York: McGraw-Hill.

Millar, C. E., and Turk, F. M. (1943) *Fundamentals of Soil Science*, New York: Wiley.

Millikan, R. A. (1895) A study of the polarization of the light emitted by incandescent solid and liquid surfaces. *Phys. Rev.* 3:177.

Minnaert, M. (1962) Planets and satellites. Vol. III, *The Solar System* (G. P. Kuiper and B. M. Middlehurst, eds.). Chicago: Univ. of Chicago Press.

Mitchell, J. M., Jr. (1961) Recent secular changes of global temperatures. *Ann. NY Acad. Sci.* 95–235.

Mitzner, K. M. (1966) Change in polarization on reflection from a tilted plane. *Radio Sci.* 1:27–29.

Monmonier, M. (1982) *Computer-assisted Cartography*. Englewood Cliffs: Prentice-Hall.

Moroz, V. I. (1964) The infrared spectrum of Mars ($\lambda 1.1 - 4.1\mu$). *Astron. Zh.* 41:350–361.

Mukai, Y., and Takeuchi, S. (1979) Estimation of primary production of vegetation in agricultural and forested areas using Landsat data. Proc. 13th Int. Symp. Remote Sensing, Ann Arbor, Mich.

Mumford, F. A., and Thompson, C. S. (1966) The stability of brucite in the weathering zone of the New Idria serpentinite. *Clays and Clay Minerals*. Proc. 14th Natl. Conf., Oxford: Pergamon Press.

Myers, V. I., and Allen, W. A. (1968) Electrooptical remote sensing methods as nondestructive testing and measuring techniques in agriculture. *Appl. Opt.* 7:1819–1838.

Nagaraja Rao, C. R. (1964) Polarization and albedo. *Nature* 204:277–278.

NASA (1964) Ranger VII photographs of the Moon, Part I (Camera A Series). NASA SP-61.

NASA (1965a) Ranger VII photographs of the Moon, Part II (Camera B Series). NASA SP-62.

NASA (1965b) Ranger VII photographs of the Moon, Part III (Camera P Series). NASA SP-63.

NASA Landsat Data Users Notes (1982) No. 23, U.S. Geological Survey, EROS Data Center, Sioux Falls, S. Dak.

NASA (1970) Tech. Brief 67-10005 and 67-10630. Image processing system. JPL Invention Rep. 30-792, January 30. Jet Propulsion Laboratory, California Institute of Technology, Pasadena, Ca.

National Research Council (1927) *International Critical Tables*. (E. W. Washburn, ed.). New York: McGraw-Hill.

National Research Council (1977) Resource sensing from space: Prospects for developing countries. Committee on Remote Sensing for Development, U.S. Dept. Commerce, Natl. Tech. Information Service, Document No. PB-264-171.

NCAR (1980), Annual Report, Fiscal Year 1980, National Center for Atmospheric Research (NCAR/AR-80, March 1981). Boulder, Co.: University Corporation for Atmospheric Research, National Center for Atmospheric Research.

Neblette, C. B. (1962) *Photography Its Materials and Processes*, 6th Ed. Princeton, N.J.: D. Van Nostrand.

Neto, A. D., and DeFreitas Pacheco, J. A. (1982) Infrared excess and line emission in stars. *Mon. Not. R. Astron. Soc.* (GB) 198(2):659–668.

Neumann, G., and Pierson, W. J., Jr. (1966) *Principles of Physical Oceanography*. Englewood Cliffs, N.J.: Prentice-Hall.

New York Times (1966) Moscow Releases 2 of Its Moon Photos. 6 February.

Newell, R. E. (1970) Stratospheric temperature change from the Mt. Agung volcanic eruption of 1963. *J. Atmos. Sci.* 27:977–978.

NOAA and USGS (1983) Landsat-4 Users Handbook. EROS Data Center, Sioux Falls, S. Dak.

O'Leary, B. T. (1967) Mars: Visible and near infrared studies and the composition of the surface. Thesis: Tech. Rep. on NASA Grant NsG101-61, *Space Sciences Laboratory Series 8*, Issue 103, University of California, Berkeley.

Olson, C. E. (1967) Accuracy of land-use interpretation from infrared imagery in the 4.5 to 5.5 micron band. *Ann. Assoc. Am. Geogr.* 57:382–388.

O'Neill, N. T., Miller, J. R., and Ahern, F. J. (1978) Radiative transfer calculations for remote sensing applications. Proc. 5th Canadian Symp. Remote Sensing.

Orlova, N. S. (1952) Radial diagrams of scattering for several materials. *Trudy Astron. Obs.* 16:166–193.

Orlova, N. S. (1962) Selected articles on light scattering and photometric relief of the lunar surface. *Astron. Zh.* 33, No. 1; NASA TTF-75.

Palmer, K. F., and Williams, D. (1975) Optical constants of sulfuric acid; application to the clouds of Venus? *Appl. Opt.* 14:208–219.

Pang, K., Ajello, J. M., Hord, C. W., and Egan, W. G. (1976) Complex refractive index of Martian dust. Mariner 9 Observations. *Icarus* 27:55–67.

Patterson, E. M., Gillette, D. A., and Stockton, B. H. (1977) Complex index of refraction between 300 and 700 nm for Saharan aerosols. *J. Geophys. Res.* 82:3153–3160.

Patterson, J., and Parsons, T. R. (1963) Distribution of chlorophyll *a* and degradation products in various marine materials. *Limnol. Oceanogr.* 8:355–356.

Pearse, C. A. (1963) *Photometry and Polarimetry of the Moon and Their Relationship to Physical Properties of the Lunar Surface*. Washington, D.C.: Bellcomm, Inc.

Perkin–Elmer (1973) Wire grid polarizers and mounting kits for infrared spectrophotometers. Bull. L-82.

Philips, A. G. D., and Hayes, D. S. (1975) Multicolor photometry and the theoretical HR diagram. Dudley Observatory Rep. No. 9.

Plass, G. N., Kattawar, G. W., and Guinn, J. A., Jr. (1975) Radiative transfer in the earth's atmosphere and ocean: influence of ocean waves. *Appl. Opt.* 14:1924–1936.

Plass, G. N., Kattawar, G. W., and Guinn, J. A., Jr. (1976) Radiance distribution over a ruffled sea: contributions from glitter, sky and ocean. *Appl. Opt.* 15:3161–3165.

Polaroid Corporation (1981) Polarizing filters. Spectral photometric data. Bull. TP115.

Polaroid Corporation (1982) Polaroid type HNP′B. Bull. TP 119A.

Pollack, J. B., Colburn, D. S., Flasar, F. M., Kahn, R., Carlston, C. E., and Pidek, D. (1979) Properties and effects of dust particles suspended in the Martian atmosphere. *J. Geophys. Res.* 84:2929–2945.

Pollack, J. B., and Sagan, C. (1965) The microwave phase effect of Venus. *Icarus* 4:62–103.

Pollack, J. B., and Sagan, C. (1967) Interpretation of surface features. Unpublished talks. Symp. Surface Mars, Goddard Institute for Space Studies, February 3–4.

Potter, J. F., and Mendlowitz (1975) On the determination of Haze levels from Landsat data. Proc. 10th Int. Symp. Remote Sensing, ERIM.

Radio Corporation of America (1963) Phototubes and photocells. Tech. Man. PT-60.

Radio Corporation of America (1966) HB-3, Vol. 3-4. Commercial Engineering, Tube Division, Harrison, N.J.

RCA Imaging Devices IMD-100 (1979) RCA Solid State Division, Electro Optics and Devices, Lancaster, Pa.

RCA Solid State Detectors SSP-100 (1981) RCA Solid State Division, Electro Optics and Devices, Lancaster, Pa.

Rahn, K. A. (1980) On the causes, characteristics and potential environmental effects of aerosols in the arctic atmosphere. Paper presented at the Conf. Arctic Ocean, London, March 11–12. Arctic Committees of Monaco and Royal Geographical Society.

Rasool, S. I., and Schneider, S. H. (1971) Atmospheric carbon dioxide and aerosols: Effect of large increases on global climate. *Science* 173:138–141.

Rateev, M. A., Gorbunova, Z. N., Lisitzyn, A. P., and Nosov, G. L. (1969) The distribution of clay minerals in the oceans. *Sedimentology* 13:21–43.

Raymont, J. E. G. (1963) *Plankton and Productivity in the Oceans.* New York: Pergamon Press.

Rea, D. G., and O'Leary, B. T. (1965) Visible polarization data of Mars. *Nature* 206:1138–1140.

Reeves, R. G. (ed.) (1975) *Manual of Remote Sensing,* Falls Church: American Society of Photogrammetry.

Reisman, E., Cumming, G., and Bartky, C. (1967) Comparison of fog scattered laser and monochromatic incoherent light. *Appl. Opt.* 6:1969–1972.

Remy-Battiau, L. (1962) Mie scattering functions for spherical particles of refractive index $m = 1.25$. Mem. 4° *Soc. R. Sc. Liege* 2:7.

Renau, J., Cheo, P. K., and Cooper, H. G. (1967) Depolarization of linearly polarized EM waves backscattered from rough metals and inhomogeneous dielectrics. *J. Opt. Soc. Am.* 57:459–466.

Richason, B. F., Jr. (1978) *Introduction to Remote Sensing of the Environment.* Dubuque: Kendall–Hunt.

Riley, G. A. (1939) Plankton studies II, The western North Atlantic, May–June 1939. *J. Marine Res.* 2:145–162.

Rogers, R. H., Peacock, K., and Shah, N. J. (1973) A technique for correcting ERTS data for solar and atmospheric effects. Paper 1-7, 3rd ERTS Symp., December.

Romanova, M. A. (1962) Air survey of sand deposits by spectral luminance. Laboratory of Aeromethods of the Academy of Sciences of the USSR; State Scientific Technical Press for Literature on Petroleum and Solid Fuels. Translation by Consultants Bureau, New York, 1964.

Rosen, J. M., and Hoffman, D. J. (1983) Balloon borne particle counter observations of the stratospheric aerosol in the 0.01–1.8 μm radius range following the eruptions of El Chichón. *Trans. AGU* 64:197.

Ruachev, V. P., and Polyanskii, V. K. (1965) Passage of polarized light through a light scattering surface. *Opt. Spectrosc.* 18:594–597.

Rubler, J. (1971) Black Light Eastern Corporation. Private communication.

Ruffine, R. S., and de Wolfe, D. A. (1965) Cross-polarized electromagnetic backscatter from turbulent plasmas. *J. Geophys. Res.* 70:4313–4321.

Russel, F. A. R., and Archibald, E. E. (1888) On the Unusual Optical Phenomena of the Atmosphere 1883–1886, including twilight effects, coronal appearances, sky haze, colored suns, moons, etc.; the eruption of Krakatoa, and subsequent phenomena, G. J. Symons (ed.). London: Royal Society, 151–463.

Rzhiga, O. N. (1967) An estimate of the physical properties of the Martian surface from the results of radio and infrared observations. *Sov. Astron.* 11:113–118.

Sadowski, F. A., and Sarno, J. E. (1976) Forest classification accuracy as influenced by multispectral scanner resolution. Rep. No. 109600-71-F. ERIM.

Sagan, C., Phaneuf, J. P., and Ihnat, M. (1965) Total reflection spectrophotometry and thermogravimetric analysis of simulated Martian surface materials. *Icarus* 4:43–61.

Saijo, Y., and Ichimura, S. (1960) Primary production in the northwestern Pacific Ocean. *J. Oceanogr. Soc. Jpn.* 16:139–145.

Salomonson, V. V., and Marlatt, W. E. (1971) Airborne measurements of reflected solar radiation over soils and vegetation. *Remote Sensing Environ.* 2:1–8.

Salomonson, V. V., Smith, P. L., Jr., Park, A. B., Webb, W. C., and Lynch, T. J. (1980) An overview of progress in the design and implementation of Landsat-D systems. *IEEE Trans. Geosci. Remote Sensing* GE-18:137–146.

Salomonson, V. V., Williams, D. L., and Barker, J. L. (1983) Information expectations from Landsat-D 1981. Proc. 15th Int. Symp. Remote Sensing ERIM 81–93.

Sandus, O. (1965) A review of emission polarization. *Appl. Opt.* 4:1634–1642.

Santa Barbara Research Center, 1972 Brochure.

Schalen, C. (1936) Über probleme der intellstellaren absorption. *Medd. Upsala Astr. Obs.* No. 64.

Schieferstein, R. H., and Loomis, W. E. (1956) Wax deposits on leaf surfaces. *Pl. Physiol.* 31:240–247.

Schmidt, E., and Eckert, E. (1935) Über die richtungsverteilung der wärmestrahlung von oberflächen (Angular distribution of thermal radiation from surfaces). *Forsch. Gebiete Ingenieurw*, 6:175–183.

Schwartz, M. (1959) *Information Transmission, Modulation and Noise.* New York: McGraw-Hill

Seiff, A., and Kirk, D. B. (1976) Structure of Mars atmosphere up to 100 kilometers from the entry measurements of Viking 2. *Science* 194:1300–1303.

Selby, J. E. A., Kneizys, F. X., Chetwynd, J. H., Jr., and McClatchey, R. A. (1977) LOWTRAN. Air Force Geophysical Laboratories Program, 1 December.

Selby, J. E. A., Kneizys, F. X., Chetwynd, J. H., Jr., and McClatchey, R. A. (1978) Atmospheric Transmittance/Radiance Computer Code LOWTRAN 4. AFGL-TR-80-0053. Environmental Res. Paper No. 626, Bedford, Ma.

Selby, J. E. A., Shettle, E. P., and McClatchey, R. A. (1976) Atmospheric Transmittance from 0.25 to 28.5 μm: Supplement LOWTRAN 3B (1976). AFGL-TR-76-0258. Environmental Res. Paper No. 587, Bedford, Ma.

Shapiro, J. (1957) Chemical and biological studies on the yellow organic acids of lake water. *Limnol. Oceanogr.* 2:161–179.

Sharanov, V. V. (1960) *The Moon* Rep. Symp. No. 14, Inst. Astron. Union, Pulkova Obs. New York: Academic Press.

Shaw, G. E. (1979) Inversion of optical scattering and spectral extinction measurements to recover aerosol size spectra. *Appl. Opt.* 18:988–993.

Shaw, G. E. (1980) Remote sensing of aerosol in free atmosphere by passive optical techniques. Rep. 1st Int. Workshop Light Absorption Aerosol Particles, Fort Collins, Colorado State University.

Shaw, G. E., and Deehr, C. S. (1975) A photoelectric coronameter for atmospheric turbidity studies, *J. Appl. Meteorol.* 14:1203–1205.

Shettle, E. P., and Fenn, R. W. (1979) Models for the Aerosols of the Lower Atmosphere and the Effects of Humidity Variations on their Optical Properties. AFGL-TR-790214. Environmental Res. Paper No. 676, Bedford, Ma.

Shurcliff, W. A. (1962) *Polarized Light.* Cambridge: Harvard University Press.

Shurcliff, W. A., and Ballard, S. S. (1964) *Polarized Light.* New York: Van Nostrand.

Singer, S. F. (ed.) (1970) *Global Effects of Environmental Pollution.* Dordrecht, Holland: Reidel Publishing Co.

Smith, L. L., Krassner, J., Egan, W. G., Hilgeman, T. W., and Selby, J. E. A. (1979) A recommended modification of LOWTRAN 4 to include first order solar scattering. SPIE 23rd International Technical Symposium, Paper 195-25.

Smith, R. A., Jones, F. E., and Chasmar, R. P. (1957) *Detection and Measurement of Infrared Radiation.* London: Oxford University Press.

Smoluchowski, M. (1910) Sur la conductibilité calorifique des corps pulvérisés. *Bull. Intern. Acad. Sci. Cracovie* (A):129–153.

Solar Energy Research Institute Review (1983) Remote Sensing Suggested as Insulation and Biomass Resource Assessment Tool. Vol. 2, pp. 1–4.

Sorensen, B. M. (ed.) (1979) Recommendations of the International Workshop on Atmospheric Correction of Satellite Observation of Sea Water Color, Ispra, Italy, March.

Sternberg, S. R. (1979) Parallel Architectures for Image Processing. Proceedings of 3rd International IEEE COMPSAC, Chicago.

Sternberg, S. R. (1980) Language and Architecture for Parallel Image Processing. Proceedings of the Conference on Pattern Recognition in Practice, Amsterdam, The Netherlands.

Stommel, H. (1958) *The Gulf Stream*. London: Cambridge University Press.

Strahler, A. H., and Xiaowen, L. (1981) An invertible coniferous forest canopy reflectance model. Proc. 15th Int. Symp. Remote Sensing, 11–15 May.

Strickland, J. D. H. (1968) A comparison of profiles of nutrient and chlorophyll concentrations taken from discrete depths and by continuous recording. *Limnol. Oceanogr.* 13:391–394.

Strickland, J. D. H., and Parsons, T. R. (1968) A practical handbook of seawater analysis. Fisheries Research Board of Canada, Ottawa.

Strong, A. E., Stowe, L. L., and Walton, C. C. (1983) Using the NOAA-7 AVHRR data to monitor El Chichón aerosol evolution and subsequent sea surface temperature anomalies. Proc. 17th Int. Symp. Remote Sensing, Ann Arbor.

Swain, P. H., and Davis, S. M. (1978) *Remote Sensing: The Quantitative Approach*. New York: McGraw-Hill.

Sytinskaya, N. N. (1959) Probable dimensions of roughness of lunar microtopography. *News Commission Phys. Planets* 1:81–84.

Tandon, S. N., and Wasson, J. T. (1968) Gallium, germanium, indium and iridium variations in a suite of L-group chondrites. *Geochim. Cosmochim. Acta.* 32:1087–1109.

Taylor, J. H., and Yates, H. W. (1956) Naval Research Laboratory Report No. 4759, published also as PB121199, May 11, 1956.

Teicholz, E., and Berry, B. J. L. (1983) *Computer Graphics and Environmental Planning*. Englewood Cliffs: Prentice-Hall.

Thomas, J. R., Myers, V. I., Heilman, M. D., and Wiegard, C. U. (1966) Factors affecting light reflectance of cotton. Proc. 4th Symp. Remote Sensing. Univ. Michigan.

Thomas, J. B., and Flight, W. F. G. (1964). Fluorescence responses of chlorophyll *in vivo* to treatment with acetone. *Biochim. Biophys. Acta* 79:500–510.

Thornbury, W. D. (1954) *Principles of Geomorphology*. New York: Wiley.

Toon, O. B., and Pollack, J. B. (1976) A global average model of atmospheric aerosols for radiative transfer calculations. *J. Appl. Meteorol.* 15:226–246.

Toon, O. B., Pollack, J. B., and Khare, B. N. (1976) The optical constants of several atmospheric aerosol species: Ammonium sulfate, aluminum oxide and sodium chloride. *J. Geophys. Res.* 81:5733–5748.

Toon, O. B., Pollack, J. B., and Sagan, C. (1977) Physical properties of the particles composing the Martian dust storm of 1971–1972. *Icarus* 30:663–696.

Tull, R. G. (1966) The reflectivity spectrum of Mars in the near-infrared. *Icarus* 5:505–514.

Turin, G. L. (1960) An Introduction to Matched Filters. IRE Transactions on Information Theory, Vol. IT6, 311.

Turner, R. E. (1975) Signature variations due to atmospheric effects. Proc. 10th Int. Symp. Remote Sensing, ERIM.

Turner, R. E., and Spencer, M. M. (1972) Atmospheric model for correction of spacecraft data. Proc. 8th Int. Symp. Remote Sensing.

Twersky, V. (1967) Multiple scattering of electromagnetic waves by arbitrary configurations. *J. Math. Phys.* 8:589–610.

Twomey, S. (1977) *Introduction to the Mathematics of Inversion in Remote Sensing and Indirect Measurements*. New York: Elsevier.

Tyler, J. E. (1964) *In situ* detection and estimation of chlorophyll and other pigments—preliminary results. *Proc. Natl. Acad. Sci. USA* 51:671–678.

Tyler, J. E., and Smith, R. C. (1967) Spectroradiometric characteristics of natural light under water. *J. Opt. Soc. Am.* 57:595–601.

Unger, S. H. (1959) Pattern Recognition and Detection. Proceedings IRE 47:173.

U.S. National Research Council (1970) *Remote Sensing with Special Reference to Agriculture and Forestry.* Washington: National Academy of Science.

U.S. Naval Oceanographic Office (1965) Publ. 700, Oceanographic Atlas of the North Atlantic Ocean, Section I, Tides and Currents.

U.S. Naval Oceanographic Office (1967) Publ. 700, Oceanographic Atlas of the North Atlantic Ocean, Section II, Physical Properties.

USDI (1970) United States Department of the Interior, Fish and Wildlife Service, Bureau of Sport Fisheries and Wildlife and Bureau of Commercial Fisheries, National Estuary Study, Vol. 3, Chesapeake Bay, Md., and Va., 65–112.

Valenzuela, G. R. (1967) Depolarization of EM waves by slightly rough surfaces. *IEEE Trans. Antennas Propag.* AP-15:552–557.

Vanderbilt, V. C. (1980) A model of plant polarization response. *Proc. Machine Processing Remotely Sensed Data, MPRSD* 80:98–108.

de Vaucouleurs, G. (1964) Geometric and photometric parameters of the terrestrial planets. *Icarus* 3:187–235.

Van de Hulst, H. C. (1957) *Light Scattering by Small Particles.* New York: Wiley.

Van Diggelen, L. (1959) Photometric properties of lunar crater floors. Recherche Astronomique de l'Observatoire d'Utrecht. Vol. XIV:No. 2.

Van Vleck, J. H., and Middleton, D. (1946) A theoretical comparison of the visual, aural, and meter reception of pulsed signals in the presence of noise. *J. Appl. Phys.* 17:9–97d.

Veverka, J. (1970) Photometric and polarimetric studies of minor planets and satellites. Ph.D. Thesis, Harvard University, Cambridge, Mass.

Veverka, J. (1971a) The physical meaning of phase coefficients. Physical Studies of Minor Planets, NASA SP-267.

Veverka, J. (1971b) Asteroid polarimetry: A progress report. Physical Studies of Minor Planets, NASA SP-267.

Veverka, J. (1971c) The polarization curve and absolute diameter of Vesta. *Icarus* 15:11–17.

Volz, F. E. (1972a) Infrared refractive index of atmospheric aerosol substances. *Appl. Opt.* 11:755–759.

Volz, F. E. (1972b) Infrared absorption by atmospheric aerosol substances. *J. Geophys. Res.* 77:1017–1031.

Von Neumann, J. (1961–63) Design of Computers, Theory of Automata and Numerical Analysis. Vol. 5 of Collected Works, (A. H. Taub, ed.). New York: Pergamon.

Warner, J. W., Jr., Hanna, W. E., Landry, R. J., Wulforst, J. P., Neeley, J. A., Holmes, R. I., and Rice, C. E. (1975) Soil Survey of Suffolk County, New York. U.S. Dept. Agriculture Soil Conservation Service, in cooperation with Cornell Agricultural Experiment Station.

Wattson, R. B., and Danielson, R. E. (1965) The infrared spectrum of the Moon. *Astrophys. J.* 142:16–22.

Weast, R. C., and Selby, S. M. (1967–1968) *Handbook of Chemistry and Physics*, 48th Ed. Cleveland, Ohio: Chemical Rubber Co.

Weber, F. P., and Olson, C. E. (1967) Remote sensing implications of changes in physiologic structure and function of tree seedlings under moisture stress. Annual Progress Report, September 30, Forestry Remote Sensing Laboratory, Berkeley, California.

Wehner, G. K., Ken Knight, C. E., and Rosenberg, D. (1963) Modification of the lunar surface by solar wind bombardment. *Planet. Space Sci.* 11:1257–1261.

Wehner, G. K., Rosenberg, D. L., and Ken Knight, C. E. (1965) Investigation of sputtering effects on the Moon's surface. Rep. No. 2792, Applied Science Division, Litton Systems, Inc., Minneapolis, Mn.

Weinstein, O. (1969) Return beam vidicon. Presentation at Symposium on Oceanography from Space Using Visible Region Sensors, December 16. Office of Naval Research, Washington, D.C.

White, P. G. (1969) Imaging spectrometers. Presentation at Symposium on Oceanography from Space Using Visible Region Sensors, December 16. Office of Naval Research, Washington, D.C.

White, W. B., and Keester, K. L. (1967) Selection rules and assignments for the spectra of ferrous iron in pyroxenes. *Am. Mineral.* 52:1508–1544.

Wiersma, D. J., and Landgrebe, D. A. (1978) The analytical design of spectral measurements for multispectral remote sensor systems. LARS Tech. Rep. 122678. Lab. Applic. Remote Sensing, West Lafayette, Ind.

Williams, R. T., and Bridges, J. W. (1964) Fluorescence of solutions: a review. *J. Clin. Pathol.* 17:371–394.

Winters, E., and Simonson, R. W. (1951) The Subsoil. *Adv. Agron.* 3:1–92.

Wintringham, W. T. (1951) Color television and colorimetry. *Proc. I.R.E.* 39:1135–1172.

Wolfe, William L. (1965) Handbook of Military Infrared Technology. Office of Naval Research; available from Superintendent of Documents, U.S. Government Printing Office.

Yaplee, B. S., Shapiro, A., Hammond, D. L., and Uliana, E. A. (1971) Ocean wave height measurements with nanosecond radar. Proc. 7th Int. Symp. Remote Sensing, Ann Arbor, Mich.

Yentsch, C. S., and Menzel, D. W. (1963) A method for the determination of phytoplankton chlorophyll and phaeophytin by fluorescence. *Deep-Sea Res.* 10:221–231.

Yentsch, C. S., and Reichert, C. A. (1962) The interrelationship between water soluble yellow substances and chloroplastic pigments in marine algae. *Botan. Marina.* 3:65–74.

Yost, E., and Wenderoth, S. (1968) In *Manual of Color Aerial Photography*, 1st Ed. (J. T. Smith, Jr., ed.). Falls Church, Va.: American Society of Photogrammetry, 451–471.

Young, L. D. Gray (1971) Interpretation of high resolution spectra of Mars-II; calculations of CO_2 abundance, rotational temperature and surface pressure. *J. Quant. Spectrosc. Radiat. Transfer* 11:1075–1086.

Younkin, R. L. (1966) A search for limonite near infrared spectral features on Mars. *Astrophys. J.* 144:809–818.

Zander, R. (1966) Spectral scattering properties of ice clouds and hoarfrost, *J. Geophys. Res.* 71:375–378.

Index